Modeling and Monitoring Extreme Hydrometeorological Events

Carmen Maftei
Transilvania University of Brasov, Romania

Radu Muntean
Transilvania University of Braşov, Romania

Ashok Vaseashta
International Clean Water Institute, USA

IGI Global
PUBLISHER of TIMELY KNOWLEDGE

A volume in the Advances in Environmental Engineering and Green Technologies (AEEGT) Book Series

Published in the United States of America by
IGI Global
Engineering Science Reference (an imprint of IGI Global)
701 E. Chocolate Avenue
Hershey PA, USA 17033
Tel: 717-533-8845
Fax: 717-533-8661
E-mail: cust@igi-global.com
Web site: http://www.igi-global.com

Library of Congress Cataloging-in-Publication Data

Names: Maftei, Carmen, 1965- editor. | Muntean, Radu, 1977- editor. | Vaseashta, A. (Ashok), editor.
Title: Modeling and monitoring extreme hydrometeorological events / edited by Carmen Maftei, Radu Muntean, Ashok Vaseashta.
Description: Hershey, PA : Engineering Science Reference, 2024. | Includes bibliographical references and index. | Summary: "This book develops a framework that will be used in the future to produce a stable, robust, innovative, and operational informational platform for the study of hydrometeorological events such as flash drought and flash floods"-- Provided by publisher.
Identifiers: LCCN 2023034570 (print) | LCCN 2023034571 (ebook) | ISBN 9781668487716 (hardcover) | ISBN 9781668487723 (paperback) | ISBN 9781668487730 (ebook)
Subjects: LCSH: Hydrometeorology. | Hydrometeorology--Remote sensing.
Classification: LCC GB2805 .M64 2024 (print) | LCC GB2805 (ebook) | DDC 551.57--dc23/eng/20231010
LC record available at https://lccn.loc.gov/2023034570
LC ebook record available at https://lccn.loc.gov/2023034571

This book is published in the IGI Global book series Advances in Environmental Engineering and Green Technologies (AEEGT) (ISSN: 2326-9162; eISSN: 2326-9170)

British Cataloguing in Publication Data
A Cataloguing in Publication record for this book is available from the British Library.

Advances in Environmental Engineering and Green Technologies (AEEGT) Book Series

Sang-Bing Tsai
Zhongshan Institute, University of Electronic Science and Technology of China, China & Wuyi University, China
Ming-Lang Tseng
Lunghwa University of Science and Technology, Taiwan
Yuchi Wang
University of Electronic Science and Technology of China Zhongshan Institute, China

ISSN:2326-9162
EISSN:2326-9170

Mission

Growing awareness and an increased focus on environmental issues such as climate change, energy use, and loss of non-renewable resources have brought about a greater need for research that provides potential solutions to these problems. Research in environmental science and engineering continues to play a vital role in uncovering new opportunities for a "green" future.

The **Advances in Environmental Engineering and Green Technologies (AEEGT)** book series is a mouthpiece for research in all aspects of environmental science, earth science, and green initiatives. This series supports the ongoing research in this field through publishing books that discuss topics within environmental engineering or that deal with the interdisciplinary field of green technologies.

Coverage

- Renewable Energy
- Waste Management
- Alternative Power Sources
- Electric Vehicles
- Contaminated Site Remediation
- Radioactive Waste Treatment
- Green Transportation
- Cleantech
- Water Supply and Treatment
- Sustainable Communities

IGI Global is currently accepting manuscripts for publication within this series. To submit a proposal for a volume in this series, please contact our Acquisition Editors at Acquisitions@igi-global.com or visit: http://www.igi-global.com/publish/.

The Advances in Environmental Engineering and Green Technologies (AEEGT) Book Series (ISSN 2326-9162) is published by IGI Global, 701 E. Chocolate Avenue, Hershey, PA 17033-1240, USA, www.igi-global.com. This series is composed of titles available for purchase individually; each title is edited to be contextually exclusive from any other title within the series. For pricing and ordering information please visit http://www.igi-global.com/book-series/advances-environmental-engineering-green-technologies/73679. Postmaster: Send all address changes to above address.

Titles in this Series

For a list of additional titles in this series, please visit:
http://www.igi-global.com/book-series/advances-environmental-engineering-green-technologies/73679

Exploring Ethical Dimensions of Environmental Sustainability and Use of AI
Hemachandran Kannan (Woxsen University, India) Raul Villamarin Rodriguez (Woxsen University, India) Zita Zoltay Paprika (Corvinus University of Budapest, Hungary) and Abejide Ade-Ibijola (University of Johannesburg, South Africa)
Engineering Science Reference • © 2024 • 426pp • H/C (ISBN: 9798369308929) • US $265.00

Perspectives on Global Biodiversity Scenarios and Environmental Services in the 21st Century
Naveen Kumar Chourasia (Government College, Bichhua, India) and Kavita Chahal (Government College, Bichhua, India)
Engineering Science Reference • © 2023 • 294pp • H/C (ISBN: 9781668490341) • US $250.00

Palynology and Human Ecology of Africa
Emuobosa Akpor Orijemie (Department of Archaeology and Anthropology, University of Ibadan, Nigeria) and Sylvester Onoriode Obigba (Palynology Laboratory, Department of Archaeology and Anthropology, University of Ibadan, Nigeria)
Engineering Science Reference • © 2024 • 300pp • H/C (ISBN: 9781668478011) • US $265.00

Nematode-Plant Interactions and Controlling Infection
Waleed Fouad Abobatta (Horticulture Research Institute, Agriculture Research Center, Egypt) and Rehab Yasin Ghareeb (Plant Protection and Biomolecular Diagnosis Department, Arid Lands Cultivation Research Institute, City of Scientific Research and Technological Applications, Egypt)
Engineering Science Reference • © 2023 • 289pp • H/C (ISBN: 9781668480830) • US $240.00

Promoting Sustainable Management Through Technological Innovation
Supriya Lamba Sahdev (ISBR Business School, India) Chitra Krishnan (Amity University, Noida, India) and Ahdi Hassan (Global Institute for Research Education and Scholarship, The Netherlands)
Engineering Science Reference • © 2023 • 293pp • H/C (ISBN: 9781668499795) • US $300.00

Environmental and Technological Aspects of Redox Processes
Gheorghe Duca (Institute of Chemistry, Moldova State University, Moldova) and Ashok Vaseashta (International Clean Water Institute, USA)
Engineering Science Reference • © 2023 • 435pp • H/C (ISBN: 9798369305126) • US $295.00

IGI Global
PUBLISHER of TIMELY KNOWLEDGE

701 East Chocolate Avenue, Hershey, PA 17033, USA
Tel: 717-533-8845 x100 • Fax: 717-533-8661
E-Mail: cust@igi-global.com • www.igi-global.com

List of Reviewers

Alina Barbulescu, *Transilvania University of Brasov, Brasov, Romania*
Erica Beilicci, *Politehnica University, Timisoara, Romania*
Nimet Bolgen, *Mersin University, Mersin, Turkey*
Constantin Buta, *Universitatea Ovidius, Constanţa, Romania*
Silvia Chelcea, *National Institute of Hydrology and Water Management, Bucharest, Romania*
Didem Demir, *Tarsus University, Mersin, Turkey*
Surik Khudaverdyan, *Armenian National Polytechnic University, Yerevan, Armenia*
Vahid Nasiri, *University of Agriculture in Krakow, Krakow, Poland*
Konstantinos Papatheodorou, *International Hellenic University, Serres, Greece*
Hafizullah Rasouli, *Kabul University, Kabul, Afghanistan*

Table of Contents

Section 2
Case Studies

Detailed Table of Contents

Hydrometeorological events are becoming more intense and are attributed to climate change from anthropogenic activities and long-term natural meteorological cycles. Such events have resulted in record droughts, rain, floods, temperatures, and even seismological activities. Recent studies have revealed that fast-moving droughts are emerging quickly, adding a new challenge to hazards for farmers and water managers. Most climate model projections show a continued precipitation decrease in several regions and a temperature increase of around 1.5 - 4.0°C through 2100. The authors posit that new drought indices can now be calculated, due to the development of satellite imaging and remote sensing approaches. Grid computing offers data-processing capability and the ability to use distributed computing resources to process the spatial data provided by satellite images. Platforms, such as Google Earth Engine, offer a collection of satellite images and the opportunity to implement algorithms to provide geo-spatial analysis. These tools can also be used for flash floods and other extreme events.

This review provides a comprehensive analysis of the use of remote sensing techniques in monitoring ecohydrological events. Ecohydrology is an interdisciplinary field that explores the interactions between ecosystems and the water cycle. Remote sensing, with its ability to capture large-scale and continuous observations, has proven to be an invaluable tool in monitoring and understanding these complex interactions. In this chapter, the authors discuss the various remote sensing platforms, sensors,

According to the World Meteorological Organization, the total economic impact of drought in Europe in the past 30 years was close to €100 billion. A recent publication by the European Environment Agency states that meteorological and hydrological droughts have become more frequent and severe in most of Europe, "being the highest in southern Europe." Significant changes in worldwide hydrography have occurred recently, as evidenced by the drying up of some lakes and a decrease in river flows. In a handbook published by WMO and Global Water Partnership, a set of indices and indicators were proposed to investigate different drought types. The data needed to calculate these indices refers to different hydrometeorological variables, and their analysis must be carried out. The indices based on remote sensing data could offer a viable alternative to improve the observation. Based on different applications realized by the authors, we propose a review of the methods used to identify hydrological drought. It suggests a methodology appropriate to the Dobrogea region in southeastern Romania.

Google Earth Engine (GEE) has emerged as a powerful platform for modeling and monitoring extreme hydrometeorological events. In recent years, GEE has been used extensively for studying floods, droughts, and other natural disasters. It offers a comprehensive suite of tools that can help researchers and practitioners better understand the complex interactions between weather, climate, and water resources. By providing access to a wealth of satellite imagery, climate data, and geospatial datasets, GEE enables users to model and monitor these events with unprecedented accuracy and efficiency. This book chapter explores the various ways in which GEE can be used for modeling and monitoring extreme hydrometeorological events, understanding hydrometeorological events and their monitoring needs, including case studies and practical examples. It's worth noting that this chapter mainly focuses on using GEE for remote sensing and geospatial data analysis into hydrometeorological modeling and monitoring.

Hydrometeorological events, such as floods and droughts, pose significant challenges to societies worldwide, causing loss of life and economic damage. Traditional methods of predicting such events often rely on statistical and physical models that are limited by their assumptions, uncertainties, and computational requirements. Machine learning (ML) techniques, with their ability to extract knowledge and insights from data, have shown great potential for improving the accuracy and lead time of hydrometeorological event prediction. This chapter reviews the use of ML for predicting hydrometeorological events focusing on flood and drought events. The chapter provides an overview of the application of ML techniques or algorithms for hydrometeorological events prediction. The chapter discusses data type, collection, and analysis for ML applications for predicting hydrometeorological events. The chapter presents case studies from different regions and highlights the benefits of ML-based approaches and the challenges. Finally, the chapter identifies future research directions.

Vahid Nasiri, University of Agriculture, Krakow, Poland
Soroor Rahmanian, German Centre for Integrative Biodiversity Research (iDiv) Halle-Jena-Leipzig, Germany
Zohreh Hashemi Aslani, University of Tehran, Iran
Carmen Maftei, Transilvania University of Brasov, Romania
Ashok Vaseashta, International Clean Water Institute, USA

Based on the normalized difference vegetation index (NDVI) time series of moderate-resolution imaging spectroradiometer products, meteorological observations, and spatial trend analysis, the authors analyze how climate change and vegetation cover have changed in the Black Sea coasts over the past 22 years. Moreover, spatial correlation analysis was used to examine the correlation between climatic factors and NDVI as a proxy of vegetation productivity. With growth rates of 0.003, 0.5°C, and 5.5 mm per year, respectively, the inter-annual variation of NDVI, temperature, and precipitation revealed a noticeably growing trend. At the seasonal time scale, a similar trend was observed, which was statistically significant in some cases. The spatial-temporal trend of NDVI showed a greening trend in the inter-annual and seasonal time scales (except autumn). The greening trend in vegetation refers to the increase in vegetation productivity and density over time. The authors conclude that the most effective factor in the vegetation greening process is an increasing trend in temperature and precipitation.

Silvia Mihaela Chelcea, National Institute of Hydrology and Water Management, Romania
Adrian Alexandru Aldea, National Institute of Hydrology and Water Management, Romania
Maria Cristina Trifu, National Institute of Hydrology and Water Management, Romania

The daily streamflow data—at 93 gauges on Romanian rivers and other 17 on the Romanian part of the Danube River, covering the period 1980 – 2020—were used to assess the characteristics of low flows, deficits, and seasonality. It was necessary to consider several indicators, such as: meanflow (MQ), mean annual minimum n-day discharges (MAMn-day), low flow percentiles (Q80, Q95), base flow index (BFI), extreme values of annual minimum 7-day discharge (AM7) and the streamflow below a fixed Q80 threshold. For assessing the occurrence and variability of low flows, rate and index of seasonality was used. The statistical computation of indices was made using the LFSTAT application packages developed in the R program. Also, in order to have an overview, at the level of the Romania, of low flows magnitude, the variability of minimum discharges on different time durations (days), the relative contribution of base flow to total river flow and for highlighting the driest year in the period 1980 - 2020, there were produced several maps with the spatial distribution of low flow indices.

profile and transported to our laboratory for characterization. The objective was to identify parameters in different locations of Nimroz and Zarang, for potential use as a water resource, since no such data exists due to years of conflict in the region. From this study, we assess the features of lithology, such as silty sand, clay with gravel, sandy silt, gravel, boulder, and clay with gravel, in addition to the groundwater quality, which is saline, due to evaporation in this region. Generally, the thickness of aquifers is about 5-25m, and the groundwater has evaporated to 5-10m depth with an annual precipitation of about 50-100 mm. It is further observed that the groundwater direction is from east to west. The results suggest that the modeling can be used for other regions.

The 2013 North Indian floods were a devastating hydrometeorological event that affected several states in northern India, primarily Uttarakhand and Himachal Pradesh. The floods were triggered by heavy rainfall combined with cloud bursts, cloudburst-induced landslides, and the melting of glaciers in the region. The event resulted in extensive flooding, landslides, and loss of life and infrastructure. This book chapter aims to explore the causes, impacts, and aftermath of the 2013 North Indian floods, highlighting both the devastation brought upon by the disaster and the remarkable resilience displayed by the affected communities. It emphasizes the need for improved disaster management strategies, sustainable development practices, and community resilience to minimize the impact of future floods and protect vulnerable regions.

Foreword

Extreme hydro-meteorological events, such as heavy precipitations or lack thereof, floods, and droughts, are becoming more intense and frequent due to climate change. The problem of extreme hydro-meteorological phenomena is intensifying in the area of changing climatic conditions and among others, it can strongly impact water quality. Urbanization alters the water cycle due to the rapid inflow of pollution from the urbanized catchment in the riverbed and river valley in contrast to agricultural and forested freshwater ecosystems. Additionally, the increasing frequency are flash floods and flash droughts pose very unsafe conditions for human beings. Although other hydrometeorological conditions are classified as extreme, this book primarily deals with floods, flash floods, droughts, and flash droughts, as all of these impact water quality for surface waters as well as for aquifers. Low water levels in rivers cause a concentration of dissolved mineral and organic components in water and hence affect the quality of water during hydrological extremes. As extreme events become more common, it is essential to conduct modeling and enhanced monitoring to push the conventional observation mechanisms to their limits. This edited book contains chapters on the mechanism, modeling, and remote monitoring of extreme hydrometeorological events. There are chapters on case studies on water quality due to extreme arid conditions.

There are 13 chapters authored by scientists and researchers who have made significant contributions to this emerging field. The context is divided into two parts. The first part of the book deals with the observation, modeling, and monitoring of extreme events. The second part consists of several case studies that highlight the importance of the subject, as the world experiences such events unfold in the future.

The book is interesting and well-written and is addressed to a broad readership. Students will discover research topics that are exciting and timely, and scientists will find a valuable tool for developing research projects. The book is also useful for teaching, providing teachers and students with essential and recent achievements in hydrometrological studies. I anticipate that future editions of this book will continue to provide the latest trends in modeling and enhanced monitoring of such events as predictive intelligence indicators to prepare for such events and mitigate the adverse impacts of such events.

I congratulate the editors and authors for this excellent work!

Ioan Tuns
Faculty of Civil Engineering, Transilvania University of Brasov, Brasov, Romania

Preface

Hydrometeorological events are becoming more intense worldwide and are attributed to climate change resulting from anthropogenic activities and long-term natural meteorological cycles. Such events have resulted in record droughts, rain, floods, temperatures, and even seismological activities. Recent research studies have revealed that fast-moving droughts are emerging relatively quickly, adding a new challenge to the list of hazards that farmers and water managers all over the world are already facing, as the actual hydroclimatic context drives temperatures to new extremes. Global Drought Observatory of Copernicus Emergency Management Services declared that the droughts have become extremely severe. The World Meteorological Organization reported that the total drought cost across Europe was about 100B Euros, in the last 30 years. Recognizing the limited progress achieved in implementing the instrument proposed in 2007, the European Commission adopted a communication "Addressing the challenge of water scarcity and droughts" and decided that the revised River Basins Management Plans should include drought risk management and other aspects concerning climate change, resulting in the Global Water Partnership and World Meteorological Organization launching an Integrated Drought Management Program in 2013. A recent European Environment Agency publication, titled "Meteorological and hydrological droughts in Europe 2020" shows that the frequency and severity of meteorological and hydrological droughts have increased significantly in most parts of Europe. The latest droughts have revealed sunken ships, abandoned villages, and ominous "hunger stones" that marked years of drought all over Europe. These "hunger stones" which were landmarks used in antiquity to predict periods of drought and food shortages have also reappeared in various parts of Europe. One of them, dating back to 1616 had the following words written on it: "If you see me, cry" (BBC News). Since many drought indices are developed yearly, at the Inter-Regional Workshop on Indices and Early Warning Systems for Drought, the participants agreed that the Standard Precipitation Index should be used to characterize meteorological drought around the world.

Most climate model projections show a continued precipitation decrease in several regions and a temperature increase between 1.5°C and 4.0°C until 2071–2100. In this context, a set of Guidelines for the preparation of a Drought Management Plan was prepared by the Global Water Partnership for Central and Eastern Europe in 2015. The principal disadvantage of the known drought indices is that they use climacteric parameters measured on "ground" meteorological stations. Many of them are region-dependent and represent a result of a huge amount of data. Furthermore, all drought indices are based on the Standardized Precipitation Index algorithm using a Gamma distribution function. If this fit is not good, the value of drought indices is not representative. The Handbook of Drought Indicators and Indices revised the most common drought indexes used around the world and classified them by their type and "ease of use". This document considers that a new generation of drought indices can now be calculated, due to the development of satellite imaging and remote sensing approaches, which provide real-time data with an accuracy range of a few hundred meters to a few kilometers. Precipitation, temperature, and even evapotranspiration could be extracted from satellite images. Through Grid computing, the resources of numerous computers can be cooperatively managed as collaboration toward a common objective, offering strong data-processing capability and the ability to use distributed computing resources to process the spatial data provided by a satellite image. A platform, such as Google Earth Engine, offers a vast collection of satellite images and the opportunity to implement algorithms/tools to provide advanced geospatial analysis.

The scientific community has been paying more and more attention lately to flash droughts than additional long-term droughts. Flash droughts, as the name suggests, are short-lived, usually lasting only a few weeks or months. But when they occur during critical growing periods, they can cause disasters. Globally, flash droughts are relatively new to science and are considered the next big climate threat and are characterized by a steep drop in rainfall, particularly during a season that normally receives plenty, along with high temperatures, radiation, heat waves, and fast winds. These changes in the weather increase evapotranspiration and lower soil moisture. New research shows that dry weather appears relatively more quickly than before, with little to no warning. Scientists have long warned that warming temperatures and shifting rainfall patterns due to climate change pose a threat to the environment, agriculture, and water management. But flash droughts are a relatively new area of research, with the term gaining usage only in the last couple of decades. The increase in their severity and frequency, though, is already being felt across the world causing extensive damage (billion $) in agriculture and water resources losses along with widespread wildfires, poor air quality, damaged ecosystems, and degraded mental health. The development of remote sensing technology over the past two decades has made it possible to identify occurrences of soil drying out quickly during droughts. This can induce drought circumstances to emerge practically out of nowhere and is the first early warning of a flash drought. It is concluded that using the advancement in drought-detecting technology and modeling tools that lead to growing awareness of the impact of flash drought is the key, to finding available and valid solutions to minimize the detrimental effects of flash droughts. The tools and methodologies under investigation can also be used for flash floods and other extreme hydrometeorological events.

As the world population continues to increase, it is quintessential to study such events to minimize the adverse consequences of such events. The book *Modeling and Monitoring Extreme Hydrometeorological Events* aims to address extreme hydrometeorological events, such as flash droughts and floods in terms of global awareness, a study using computer modeling, remote sensing, artificial intelligence, and selected case studies. The scope of this book has been limited to droughts and floods, as these studies will serve as the basis and models for other extreme events in nature. It is anticipated that such a book will serve as a guide to many researchers, practitioners, students, and policymakers to find ways to slow the onset of such events. It is impossible to stop such events, however, a careful study will help us plan for contingencies, develop policies to change the trajectory of extreme hydrometeorological events, and find alternatives to deal with such events.

The objective of the book is to develop a framework that will be used in the future to produce a stable, robust, innovative, and operational informational platform for the study of hydrometeorological events such as flash drought and flash floods, risk assessment based on a multicriteria evaluation method by integrating spatial remote sensing techniques for providing data necessary for indices computation, modeling using Bayesian and other computational platforms, data analytics and artificial intelligence tools to develop early warning indicators. A few regional case studies are included to show the results accomplished so far and a path forward.

A few specific objectives are to:

- Improve the knowledge of drought monitoring and environmental protection focused on the flash drought process.
- Validate the existing models and explore new directions for developing appropriate tools for flash drought risk assessment.
- Validate, homogenize, and stabilize dataset input for flash drought monitoring.
- Produce accepted and validated methods for flash drought assessment and flash drought risk assessment based on multicriteria evaluation method. -
- Foster scientific exchanges among scientists.
- Foster the theme "Tools for water management - in the context of hydroclimatic extreme events".
- Develop an operational tool for flash drought risk assessment based on a multicriteria evaluation method, to integrate remote sensing techniques for providing data necessary for the drought indices computation.
- Improve the knowledge of the drought methodology and risk assessment, which integrates spatial remote sensing techniques for providing data necessary for the drought indices computation.
- Address knowledge gaps in our understanding of the causes of water scarcity, drought events, and seasonal variability in climate impacts to develop adaptation and mitigation measures, taking climate change into account.
- Develop and demonstrate innovative (or improved) societally acceptable adaptation and mitigation strategies to cope with extreme hydro-climatic events.

The principal objective of this book is to a set of drought maps that could improve the operational capacity of competent authorities and contribute to the preparation of the Drought Management Plan. The authors of some chapters have analyzed available data sets using data analytics and commercially available artificial intelligence software tools to develop adaptation and mitigation strategies to cope with hydroclimatic extreme events in the future. The book consists of 13 chapters, written by experts in their respective fields, organized into two sections: Modelling and Monitoring, and Selected Case Studies. The first chapter on monitoring and modeling of extreme hydrometeorological events by Maftei, Muntean, and Vaseashta lays the framework for this book and describes several modeling and simulation (MODSIM), and monitoring capabilities for geospatial intelligence. The next chapter by Nasiri, Sarli, and Afshari provides a review of ecohydrological events using remote sensing techniques. Ecohydrology is a relatively new interdisciplinary field that explores the interactions between ecosystems and the water cycle. The next chapter by Beilicci and Beilicci provides a review of flash floods and provides the identification of flash flood causes, their negative effects on environmental factors, and to study of modeling with advanced hydro-informatic tools. Flash flood early warning systems (FFEWS) have been developed to mitigate the negative impacts of flash floods and other extreme events. A chapter by Msigwa, Makinde, and Mwaijengo provides a critical review of their effectiveness in reducing disaster risks. A chapter by Dobrica, Maftei, Barbulescu, and Serban titled "Analysis Methods for Hydrological Drought" provides a comprehensive review of the methods used to identify the hydrological drought and suggests a methodology specific to the Dobrogea region in Romania. As mentioned earlier, remote sensing plays a major role in monitoring hydrometeorological events, a chapter by Hazaymeh and Zeitoun describes the use of Google Earth Engine (GEE) as a platform for modeling and monitoring extreme hydrometeorological events. Recently GEE has been used extensively for studying floods, droughts, and other natural disasters using GEE's comprehensive suite of tools that can help researchers and practitioners better understand the complex interactions between weather, climate, and water resources. The use of artificial intelligence, deep learning, and machine learning has significantly improved our understanding of the vast amount of data generated by systems around us by deciphering patterns, identifying a specific feature or even computing relevant sections from a vast amount of data. Along these lines, to obtain a specific result Msigwa, and Makinde describe in their chapter the application of machine learning techniques for the prediction of hydrometeorological events. Based on the normalized difference vegetation index (NDVI) time series of moderate-resolution imaging spectroradiometer products, meteorological observations, and spatial trend analysis, Nasiri, Rahmanian, Aslani, Maftei, and Vaseashta analyze how climate change and vegetation cover have changed in the Black Sea coasts over the past 22 years. It is concluded that the most effective factor in the vegetation greening process is an increasing trend in temperature and precipitation. In the following chapter, Chelcea, Aldea, and Trifu discuss an assessment of low flow and water deficits on the Danube and Romanian rivers during 1980 – 2020. Realizing that the Hydro-meteorological events result from the interactions between the Earth's atmosphere and the water cycle, it encompasses a wide range of phenomena. Hence Nabi, Kundan, Verma, and Kumar, in their chapter describe tools and techniques that are used in the modeling of hydrometeorological events. In the second part of the book, some examples of such extreme events are presented such as a chapter by Rasouli and Vaseashta that describes an artificial recharge to the groundwater resources using rainfall water collection in Afghanistan, which is known for its extreme arid condition. In yet another chapter the same authors model the groundwater features due to extreme arid conditions in the Nimroz Province in Afghanistan. Since the impact of such extreme events is catastrophic, the last chapter by Nabi, Kundan, Verma, and Kumar provides a case study of the 2013 North Indian floods that affected several states in northern India, primarily Uttarakhand and Himachal Pradesh. The floods were triggered by heavy rainfall combined with cloud bursts, cloudburst-induced landslides, and the melting of glaciers in the region and resulted in extensive flooding, landslides, and loss of life and infrastructure. The event resulted in policy implementation to improve disaster management strategies, sustainable development practices, and community resilience to minimize the impact of future floods and protect vulnerable regions.

The expected result is the accumulation of new knowledge on modeling and monitoring extreme hydrometeorological events, the remote sensing methodologies, theoretical disclosure and justification of key mechanisms, and the discovery of new aspects of hydrometeorological events, all of which have a major impact on us due to need driven by climate change, to find effective solutions. The target audience of this book will consist of professionals and researchers working in the fields of Civil Engineering, Environmental Sciences, Computer Modeling, Geospatial Sciences, Remote Sensing, and Ecological Sciences. The book will also be useful to city planners, policymakers, and various other related disciplines, for example, Environment, Administrative Sciences and Management, Education, Sociology, and topics related to public policy and decision support. In addition, the book will provide information and support to executives involved in managing expertise, knowledge, information, and policy development in different types of scientific and educational communities. In addition, the book will be very useful for many researchers, undergraduate /postgraduate students, scientists, policymakers, industry professionals, and practitioners, and some chapters can be adopted as text supplements for independent study. We hope that the information provided by the authors of this book guides them in their research, discoveries, and many directions, but undiscovered. The book will be very beneficial for creating confidence and facilitating cross-border communication among forecasters.

Finally, we would like to recognize many contributors for their excellent contributions, as they have been working with passion in this field for a long time. We also thank our reviewers who have tirelessly helped the book with their expert and objective comments. Such observations are essential to ensure that any technical gaps are bridged, and chapters appear on an ongoing basis. The editors recognize the publisher of IGI for his guidance and help with the book. We certainly hope that many researchers, students, and scientists benefit from these chapters for their ongoing investigations to seek answers from these chapters. We also look forward to the opportunity for many readers to become contributors to our later editions. Of course, efforts to prepare books invariably consume a lot of time away from our families and pets. One of the editors (AV) owes a lot of missed playtimes to his very cute dog, Isabella, who is an avid frisbee and size 3 soccer ballplayer.

Disclaimer: The opinions, recommendations, and proposals presented in different chapters represent the findings, views, and opinions of their authors and may not coincide with the position(s) of the editors. The editors do not promote, endorse, or express their dissent against any product and/or company mentioned in the chapters. Although due precautions are exercised before publication, the editors, however, do not take any responsibility for any infringement of any copyrighted content by the authors of either their own or from elsewhere.

Carmen Maftei
Transilvania University of Brasov, Romania

Radu Muntean
Transilvania University of Brasov, Romania

Ashok Vaseashta
International Clean Water Institute, USA

Section 1
Modelling and Monitoring

Chapter 1
Monitoring and Modelling of Extreme Hydrometeorological Events

Carmen Maftei
Transilvania University of Brasov, Romania

Radu Muntean
Transilvania University of Brasov, Romania

Ashok Vaseashta
https://orcid.org/0000-0002-5649-0067
International Clean Water Institute, USA

ABSTRACT

Hydrometeorological events are becoming more intense and are attributed to climate change from anthropogenic activities and long-term natural meteorological cycles. Such events have resulted in record droughts, rain, floods, temperatures, and even seismological activities. Recent studies have revealed that fast-moving droughts are emerging quickly, adding a new challenge to hazards for farmers and water managers. Most climate model projections show a continued precipitation decrease in several regions and a temperature increase of around 1.5 - 4.0°C through 2100. The authors posit that new drought indices can now be calculated, due to the development of satellite imaging and remote sensing approaches. Grid computing offers data-processing capability and the ability to use distributed computing resources to process the spatial data provided by satellite images. Platforms, such as Google Earth Engine, offer a collection of satellite images and the opportunity to implement algorithms to provide geo-spatial analysis. These tools can also be used for flash floods and other extreme events.

DOI: 10.4018/978-1-6684-8771-6.ch001

INTRODUCTION

Hydrometeorological events are becoming more intense worldwide and are attributed to climate change resulting from anthropogenic activities and long-term natural meteorological cycles. Such events have resulted in record droughts, rain, floods, temperatures, and even seismological activities. Recent research studies have revealed that fast-moving droughts are emerging relatively quickly, adding a new challenge to the list of hazards that farmers and water managers all over the world are already facing. Most climate model projections show a continued precipitation decrease in several regions and a temperature increase between 1.5°C and 4.0°C until 2071–2100. In this context, a set of Guidelines for the preparation of a Drought Management Plan was prepared by the Global Water Partnership for Central and Eastern Europe in 2015. The principal disadvantage of such drought indices is that they use climacteric parameters measured on "ground" meteorological stations. We posit that a new generation of drought indices can now be calculated, due to the development of satellite imaging and remote sensing approaches, which provide real-time data with an accuracy range of a few hundred meters to a few kilometers. Precipitation, temperature, and even evapotranspiration could be extracted from satellite images. Through Grid computing, the resources of many computers can be cooperatively managed as collaboration toward a common objective, offering strong data-processing capability and the ability to use distributed computing resources to process the spatial data provided by a satellite image. A platform, such as Google Earth Engine, offers a vast collection of satellite images and the opportunity to implement algorithms/tools to provide advanced geospatial analysis. The scientific community has been paying more and more attention lately to flash droughts since they can cause disasters. Globally, flash droughts are relatively new to science and are considered the next big climate threat and are characterized by a steep drop in rainfall, particularly during a season that normally receives plenty, along with high temperatures, radiation, heat waves, and fast winds. These changes in the weather increase evapotranspiration and lower soil moisture. New research shows that dry weather appears relatively more quickly than before, with little to no warning. Scientists have long warned that warming temperatures and shifting rainfall patterns due to climate change pose a threat to the environment, agriculture, and water management. But flash droughts are a relatively new area of research, with the term gaining usage only in the last couple of decades. The development of remote sensing technology over the past two decades has made it possible to identify occurrences of soil drying out quickly during droughts. It is concluded that using the advancement in drought-detecting technology and modeling tools that lead to growing awareness of the impact of flash drought is the key, to finding available and valid solutions to minimize the detrimental effects of flash droughts. The tools and methodologies under investigation can also be used for flash floods and other extreme hydrometeorological events. It is impossible to stop such events, however, a careful study will help us plan for contingencies, develop policies to change the trajectory of extreme hydrometeorological events, and find alternatives to deal with such events. We aim to develop an operational tool for flash drought risk assessment based on a multicriteria evaluation method, which integrates spatial remote sensing techniques for providing data necessary for the drought indices computation, improving the knowledge of the drought methodology and risk assessment, addressing knowledge gaps in our understanding of the causes of water scarcity, drought events, and seasonal variability in climate impacts to develop adaptation and mitigation measures, and developing and demonstrating innovative and societally acceptable adaptation and mitigation strategies to cope with extreme hydro-climatic events.

EXTREME HYDROMETEOROLOGICAL EVENTS – BRIEF OVERVIEW AND RECENT OCCURRENCES

Water is crucial for the existence of all living beings on our planet. However, an excess or insufficiency of water can lead to significant issues. According to the International Centre of Excellence on Risk Interconnectivity and Governance on Weather, 2022, there are two principal groups of hazards: natural and anthropic. Considering natural hazards, this one is subdivided into five more levels; in addition, for each level, several main events are assigned (Figure 1).

Figure 1. Classification of hazard type according to International Centre of Excellence on Risk Interconnectivity and Governance on Weather, 2022

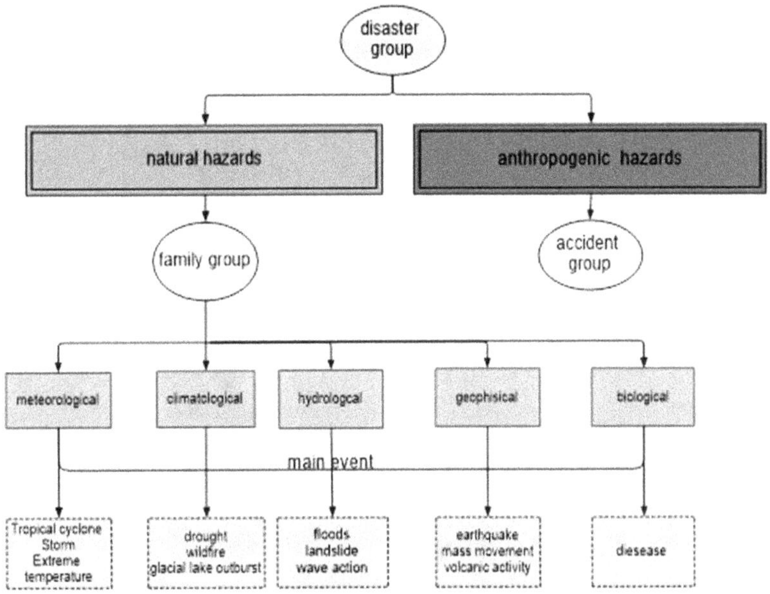

The principal natural hazards related to water are floods, drought, storms, landslides, and epidemics (waterborne diseases). Complex and interconnected connections frequently exist among hazards; for example, an earthquake might trigger a tsunami, leading to subsequent flooding in coastal areas. According to (IRDR - the International Centre of Excellence on Risk Interconnectivity and Governance on Weather, 2022), we provide here several definitions of natural hazards related to water as follows.

Floods – "*A general term for the overflow of water from a stream channel onto normally dry land in the floodplain (riverine flooding), higher-than-normal levels along the coast and in lakes or reservoirs (coastal flooding) as well as ponding of water at or near the point where the rain fell (flash floods).*"

Drought – "*An extended period of unusually low precipitation that produces a shortage of water for people, animals and plants.*"

Landslides – "*the movement of a mass of rock, debris, or earth down a slope*"

Storms – a type of meteorological hazard related to a violent atmospheric disturbance; if the disturbance is "*generated by the heating of air and the availability of moist and unstable air masses*" the hazard is called a "*convective storm*". "An abnormal rise in sea level generated by a tropical cyclone or other intense storms" is known as a *surge storm*.

The frequency of extreme weather and water events has been on the rise across the globe resulting in millions of fatalities and impeding social and economic progress. Based on different databases, the WMO (World Meteorological Organization) published an updated work related to mortality and economic losses caused by climate and water extremes. According to this work (WMO, 2023a), spanning the 1970-2021 period, floods (general floods, flash floods, coastal floods, riverine floods) have most often been disasters reported in terms of frequency (45%) and economic losses in the United States (38%). In terms of reported death, the tropical cyclone was the most frequent hazard with 38% followed by drought at 35% and flood with 16% (WMO, 2023a). In Europe, from all natural hazards, floods were the primary cause of economic losses between 1970 and 2021. These catastrophes resulted in 166,492 fatalities and economic damage of US$ 562.0 Billion. Figure 2 presents the spatial distribution of the economic losses of Europe in the period 1970 – 2021. All the countries were affected, however, France, Germany, and Italy reported losses of more than 15.1 billion US dollars.

Figure 2. Economic losses in Europe between 1970-2021
(WMO, 2023a)

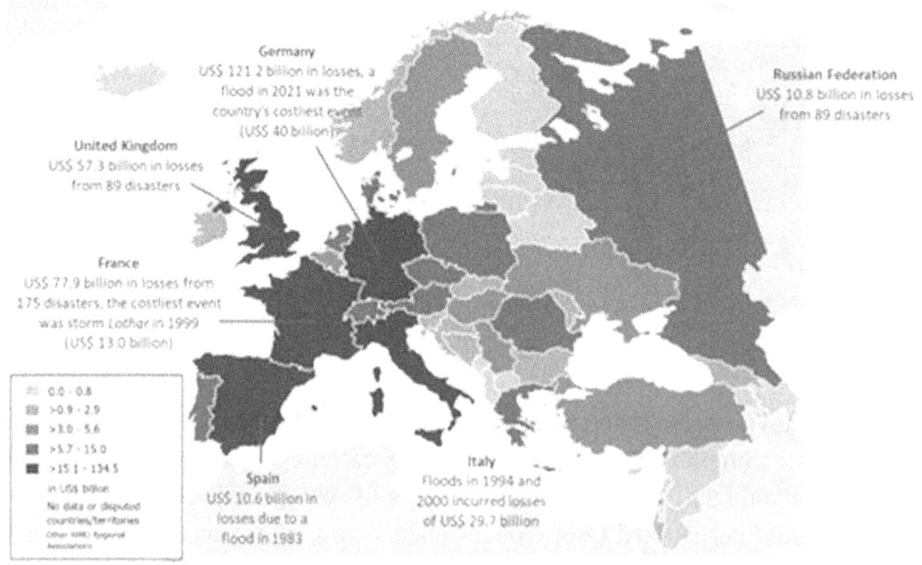

Climate change is a driver that triggers extreme events. In 1979, after the first Conference of World Climate organized by WMO in collaboration with other organizations interested in climate as UNESCO (The United Nations Educational, Scientific and Cultural Organization), FAO (Food and Agriculture Organization), WHO (The World Health Organization), and UNEP (The United Nations Environment Program), as the organizers, issued a statement in which they drew attention to climate change in the following manner: "*To foresee and prevent potential man-made changes in climate that might be adverse to the well-being of humanity.*"

Based on the Declaration, WMO established the World Climate Program. Among the four components of the Program, one referred to climate change and variability of climate, known today as the World Climate Research Program. All Member States were encouraged to establish their programs and strategies related to climate change. At the tenth meteorology congress in May 1987, many of the participating delegations requested that the WMO provide relevant scientific information about human-induced climate changes. This is how the idea of establishing an intergovernmental body under the auspices of the United Nations was born, to conduct pertinent scientific assessments on climate changes and their effects. Starting in 1990, the IPPC (Intergovernmental Panel on Climate Change) provides time-to-time assessment reports. In March 2023, the Sixth Assessment Report (AR6) was completed and the last report (Calvin et al., 2023) shows that global warming is undoubtedly being driven by rising greenhouse gas (GHG) emissions, which have been increasing from 2010 to 2019. The increase of global temperature by 1.1°C spanning 2011-2020 compared to the 1850-1990 period is the principal consequence of increased GHG emissions. Figure 3 presents a synthesis of reported changes in heat extremes, heavy precipitation, and agricultural and ecological drought and the confidence level that humans have contributed to the changes observed in different parts of the world. From 2010 to 2020, regions with significant vulnerability experienced a 15 times increase in human fatalities resulting from floods, droughts, and storms (Calvin et al., 2023).

Facing these challenges, the UN-water (an organization established in 2013, as the successor of the United Nation's Inter-Secretariat Group for Water Resources) tried to develop a water security policy that represents one of the cornerstones of environmental protection. In 2013 UN-Water defined water security policy as "*the capacity of a population to safeguard sustainable access to adequate quantities of acceptable quality water for sustaining livelihoods, human well-being, and socio-economic development, for ensuring protection against water-borne pollution and water-related disasters, and for preserving ecosystems in a climate of peace and political stability*." In 2018 UN-Water produced the first report related to water resources and sanitation, a report well-known as Sustainable Development Goal 6 (SDG 6). SDG 6 offers a set of indicators related to clean water and sanitation through which the Member States could elaborate regulations and policies. The platform of SDG 6 (https://www.unwater.org/our-work/sdg-6-data-portal) offers all the information and data needed. The website of UN-Water also offers information about disasters related to water such as floods, landslides, drought, storms, waterborne diseases, etc.

To assist the Member States with the world in meeting the challenges generated by the drought, the WMO and Global Water Partnership (GWP) started 2013 a program called the Integrated Drought Management Program (IDMP) in support of SDG 6.5. (Water Resources Management). According to this program, the partners produce several documents (Fatulová et al., 2015; WMO & GWP, 2014; World Meteorological Organization -WMO & Global Water Partnership (GWP), 2016). Those publications will contribute to providing a better understanding of drought phenomena, drought assessment and monitoring (Vaseashta et al. 2021), early warning, drought risk management, response, and preparedness, etc.

Similarly, WMO and GWP launched in 2003 the program Integrated Flood Management through the Associated Program on Flood Management (APFM) (https://www.floodmanagement.info/). The concept of Integrated Flood Management is very clearly stipulated in WMO report #1047 (Associated Programme on Flood Management, 2009). In line with the proposed concept, the river basin/catchment/watershed is the keystone of water resources management, which integrates all the interactions between the water cycle, land, and land cover. Six key elements are the fundament of flood management of which the most notable are: "manage the water cycle as a whole and adopt an integrated hazard management

Figure 3. The impact of climate change
(Calvin et al., 2023)

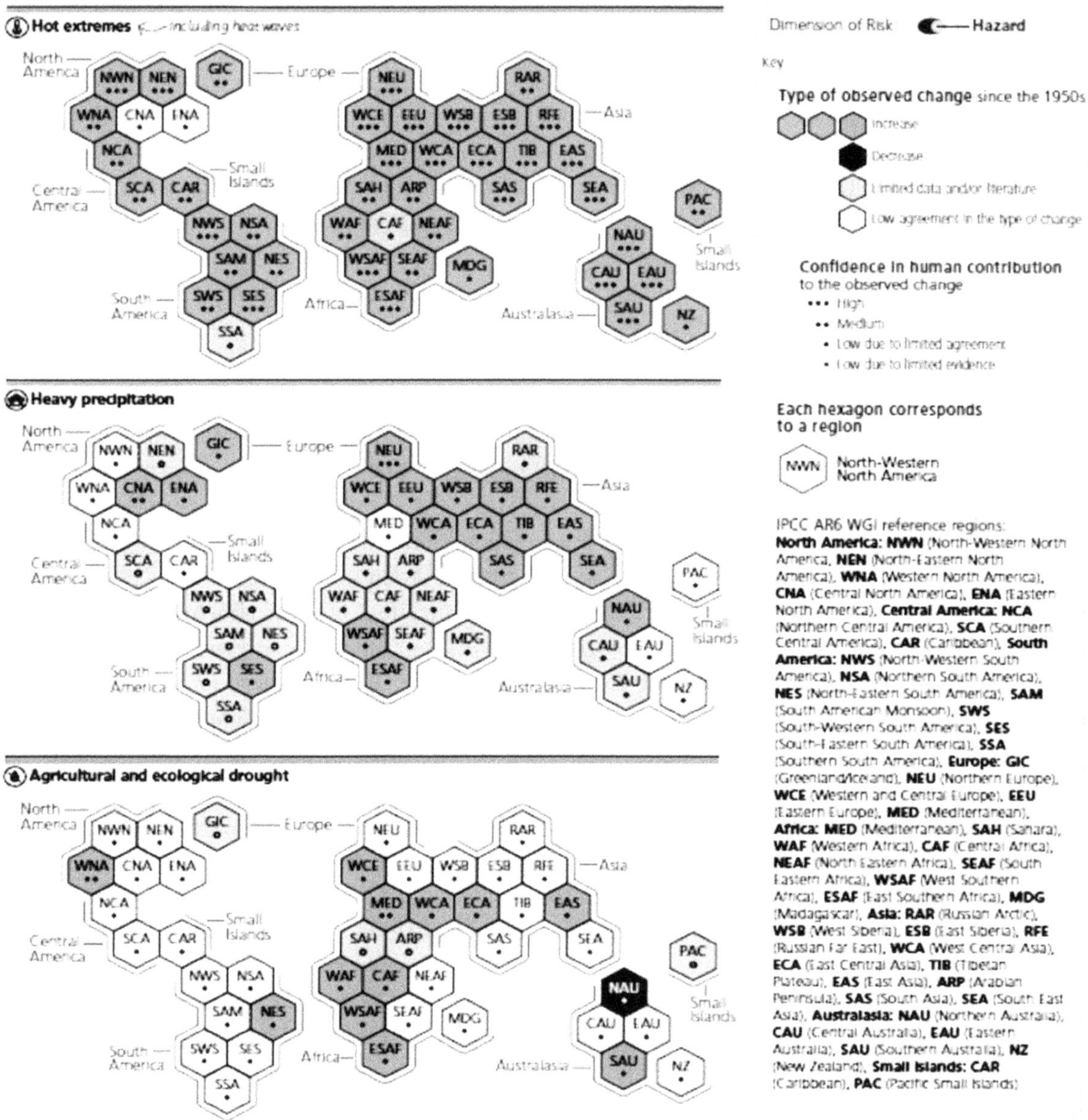

approach". (Associated Programme on Flood Management, 2009). While the Integrated Flood Management Concept Paper contains the essential idea of the Integrated Flood Management (IFM) strategy, the Policy Series books are devoted to addressing various aspects of flood management policy. These reports include considerations spanning economic, environmental, legal, institutional, and most social aspects and are readily accessible on the website of WMO at URL (https://www.floodmanagement.info/flood-management-policy-series/).

EU reacted starting in 2000 by elaborating and adopting of Water Framework Directive (Directive 2000/60/EC, Water Framework Directive, 2000) which represents the primary legislation related to water. Article 4 of the Directive outlines the WFD's primary goals: to prevent deterioration and maintain a good status (chemical and ecological) of water bodies. An integrated management system based on river basin/catchment/watershed, which represents an innovative way to explore the water budget, is one of the main goals of WFD. Each Member State shall provide a River Basin Management Plan (RBMP) and a set of measures through which the water bodies could be protected and restored. The third draft of RBMP (2022-2027) is reported under public consultation. Figure 4 presents the structure of European environmental-related water policies.

Figure 4 shows that the EU also provides specific policy recommendations related to floods (the Flood Directive was adopted in 2007), drinking water, water reuse, etc. The data and information related to water, and environmental policies, are provided through the WISE (Water Information System for Europe) platform launched in 2007 (https://water.europa.eu/about). The main objective of the Floods Directive (2007/60/EC) was to establish a framework for the evaluation and management of flood risks to minimize the adverse impact of floods on people's health, economic activity, the environment, and cultural heritage in the European Union. In contrast to the IFM, the Flood European Directive requires Member States to identify areas within watersheds with high potential flood risk and to create flood hazard and risk maps. Flood hazard maps offer insights into the extent of flood impact, encompassing both the probability and scale of a potential flood occurrence. Based on this information, Member States prepared and published the Flood Risk Management Plans (FRMP). The second FRMP is reported and is under public consultation.

Figure 4. The Water Information System for Europe (WISE) - a partnership between the European Commission and the European Environment Agency (EEA)

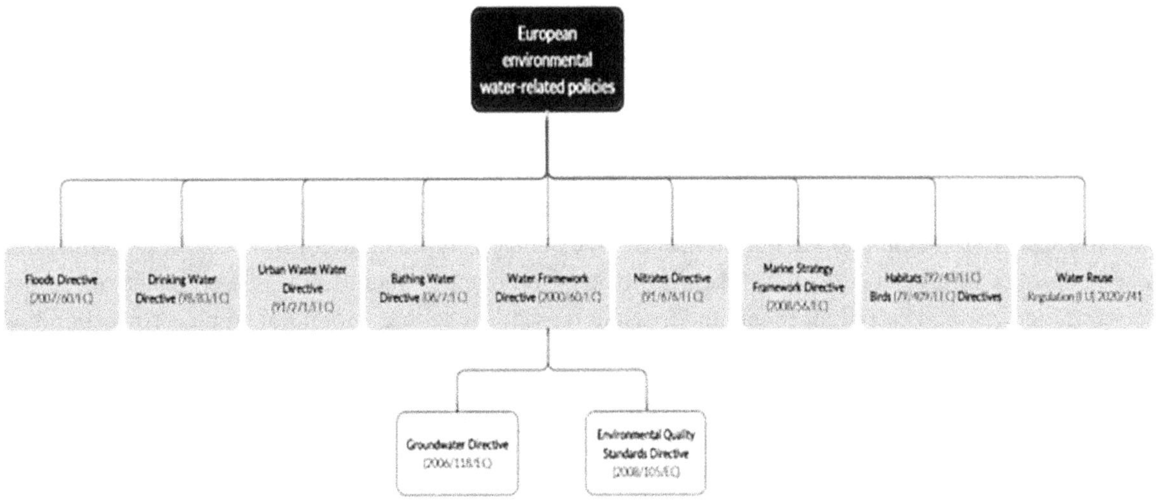

Concerning drought, the EU considered that WFD provides an adequate framework. Acknowledging the threat posed by droughts and anticipating their occurrence in the upcoming years, the European Commission approved the Communication "*Addressing the challenge of water scarcity and droughts*" in 2007. In this communication, the European Commission proposed some measures related to the management improvement of drought risk by developing the drought risk management plan at the national level, developing a drought observatory, and developing additional water infrastructure (*COM(2007)414 - Addressing the Challenge of Water Scarcity and Droughts in the EU - EU Monitor*, n.d.). The EU Commission recognized the slow implementation of the 2007 proposal's instrument through COM/2007/414 and adopted a plan to safeguard the water resources (European Commission, COM/2012/0673, 2012). Accordingly, the Drought Risk Management Plan will be introduced in the Revised Basin Management Plan (2nd RBMP).

Landslides result from many causes, such as, (i): excess rainfall, (ii): earthquakes, and (iii): volcanic activity. However, for this chapter, only landslides influenced by water will be considered to limit the scope under consideration. The challenge to the strategy of soil protection was proposed by the EU Commission to the EU Parliament and Council in 2006 COM (2006) 231, which represents a proposal for the EU Directive (COM (2006) 232). In 2012, the EU Commission published a report related to the implementation of the strategy (EU, 2012) but after eight years, the Commission withdrew the proposal. Even if the strategy is much more qualitative, the main positive element of the proposal is to introduce the notion related to the "area of risk". The principal limit of this proposal refers to the fact that any indicators or guidelines for monitoring landslides or other soil threats already exist. Moreover, some reports describe the status of soil threat or soil risk evaluation that currently exists. In a recent study (Van Beek & Tóth, 2012), the authors define landslide, the five principal movements of soil, data collection methods, and principal methods of risk assessment. To conclude, a harmonized methodology to monitor landslides and delineate the landslide-prone area is far from complete for Europe. This is still a controversial topic due to the "terms" that are used to define the hazard. A new proposal related to soil was launched by the EU in 2023, with a key outcome of the EU biodiversity strategy which has as its main objective the development of a monitoring soil system.

EXTREME EVENTS MONITORING TOOLS

This section is dedicated to monitoring extreme events. According to WMO (WMO, 2023b), the monitoring activity refers to the intensity and spatial-time distribution of such events. Starting with this widely accepted principle, the monitoring activity is based on detecting the following characteristics of extreme events: (i) magnitude, (ii) duration, and (iii) extension. The magnitude refers to the extremity and practically this parameter determines the departure of a threshold. Duration is related to the difference between the start and end of an event. and extension is in line with a geographical area corresponding to a measurement station.

Generally, extreme events are difficult to measure but the factors influencing them can be. For example, natural floods could be detected by measuring the level/discharge of a river. In this case, the river gauge station provides the necessary information. Drought could not be directly measured, but this phenomenon is a consequence of lack of precipitation, etc. Knowing the factors which influence the natural hazards, a relationship between them could be determined and a set of indicators could be calculated.

In this chapter, we do not want to define each type and subtype of extreme events, but only to draw attention to the fact that it is important to know the factors they depend on and thus understand how they can be monitored. Several information will be provided as examples.

Drought as a natural phenomenon is typically for all climates but its development is slow and insidious making difficult its detection. For this reason, the researchers and all the stakeholders involved in drought management detect a set of specific indicators for monitoring each category of drought:

meteorological, agricultural, hydrological, or socioeconomic. To monitor meteorological drought some specific climate parameters could be used such as precipitation, temperature, etc. WMO recommends SPI (Standardized Precipitation Index) as an indicator to detect and monitor meteorological drought. Hydrological drought indicators are based on discharge values. Agricultural drought indicators are related generally to soil moisture. Minimum In-stream Water Requirements (MIWR) indicators are generally used for socioeconomic drought monitoring. As we mentioned above, GWP and WMO provide technical information about drought indicators in different publications such as: "*Handbook of Drought Indicators and Indices*" (World Meteorological Organization -WMO & Global Water Partnership - GWP, 2016) or "*Guidelines for Preparation of the Drought Management Plans: development and implementation of risk-based Drought Management Plans in the context of the EU Water Framework Directive - as part of the River Basin Management Plans*" (Fatulová et al., 2015).

It is known that there are four major types of floods: riverine (also known as fluvial or stream floods), flash floods, coastal floods (storm surge), and inland floods. Fluvial (riverine) floods are related to the substantial increase of river water levels which overflow its banks. Generally, flash floods are caused by a high rate of precipitation. If the precipitation intensity exceeds the infiltration capacity of the soil, the water in excess overflows on the slope. The monitoring of flash floods is based on a rainfall-runoff relationship reliant on two indicators: FFG Flash Flood Guidance and FFT – Flash Flood Threat.

For those two extreme events, it is important to know how and from where the data could be collected. Since this situation is a widespread phenomenon, climate and water extremes were among the first to come under scrutiny. When WMO was established in March 1950 as a specialized agency of the United Nations a new era began, an era dedicated to international cooperation in the areas of meteorology, hydrology, and its interaction with other fields of geosciences. In line with its mandate, WMO organizes in collaboration with other agencies such as the World Integrated Observation Network (WIGOS) to offer a rapid exchange dataset related to weather, climate, and water. The first edition of the WIGOS manual was approved at the 16th World Meteorological Congress (Organization (WMO) & World Meteorological Organization (WMO), 2021).

The monitoring and data management activities are supported by solid research and training related to weather, climate, and water. In view of this WMO created a sponsored website – World Weather Information Services (WWIS) – which provides weather information delivered by National Agencies. The data can be downloaded following the guide (https://worldweather.wmo.int/en/dataguide.html). Climate observations are a part of the Climate Data Management System (CDMS) existing in which country and governed by the National Meteorological and Hydrological Services (NMHSs) which are memberships of WMO. WMO provides information about terms, standards, and procedures in their technical regulations which are available on the WMO website at Technical<Basic Documentation< Technical regulations address (Organization (WMO), n.d.-a). Hydrological observations are also governed by the National Agencies (NMHSs). To improve the accessibility and sharing of hydrological data observations and/or practice WMO created in 1981 a system known as Hydrological Operational Multipurpose System (Water Resources Council (U.S.), 1981). As part of this, the gauge network systems measure the principal element which could be directly used to delineate the flood area and to estimate the discharge/flow. This parameter refers to the water level of water bodies. WMO provides several technical documents related to hydrological practice: Guidelines for Hydrological Practice (Organization (WMO), n.d.-b).

It is important to know that the National Agencies have additional data that is not available internationally (WMO, 2023b).

In accordance with the endorsement of the current EU Strategy on Climate Change Adaptation (European Commission, COM/2021/82, 2021), the European Drought Observatory is a service led by the Joint Research Center which was launched in 2022 This platform represents a collection of information related to drought achieved from different data sources like precipitation or satellite measurements. The information is provided as maps or reports of drought indicators (https://edo.jrc.ec.europa.eu/edov2/php/index.php?id=1000). Similarly, for North America exist US Drought Monitor (https://droughtmonitor.unl.edu) a product developed by specialist from the National Drought Mitigation Center - NDMC (University of Nebraska-Lincon), the National Oceanic and Atmospheric Administration (NOAA) and the U.S. Department of Agriculture in 1999. This platform offers a lot of information about drought as a phenomenon (including educational documentation), maps, and data sets.

EFAS system (European Flood Awareness System) was realized to ensure the preparation of flood measures across Europe. Unfortunately, the EFAS is designated for the transnational river basins. Using the meteorological datasets and/or meteorological predictions, and land cover data the hydrological model LISFLOOD calculates the water balance at daily or 6-hour time step and provides a set of maps and graphs related to flood indicators, flood forecasting, or flood impact. The EFAS platform (https://www.efas.eu/en) offers the possibility of downloading the model data output, map viewer, and training.

US Geological Survey has a website dedicated to flood information at the following address https://www.usgs.gov/mission-areas/water-resources/science/usgs-flood-information#web-tools. According to this website, the US Geological Survey (USGS) offers a set of tools through the population, and all those interested are informed about the flood conditions. Water Watch tool (https://waterwatch.usgs.gov/) offers information about current streamflow in real time or daily/weekly/monthly, flood maps and extreme flow conditions, past flow. Data could be downloaded as a shapefile or table. Some statistics could also be downloaded. Each country, based on its policy concerning climatological and hydrological data, provides researchers with the opportunity to access data to conduct their research.

Natural hazards and risk studies may be compromised in their precision and reliability by the lack of meteorological and hydrological measurements (especially in ungauged catchments), low station density, short data histories, data gaps, and restricted data access. To address these issues, finding alternate data sources and modeling approaches is crucial. Remote sensing techniques could be a solution because they offer the possibility to monitor and record in real-time the Earth's Physical characteristics from a distance. There are four platform categories as support to the sensors: terrestrial platform, drone (airborne platform without pilot), airborne with the pilot (planes, etc.), and spatial platforms (satellites). The last one presents some advantages such as stability and covering a large area, data processing, and analyzing.

- Part of WIGOS, the WMO Space Programme (https://public.wmo.int/en/programmes/wmo-space-programme) was created to offer remote sensing data and products related to weather, climate, and water to all Member States (Organization (WMO) & World Meteorological Organization (WMO), 2020). WMO provides also virtual training activities and education in remote sensing techniques and guidance on the usability of satellite data (WMO, 2017, 2018).
- The Copernicus EU Space program - https://www.copernicus.eu/en. Copernicus, previously known as GMES (Global Monitoring for Environment and Security), is the European Union's Earth observation program. The European Commission manages the Programme. It is implemented in partnership with the Member States, the European Space Agency (ESA), the European Organization for the Exploitation of Meteorological Satellites (EUMETSAT), the European Centre for Medium-Range Weather Forecasts (ECMWF), EU Agencies, and Mercator Ocean.

Vast amounts of global data from satellites and ground-based, airborne, and seaborne measurement systems provide information to help service providers, public authorities, and other international organizations improve European citizens' quality of life and beyond. The information services provided are free and openly accessible to users.

- USGS Earth Explorer - https://earthexplorer.usgs.gov/ https://www.usgs.gov/educational-resources/earth-explorer provides science about the natural hazards that threaten lives and livelihoods, the water, energy, minerals, and other natural resources. The USGS scientists develop new methods and tools to enable timely, relevant, and useful information about the Earth and its processes so that policymakers and scientists can make better decisions by deep understanding to address complex environment, natural resources, and public health and safety issues.
- Google Earth Engine - https://earthengine.google.com/ - is a platform that offers a catalog of remote sensing data free of charge for academic users. But this platform is more than a simple catalog is a computing platform that offers users the possibility to run their one analysis using Google's infrastructure and the Code editor (a web-design IDE for running edit and writing scripts).

As monitoring tools evolve, it has led to a field of Geospatial Intelligence (GEOINT) - typically referred to as remotely sensing human activity on earth by the exploitation and analysis of imagery, signals, or signatures with geospatial information. GEOINT describes, assesses, and visually depicts physical features and geographically referenced activities on the Earth and consists of imagery, imagery intelligence (IMINT), and geospatial information. The subject provides the ability to process and make decisions based on spatial data which consists of a set of information that describes locations in space. With the four platform categories, described above, the subject is growing exponentially. A detailed description is beyond the scope of this chapter, except to briefly describe some of the monitoring tools that are used in support of GEOINT and IMINT. Remote sensing and geographic information systems (GIS) software play a key role in geospatial intelligence. Several specialized software, such as ENVI and ArcGIS spatial analyst or defense mapping can leverage images from satellites and drones, as well as spatial vector data. Google Earth Engine, as described above, also provides basic functionality for GEOINT purposes. Additionally, it is possible to use many open-source GIS software to draw GEOINT intuitions, including QGIS and GRASS. Combining imaging with feature extraction allows satellite imagery to see where things are on Earth in order to provide contextual foundations. However, analyzing, synthesizing, and communicating information in an organized way to develop comprehensive intelligence also requires the use of Artificial intelligence (AI) (Vaseashta, 2022). Algorithm use of AI is revolutionizing GEOINT through the automatic classification and detection of objects from satellite imagery. Hence, GEOINT can support disaster prevention by developing models for determining the probability of an event happening in the future and providing early warning indicators. The field is growing since it is a discipline-specific practice. There is no single software that offers all of the spatial intelligence functionality and hence it becomes resource intensive. Considering its use in defense, disaster preparedness, wide-spectrum disease surveillance through the use of bioinformatics (Baldwin et al., 2019), real estate planning, census surveys, migration pattern analysis, etc., it is essential to explore this subject in much more detail and depths.

OVERVIEW OF MODELING AND SIMULATION TOOLS

A model is a simple representation of a physical process (Petrescu (Maftei), 2002). The representation can be mathematical or material. In the first case, the model is an ensemble of equations that describe the behavior of the modeled system. A material model, also named a physical model (Chow et.al., 1988), is a representation of a system (e.g., hydrological system) using another physical system. Generally, material models include the scale model and the analog model. The scale model represents the system on a reduced scale. The best-known scale model includes laboratory models. The analog model is a representation based on the similarity between two physical systems (e.g., water flow in pipes versus the flow of electrical current). Several modeling and simulation (MODSIM) tools and their basis for several events are described as follows:

Drought (Including Flash Drought)

Drought modeling is based on several drought indicators. There are more than 100 indices to calculate drought (Sundararajan et al., 2021) which are developed according to the type of drought (agricultural, meteorological, hydrological, socio-economical) and the data availability. WMO offers a detailed handbook related to drought indicators classified based on ease of use (World Meteorological Organization -WMO & Global Water Partnership - GWP, 2016). Drought is one of the most devastating natural hazards that impact any region's agricultural system, water availability, and economy. This handbook also offers the software used to calculate indicators.

Studies suggest that drought events are traditionally long-term and last for several months or years. However, in recent decades, there has been a growing interest in short-term drought events due to their rapid, variable, and adverse impacts, especially on agriculture. These short-term events, known as - Flash Drought (FD), are characterized by a sudden onset and rapid intensification due to high temperatures and very low levels of precipitation, hence a decline in soil moisture (Mahto & Mishra, 2020; Otkin et al., 2018). However, due to a lack of consistent definitions, compounded with fragmented documentation and assessments, and limited progress toward improved prediction models. Some studies have included hydrometeorological variables such as Surface Air Temperature, Precipitation, Evapotranspiration, Soil Moisture, and Potential Evapotranspiration as some of the primary indicators. Intuitively, simultaneous changes in various other key indicators provide an efficient definition of FD. This would include several normalized drought indices, viz. Standardized Precipitation Index, Standardized Precipitation Evapotranspiration Index, Evaporative Demand Drought Index, Evaporative Stress Index, Normalized Difference Vegetation Index, Standardized Evaporative Stress Ratio, and Remotely Sensed Drought Indices. The FD capturing ability of such drought indices-based methods is limited by their ineffectiveness in including multicollinearity and non-linear relations between the independent and dependent variables. However, traditional drought indices tend to overestimate (or underestimate) the Frequency of Occurrence and spatial extent of FD. As an alternative, physical models, such as observational data and the Variable Infiltration Capacity model are used to simulate soil moisture. However, the efficiency of such modeling efforts is limited by a lack of good quality input data and due to insufficient information on underlying complex physical processes. Hence, Machine learning and Deep Learning methods have evolved to address the issue of multi-dimensional relations efficiently and obtain information from complex datasets (Vaseashta, 2022). Within the context of this review, machine learning applies computational algorithms for classification, prediction, clustering, and pattern recognition in a target data set by transforming, sorting, and splitting the input data set. As a standalone definition, deep learning itself implies sophisticated machine learning and recognition algorithms with multiple hierarchical layers to adapt to complex functions.

Climatology

It is normally defined as the science that deals with climates and their phenomena. Climate models are sophisticated computer algorithms that simulate weather patterns over time. By executing these simulations, climate models can estimate the Earth's average weather patterns—the climate—under different conditions. These climate models, depending upon the input, forecast how the climate conditions might change, going forward. With the steady increase in population, increased transportation, and use of plastics, there is a concern as to how such use of technology is likely to add greenhouse gases to the atmosphere and change the basic conditions of our planet.

The subject is quite complex due to the vast interplay of parameters and a comprehensive coverage is beyond the scope of this chapter. There are sufficiently detailed reports available in the literature, viz. at the fifth Session of the Commission for Climatology (Geneva, 1969), Flohn et al. drew attention to the requirement of developing atmospheric physical-mathematical models that are sufficiently accurate to simulate all of climate change's effects (Flohn, 1970). Another excellent review of climate models is presented by Flato et al. (*WG1AR5_Chapter09_FINAL.Pdf*, n.d.).

For the sake of continuity, it should be pointed out here that the dynamics of the climate system are governed by physical principles, viz. conservation of air mass, conservation of water mass, conservation of energy, conservation of momentum of air in three directions, and the ideal gas law applied to air. Furthermore, climate models are described by equations, which constrain the following variables: air temperature, pressure, density, water vapor content, and wind magnitude in three directions. By solving these equations, climate models can simulate all these variables in three dimensions and in time. Other variables that affect the Earth's climate are hard to model directly. Clouds are much smaller than the smallest unit of distance in a typical climate model, so the model cannot "simulate" individual clouds, but taken together they have big effects on the Earth's temperature. For these factors, climate models use "parameterizations," or simplified equations that behave roughly the same as the real thing. Rain, snow, and evaporation are other physical processes that should be "parameterized" in climate models. These are some important features of the Earth's climate, so getting the parameterizations right is an essential part of designing a good climate model. However, the inclusion of parameters that can forecast how climate might vary with human resources consumption is very complex and include socio-economic factors and parameters from local geographical location. Such individual models need to be integrated to develop into a larger climate model.

Modeling Floods (Including Flash Flood)

Considering floods, the principal issue is to evaluate the discharge values related to floods and to create hazard and risk maps. In a study performed in 1988 by the US Geology Survey (Hill et al., 1988.) the methodologies to delineate the flood extent are grouped into 3 categories according to the accuracy level: comprehensive, intermediate, and approximate (Figure 5). The first category can include detailed and historical methods.

Figure 5. Synoptic of USGS method to map the flood extent

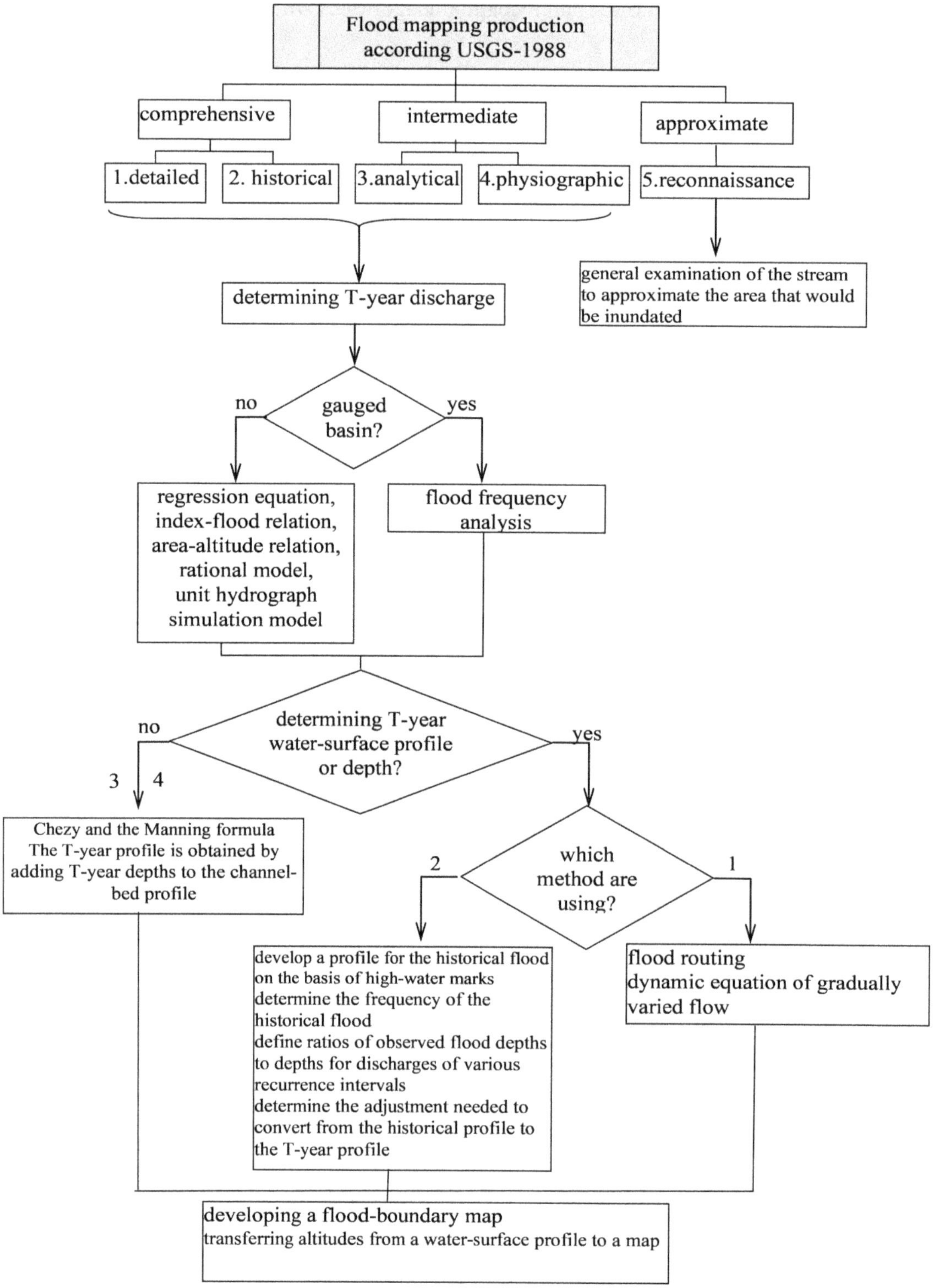

The analytical and physiographic are intermediate methods and the reconnaissance method is approximate. All these methods (without the reconnaissance method) require the determination of the maximum discharge with a given return period (T-years discharge). Once the peak discharge is obtained, the next basic steps may be performed, i.e., the determination of a water-surface profile and/and water depth and developing a flood-boundary map (for the T-year discharge calculated). The evaluation of T-year discharge is the subject of flood frequency analysis. Flood frequency analysis uses different approaches which depend on the available data. For gauged watersheds, an empirical distribution, or theoretical distribution, can be computed. In the US the logPearson III distribution is proposed by the US Water Resource Council (Water Resources Council (U.S.), 1977, 1981). For ungauged watersheds different types of models may be used: regression equation (Thomas & Benson, 1970) index-flood relation (Butler et al., 1964; Dalrymple, 1960; Kennon, 1955), area-altitude (Moore, 1976), rational model (Mulvaney, 1851), unit hydrograph (Sherman, 1932a) and simulation model (Crawford & Linsley, 1966).

To obtain the water surface profile, the detailed method proposes two methods: flood routing (Carter & Godfrey, 1960; Cunge, 2010) and the dynamic equation of gradually varied flow (Chow, 1959). In the analytical method, the Chezy and the Manning formula (Chow, 1959) are proposed. The historical method's purpose is to develop a profile for the historical flood based on high-water marks and determine some adjustments needed to convert from the historical profile to the T-year profile (Figure 6). The National Flood Insurance Program (NFIP) of the Federal Emergency Management Agency (FEMA, 2022)considers that the most accurate and widely used method is: (1) to calculate the flood discharge for differing flood frequencies, magnitudes, or recurrence intervals and (2) to determine water surface profiles/elevation.

Figure 6. Synoptic of FEMA method

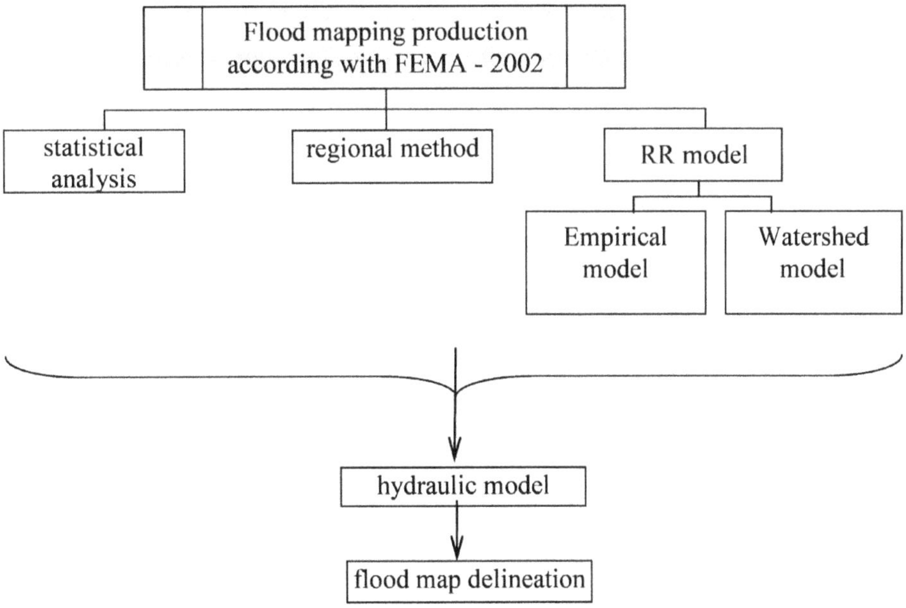

According to the Flood Directive, (Directive 2000/60/EC, Water Framework Directive, 2000), flood hazard assessment must be evaluated for three different hazard levels: frequent events (10 years Return Period), medium probability events (100 years Return Period), and extreme events (1000 years RP). For each scenario, the hazard map must provide the following results: flood extent, water surface depth/ level, and water velocity - if it is relevant.

In the report, Integrated Flood Management Tools Series: Flood Mapping (Partnership (GWP) & Organization (WMO), 2013), the different tasks necessary to perform a flood hazard map are explained. The first step is the determination of discharge values and hydrographs, and the second is the determination of flood-prone areas. The methods used to obtain discharge and hydrographs are (1) flood frequency analysis - observed and simulated; (2) a physically based hydrological model. Whatever method selected is obtained from the hydrograph with a given return period. The following return period are proposed: frequent event (Q10); rare event (Q20 to Q30); very rare event (Q50 to Q100); extreme event (Q300, Q500 or Q1000). Concerning the determination of flood-prone areas, the same report proposes three different approaches to developing flood hazard maps: (i) a historic approach based on past flood events, (ii) a geomorphologic approach: floods and flows leave distinct marks in the landscape. All the information provided from various sources (remote sensing images, orthophotos, field observations) may be superposed to create a very comprehensive tool that can describe the evolution of the river, and (iii): Modeling approach: hydraulic models are applied to simulate floods.

The historic and geomorphologic approaches are mostly used in the preliminary stage, whereas the modeling approach is mainly used in the detailed mapping stage. To conclude the steps necessary to develop a study regarding flood hazards are: (i): collecting data (ii): estimating discharge for different return periods, (iii): using 1D or 2D hydrodynamic models to calculate the flood parameters, (iv): delineation of the flooded area by combining the flood parameters (water level) with Digital Elevation Models (DEM).

Collecting data is the first stage and maybe the most important and difficult step. As we stated many organisms could provide data sets, from national agencies to world organizations. Creating a dataset, in a statistical sense, is a process during which the occurrence of errors (systematic and/or random) is possible. In situ, or statistical tests are used to control the time series data.

The most applied techniques used to define discharge for different return periods are (a) Statistical analysis of stream-flow records, the method used for gauged basin/stations; and the discharge frequency analysis can be used to fit the best probability distribution. Computer programs for performing the analyses are available from the U.S. Army Corps of Engineers (USACE) HEC-FFA Frequency (USACE, 1992) and the USGS PEAKFQ, Annual Flood Frequency Analysis Using Bulletin 17B Guidelines (John F. England et al., 2019). Guidance on frequency analysis can also be found in USACE Engineering Manual No.1110-2-1415 (U.S. Army Corps of Engineers, 1993) or (Partnership (GWP) & Organization (WMO), 2013). There are several other open-source (e.g., Hydrogonom - http://hydrognomon.org/) or proprietary software options available in the market; if the watershed is ungauged, there are other available data (rainfall, topographic, etc) that can be used with the regional model, with the index-flood relation and with the area-altitude relation; and (c) rainfall-runoff models also called hydrological mathematical models which transform the rainfall into a runoff through a set of relationships between the factors which control the hydrological process (water, soil, land use, climate). There are a couple of hydrological models that can be used to estimate discharge data, and these include: i) the empirical model which consists of the rational model and the unit hydrograph (Sherman, 1932b), and ii) the watershed model may be included; this model can be classified according to the type of simulation into two categories: single or continuous event.

There is software available to apply the empirical model, provided by various sources including the U.S. Department of Agriculture. HEC-HMS (https://www.hec.usace.army.mil/software/hec-hms/), and MIKE 11 (https://www.mikepoweredbydhi.com/products/mike-11). The most widely applied hydrological models in Europe are the HBV- Hydrologiska Byråns (Bergström, 2019) proposed by Swedish Meteorological and Hydrologic Institute (SHMI) and used in 40 countries, the LISFLOOD (https://ec-jrc.github.io/lisflood-model/) and the TOPKAPI model (http://www.progea.net/prodotti.php?c=Software&p=TOPKAPI). Once the discharge for a return period is obtained, 1D or 2D hydraulic models are used in order to obtain the parameters of the flood. These models provide the water surface which is assessed by using each of: (i) normal-depth calculations using Manning's Equation and (ii) highway culvert nomographs from Hydraulic Design of Highway Culverts (https://www.fhwa.dot.gov/engineering/hydraulics/library_arc.cfm?pub_number=7&id=13) and (iii) software Quick-2, which may be used to compute critical depth and normal depth. There is a wide range of tools for this modeling.

Some examples are mentioned here, such as Mike11 (https://www.mikepoweredbydhi.com/products/mike-11), HEC-RAS (https://www.hec.usace.army.mil/software/hec-ras/). Finally, the delineation of flood limits could be done. FEMA recommended the implementation of a GIS tool to create cross-section and structure data for the HEC-RAS program to perform water surface elevation. Today, the HEC-GeoRAS (*HEC-GeoRAS and ArcGIS*) tool exists as a component of ArcGIS and allows the design of the components of the topographical model and the automated export to HEC-RAS and then imports the results of the hydraulic simulation into ArcGIS. In recent work (Maftei & Papatheodorou, 2016) a detailed review of hydrological models used in flood modeling is presented, including a proposed methodology used and examples.

CONCLUSION AND POLICY RECOMMENDATIONS

The chapter presents various challenges presented by extreme hydrometeorological events. It is now considered a global priority of scientific research and is well integrated in the field of Environmental and climate change, societal impact, and policy. Among the solutions presented above include the need to improve knowledge and data collection about extreme events by developing real-time information systems, with early warnings and better management of such events. Due to multi-parametric and complex dynamics, we suggest the use of the Grid Computing technique to process the remote sensing data which will provide the computational power of several computers (Grid cluster) and, hence, distribute the computational effort and minimize the processing time. Furthermore, the adoption of Open-Source GIS tools combined with R software, Grid Computing Technique, and the development of support for potential users will make the adoption and use of this platform by several users. We further posit the formation of Prevention, Preparedness, and Response actions, to provide Environmental Hazard and Risk analysis associated with each extreme event. In addition, formation of the development of a harmonization framework for data, information, and methodologies and a reliable extreme event Early Warning System for Emergency Response to increase Response efficiency. We also propose the formation of a framework that will be used in the future to produce a stable, robust, innovative, and operational tool for all extreme events – may it be floods, flash floods, or flash droughts. The strategy will include risk assessment based on a multicriteria evaluation method, which integrates spatial remote sensing techniques for providing data necessary for extreme event indices computation. The specific objectives are: (i) to improve the knowledge of extreme event monitoring and environmental protection, (ii) to validate the existing models, and to explore new directions for developing appropriate tools for composite risk assessment. (iii) to validate, homogenize, and stabilize datasets input for extreme event monitoring, (iv) to produce accepted and validated methods for extreme event assessment and risk assessment based on multicriteria evaluation method; and (v) to foster scientific exchanges among scientists. This strategy is based on the following three pillars.

I. Resilience, adaptation, and mitigation to hydroclimatic extreme events
   - Addressing knowledge gaps in our understanding of the causes of water scarcity, drought events, and seasonal variability in climate impacts to develop climate change adaptation and mitigation measures 10%
   - Developing and demonstrating innovative (or improved) societally acceptable adaptation and mitigation strategies to cope with hydro-climatic extreme events and their increase in length and duration. This includes floods and droughts, although is not limited to them, adopting a regional and/or a catchment scale-wide approach. Both surface and groundwater can be included.
   - Improving resilience and adaptation capacity of water infrastructure (e.g., industrial water facilities, urban networks, wastewater treatment facilities, stormwater management systems) to hydroclimatic extreme events.

II. Tools for water management - in the context of hydroclimatic extreme events.
   - Developing tools (e.g., multi-risk approach, decision support tools, monetary/non-monetary costs valuation) to support the design and implementation of strategies for adaptation and mitigation to
   - hydro-climatic extreme events, especially floods (including "flash floods"), heat waves, and droughts in a catchment to sea.
   - Generating new methodologies, tools, and models for water resources assessment/modeling for water bodies in scarcely monitored /data-scarce areas.
   - "Smartening the water system" and use of innovative digitalization, including improved/new sensors, models, communications, and computing technologies.

III. Improved water governance in the context of hydroclimatic extreme events and international contexts
   - Undertaking an analysis and developing robust Governance models for the management of water resources in the context of extreme events, is critical in increasing the decision-making capacity of institutions and the involvement of citizens. This should include ways of improving coordination for water managers to increase our capacity to reduce our vulnerability to extreme events, as well as effectively responding to them addressing and encouraging international cooperation in the field of water, including management of transboundary water resources and contribution to Water Diplomacy.

REFERENCES

Associated Programme on Flood Management (Ed.). (2009). Integrated flood management: Concept paper. World Meteorological Organization: Associated Programme on Flood Management. APFM.

Baldwin, J., Noorali, S., & Vaseashta, A. (2019, July). *Wide Spectrum Bio-threats Identification and Classification*. Conference: Chemical, Biological, Radiological, and Nuclear (CBRN) Defense - Modernizing the Future Fight: Accelerate & Adapt, Delaware.

Bergström, S. (2019). *HBV (Hydrologiska Byråns Vattenbalansavdelning), Model Item, OpenGMS*. GeoModeling. https://geomodeling.njnu.edu.cn/modelItem/5ad110bc-5655-438d-908d-ca8c6452c95f

Butler, E., Reid, J. K., & Berwick, V. K. (1964). Magnitude and frequency of floods in the United States: Part 10, the Great Basin. In *Water Supply Paper*. U.S. Government Printing Office. (Original work published 1684), doi:10.3133/wsp1684

Calvin, K., Dasgupta, D., Krinner, G., Mukherji, A., Thorne, P. W., Trisos, C., Romero, J., Aldunce, P., Barrett, K., Blanco, G., Cheung, W. W. L., Connors, S., Denton, F., Diongue-Niang, A., Dodman, D., Garschagen, M., Geden, O., Hayward, B., Jones, C., & Péan, C. (2023). IPCC, 2023: Climate Change 2023: Synthesis Report. Intergovernmental Panel on Climate Change (IPCC). doi:10.59327/IPCC/AR6-9789291691647

Carter, R. W., & Godfrey, R. G. (1960). *Storage and flood routing: Manual of hydrology Part 3*. Research Gate.

Chow, V. T. (1959). *Open Channel Hydraulics*. McGraw-Hill.

Crawford, N. H., & Linsley, R. K. (1966). *Digital simulation in hydrology: Stanford watershed model IV*. Dept. of Civil Engineering, Stanford University.

Cunge, J. (2010). On The Subject of A Flood Propagation Computation Method (Muskingum Method). *Journal of Hydraulic Research, April 1969*, 205–230. doi:10.1080/00221686909500264

Dalrymple, T. (1960). Flood-frequency analyses, Manual of Hydrology: Part 3. In Water Supply Paper (1543-A). U.S. G.P.O., doi:10.3133/wsp1543A

EU. (2012). *COM (2012)46—Implementation of the Soil Thematic Strategy and ongoing activities—EU monitor*. Eumonitor. https://www.eumonitor.eu/9353000/1/j9vvik7m1c3gyxp/viwytay5yfzt

European Commission. (2012). *COM/2012/0673, COM/2012/0673*. Europea. https://eur-lex.europa.eu/legal-content/EN/TXT/?uri=CELEX%3A52012DC0673&qid=1692729471689 Directive 2000/60/EC,

Fatulová, E., Majerčáková, O., Houšková, B., Bardarska, G., Alexandrov, V., Kulířová, P., Gayer, J., Molnár, P., Fiala, K., Tamás, J., Zupan, M., Sušnik, A., Gregorič, G., & Bokal, S. (2015). *Guidelines for the preparation of the Drought Management Plans: Development and implementation of risk-based Drought Management Plans in the context of the EU Water Framework Directive - as part of the River Basin Management Plans*. Global Water Partnership Central and Eastern Europe.

FEMA. (2022). *Guidance for Flood Risk Analysis and Mapping*. General Hydrologic Considerations.

Flohn, H. (1970). Climatology—Descriptive or Physical Science? *World Meteorological Organization Bulletin*, *19*(4), 223–229.

Hill, G. W., Hales, T. A., & Aldridge, B. N. (n.d.). *Flood Hydrology Near Flagstaff, Arizona*. Research Gate.

IRDR International Centre of Excellence on Risk Interconnectivity and Governance on Weather. (2022). *Peril Classification and Hazard Glossary* (IRDR DATA Publication No. 1; p. 28). International Science Council. doi:10.24948/2022.05

John, F. England, Cohn, T. A., Faber, B. A., Stedinger, J. R., Jr, W. O. T., Veilleux, A. G., Kiang, J. E., & Robert R. Mason, J. (2019). Guidelines for determining flood flow frequency—Bulletin 17C. In Techniques and Methods (4-B5). U.S. Geological Survey. doi:10.3133/tm4B5

Kennon, F. W. (1955). Magnitude and frequency of summer floods in western New Mexico and eastern Arizona. In *Open-File Report* (pp. 55–82). U.S. Geological Survey. doi:10.3133/ofr5582

Maftei, C., & Papatheodorou, K. (2016). Mathematical Models Used for Hydrological Floodplain Modeling. In *Extreme Weather and Impacts of Climate Change on Water Resources in the Dobrogea Region* (pp. 69–100). IGI Global.

Mahto, S. S., & Mishra, V. (2020). Dominance of summer monsoon flash droughts in India. *Environmental Research Letters, 15*(10), 104061. doi:10.1088/1748-9326/abaf1d

Moore, D. O. (1976). *Estimating peak discharge from small drainages in Nevada according to basin areas within elevation zones* (Vol. 3). Nevada Highway Department Hydrologic.

Mulvaney, T. J. (1851). On the use of self-registering rain and flood gauges in making observations of the relations of rainfall and flood discharges in a given catchment. *Proceedings of the Institution of Civil Engineers of Ireland, 4*, 18–33.

World Meteorological Organization (WMO). (2020). *Vision for the WMO Integrated Global Observing System in 2040* (2019 edition). WMO.

World Meteorological Organization (WMO). (2021). *Manual on the WMO Integrated Global Observing System (WMO-No. 1160): Annex VIII to the WMO Technical Regulations* (2021 edition). WMO.

WMO. (n.d.-a). *Climate Data Management System Specifications*. WMO. https://library.wmo.int/records/item/51447-climate-data-management-system-specifications

WMO. (n.d.-b). *Guide to Hydrological Practices, Volume I*. WMO. https://library.wmo.int/records/item/35804-guide-to-hydrological-practices-volume-i?offset=2

Otkin, J. A., Svoboda, M., Hunt, E. D., Ford, T. W., Anderson, M. C., Hain, C., & Basara, J. B. (2018). Flash Droughts: A Review and Assessment of the Challenges Imposed by Rapid-Onset Droughts in the United States. *Bulletin of the American Meteorological Society, 99*(5), 911–919. doi:10.1175/BAMS-D-17-0149.1

Partnership (GWP). G. W., & Organization (WMO), W. M. (2013). *Flood Mapping*. WMO. https://library.wmo.int/records/item/37083-flood-mapping

Petrescu (Maftei). C. (2002). *Etudes concernant les écoulements superficiels modélisation spatialisée de l'écoulement sur le bassin versant de Voinesti-Roumanie* [PhD Thesis, Université de Montpellier II].

Sherman, I. (1932a). Streamflow from Rainfall by Unit-Graph Method. *Engineering News Record, 108*, 501–505.

Sherman, l., (1932b). The relation of hydrographs of runoff to size and character of drainage basins. *Eos, Transactions American Geophysical Union, 13*(1), 332–339. doi:10.1029/TR013i001p00332

Sundararajan, K., Garg, L., Srinivasan, K., Bashir, A., Kaliappan, J., Ganapathy, G., Selvaraj, S., & Meena, T. (2021). A Contemporary Review on Drought Modeling Using Machine Learning Approaches. *Computer Modeling in Engineering & Sciences*, *128*(2), 447–487. doi:10.32604/cmes.2021.015528

Thomas, D. M., & Benson, M. A. (1970). Generalization of streamflow characteristics from drainage-basin characteristics. In *Water Supply Paper*. U.S. Govt. Print. Off. (Original work published 1975), doi:10.3133/wsp1975

U.S. Army Corps of Engineers. (1993). *Hydrologic Frequency Analysis*. US Army.

USACE. (1992). HEC-FFA: Flood Frequency Analysis. User's Manual [HEC-FFA Frequency]. US Army Corps of Engineers, Hydrologic Engineering Center.

Van Beek, C., & Tóth, G. (2012). *Risk assessment methodologies of soil threats in Europe : status and options for harmonization for risks by erosion, compaction, salinization, organic matter decline, and landslides*. Publications Office of the European Union. https://data.europa.eu/doi/10.2788/47096

Vaseashta, A. (2022). Nexus of Advanced Technology Platforms for Strengthening Cyber-Defense Capabilities. In I. Djugumanov & M. Hadji-Janev (Eds.), *NATO Science for Peace and Security Series – E: Human and Societal Dynamics*. IOS Press. doi:10.3233/NHSDP220003

Vaseashta, A., Duca, G., & Travin, S. (Eds.). (2022). *Handbook of Research on Water Sciences and Society* (Vols. 1–2). IGI Global., doi:10.4018/978-1-7998-7356-3

Water Resources Council (U.S.) (Ed.). (1977). Guidelines for determining flood flow frequency (Rev. June 1977). Water Resources Council, Hydrology Committee.

Water Resources Council (U.S.). (1981). *Guidelines for determining flood flow frequency* (Rev September 1981). Water Resources Council, Hydrology Committee. https://library.wmo.int/records/item/35804-guide-to-hydrological-practices-volume-i

WMO. (2017). *Guidelines on Best Practices for Achieving User Readiness for New Meteorological Satellites*. WMO. https://library.wmo.int/doc_num.php?explnum_id=3553

WMO. (2018). *Guidelines on Satellite Skills and Knowledge for Operational Meteorologists*. WMO. https://library.wmo.int/doc_num.php?explnum_id=5799

WMO. (2023a). *Atlas of Mortality and Economic Losses from Weather, Climate, and Water Extremes (1970–2019)*. WMO. https://storymaps.arcgis.com/stories/8df884dbd4e849c89d4b1128fa5dc1d6

WMO. (2023b). *Guidelines on the Definition and Characterization of Extreme Weather and Climate Events*. World Meteorological Organization. https://library.wmo.int/records/item/58396-guidelines-on-the-definition-and-characterization-of-extreme-weather-and-climate-events

WMO, & GWP. (2014). *National Drought Management Policy Guidelines: A template for action*. MWO. https://www.droughtmanagement.info/literature/IDMP_NDMPG_en.pdf

World Meteorological Organization -WMO, & Global Water Partnership - GWP. (2016). *Handbook of Drought Indicators and Indices* WMO. https://public.wmo.int/en/resources/library/handbook-of-drought-indicators-and-indices

Chapter 2
A Review of Monitoring Ecohydrological Events Using Remote Sensing

Vahid Nasiri
Agriculture University in Krakow, Poland

Reza Sarli
Faculty of Forestry, University of Agriculture in Krakow, Poland

Samaneh Afshari
University of Tehran, Iran

ABSTRACT

This review provides a comprehensive analysis of the use of remote sensing techniques in monitoring ecohydrological events. Ecohydrology is an interdisciplinary field that explores the interactions between ecosystems and the water cycle. Remote sensing, with its ability to capture large-scale and continuous observations, has proven to be an invaluable tool in monitoring and understanding these complex interactions. In this chapter, the authors discuss the various remote sensing platforms, sensors, and techniques employed to monitor ecohydrological events, including vegetation dynamics, water availability, and land-use changes. The authors also examine the challenges and future prospects in the field, highlighting the potential for advancing our understanding of ecohydrological processes through remote sensing.

INTRODUCTION: DEFINITION AND SIGNIFICANCE OF ECOHYDROLOGY AND EXTREME ECOHYDROLOGICAL EVENTS

All life on Earth depends on water as a fundamental resource, and the quantity and quality of water directly affect the resilience, structure, and function of ecosystems (Falkenmark and Wang, 2021). Ecohydrology strives to advance our understanding of how changes in hydrological conditions impact the health and operation of ecological systems by examining the intricate connections between water and

DOI: 10.4018/978-1-6684-8771-6.ch002

ecosystems (Cao et al., 2022). In order to address pressing environmental issues including water scarcity, land degradation, climate change, and biodiversity loss, ecohydrology is extremely important (Lindsay et al., 2021). It facilitates the development of sustainable water management strategies and conservation initiatives by providing insightful information about the effects of human activity on water resources and ecosystem health (Makanda et al., 2022). Additionally, the ecohydrological study advances our knowledge of ecosystem services like water purification, flooding prevention, and habitat provision by emphasizing the critical function of healthy, functioning ecosystems in promoting human well-being (Li et al., 2022).

Extreme ecohydrological events are instances that dramatically depart from predicted or normal hydrological and ecological circumstances (Moore et al., 2015). On ecosystems, water resources, and human societies, these occurrences may have significant effects. They tend to take on a variety of forms, including floods, droughts, wildfires, storms, and heatwaves, and are frequently distinguished by their strength, duration, or frequency (Seneviratne et al., 2012). They are influenced by various factors, including climate change, land-use changes, and natural climate variability (Duveiller et al., 2018). Monitoring and understanding these events using remote sensing, hydrological modeling, and other tools can help in assessing their impacts, developing early warning systems, and implementing effective mitigation and adaptation strategies (Kumar et al., 2021). By integrating ecohydrological research and monitoring with proactive management approaches, societies can strive to reduce the risks and enhance the resilience of ecosystems and water resources in the face of extreme events.

ROLE OF REMOTE SENSING IN MONITORING ECOHYDROLOGICAL EVENTS

Remote sensing plays a pivotal role in ecohydrological event monitoring through the acquisition of comprehensive and continuous observations of the Earth's surface remotely. It facilitates the retrieval of data pertaining to diverse ecohydrological parameters (Laiolo et al., 2016), encompassing vegetation dynamics (Stančikaitė et al., 2021), water availability (Huang et al., 2021), and land-use alterations (Bhattacharya et al., 2020), across extensive spatial domains and diverse temporal scales. Here are the key roles of remote sensing in monitoring ecohydrological events:

Large-scale monitoring: Remote sensing platforms, including satellites, airborne instruments, and aerial platforms, afford a synoptic perspective of the Earth's surface, facilitating the comprehensive monitoring of ecohydrological events across extensive geographic extents (Gomes et al., 2020). This capability proves especially valuable for assessing the magnitude and intensity of phenomena such as floods, droughts, and wildfires, which often transcend regional and national boundaries (Pamela et al., 2023). By enabling the detection and mapping of these events, remote sensing expedites timely response and efficient management strategies.

Continuous monitoring: Remote sensing enables continuous monitoring of ecohydrological processes, capturing changes and trends over time (Wang et al., 2023). This is essential for understanding the dynamics of ecosystems and water resources. For example, remote sensing can track the progression of vegetation growth (Dube et al., 2022), assess variations in soil moisture content (Wang et al., 2022), and monitor the filling and depletion of water bodies (Walker et al., 2016). Continuous monitoring provides valuable information for ecological studies, water resource management, and early warning systems for extreme events.

Multispectral and hyperspectral analysis: Remote sensing sensors can capture data across different spectral bands, ranging from visible to thermal infrared and microwave regions. This allows for the analysis of specific ecohydrological parameters, such as vegetation indices (Lemenkova and Debeir, 2023), land cover classification (Torabi et al., 2023), and surface temperature (Pourshanbeh et al., 2022). The multispectral and hyperspectral analysis provides insights into vegetation health (Sousa and Small, 2022), water availability (Hasson et al., 2019), and land-use changes (Darvishi and Arekhi, 2023), facilitating the assessment of ecosystem condition and hydrological processes (Zhang et al., 2023).

Integration with hydrological models: Remote sensing data can be integrated with hydrological models to improve the understanding and prediction of ecohydrological events. By assimilating remote sensing observations into models, researchers can calibrate and validate model parameters, estimate key variables (e.g., evapotranspiration, runoff), and enhance the accuracy of predictions. This integration helps in better managing water resources, assessing the impacts of land-use changes, and supporting decision-making for sustainable water management (García et al., 2023).

Early warning and disaster response: Remote sensing plays a vital role in providing early warning of ecohydrological events and supporting disaster response efforts (Panwar and Sen, 2019). For instance, satellite-based rainfall estimates can aid in predicting floods, while satellite imagery can help identify drought-affected areas and assess wildfire risks (Samansiri et al., 2022). Remote sensing data can also assist in post-disaster damage assessment (Yamazaki and Matsuoka, 2007), mapping affected areas (Kushiyama and Matsuoka, 2019), and supporting relief (Guo et al., 2021), and recovery efforts (Kaplow, 2018).

Long-term trend analysis: Remote sensing data archives enable the analysis of long-term trends in ecohydrological events, allowing researchers to study the effects of climate change, land-use changes, and other drivers on ecosystems and water resources (Roshan et al., 2021). By examining historical satellite imagery and other remote sensing datasets, scientists can identify shifts in vegetation patterns (Beamish et al., 2020), changes in water availability (Schwank et al., 2014), and alterations in landscape dynamics (Van der Plas et al., 2019), aiding in understanding and managing the impacts of environmental change.

REMOTE SENSING PLATFORMS AND SENSORS

Remote sensing platforms encompass a range of vehicles and devices specifically designed for the acquisition of data pertaining to the Earth's surface and atmosphere, employing a distance-based approach (Jafarbiglu and Pourreza, 2022). These platforms exhibit diversity in their technological foundations, drawing upon satellite systems, aircraft, drones, and ground-based sensors. Equipped with a multitude of sensor types, these platforms possess the capability to effectively capture and record intricate details concerning the target area (Liu, 2021). Here are some common remote sensing platforms and sensors:

Satellites: Satellites orbiting the Earth provide a global perspective and are commonly used for remote sensing. They can be classified into two categories: passive and active sensors. Passive sensors measure natural energy (radiation) emitted or reflected by the Earth's surface and atmosphere. Examples include optical sensors, such as multispectral and hyperspectral sensors, which capture data in different wavelength ranges (visible, near-infrared, and thermal infrared) to analyze vegetation, land cover, and atmospheric conditions. Active sensors emit energy (e.g., microwaves or lasers) and measure the reflected or backscattered signals. Synthetic Aperture Radar (SAR) is a commonly used active sensor that can penetrate clouds and capture high-resolution images of the Earth's surface (Mondini et al., 2021). Passive sensors offer a cost-effective solution for continuous monitoring as they rely on solar radiation and can operate day and night. However, they face limitations (Naseem et al., 2016). Cloud cover or nighttime conditions can hinder their effectiveness, and their ability to penetrate through obstacles like clouds or dense vegetation is limited. Additionally, passive sensors might struggle with depth estimation in water bodies during events like floods (Van Dijk et al., 2016). On the other hand, active remote sensing devices emit their own energy sources, such as radar or lidar, and record the reflected signal (Lopacet al., 2022). They excel in all weather conditions, penetrating through clouds, vegetation, and water, providing valuable data even when visibility is poor. Active sensors offer higher-resolution imaging, allowing for the detection of fine details in flood-prone areas (Muhadi et al., 2022). They can estimate water depth accurately, a critical factor in flood monitoring. However, their complexity and higher cost can limit their widespread deployment. Signal interference and a narrower field of view compared to passive sensors are also considerations in their use for flood monitoring (Jakovljević and Govedarica, 2018). Combining the strengths of both types of sensors could offer a more comprehensive solution. Passive sensors provide continuous global coverage, while active sensors offer detailed and accurate data in adverse conditions. Integrating data from both could significantly enhance our ability to monitor and manage hydrological events like floods more effectively. The study by Bhaga et al. (2023) builds upon this foundation, utilizing Landsat-8 OLI and Sentinel-2 MSI data to characterize and monitor drought impacts on surface water resources. The inclusion of multiple multispectral indices, such as the Normalized Different Vegetation Index (NDVI), VCI (Vegetation Condition Index), NDWI (Normalized Different Water Index), MNDWI (Modified Normalized Different Water Index), and Land Surface Water Index (LSWI), adds depth to the analysis, allowing for a comprehensive evaluation of suitable methods for surface water detection and drought monitoring. The study conducted by Benzougagh et al. (2022) indicates that the combination of Sentinel-2 and Landsat-8 data provides an effective approach for monitoring water surface areas and assessing drought risks with precision. The use of remote sensing tools and practical software enhances the understanding of the impacts of drought on various sectors, including water resources, agriculture, ecology, and socio-economic aspects.

Aircraft: Aircraft-based remote sensing platforms offer high spatial resolution and flexibility in data acquisition. They can carry a range of sensors, including those similar to satellite-based sensors. Additionally, aerial platforms can use specialized sensors such as LIDAR (Light Detection and Ranging) to capture detailed elevation data by emitting laser pulses and measuring their return time (You et al., 2021). Their adaptability in terms of choosing and deploying sensors allows for customized strategies, maximizing observations for certain goals, and improving the accuracy of the data gathered during these occurrences (Wang, 2015). Furthermore, their quick mobilization allows for the collection of data in real-time or almost real-time, which speeds up evaluations and response preparation (Chien et al., 2019). These platforms nevertheless come with certain restrictions. Because of its relatively lower coverage area compared to satellite-based systems, it takes several flights to cover large areas affected by hydrological events. Furthermore, the prices associated with fuel, maintenance, and personnel can be exorbitant, which may restrict the quantity or extensive application of continuous monitoring (Matese et al., 2015). Additionally, aircraft operations are weather-dependent and might be impacted by unfavorable conditions that could prevent flights during extreme weather. In short, their detailed observations aid in emergency response, impact assessment, and resource allocation. Balancing the cost, coverage limitations, and operational complexities against the advantages they offer is essential to effectively utilizing aircraft-based platforms for monitoring floods, droughts, and other hydrological events. Integration of data from aircraft-based systems with other remote sensing methods enhances the comprehensive understanding of these events, facilitating better-informed decision-making and response strategies.

Drones (Unmanned Aerial Vehicles - UAVs): Drones have gained popularity in recent years due to their accessibility, maneuverability, and affordability. They are equipped with various sensors, including optical sensors, thermal sensors, LIDAR, and multispectral sensors. Drones allow for localized and on-demand data collection, making them suitable for smaller-scale mapping and monitoring tasks (Mohsan et al., 2023). Drones, with their rapid deployment capabilities, provide swift decision-making for disaster response and mitigation measures by providing real-time or near-real-time data collecting during present hydrological events. Their cost-effectiveness, especially for concentrated evaluations in limited areas, makes them a feasible and easily obtainable choice for accurate observation. Drones nevertheless have restricted operational range. When compared to satellite or manned aircraft platforms, their short flying duration and range limit the coverage areas. During emergencies, quick deployment may be hampered by weather sensitivity and regulatory restrictions, such as licenses and regulations pertaining to airspace. Payload restrictions also limit the kinds and quantity of sensors that may be carried, which may have an effect on the variety and complexity of the data gathered. Notwithstanding these disadvantages, drones are essential for delivering accurate, localized data that is needed for efficient reaction and management strategies in the event of hydrological events. Combining data from drones with other remote sensing techniques improves the overall knowledge required for well-informed decisions and focused interventions. For example, the study conducted by Otsu et al. (2019) focused on the growing threat of the pine processionary moth to Mediterranean forests, specifically in Catalonia, Spain, following a recent outbreak. Researchers assessed damage severity by analyzing the maximum defoliation period using Landsat 8 imagery. They employed the Moisture Stress Index (MSI), derived from the difference in vegetation index (dVI) and calibrated with UAV imagery, as a promising method to estimate defoliation severity in areas with limited ground-truth data. The study emphasizes the potential of UAVs as a cost-effective means to obtain ground-truth data for monitoring forest health. Future research may explore integrating UAV images with satellite data to validate model predictions and enhance the development of ecosystem service tools for assessing forest health.

Ground-based Sensors: These sensors are deployed on the Earth's surface to collect data at specific locations. Examples include weather stations, soil moisture sensors, and seismic sensors. These sensors provide detailed, localized information and can be used for specific monitoring purposes (Zhang et al., 2013). Ground-based sensors, with their precision and localized measuring capabilities, are essential for monitoring hydrological events such as droughts and floods. These sensors provide extremely precise data gathering at certain locations, offering comprehensive insights into soil moisture content, water levels, and other vital hydrological factors. Their ability to transmit data in real-time or almost real-time facilitates prompt answers and interventions during ongoing events, which supports prompt decision-making for efficient management. Their limited coverage area around the installation point, on the other hand, makes many sensors necessary for thorough monitoring of broader territories, which raises the costs of deployment and maintenance. Challenges also include the need for constant infrastructure maintenance and calibration, as well as the possibility of damage or sabotage. Despite these limitations, when combined with information from other monitoring platforms for an expanded perspective, ground-based sensors become a vital component of recognizing and reacting to hydrological events due to their dependability in challenging circumstances and capacity to provide accurate, localized data. In this context, the study conducted by Gallant et al. (2018) addresses the challenges of assessing climate-related ecological changes across different spatiotemporal scales, particularly in wetland-rich areas of the U.S. Midwest. The researchers integrated data from ground- and satellite-based sensors, examining the relationship between temperature, precipitation, and various ecological variables measured both on the ground at individual wetlands and via satellite sensors within surrounding landscape blocks. While responses at the landscape block level generally mirrored ground-level changes, the latter were more dynamic, and crucial ecological conditions for biota were not always apparent in block-level measurements. The study evaluates the decisions made in applying remotely sensed data and underscores the value of integrating observations across scales, sensors, and disciplines for a comprehensive understanding of climate-related ecological changes.

The comparison of platforms and sensors for ecohydrological monitoring should consider the specific research objectives, spatial and temporal scales, and resource constraints. Integrating multiple platforms and sensors can provide a comprehensive assessment of ecohydrological dynamics, allowing for more informed decision-making in water resource management and ecosystem conservation. In the next sections, the review focuses on the paramount significance of extreme ecohydrological events, examining their characteristics and impacts. Subsequently, it delves into the utilization of remote sensing platforms and sensors for effective monitoring and assessment of these events.

MONITORING VEGETATION DYNAMICS AND EXTREME ECOHYDROLOGICAL EVENTS

Vegetation is a highly sensitive indicator of environmental changes, encompassing variations in temperature, precipitation, and water availability (Nasiri et al., 2023). By systematically monitoring vegetation dynamics, which include changes in leaf area, canopy cover, and biomass, early warning signs of ecohydrological events can be detected (Xie et al., 2019). Notably, abrupt reductions in vegetation productivity or the occurrence of browning phenomena indicate the presence of drought conditions or water stress (Seleiman et al., 2021). By closely observing these vegetation parameters, it becomes possible to discern these early indications and proactively respond to potential ecohydrological events (Fang et al., 2022). In the context of droughts, water-stressed plants exhibit evident signs such as reduced growth rates, wilting, leaf loss, or premature senescence (Quiroga et al., 2020). Monitoring techniques such as satellite imagery, remote sensing, or ground-based measurements allow for tracking these vegetation changes and assessing the extent and severity of drought events. This comprehensive approach enables a more precise understanding of drought dynamics, facilitating effective management strategies and resource allocation to mitigate the impacts on ecosystems and water resources (Aroca et al., 2007). Vegetation dynamics also contribute to assessing flood risks, as dense vegetation like forests or wetlands can act as natural buffers, slowing down water movement and reducing runoff. Monitoring changes in vegetation cover and density provides valuable insights into the effectiveness of these natural flood control measures (Chen et al., 2020). Furthermore, vegetation conditions, particularly the presence of dry and combustible fuels, play a crucial role in determining the risk of wildfires (Nasiri et al., 2022a). Monitoring parameters such as vegetation greenness, fuel moisture content, and vegetation flammability aids in identifying areas prone to extreme fire events (Rafaqat et al., 2022). This information is invaluable for fire management agencies to prioritize prevention measures, allocate resources effectively, and issue timely warnings. Additionally, vegetation dynamics indicate the overall health and resilience of ecosystems. Sudden changes or declines in vegetation cover, such as forest dieback or desertification, serve as indicators of ecological disturbances or shifts within ecosystems (Sample et al., 2022). By monitoring these changes, a better understanding of the drivers behind ecosystem degradation is gained, leading to appropriate measures for ecosystem restoration and protection.

Time series spectral indices, particularly vegetation indices, are widely recognized and essential tools for monitoring vegetation dynamics and assessing the influence of drought conditions on vegetation (Van Hoek et al., 2016). These indices, such as the Normalized Difference Vegetation Index (NDVI) or Enhanced Vegetation Index (EVI), quantify the photosynthetic activity and greenness of vegetation (Telesca et al., 2022). Calculated from remotely sensed data, such as satellite and aerial imagery, they provide valuable information on vegetation vigor, health, and productivity. Anomalies or reductions in vegetation indices serve as indicators of stress, decreased biomass, or alterations in vegetation cover, offering valuable insights into the impact of drought or other environmental factors on vegetation (Candiago et al., 2015). Drought indices, on the other hand, are numeric indicators designed to capture the severity and spatial extent of drought conditions (Vicente-Serrano et al., 2015). These indices integrate various meteorological and hydrological variables, including precipitation, temperature, soil moisture, and evapotranspiration. Commonly used drought indices include the Standardized Precipitation Index (SPI), Palmer Drought Severity Index (PDSI), and Vegetation Drought Response Index (VegDRI) (Lee et al., 2021). By correlating drought indices with time series vegetation indices, researchers can establish the relationship between vegetation dynamics and drought conditions. This correlation aids in the identification of periods characterized by vegetation stress, evaluation of the impacts of drought on vegetation health and productivity, and comprehension of vegetation responses to evolving water availability. In previous studies, for time series analysis of vegetation, several satellite imagery datasets are commonly used.

Moderate Resolution Imaging Spectroradiometer (MODIS) sensors onboard the NASA Terra (https://terra.nasa.gov/about/terra-instruments/modis) and Aqua (https://oceancolor.gsfc.nasa.gov/data/aqua/) satellites provide moderate spatial resolution data (250 m to 1 km) with a daily revisit time. MODIS offers a wide range of spectral bands, including the red and near-infrared bands used to calculate vegetation indices like NDVI. MODIS data is widely used for long-term vegetation monitoring and trend analysis. The spatial resolutions that MODIS sensors usually provide range from 250 meters to 1 kilometer, which may not be sufficient to detect small-scale or fine-scale changes associated with ecohydrological events. This limitation may have an impact on localized flood or drought effects, subtle changes in the health of the vegetation, and the detection and detailed characterization of smaller water bodies. As such, accurate monitoring of particular, localized effects may prove difficult (Wang et al., 2016; Kuenzer et al., 2015). Moreover, MODIS's daily observations can be greatly impacted by cloud cover. Daily observations may not be available in areas where cloud cover or atmospheric interference occur frequently. It becomes difficult to monitor continuously, which causes gaps in the dataset and makes it more difficult to track daily changes in ecohydrological events in these areas (Almamalachy et al., 2019). Addressing these limitations often involves using a combination of satellite sensors with varying spatial resolutions, integrating data from multiple sources, and employing advanced modeling or data fusion techniques to enhance the monitoring and understanding of ecohydrological events at different scales. For instance, Son et al. (2019) leveraged a fusion technique using multi-temporal MODIS-Landsat satellite imagery to evaluate cropland drought in Central America. Their study showcased the fusion's effectiveness by comparing the results with reference Landsat data. This comparison demonstrated a close agreement between the fused data and two products, indicating that this fusion approach could yield dependable information regarding drought based on MODIS sensor data, albeit at a Landsat-equivalent spatial resolution of 30 meters.

Landsat: The Landsat series of satellites, such as Landsat 7 and Landsat 8, provide moderate to high spatial resolution (30 m) multispectral data. The Landsat program has a rich historical archive of satellite imagery that spans several decades. Landsat satellites have been capturing Earth's surface imagery since the launch of Landsat 1 in 1972, with subsequent missions including Landsat 2, Landsat 3, Landsat 4, Landsat 5, Landsat 7, and the latest mission, Landsat 8. Landsat 9 in late 2021 (https://www.usgs.gov/landsat-missions/landsat-data-access). Their spectral diversity and open, free data access make them more useful for the environmental management, policy, and research domains. Nevertheless, one of Landsat's drawbacks is its temporal resolution, which is only 8 to 16 days, which may make it difficult to record abrupt or short-lived changes in ecohydrological events. Furthermore, the accuracy of localized monitoring may be impacted by its 30-meter resolution's inability to capture fine-scale variations. In certain situations, cloud cover, atmospheric interference, and significant data processing needs make it even less effective. Landsat's extensive historical data and moderate spatial resolution, despite these drawbacks, continue to be essential for comprehending long-term ecohydrological trends and significant environmental changes (Yang et al., 2021).

Sentinel-2: The Sentinel-2 satellites, operated by the European Space Agency, offer multispectral imagery with high spatial resolution (ranging from 10 to 20 meters) and a revisit time of 5 days for global coverage (https://sentinel.esa.int/web/sentinel/missions/sentinel-2). Equipped with sensors that capture various spectral bands, including red-edge and near-infrared wavelengths, Sentinel-2 is well-suited for scientific applications such as vegetation monitoring, land cover classification, and the calculation of vegetation indices (Nasiri et al., 2022b). Its high-resolution data enables detailed analysis of vegetation dynamics, including the assessment of vegetation health, mapping of land surface characteristics, and identification of vegetation stress indicators. Nevertheless, despite its advantages, Sentinel has limitations such as handling massive data volumes and intricate processing needs that necessitate high computational power for analysis and storage. Cloud cover is still a constraint on optical sensors, which can make it more difficult to acquire data in areas where bad weather or extended cloud cover is common (Traganoset al., 2018).

Advanced Very High-Resolution Radiometer (AVHRR): AVHRR (Advanced Very High-Resolution Radiometer) instruments are deployed on multiple NOAA (National Oceanic and Atmospheric Administration) satellites, delivering data with a moderate spatial resolution of approximately 1 kilometer and a daily revisit time. AVHRR data has been extensively employed for long-term vegetation monitoring and climate studies. With its consistent and frequent observations, AVHRR enables the analysis of vegetation dynamics and the detection of changes over extended time periods. The data obtained from AVHRR supports investigations into vegetation phenology, vegetation cover mapping, and the assessment of vegetation responses to environmental factors. Moreover, AVHRR data plays a crucial role in climate studies by contributing to the analysis of land surface temperature, the monitoring of land-ocean interactions, and the evaluation of climate-driven impacts on vegetation ecosystems.

PlanetScope: PlanetScope is a commercial satellite constellation that provides high-resolution imagery (3 to 5 m) with a high revisit time (https://earth.esa.int/eogateway/missions/planetscope). The frequent revisit allows for capturing short-term vegetation dynamics and monitoring changes at a finer spatial scale. This high temporal resolution is crucial for tracking quickly changing events almost every day, enabling quick reactions and flexible management techniques. Its quick data availability also enables near-real-time monitoring, which facilitates prompt decision-making in environmental emergencies. While using PlanetScope for long-term trend analysis or monitoring areas with persistent cloud cover, users should be aware of its limitations in historical archives, potential cost implications for extensive coverage, and susceptibility to atmospheric interference and cloud cover.

The selection of satellite imagery for time series analysis of vegetation is contingent upon various factors, including the desired spatial resolution, temporal frequency, and specific research or monitoring objectives. Each satellite sensor possesses unique strengths and limitations, necessitating careful consideration of these factors during the dataset selection process.

ASSESSING WATER AVAILABILITY

The assessment of water availability and extreme ecohydrological events entails a comprehensive examination of the intricate interplay among water resources, ecosystems, and human activities (Elias et al., 2015). This scientific assessment assumes paramount significance as it serves to elucidate the present condition of water resources, prognosticate future water availability, and effectively address water-related risks and challenges (Guo et al., 2019). The integration of remote sensing data with hydrological models and ground-based measurements facilitates water availability assessments by harnessing the spatial and temporal coverage offered by remote sensing technology (Wu et al., 2023). This integration enables a more extensive and precise monitoring of water resources, which, in turn, bolsters the efficacy of water management practices and decision-making processes. Remote sensing can be used for assessing water availability in several ways:

Surface Water Mapping

Remote sensing is extensively used for surface water mapping due to its ability to provide wide-scale coverage and repetitive observations. Each remote sensing sensor has its strengths and limitations, and the choice depends on the specific requirements of the surface water mapping study, including spatial resolution, coverage area, temporal frequency, and data accessibility. Integrating data from multiple sensors and platforms often provides a more comprehensive understanding of surface water dynamics. Optical sensors installed on satellites like Landsat, Sentinel-2, and MODIS capture data across various spectral bands, encompassing the visible and near-infrared regions. These sensors are well-suited for surface water mapping owing to their capability to discern water bodies from adjacent land areas by exploiting distinctive reflectance properties. Based on optical sensors, different spectral indices can be extracted for mapping water bodies:

- Normalized Difference Water Index (NDWI): The NDWI is calculated using the near-infrared (NIR) and green or blue spectral bands. NDWI is sensitive to the presence of water, and higher values indicate the presence of water bodies.
- Modified Normalized Difference Water Index (MNDWI): The MNDWI is similar to NDWI but utilizes the green and mid-infrared (MIR) spectral bands. MNDWI is effective in differentiating water bodies from other land cover types (Laonamsai et al., 2023).
- Normalized Difference Vegetation Index (NDVI): Although primarily used for vegetation assessment, NDVI can also aid in surface water mapping. NDVI is calculated using the NIR and red spectral bands. Water bodies typically exhibit low NDVI values due to their low reflectance in the NIR band.

- Floating Algae Index (FAI): The FAI is designed to detect and monitor floating algae on water surfaces. It uses the red and blue spectral bands. High FAI values indicate the presence of floating algae on the water surface (Muzhoffar et al., 2023).
- Water Ratio Index (WRI): The WRI utilizes the NIR and shortwave infrared (SWIR) spectral bands and is calculated as NIR/SWIR. WRI is sensitive to water bodies and can help identify and map them accurately (Laonamsai et al., 2023).
- Blue Normalized Difference Index (BNDI): The BNDI utilizes the blue and green spectral bands. BNDI can assist in differentiating water bodies from other land cover types, particularly in coastal areas (Huang et al., 2016).

In addition to optical sensors, synthetic aperture radar (SAR) sensors, such as those on Sentinel-1 and RADARSAT satellites, use microwave frequencies to measure backscattered signals from the Earth's surface (Ayehu et al., 2020). SAR is particularly useful for surface water mapping because microwaves can penetrate clouds and vegetation, allowing for all-weather and day-night imaging capabilities (El Hajj et al., 2019). SAR measures the strength of backscattered radar signals, which are influenced by the surface properties of water bodies. Calibrated SAR backscatter intensity can be used to identify and differentiate water bodies from surrounding land areas (Kseňak et al., 2022). For example, SAR data possesses the ability to capture surface texture variations associated with water bodies. Through the application of texture analysis techniques, such as the utilization of the gray-level co-occurrence matrix (GLCM), it becomes possible to quantitatively assess textural characteristics. This facilitates the discrimination and delineation of water bodies, enhancing the accuracy of their identification (Bayanudin et al., 2016). Also, SAR data can be utilized to derive polarimetric indices that are specifically tailored for water body mapping purposes. Notable examples encompass the Water Cloud Model (WCM) and the Water Cloud Similarity (WCS) index. These indices leverage the polarimetric responses exhibited by water bodies, thereby augmenting their detection and classification capabilities (Li et al., 2018).

Soil Moisture Estimation

Radar satellites are extensively utilized for soil moisture estimation due to their capacity to penetrate through vegetation and clouds (Singh et al., 2020). Specifically, radar sensors operating in the microwave frequency range are exceptionally suitable for this purpose. In contrast, optical satellites rely on sunlight reflection and are not as proficient in directly measuring soil moisture content (Zribi et al., 2020). Nevertheless, optical satellite data remains valuable as it can furnish pertinent information concerning vegetation health and land surface conditions, which indirectly contribute to soil moisture estimation through derived parameters such as vegetation indices (Rasheed et al., 2022). Consequently, while radar satellites are the primary preference for soil moisture estimation, the amalgamation of radar and optical satellite data yields a more comprehensive understanding of soil moisture dynamics (He et al., 2014). Several radar satellites are commonly employed for soil moisture estimation. The Soil Moisture Active Passive (SMAP) mission by NASA combines active (radar) and passive (radiometer) microwave measurements, operating in the L-band frequency, to provide high-resolution soil moisture data globally (Brown et al., 2013). The Sentinel-1 constellation, operated by the European Space Agency (ESA), utilizes C-band synthetic aperture radar (SAR) with dual-polarization capabilities, enabling frequent global coverage. ALOS-2, operated by JAXA, carries a fully polarimetric L-band SAR sensor, offering valuable data for soil moisture estimation by capturing vegetation and soil moisture variations. RADARSAT-2, operated by the Canadian Space Agency (CSA), features a C-band SAR sensor and provides high-resolution radar imagery, widely utilized for soil moisture estimation (Moon et al., 2010). Additionally, the twin satellites TerraSAR-X and TanDEM-X, operated by the German Aerospace Center (DLR), employ X-band SAR sensors, offering high-resolution capabilities and dual-polarization options suitable for soil moisture estimation (Buckreuss et al., 2018).

Evapotranspiration Monitoring

Evapotranspiration (ET) monitoring is of paramount importance for comprehending the water cycle and effectively managing water resources. In this regard, remote sensing techniques emerge as indispensable tools for estimating and monitoring ET across extensive geographical regions (Wanniarachchi et al., 2022). The utilization of remote sensing for ET monitoring encompasses several key approaches.

- Thermal Infrared Remote Sensing: Land surface temperature (LST) is measured using thermal infrared sensors on satellites like Landsat, MODIS, and Sentinel-3. By combining LST data with meteorological information and vegetation indices such as NDVI, the energy balance equation can be applied to estimate ET (Roshan et al., 2022).
- Vegetation Indices: Optical satellite imagery allows the derivation of vegetation indices such as NDVI and EVI, offering insights into vegetation growth and vigor. By integrating vegetation indices with ET models, the estimation of ET based on vegetation transpiration processes becomes feasible.
- Surface Energy Balance Models: Remote sensing data, including land surface temperature, vegetation indices, and solar radiation, can be integrated into surface energy balance models like SEBAL or TSEB. These models calculate ET by considering the energy fluxes at the land-atmosphere interface (Andreu et al., 2018).
- Microwave Remote Sensing: Microwave sensors on satellites like SMAP and AMSR provide information on soil moisture content. Since soil moisture is closely related to ET, integrating microwave data with other variables enables indirect estimation of ET (Zhu et al., 2022).
- Water Balance Modeling: Remote sensing data facilitates the derivation of essential inputs for water balance models used in ET estimation. Parameters such as precipitation, temperature, vegetation cover, and soil moisture, obtained from remote sensing, enhance the accuracy of water balance modeling.
- Evapotranspiration Product Services: Several organizations and research groups offer ET products derived from remote sensing data, such as the MODIS Evapotranspiration (MOD16) product. These pre-processed ET estimates serve as valuable resources for monitoring water resources.

Flood Monitoring and Prediction

Flood monitoring and prediction are crucial components of effective disaster management and mitigation strategies. Remote sensing techniques offer valuable capabilities for these purposes. Optical sensors on satellites, such as Landsat and Sentinel-2, provide high-resolution imagery to identify and monitor flood-affected areas, enabling flood extent mapping through pre- and post-flood image comparisons. Synthetic Aperture Radar (SAR) sensors, like those on Sentinel-1 and RADARSAT-2 satellites, are particularly useful as they can penetrate clouds and provide continuous monitoring, detecting changes in backscattering properties caused by floodwaters. Digital Elevation Models (DEMs) derived from remote sensing data aid in floodplain mapping and risk assessment by accurately representing the elevation of flood-prone areas.

Remote sensing-based rainfall estimation techniques, such as those employed by the Tropical Rainfall Measuring Mission (TRMM) and Global Precipitation Measurement (GPM) satellites, provide valuable data for flood prediction by monitoring rainfall patterns and intensities (Xu et al., 2019). This information, along with remote sensing data on soil moisture and vegetation cover, can be assimilated into hydrological models to simulate and forecast flood events, contributing to early warning systems (Li et al., 2016). Furthermore, remote sensing data enables the generation of flood inundation maps, illustrating the spatial extent and depth of floodwaters, which are essential for emergency response planning and evacuation strategies.

CONCLUSION

In this review, we present a comprehensive overview of the use of remote sensing in monitoring ecohydrological events. By leveraging the capabilities of remote sensing platforms and sensors, researchers and practitioners can gain valuable insights into the complex interactions between ecosystems and the water cycle. Although challenges and limitations exist, ongoing advancements in technology and data processing techniques provide immense opportunities for further exploration in this field. The integration of remote sensing with hydrological models and the adoption of innovative approaches, such as artificial intelligence and data fusion, hold significant promise for improving our understanding of ecohydrological processes and supporting sustainable water resource management in the future.

REFERENCES

Almamalachy, Y. S., Al-Quraishi, A. M. F., & Moradkhani, H. (2019). Agricultural drought monitoring over Iraq utilizing MODIS products. In Springer water (pp. 253–278). Springer. doi:10.1007/978-3-030-21344-2_11

Andreu, A., Kustas, W., Polo, M., Carrara, A., & González-Dugo, M. (2018). Modeling Surface Energy Fluxes over a Dehesa (Oak Savanna) Ecosystem Using a Thermal Based Two-Source Energy Balance Model (TSEB) I. *Remote Sensing (Basel)*, *10*(4), 567. doi:10.3390/rs10040567

Aroca, R., Porcel, R., & Ruiz-Lozano, J. M. (2007). How does arbuscular mycorrhizal symbiosis regulate root hydraulic properties and plasma membrane aquaporins in Phaseolus vulgaris under drought, cold, and salinity stresses? *The New Phytologist*, *173*(4), 808–816. doi:10.1111/j.1469-8137.2006.01961.x PMID:17286829

Ayehu, G., Tadesse, T., Gessesse, B., Yigrem, Y. M., & Melesse, A. (2020). Combined Use of Sentinel-1 SAR and Landsat Sensors Products for Residual Soil Moisture Retrieval over Agricultural Fields in the Upper Blue Nile Basin, Ethiopia. *Sensors (Basel)*, *20*(11), 3282. doi:10.339020113282 PMID:32526894

Bayanudin, A. A., & Jatmiko, R. H. (2016). Orthorectification of Sentinel-1 SAR (Synthetic Aperture Radar) Data in Some Parts of South-eastern Sulawesi Using Sentinel-1 Toolbox. *IOP Conference Series. Earth and Environmental Science*, *47*, 012007. doi:10.1088/1755-1315/47/1/012007

Beamish, A., Raynolds, M. K., Epstein, H., Frost, G. V., Macander, M. J., Bergstedt, H., Bartsch, A., Kruse, S., Miles, V., Tanis, C. M., Heim, B., Fuchs, M., Chabrillat, S., Shevtsova, I., Verdonen, M., & Wagner, J. (2020). Recent trends and remaining challenges for optical remote sensing of Arctic tundra vegetation: A review and outlook. *Remote Sensing of Environment*, *246*, 111872. doi:10.1016/j.rse.2020.111872

Benzougagh, B., Meshram, S. G., Fellah, B. E., Masters, M., Dridri, A., Sadkaoui, D., Mimich, K., & Khedher, K. M. (2022). Combined use of Sentinel-2 and Landsat-8 to monitor water surface area and evaluated drought risk severity using Google Earth Engine. *Earth Science Informatics*, *15*(2), 929–940. doi:10.100712145-021-00761-9

Bhaga, T. D., Dube, T., Shekede, M. D., & Shoko, C. (2023). Investigating the effectiveness of Landsat-8 OLI and Sentinel-2 MSI satellite data in monitoring the effects of drought on surface water resources in the Western Cape Province, South Africa. *Remote Sensing Applications: Society and Environment*, *32*, 101037. doi:10.1016/j.rsase.2023.101037

Bhattacharya, R. K., Chatterjee, N. D., & Das, K. (2020). An integrated GIS approach to analyze the impact of land use change and land cover alteration on groundwater potential level: A study in Kangsabati Basin, India. *Groundwater for Sustainable Development*, *11*, 100399. doi:10.1016/j.gsd.2020.100399

Brown, M. E., Escobar, V., Moran, S., Entekhabi, D., O'Neill, P. E., Njoku, E. G., Doorn, B., & Entin, J. K. (2013). NASA's Soil Moisture Active Passive (SMAP) Mission and Opportunities for Applications Users. *Bulletin of the American Meteorological Society*, *94*(8), 1125–1128. doi:10.1175/BAMS-D-11-00049.1

Buckreuss, S., Schättler, B., Fritz, T., Mittermayer, J., Kahle, R., Maurer, E., Böer, J., Bachmann, M., Mrowka, F., Schwarz, E., Breit, H., & Steinbrecher, U. (2018). Ten Years of TerraSAR-X Operations. *Remote Sensing (Basel)*, *10*(6), 873. doi:10.3390/rs10060873

Candiago, S., Remondino, F., De Giglio, M., Dubbini, M., & Gattelli, M. (2015). Evaluating Multispectral Images and Vegetation Indices for Precision Farming Applications from UAV Images. *Remote Sensing (Basel)*, *7*(4), 4026–4047. doi:10.3390/rs70404026

Cao, Z., Wang, S., Luo, P., Xie, D., Zhu, W. (2022). Watershed Ecohydrological Processes in a Changing Environment: Opportunities and Challenges. *Water, 14*(9), 1502. Re. doi:10.3390/w14091502

Chen, X., Zhao, P., Ouyang, L., Zhu, L., Ni, G., & Schäfer, K. V. R. (2020). Whole-plant water hydraulic integrity to predict drought-induced Eucalyptus urophylla mortality under drought stress. *Forest Ecology and Management*, *468*, 118179. doi:10.1016/j.foreco.2020.118179

Chien, S., McLaren, D., Doubleday, J., Tran, D., Tanpipat, V., & Chitradon, R. (2019). Using taskable remote sensing in a sensor web for Thailand flood monitoring. *Journal of Aerospace Information Systems*, *16*(3), 107–119. doi:10.2514/1.I010672

Darvishi, Y., & Arekhi, S. (2023). Analysis of Spatial Changes in the Landscape of Cites (Case Study: Kermanshah City). *Landscape Ecology*, *1*(1), 79–95. doi:10.22034/el.2023.170942

Dube, T., Shekede, M. D., & Massari, C. (2022). Remote Sensing for Water Resources and Environmental Management. *Remote Sensing (Basel)*, *15*(1), 18. doi:10.3390/rs15010018

Duveiller, G., Hooker, J., & Cescatti, A. (2018). the mark of vegetation change on Earth's surface energy balance. *Nature Communications*, *9*(1), 679. doi:10.103841467-017-02810-8 PMID:29463795

El Hajj, M. M., Baghdadi, N., Zribi, M., & Bazzi, H. (2017). Synergic Use of Sentinel-1 and Sentinel-2 Images for Operational Soil Moisture Mapping at High Spatial Resolution over Agricultural Areas. *Remote Sensing (Basel)*, *9*(12), 1292. doi:10.3390/rs9121292

Elias, E. H., Rango, A., Steele, C. M., Mejia, J. F., & Smith, R. (2015). Assessing climate change impacts on water availability of snowmelt-dominated basins of the Upper Rio Grande basin. *Journal of Hydrology. Regional Studies*, *3*, 525–546. doi:10.1016/j.ejrh.2015.04.004

Falkenmark, M., & Wang-Erlandsson, L. (2021). A water-function-based framework for understanding and governing water resilience in the Anthropocene. *One Earth*, *4*(2), 213–225. doi:10.1016/j.oneear.2021.01.009

Fang, Q., Wang, G., Zhang, S., Peng, Y., Xue, B., Cao, Y., & Shrestha, S. (2022). A novel ecohydrological model by capturing variations in climate change and vegetation coverage in a semi-arid region of China. *Environmental Research*, *211*, 113085. doi:10.1016/j.envres.2022.113085 PMID:35307372

Gallant, A. L., Sadinski, W. J., Brown, J. F., Senay, G. B., & Roth, M. F. (2018). Challenges in Complementing Data from Ground-Based Sensors with Satellite-Derived Products to Measure Ecological Changes in Relation to Climate—Lessons from Temperate Wetland-Upland Landscapes. *Sensors (Basel)*, *18*(3), 880. doi:10.339018030880 PMID:29547531

García-Galiano, S. G., Cánovas-García, F., & Giraldo-Osorio, J. D. (2023). Editorial for Special Issue: "Remote Sensing of Hydrological Processes: Modelling and Applications.". *Remote Sensing (Basel)*, *15*(5), 1466. doi:10.3390/rs15051466

Gomes, V., Queiroz, G., Ferreira, K. (2020). An Overview of Platforms for Big Earth Observation Data Management and Analysis. *Remote Sensing, 12*(8), 1253. MDPI AG. . doi:10.3390/rs12081253

Guo, A., Chang, J., Wang, Y., Huang, Q., Guo, Z., & Li, Y. (2019). Uncertainty analysis of water availability assessment through the Budyko framework. *Journal of Hydrology (Amsterdam)*, *576*, 396–407. doi:10.1016/j.jhydrol.2019.06.033

Guo, L., Sun, X., Fu, P., Shi, T., Dang, L., Chen, Y., & Zeng, C. (2021). Mapping soil organic carbon stock by hyperspectral and time-series multispectral remote sensing images in low-relief agricultural areas. *Geoderma*, *398*, 115118. doi:10.1016/j.geoderma.2021.115118

Hamid, J., & Alireza, P. (2022). A comprehensive review of remote sensing platforms, sensors, and applications in nut crops. *Computers and Electronics in Agriculture*, *197*, 106844. doi:10.1016/j.compag.2022.106844

Hamid, J., & Alireza, P. (2022). A comprehensive review of remote sensing platforms, sensors, and applications in nut crops. *Computers and Electronics in Agriculture*, *197*, 106844. doi:10.1016/j.compag.2022.106844

Hasson, S., Saeed, F., Böhner, J., & Schleussner, C.-F. (2019). Water availability in Pakistan from Hindukush-Karakoram-Himalayan watersheds at 1.5ºC and 2ºC Paris Agreement Targets. *Advances in Water Resources*, *131*, 103365. doi:10.1016/j.advwatres.2019.06.010

He, B., Xing, M., & Bai, X. (2014). A Synergistic Methodology for Soil Moisture Estimation in an Alpine Prairie Using Radar and Optical Satellite Data. *Remote Sensing (Basel)*, *6*(11), 10966–10985. doi:10.3390/rs61110966

Huang, Z., Liu, X., Jin, M., Ding, C., Jiang, J., & Wu, L. (2016). Deriving the Characteristic Scale for Effectively Monitoring Heavy Metal Stress in Rice by Assimilation of GF-1 Data with the WOFOST Model. *Sensors (Basel)*, *16*(3), 340. doi:10.339016030340 PMID:26959033

Huang, Z., Yuan, X., & Liu, X. (2021). The key drivers for the changes in global water scarcity: Water withdrawal versus water availability. *Journal of Hydrology (Amsterdam)*, *601*, 126658. doi:10.1016/j.jhydrol.2021.126658

Jakovljević, G., & Govedarica, M. (2018). Water body extraction and flood risk assessment using LIDAR and open data. In Climate change management (pp. 93–111). doi:10.1007/978-3-030-03383-5_7

Kseňak, Ľ., Pukanská, K., Bartoš, K., & Blišťan, P. (2022). Assessment of the Usability of SAR and Optical Satellite Data for Monitoring Spatio-Temporal Changes in Surface Water: Bodrog River Case Study. *Water (Basel)*, *14*(3), 299. doi:10.3390/w14030299

Kuenzer, C., Klein, I., Ullmann, T., Foufoula-Georgiou, E., Baumhauer, R., & Dech, S. (2015). Remote Sensing of River Delta inundation: Exploiting the potential of coarse Spatial Resolution, Temporally-Dense MODIS Time Series. *Remote Sensing (Basel)*, *7*(7), 8516–8542. doi:10.3390/rs70708516

Kumar, P., Debele, S. E., Sahani, J., Rawat, N., Marti-Cardona, B., Alfieri, S. M., & Zieher, T. (2021). An overview of monitoring methods for assessing the performance of nature-based solutions against natural hazards. *Earth-Science Reviews*, *217*, 103603. doi:10.1016/j.earscirev.2021.103603

Kushiyama, Y., & Matsuoka, M. (2019). Time Series GIS Map Dataset of Demolished Buildings in Mashiki Town after the 2016 Kumamoto, Japan Earthquake. *Remote Sensing (Basel)*, *11*(19), 2190. doi:10.3390/rs11192190

Laiolo, P., Gabellani, S., Campo, L., Silvestro, F., Delogu, F., Rudari, R., & Puca, S. (2016). Impact of different satellite soil moisture products on the predictions of a continuous distributed hydrological model. *International Journal of Applied Earth Observation and Geoinformation*, *48*, 131–145. doi:10.1016/j.jag.2015.06.002

Laonamsai, J., Julphunthong, P., Saprathet, T., Kimmany, B., Ganchanasuragit, T., Chomcheawchan, P., & Tomun, N. (2023). Utilizing NDWI, MNDWI, SAVI, WRI, and AWEI for Estimating Erosion and Deposition in Ping River in Thailand. *Hydrology*, *10*(3), 70. doi:10.3390/hydrology10030070

Lee, S.-J., Kim, N., & Lee, Y. (2021). Development of Integrated Crop Drought Index by Combining Rainfall, Land Surface Temperature, Evapotranspiration, Soil Moisture, and Vegetation Index for Agricultural Drought Monitoring. *Remote Sensing (Basel)*, *13*(9), 1778. doi:10.3390/rs13091778

Lemenkova, P., & Debeir, O. (2023). Multispectral Satellite Image Analysis for Computing Vegetation Indices by R in the Khartoum Region of Sudan, Northeast Africa. *Journal of Imaging*, *9*(5), 98. doi:10.3390/jimaging9050098 PMID:37233317

Li, J., & Wang, S. (2018). Using SAR-Derived Vegetation Descriptors in a Water Cloud Model to Improve Soil Moisture Retrieval. *Remote Sensing (Basel)*, *10*(9), 1370. doi:10.3390/rs10091370

Li, P., Li, D., Sun, X., Chu, Z., Xia, T., Zheng, B. (2022). Application of Ecological Restoration Technologies for the Improvement of Biodiversity and Ecosystem in the River. *Water, 14*(9), 1402. MDPI AG. . doi:10.3390/w14091402

Liu, J. G. (2013). *REMOTE SENSING | Passive Sensors*. Reference Module in Earth Systems and Environmental Sciences. doi:10.1016/B978-0-12-409548-9.02956-0

Lopac, N., Jurdana, I., Brnelić, A., & Krljan, T. (2022). Application of laser systems for detection and ranging in the modern road transportation and maritime sector. *Sensors (Basel)*, *22*(16), 5946. doi:10.339022165946 PMID:36015703

Makanda, K., Nzama, S., Kanyerere, T. (2022). Assessing the Role of Water Resources Protection Practice for Sustainable Water Resources Management: A Review. *Water, 14*(19), 3153. MDPI AG. . doi:10.3390/w14193153

Matese, A., Toscano, P., Di Gennaro, S. F., Genesio, L., Vaccari, F. P., Primicerio, J., Belli, C., Zaldei, A., Bianconi, R., & Gioli, B. (2015). Intercomparison of UAV, aircraft, and satellite remote sensing platforms for precision viticulture. *Remote Sensing (Basel)*, *7*(3), 2971–2990. doi:10.3390/rs70302971

Mohsan, S. A. H., Othman, N. Q. H., Li, Y., Alsharif, M. H., & Khan, M. A. (2023). Unmanned aerial vehicles (UAVs): Practical aspects, applications, open challenges, security issues, and future trends. *Intelligent Service Robotics*, *16*, 109–137. doi:10.100711370-022-00452-4 PMID:36687780

Mondini, A. C., Guzzetti, F., Chang, K.-T., Monserrat, O., Martha, T. R., & Manconi, A. (2021). Landslide failures detection and mapping using Synthetic Aperture Radar: Past, present and future. *Earth-Science Reviews*, *216*, 103574. doi:10.1016/j.earscirev.2021.103574

Moon, W. M., Staples, G., Kim, D., Park, S.-E., & Park, K.-A. (2010). RADARSAT-2 and Coastal Applications: Surface Wind, Waterline, and Intertidal Flat Roughness. *Proceedings of the IEEE*, *98*(5), 800–815. doi:10.1109/JPROC.2010.2043331

Moore, G., McGuire, K., Troch, P., Barron-Gafford, G. (2015). Ecohydrology and the Critical Zone: Processes and Patterns across Scales. *Developments in Earth Surface Processes*, 239–266. . doi:10.1016/B978-0-444-63369-9.00008-2

Muhadi, N. A., Abdullah, A. F., Bejo, S. K., Mahadi, M. R., & Mijic, A. (2020). The Use of LiDAR-Derived DEM in Flood Applications: A review. *Remote Sensing (Basel)*, *12*(14), 2308. doi:10.3390/rs12142308

Muzhoffar, D. A. F., Sakuno, Y., Taniguchi, N., Hamada, K., Shimabukuro, H., & Hori, M. (2023). Automatic Detection of Floating Macroalgae via Adaptive Thresholding Using Sentinel-2 Satellite Data with 10 m Spatial Resolution. *Remote Sensing (Basel)*, *15*(8), 2039. doi:10.3390/rs15082039

Naseem, B., Ajami, H., Liu, Y., Cordery, I., & Sharma, A. (2016). Multi-objective assessment of three remote sensing vegetation products for streamflow prediction in a conceptual ecohydrological model. *Journal of Hydrology (Amsterdam)*, *543*, 686–705. doi:10.1016/j.jhydrol.2016.10.038

Nasiri, V., Beloiu, M., Darvishsefat, A.A., Griess, V.C., & Maftei, C., Waser, L.T. (. (2023). Mapping tree species composition in a Caspian temperate mixed forest based on spectral-temporal metrics and machine learning. *International Journal of Applied Earth Observation and Geoinformation, 103154*. doi:10.1016/j.jag.2022.103154

Nasiri, V., Sadeghi, S. M. M., Bagherabadi, R., Moradi, F., Deljouei, A., & Borz, S. A. (2022b). Modeling wildfire risk in western Iran based on the integration of AHP and GIS. *Environmental Monitoring and Assessment*, *194*(9), 644. doi:10.100710661-022-10318-y PMID:35930117

Nasiri, V., Sadeghi, S. M. M., Moradi, F., Afshari, S., Deljouei, A., Griess, V. C., Maftei, C., & Borz, S. A. (2022a). The Influence of Data Density and Integration on Forest Canopy Cover Mapping Using Sentinel-1 and Sentinel-2 Time Series in Mediterranean Oak Forests. *ISPRS International Journal of Geo-Information*, *11*(8), 423. doi:10.3390/ijgi11080423

Pamela, S. F., Hyun-Han, K., Meththika, V., & Joo-Heon, L. (2023). Modeling, challenges, and strategies for understanding impacts of climate extremes (droughts and floods) on water quality in Asia: A review. *Environmental Research*, *225*, 115617. doi:10.1016/j.envres.2023.115617 PMID:36871941

Panwar, V., & Sen, S. (2019). Economic impact of natural disasters: An empirical re-examination. *Margin - the Journal of Applied Economic Research*, *13*(1), 109–139. doi:10.1177/0973801018800087

Pourshanbeh, Y., Khoorani, A., & Jafari, L. (2022). The Impact of Climate Change on Jiroft Spring Wheat Yield Using Radiative Forcing Scenarios. *Land Ecology, 1*(1), 1-13. https://dorl.net/dor/20.1001.1.28211782.1401.1.1.1.3

Quiroga, G., Erice, G., Aroca, R., Zamarreño, Á. M., García-Mina, J. M., & Ruiz-Lozano, J. M. (2020). Radial water transport in arbuscular mycorrhizal maize plants under drought stress conditions is affected by indole-acetic acid (IAA) application—. *Journal of Plant Physiology*, *246-247*, 153115. doi:10.1016/j.jplph.2020.153115 PMID:31958683

Rafaqat, W., Iqbal, M., Kanwal, R., & Weiguo, S. (2022). Evaluation of Wildfire Occurrences in Pakistan with Global Gridded Soil Properties Derived from Remotely Sensed Data. *Remote Sensing (Basel)*, *14*(21), 5503. doi:10.3390/rs14215503

Rasheed, M. W., Tang, J., Sarwar, A., Shah, S., Saddique, N., Khan, M. U., Imran Khan, M., Nawaz, S., Shamshiri, R. R., Aziz, M., & Sultan, M. (2022). Soil Moisture Measuring Techniques and Factors Affecting the Moisture Dynamics: A Comprehensive Review. *Sustainability (Basel)*, *14*(18), 11538. doi:10.3390u141811538

Roshan, G., Sarli, R., & Grab, S. W. (2021). The case of Tehran's urban heat island, Iran: Impacts of urban "lockdown" associated with the COVID-19 pandemic. *Sustainable Cities and Society*, *75*, 103263. doi:10.1016/j.scs.2021.103263 PMID:36568531

Sadek, M., & Li, X. (2019). Low-Cost solution for assessment of urban flash flood impacts using Sentinel-2 satellite images and fuzzy analytic hierarchy process: A case study of Ras Ghareb City, Egypt. *Advances in Civil Engineering*, *2019*, 1–15. doi:10.1155/2019/2561215

Samansiri, S., Fernando, T., & Ingirige, B. (2022). Advanced Technologies for Offering Situational Intelligence in Flood Warning and Response Systems: A Literature Review. *Water (Basel)*, *14*(13), 2091. doi:10.3390/w14132091

Sample, M., Thode, A. E., Peterson, C., Gallagher, M. R., Flatley, W., Friggens, M., Evans, A., Loehman, R., Hedwall, S., Brandt, L., Janowiak, M., & Swanston, C. (2022). Adaptation Strategies and Approaches for Managing Fire in a Changing Climate. *Climate (Basel)*, *10*(4), 58. doi:10.3390/cli10040058

Schwank, J., Escobar, R., Girón, G. H., & Morán-Tejeda, E. (2014). Modeling of the Mendoza River watershed as a tool to study climate change impacts on water availability. *Environmental Science & Policy*, *43*, 91–97. doi:10.1016/j.envsci.2014.01.002

Seleiman, M. F., Al-Suhaibani, N., Ali, N., Akmal, M., Alotaibi, M., Refay, Y., Dindaroglu, T., Abdul-Wajid, H. H., & Battaglia, M. L. (2021). Drought Stress Impacts on Plants and Different Approaches to Alleviate Its Adverse Effects. *Plants*, *10*(2), 259. doi:10.3390/plants10020259 PMID:33525688

Seneviratne, S. I., Nicholls, N., Easterling, D., Goodess, C. M., Kanae, S., Kossin, J., Luo, Y., Marengo, J., McInnes, K., Rahimi, M., Reichstein, M., Sorteberg, A., Vera, C., & Zhang, X. 2012: Changes in climate extremes and their impacts on the natural physical environment. In: Managing the Risks of Extreme Events and Disasters to Advance Climate Change Adaptation. Cambridge University Press. doi:10.1017/CBO9781139177245.006

Singh, A., Gaurav, K., Meena, G. K., & Kumar, S. (2020). Estimation of Soil Moisture Applying Modified Dubois Model to Sentinel-1; A Regional Study from Central India. *Remote Sensing (Basel)*, *12*(14), 2266. doi:10.3390/rs12142266

Son, N. T., Chen, C. F., Chen, C. R., Masferrer, M. G. M., & Recinos, L. E. M. (2018). Multitemporal Landsat-MODIS fusion for cropland drought monitoring in El Salvador. *Geocarto International*, *34*(12), 1363–1383. doi:10.1080/10106049.2018.1489421

Sousa, D., & Small, C. (2023). Which Vegetation Index? Benchmarking Multispectral Metrics to Hyperspectral Mixture Models in Diverse Cropland. *Remote Sensing (Basel)*, *15*(4), 971. doi:10.3390/rs15040971

Stančikaitė, M., Zernitskaya, V., Kluczynska, G., Valūnas, D., Gedminienė, L., Uogintas, D., & Šeirienė, V. (2021). The Lateglacial and Early Holocene vegetation dynamics: New multi-proxy data from the central Belarus (Vol. 630). Quaternary International. , PP 121-136. doi:10.1016/j.quaint.2021.05.004

Stringer, L. C., Mirzabaev, A., Benjaminsen, T. A., Harris, R. M. B., Jafari, M., Lissner, T. K., Stevens, N., & Tirado-von der Pahlen, C. (2021). Climate change impacts on water security in global drylands. *One Earth*, *4*(6), 851–864. doi:10.1016/j.oneear.2021.05.010

Telesca, L., Aromando, A., Faridani, F., Lovallo, M., Cardettini, G., Abate, N., Papitto, G., & Lasaponara, R. (2022). Exploring Long-Term Anomalies in the Vegetation Cover of Peri-Urban Parks Using the Fisher-Shannon Method. *Entropy (Basel, Switzerland)*, *24*(12), 1784. doi:10.3390/e24121784 PMID:36554193

Torabi, O., & Karimi, N. shesh angosht, S., Rashtbari, M., & Sarbazvatan, A. H. (2023). Investigating the effect of radar images in classifying land use classes in machine learning-based algorithms. *Land Ecology, 2*(2), 137-150. https://dorl.net/dor/20.1001.1.28211782.1402.1.1.4.0

Traganos, D., Aggarwal, B., Poursanidis, D., Topouzelis, K., Chrysoulakis, N., & Reinartz, P. (2018). Towards Global-Scale Seagrass Mapping and Monitoring using Sentinel-2 on Google Earth Engine: The case study of the Aegean and Ionian Seas. *Remote Sensing (Basel), 10*(8), 1227. doi:10.3390/rs10081227

Van der Plas, G. W., De Cort, G., Petek-Sargeant, N., Wuytack, T., Colombaroli, D., Lane, P. J., & Verschuren, D. (2019). Distinct phases of natural landscape dynamics and intensifying human activity in the central Kenya Rift Valley during the past 1300 years. *Quaternary Science Reviews, 218*, 91–106. doi:10.1016/j.quascirev.2019.06.009

Van Dijk, A. I. J. M., Brakenridge, G. R., Kettner, A. J., Beck, H. E., De Groeve, T., & Schellekens, J. (2016). River gauging at a global scale using optical and passive microwave remote sensing. *Water Resources Research, 52*(8), 6404–6418. doi:10.1002/2015WR018545

Van Hoek, M., Jia, L., Zhou, J., Zheng, C., & Menenti, M. (2016). Early Drought Detection by Spectral Analysis of Satellite Time Series of Precipitation and Normalized Difference Vegetation Index (NDVI). *Remote Sensing (Basel), 8*(5), 422. doi:10.3390/rs8050422

Vicente-Serrano, S., Cabello, D., Tomás-Burguera, M., Martín-Hernández, N., Beguería, S., Azorin-Molina, C., & Kenawy, A. (2015). Drought Variability and Land Degradation in Semiarid Regions: Assessment Using Remote Sensing Data and Drought Indices (1982–2011). *Remote Sensing (Basel), 7*(4), 4391–4423. doi:10.3390/rs70404391

Walker, D., Forsythe, N., Parkin, G., & Gowing, J. (2016). Filling the observational void: Scientific value and quantitative validation of hydrometeorological data from a community-based monitoring programme. *Journal of Hydrology (Amsterdam), 538*, 713–725. doi:10.1016/j.jhydrol.2016.04.062

Wang, J., Ling, Z., Wang, Y., & Zeng, H. (2016). Improving the spatial representation of soil moisture by integration of microwave observations and the temperature–vegetation–drought index derived from MODIS products. *Isprs Journal of Photogrammetry and Remote Sensing, 113*, 144. doi:10.1016/j.isprsjprs.2016.01.009

Wang, S., Li, R., Wu, T., Zhao, L., Wu, X., Hu, G., Yao, J., Ma, J., Liu, W., Jiao, Y., Xiao, Y., Yang, S., Shi, J., & Qiao, Y. (2022). Evaluating the Impact of Soil Enthalpy upon the Thawing Process of the Active Layer in Permafrost Regions of the Qinghai–Tibet Plateau Using CLM5.0. *Remote Sensing (Basel), 15*(1), 249. doi:10.3390/rs15010249

Wang, X., Rodríguez, D. A., & Mahendra, A. (2021). Support for market-based and command-and-control congestion relief policies in Latin American cities: Effects of mobility, environmental health, and city-level factors. *Transportation Research Part A, Policy and Practice, 146*, 91–108. doi:10.1016/j.tra.2020.12.004 PMID:34295022

Wang, Y. (2015). Advances in remote sensing of flooding. *Water (Basel), 7*(11), 6404–6410. doi:10.3390/w7116404

Wang, Y., Zhang, F., Li, X., Johnson, V. C., Tan, M. L., Kung, H.-T., Shi, J., Bahtebay, J., & He, X. (2023). Methodology for Mapping the Ecological Security Pattern and Ecological Network in the Arid Region of Xinjiang, China. *Remote Sensing (Basel)*, *15*(11), 2836. doi:10.3390/rs15112836

Wanniarachchi, S., & Sarukkalige, R. (2022). A Review on Evapotranspiration Estimation in Agricultural Water Management: Past, Present, and Future. *Hydrology*, *9*(7), 123. doi:10.3390/hydrology9070123

Wu, X., Lu, G., & Wu, Z. (2023). Remote Sensing Technology in the Construction of Digital Twin Basins: Applications and Prospects. *Water (Basel)*, *15*(11), 2040. doi:10.3390/w15112040

Xie, Z., Huete, A., Cleverly, J., Phinn, S., McDonald-Madden, E., Cao, Y., & Qin, F. (2019). Multi-climate mode interactions drive hydrological and vegetation responses to hydroclimatic extremes in Australia. *Remote Sensing of Environment*, *231*, 111270. doi:10.1016/j.rse.2019.111270

Xu, X., Li, J., & Tolson, B. A. (2014). Progress in integrating remote sensing data and hydrologic modeling. *Progress in Physical Geography*, *38*(4), 464–498. doi:10.1177/0309133314536583

Yamazaki, F., & Matsuoka, M. (2007). Remote sensing technologies in post-disaster damage assessment. *Journal of Earthquake and Tsunami*, *1*(3), 193–210. doi:10.1142/S1793431107000122

Yang, Y., Anderson, M. C., Gao, F., Wood, J. D., Gu, L., & Hain, C. (2021). Studying drought-induced forest mortality using high spatiotemporal resolution evapotranspiration data from thermal satellite imaging. *Remote Sensing of Environment*, *265*, 112640. doi:10.1016/j.rse.2021.112640

You, J.-H., Oh, S., Park, J.-E., Song, H., & Kim, Y.-K. (2021). A novel LiDAR sensor alignment inspection system for automobile productions using 1-D photodetector arrays. *Measurement*, *183*, 109817. doi:10.1016/j.measurement.2021.109817

Zhang, C., Wang, K., Yue, Y., Qi, X., & Zhang, M. (2023). Assessing Regional Ecosystem Conditions Using Geospatial Techniques—A Review. *Sensors (Basel)*, *23*(8), 4101. doi:10.339023084101 PMID:37112442

Zhang, H., Lan, Y., Suh, C. P.-C., Westbrook, J., Clint Hoffmann, W., Yang, C., & Huang, Y. (2013). Fusion of remotely sensed data from airborne and ground-based sensors to enhance detection of cotton plants. *Computers and Electronics in Agriculture*, *93*, 55–59. doi:10.1016/j.compag.2013.02.001

Zribi, M., Albergel, C., & Baghdadi, N. (2020). Editorial for the Special Issue "Soil Moisture Retrieval using Radar Remote Sensing Sensors.". *Remote Sensing (Basel)*, *12*(7), 1100. doi:10.3390/rs12071100

Chapter 3
Flash Floods:
Causes, Effects, and Modeling Possibilities With Advanced Hydroinformatic Tools

Erika Beata Maria Beilicci
Politehnica University, Timisoara, Romania

Robert Florin Beilicci
Politehnica University, Timisoara, Romania

ABSTRACT

The objectives of the chapter are the identification of flash flood causes, their negative effects on environmental factors, and to study of modeling with advanced hydroinformatic tools. The authors analyzed the following models: DUFLOW, developed by the International Institute for Hydraulic and Environmental Engineering, Delft, The Rijkswaterstaat (Public Works Department), Tidal Water Division, The Hague, The Delft University of Technology The Netherlands; Hydrologic Engineering Center's Hydrological Modeling System runoff model, developed for the U.S. Army Corps of Engineers; Water Erosion Prediction Project, developed by USDA Forest Service, Agricultural Research Service, Natural Resources Conservation Service, Department of Interior's Bureau of Land Management, and Geological Survey from USA and MIKE11 by DHI, developed by Danish Hydraulic Institute, rainfall-runoff module. The authors conducted case studies of these models on different small hydrographic basins, located in west Romania, in the Banat region. Some possible measures to reduce the negative effects of flash floods are listed.

INTRODUCTION

Flash Floods Causes

Flash floods are becoming more frequent nowadays. For flash floods study are important the following aspects are: causes, concentration time of runoff, maximum water and solid discharge, and maximum water level in different cross sections of valleys. Their occurrence is due, in specially, to both climate

DOI: 10.4018/978-1-6684-8771-6.ch003

change and anthropogenic causes: heavy rain, massive deforestation, irrational exploitation of sloping lands, and absence of small watershed arrangements with runoff and soil erosion control works. The rains, and especially the torrential ones, consist of the fall of large amounts of precipitation in a very short time so that the infiltration capacity of the soil is quickly exceeded and almost the entire amount of water that falls flows towards the network of valleys, generating floods. Also, flash floods may be caused by heavy rain, associated with meltwater from ice or snow flowing over ice sheets or snowfields. Flash floods may occur also, after the collapse of a natural ice or debris dam, or man-made hydrotechnical structures, such as dams or dikes. Flash floods can occur under several types of conditions: when precipitation falls rapidly on saturated soil or dry soil that has poor infiltration capacity. The runoff collects in gullies and streams and often forms a fast-flowing front of water and debris.

Among the factors that favor rapid flooding is the inadequate maintenance of channel networks designed to collect and transport water from precipitation [Beilicci & Beilicci, 2015a]

A flood can be considered a flash flood if the area of the catchment does not exceed 200 km^2, the time of concentration of runoff is less than 6 hours, the rain duration is less than the time of concentration, and the rain height exceeds 100 mm. [Drobot, 2007]

A flash flood is a rapid inundation of low-lying areas. Flash floods are characterized by the short duration of the rising and falling branches of the flood hydrograph, tens of minutes or hours. Flash floods occur predominantly on specific watercourses draining small catchments (up to a few hundred km^2) characterized by high slopes, pronounced landscape fragmentation, low forest cover, and a high rate of impervious surfaces. They are generated by torrential rains with high intensities (high height and short duration), often of the convective type with short durations (of 2-6 hours), which exceed the concentration time of the basin. The speed of the water flow is very high, it causes a strong erosion of the bed of the river and causes the dislocation and entrainment of alluvium of different sizes. Such floods are generally localized but destructive, which is why they are called "brutal floods". [National Oceanic and Atmospheric Administration, 2000], [Drobot, 2007]

Flash Floods Effects

Flash flood effects are multiple and complex: social, economic, and environmental.

Social effects are loss of human life, health effects - illness, psychological effects, destruction and damage to homes and family farms, impairment of educational, cultural, sports activities, damage to transport infrastructure, and deterioration of objectives and social-cultural institutions.

Economic effects are direct and indirect damage, damage to multiple items with economic importance, loss due to malfunctioning of public institutions and economic entities, as a result of the reduction or temporary interruption of work caused by lack of manpower, raw materials, interruption of power supply, etc., payment of material and human goods insurances.

Environmental effects are pollution of flooded areas, changes in the physic-chemical and bacteriological properties of water courses, silting of reservoirs, bio-edaphic effects, and reduction in the stability of slopes: landslides, and collapses.

Material and social vulnerability constitutes an important factor that can amplify the effects of flash floods. The material vulnerability is due to the location of human settlements in areas exposed to the risk of flooding, and the practice of economic activities in these areas. [https://www.academia.edu]

FLASH FLOODS MODELLING POSSIBILITIES WITH HYDROINFORMATIC TOOLS

The need for rainfall-runoff modeling is to provide data on flood characteristics downstream of small catchment closure sections (discharges, levels). Based on this data, the necessary measures can be established to protect the population and socio-economic assets in the flash flood area, namely measures to reduce the maximum water level: afforestation, surface runoff control, hydrotechnical measures, and non-structural measures, such as the refusal to issue building permits in the potential flood affected area.

Also, data from rainfall-runoff simulation can be used as input for hydrodynamic modeling of the main watercourse, allowing the modeling of flood wave propagation in the main rivers of the watersheds. [Beilicci & Beilicci, 2015a]

To obtain results with satisfactory accuracy, it is necessary to use advanced hydroinformatics tools to model the phenomenon of rapid floods.

Hydroinformatics is the integration of computational fluid dynamics and of artificial intelligence. Computational hydraulics is the fusion of numerical methods of applied mathematics, hydrodynamics, and hydraulics. Hydroinformatics is about making the best use of information technologies to manage water in the environment. [https://en.wikipedia.org/]

Flash floods are difficult to predict because they are not always caused by meteorological phenomena alone. Flash floods occur when favorable meteorological, hydrological, and soil conditions are met simultaneously. Although the main cause of flash floods is torrential rain, a certain amount and duration of rainfall may or may not lead to a flash flood, depending on the hydrological and soil characteristics of the watershed in which it falls. Variables include: how fast the storm is moving, how intense the rainfall is, whether torrential rains repeatedly affect the same area, the percentage of rainfall that becomes surface runoff (and where it runs off), the net rainfall, how porous the soil is (infiltration rate) and initial moisture, the amount, quality and type of vegetation covering the ground, how much of the land surface is impervious and will not absorb water, if there are canals/ valleys to efficiently transport water from the area, how steep it is the terrain, the depth and condition of the snow cover, the temperature value that can cause a rapid melting of snow and ice. [https://www.nssl.noaa.gov]

By using advanced hydroinformatics tools, different scenarios can be built that lead to the production of flash floods, taking into account all the above factors.

Among the multitude of existing hydroinformatic tools, the following will be briefly presented: DUFLOW, developed by the International Institute for Hydraulic and Environmental Engineering (IHE) Delft, The Rijkswaterstaat (Public Works Department), Tidal Water Division, The Hague, The Delft University of Technology The Netherlands; Hydrologic Engineering Center's Hydrological Modeling System (HEC-HMS) runoff model, developed for the U.S. Army Corps of Engineers; WEPP (Water Erosion Prediction Project), developed by USDA Forest Service, Agricultural Research Service, Natural Resources Conservation Service, Department of Interior's Bureau of Land Management and Geological Survey from USA and MIKE by DHI, developed by Danish Hydraulic Institute, rainfall-runoff module (UHM and NAM methods).

DUFLOW

As in many water management problems, the runoff from catchment areas is important, a simple rainfall-runoff relation is part of the model set-up. The selected numerical scheme allows for a rather large time step in the computation and for choosing different lengths of the elementary sections. Precipitation generating an additional inflow to the nodes of the valley's network:

$$Q_{add} = R \bullet 0.001 \bullet A \bullet 10000 \bullet c/(24 \bullet 3600) \quad (1)$$

where: Q_{add} is additional inflow to the node (m^3/s); R is precipitation intensity (mm/day); A is catchment area (ha); c is runoff factor.

DUFLOW is based on the one-dimensional partial differential equation that describes non-stationary flow in open channels. These equations, which are the mathematical translation of the laws of conservation of mass and momentum, read:

$$\frac{\partial B}{\partial t} + \frac{\partial Q}{\partial x} = 0 \quad (2)$$

$$\frac{\partial Q}{\partial t} + gA\frac{\partial H}{\partial x} + \frac{\partial(\alpha Q v)}{\partial x} + \frac{g|Q|Q}{C^2 AR} = b\gamma w^2 \cos(\Phi - \phi) \quad (3)$$

while the relation:

$$Q = v \cdot A \quad (4)$$

holds and where: t time [s]; x distance as measured along the channel axis [m]; H(x, t) water level concerning reference level [m]; v(x, t) mean velocity (averaged over the cross-sectional area) [m/s]; Q(x, t) discharge at location x and at time t [m^3/s]; R(x, H) hydraulic radius of cross-section [m]; A(x, H) cross-sectional flow area [m^2]; b(x, H) cross-sectional flow width [m]; B(x, H) cross-sectional storage area [m^2]; g acceleration due to gravity [m/s^2]; C(x, H) coefficient of De Chézy [$m^{1/2}/s$]; w(t) wind velocity [m/s]; Φ(t) wind direction in degrees [degrees]; φ(x) direction of channel axis in degrees, measured clockwise from the north [degrees]; (x) wind conversion coefficient [-]; α correction factor for non-uniformity of the velocity distribution in the advection term, defined as:

$$\alpha = \frac{A}{Q^2}\int v(y,z)^2 dydz \quad (5)$$

where the integral is taken over the cross-section A. [m^2]

Equations (2) and (3) are discretized in space and time using the four-point implicit Preissmann scheme.

For a unique solution to the set of equations, additional conditions have to be specified at the physical boundaries of the network and the sections defined as hydraulic structures. The user-defined conditions at the physical boundaries may be specified as levels, discharges, or a relation between both, for instance, a (tidal) elevation H, a discharge Q, or a so-called QH relation. At internal junctions, the (implicit) condition states that the water level is continuous over such a junction node and that the flows towards the junction are in balance since continuity requires:

$$\sum_{j=1}^{n} Q_{j,i} + q_i = 0 \tag{6}$$

where: i indication for the junction node; $Q_{j,i}$ is discharge from node j to node; q_i is additional or lateral flow to node i.

The above equations are solved at each time step. They are transformed into a system of (linear) equations for the water levels. After the water levels are computed using a standard solution method for systems of linear equations, the discharges are found by substituting the computed water levels into equations.

Equation (6) is not used in nodes where a water level is prescribed as a boundary condition. In such a node no equation is needed because the water level is already known. Discharge boundary conditions are taken into account as the additional flow qi.

To start the computations, initial values for H and Q are required. These initial values must be provided by the user; they may be historical measurements, obtained from former computations or just a first reasonable guess. Additionally, wind stress and rainfall conditions can be specified.

The results from the simulation with DUFLOW are: the variation of water levels and water discharges in each node of network. The DUFLOW model, because is one dimensional, has the same limitations: water velocity is averaged vertically and horizontally; not can be used to simulate multi-layer flow; consider the different substance concentrations constant in the whole area of cross-section. Another unfavorable aspect is that the model considers that the rains are produced simultaneously in all nodes of the network and then rains can be input variable in time. [IHE, 1995]

HEC-HMS

HEC-HMS is developed to simulate the rainfall-runoff processes of dendritic watershed systems. It is designed to be applicable for solving different problems: large river basin water supply and flood hydrology to small urban or natural watershed runoff. The results of modeling (hydrographs) can be used for studies of water availability, urban drainage, flow forecasting, future urbanization impact, reservoir spillway design, flood damage reduction, flood plain regulation, wetlands hydrology, and hydrotechnical systems operation.

For precipitation-runoff-routing simulation, the program provides the following components:

- Precipitation-specification options that can describe an observed (historical) precipitation event, a frequency-based hypothetical precipitation event, or an event that represents the upper limit of precipitation possible at a given location.
- Loss models that can estimate the volume of runoff, given the precipitation and properties of the watershed.

- Direct runoff models that can account for overland flow, storage, and energy losses as water runs off a watershed and into the stream channels.
- Hydrologic routing models that account for storage and energy flux as water moves through stream channels.
- Models of naturally occurring confluences and bifurcations.
- Models of water-control measures, including diversions and storage facilities.
- A distributed runoff model for use with distributed precipitation data, such as the data available from weather radar.
- A continuous soil-moisture-accounting model used to simulate the long-term response of a watershed to wetting and drying.
- An automatic calibration package that can estimate certain model parameters and initial conditions, given observations of hydro-meteorological conditions. [US Army Corps of Engineers, 2000]

HEC-HMS computes runoff volume by computing the volume of water that is intercepted, infiltrated, stored, evaporated, or transpired and subtracting it from the precipitation. Interception and surface storage are intended to represent the surface storage of water by trees or grass, local depressions in the ground surface, cracks and crevices in parking lots or roofs, or a surface area where water is not free to move as overland flow. Infiltration represents the movement of water to areas beneath the land surface. [US Army Corps of Engineers, 2000]

HEC-HMS model components are used to simulate hydrologic response in a watershed, these components are: basin models (represent the physical watershed, developed by adding and connecting different hydrological elements – sub-basin, reach, junction, source, sink, reservoir, diversion; hydrological element use mathematical models to describe physical processes in the watershed); meteorological models (calculates the precipitation input required by a sub-basin element); control specifications (set the period of a simulation run – starting and ending simulation period, and computation time step); input data (time-series data, paired data, gridded date – often required as parameter or boundary conditions in basin and meteorological models). A simulation calculates the rainfall-runoff response in the basin model given input from the meteorological model. [US Army Corps of Engineers, 2010]

WEPP

The USDA Water Erosion Prediction Project erosion model represents a new generation technology for estimating soil erosion on and sediment delivery from hillslope profiles and small watersheds. The WEPP erosion model simulate the erosion processes of detachment and transport by raindrop impact on inter-rill areas, detachment, transport, and deposition by overland flow in rill channels, detachment, transport, and deposition by concentrated flow in channels, and deposition in impoundments (drop spillway, perforated riser, two sets of identical culverts, emergency spillway or open channel, rock-fill check dam, filter fence or a straw bale check dam. The continuous simulation model also includes components that mimic climate, surface and subsurface hydrology, winter processes, irrigation, plant growth, and residue decomposition. The WEPP computer program calculates spatial and temporal distributions of soil loss, as well as sediment delivery and sediment particle characteristics.

Also included as part of the WEPP erosion prediction system are user interface programs, input file building programs, a climate database, a soil database, a crop parameter database, and a tillage implement database. These additional programs and databases make the WEPP model a very powerful tool for users involved in natural resource conservation and environmental assessment. [USDA-ARS, 1995]

The WEPP model may be used in both hillslope and watershed applications. The model is a distributed parameter, continuous simulation, erosion prediction model, implemented as a set of computer programs for personal computers. The hillslope component of the WEPP erosion model requires a minimum of four input data files to run: climate file, slope file, soil file, and plant/management file. The watershed component requires a minimum of seven input data files: each hillslope information file, structure file, slope file, soil file, management file, climate file, and channel file. WEPP considers the hillslope to consist of numerous parallel rills; the surface erosion occurs on inter-rill surfaces and the dislocated soil particles are transported downhill by rill flow (rill erosion is also considered in calculus).

The model has two basic components:

- Erosion component - rill erosion is divided into two parts, detachment and deposition. Rill detachment is predicted when the flow shear stress exceeds the critical shear stress of the soil and when the sediment load is below the calculated sediment transport capacity. Deposition in a rill will be predicted when the sediment load is greater than the sediment transport capacity.
- The hydrology component produces four inputs for the erosion component: peak runoff rate, effective runoff duration, effective rainfall intensity, and effective rainfall duration.

WEPP produces many different kinds of output, in various quantities, depending upon the user's needs. The most basic output contains the runoff and erosion summary information, which may be produced on a storm-by-storm, monthly, annual, or average annual basis. The time–integrated estimates of runoff, erosion, sediment delivery, and sediment enrichment are contained in this output, as well as the spatial distribution of erosion on the hillslopc. [USDA-ARS, 1995]

MIKE11 by DHI

MIKE 11 is a hydroinformatic tool that simulates discharge and water level, water quality, and sediment transport in rivers, flood plains, irrigation canals, reservoirs, and other inland water bodies. MIKE 11 is a 1-dimensional river model. MIKE11 has long been known as a software tool with advanced interface facilities.

The computational core of MIKE 11 is a hydrodynamic simulation engine, and this is complemented by a wide range of additional modules and extensions covering almost all conceivable aspects of river modeling.

The most important modules of MIKE11 are:

- HD module: it provides a fully dynamic solution to the complete nonlinear 1-D Saint-Venant equations, diffusive wave approximation and kinematic wave approximation, Muskingum method, and Muskingum-Cunge method for simplified channel routing. It can automatically adapt to subcritical flow and supercritical flow. It can simulate standard hydraulic structures such as weirs, culverts, bridges, pumps, energy loss, and sluice gates.

- RR module: it is a rainfall-runoff module, including the unit hydrograph method (UHM), a lumped conceptual continuous hydrological model, and a monthly soil moisture accounting model. It includes an auto-calibration tool to estimate model parameters based on statistical data of comparison of simulated water levels/discharges and observations.
- AD module: it is an advection-dispersion module. It simulates the transport and spreading of conservative pollutants and constituents as well as heat with linear decay.
- ST/GST module: it is a non-cohesive sediment module. It simulates transport, erosion, and deposition of non-cohesive and graded non-cohesive sediments, including simulations of river morphology.
- ACS module: it is a cohesive sediment module. It has a 3-layer bed description, including quasi-2D erosion.
- ECO Lad module: it is ecological modeling. It can simulate BOD/DO, Ammonia, Nitrate, Eutrophication, Heavy metal, and Wetlands. It includes standard templates that are well-documented and have been used extensively in numerous applications worldwide. Based on a pre-defined process template, one can develop his/her templates.

MIKE11 has been used in hundreds of applications around the world. Its main application areas are flood analysis and alleviation design, real-time flood forecasting, dam break analysis, optimization of reservoir and canal gate/structure operations, ecological and water quality assessments in rivers and wetlands, sediment transport and river morphology studies, and salinity intrusion in rivers and estuaries. [DHI, 2011]

The UHM methods of the Rainfall-Runoff module of MIKE11 simulate the runoff from single storm events by the use of the unit hydrograph techniques and constitute an alternative to the NAM model for flood simulation in areas where no streamflow records are available or where unit hydrograph techniques have already been well established. A unit hydrograph is a hypothetical unit response of the watershed to a unit input of rainfall. The module calculates simultaneously the runoff from several catchments and includes facilities for the presentation and extraction of the results. The output from the module can further be used as lateral inflow to the advanced hydrodynamic module in MIKE11.

The basic equations of the model are the continuity equation and momentum equation:

$$A\frac{dS_i}{dt} = f \cdot r_{rave} \cdot A - Q_i \tag{7}$$

$$S_i = K \cdot Q_i^p \tag{8}$$

where: S_i apparent basin storage; Q_i direct runoff from a basin with consideration of the time of delay (except for base flow); T_i time of delay; K,p constants; f coefficient of inflow; rave mean rainfall of river basin; A catchment area.

Input data requirements: area of the sub-catchment (GIS); hydraulic length (GIS); land use and percentage area of individual categories; soil hydrologic groups (A-D) and percentage area of individual categories; average slope in the catchment (GIS); observed values: rainfall and runoff (for calibration).

In the unit hydrograph module, the excess rain is calculated assuming that the losses to infiltration can be described as a fixed initial and constant loss, a proportional loss (the rational method), or by the SCS curve number method.

The excess rainfall is routed to the river by unit hydrograph methods. The module includes the SCS-dimensionless hydrographs as well as facilities for establishing and managing databases with user-defined unit hydrographs and time series of recorded rainfall and stream flow.

The NAM hydrological model simulates the rainfall-runoff processes occurring at the catchment scale. NAM forms part of the rainfall-runoff (RR) module of the MIKE11 river modeling system. The rainfall-runoff module can either be applied independently or used to represent one or more contributing catchments that generate lateral inflows to a river network.

A mathematical hydrological model like NAM is a set of linked mathematical statements describing, in a simplified quantitative form, the behavior of the land phase of the hydrological cycle. NAM represents various components of the rainfall-runoff process by continuously accounting for the water content in four different and mutually interrelated storages. Each storage represents different physical elements of the catchment. NAM can be used either for continuous hydrological modeling over a range of flows or for simulating single events. The NAM model can be characterized as a deterministic, lumped, conceptual model with moderate input data requirements. The NAM model is a well-proven engineering tool that has been applied to several catchments around the world, representing many different hydrological regimes and climatic conditions. [DHI, 2011]

A critical observation about the WEPP and DUFLOW models is that they can only be applied to small river basins, and the models can only use a single file of climate characteristics, i.e. the same type of rain or torrential rain over the whole river basin (same intensity and duration). For larger hydrographic basins, it is necessary to divide them into smaller sub-basins, each sub-basin having the same precipitation characteristics, but different association times; to obtain the hydrographs of flows and levels and the variation of the concentration of suspended materials in the section exiting the basin, it is necessary to integrate the discharge over the entire hydrographic basin.

The choice of utilized model for flash flood simulation must be based on the availability of topographic surveys data, geomorphological data, climatic data, soil data, data about soil erosion and degradation, hydrological data, hydrogeological data, geotechnical data, vegetation, and socio–economic data. For the improvement and calibration of models, it is necessary to perform measurements immediately after flash floods in affected areas. [Beilicci & Beilicci, 2015a]

Based on the results of the simulation of the rainfall-runoff phenomenon (especially in the areas where the probability of the occurrence of flash floods is high, with one or more of the programs presented previously, in different scenarios of the production of torrential rains), the necessary measures to be implemented can be established and the hydrotechnical works needed to be carried out to reduce the negative effects of these rapid floods. All the decision-makers (local/national authorities), respectively the users/owners of the affected lands, must be involved and consulted, in addition to the specialists, to develop sustainable management plans for the hydrographic basins subject to the risk of flash floods.

The measures that can be implemented and the works that can be carried out are presented in the next section.

POSSIBLE MEASURES TO REDUCE THE NEGATIVE EFFECTS OF FLASH FLOODS

Preventive measures (for prevention, protection, and preparation) are focused on the prevention/reduction of the flash flood-generated potential damage through:

- Avoidance of the construction of houses, social, cultural, and/or economic objectives in the potential flooding zones, by mentioning in the land use planning documentation the data regarding the previous flood effects; adaptation of future developments to the flood risk conditions;
- Promotion of adequate use of land and forest and agricultural land;
- Achievement of structural protection measures, inclusively in the areas of bridges and footbridges;
- Achievement of structural protection measures (control of minor streambed uses, elaboration of watershed plans for flood risk reduction and of measure plans, introduction of insurance systems, etc.);
- Detailed identification, geographical delimitation of areas of natural flood risk on the territory of the administrative units, the introduction of these areas in the general locality plans, and provision in the locality planning regulations of the specific measures aimed at preventing and diminishing the flood risk, construction building, and land use;
- Implementation of flood forecast, warning, and alarm systems;
- Maintenance of the existent infrastructure aimed at assuring protection against floods and river streambeds;
- Execution of protection works against river streambed erosion in the areas of the existing bridges and footbridges;
- Communication with the population and its education as regards flood risk and the means of action in emergencies. [Government Decision, 2005]
- Construction, exploitation, and maintenance of works to collect, retain, or transport water from surface runoff in a directed manner on hillslope.

During floods, the following expeditious measures are to be taken:

- Detection of potential flood wave and flood formation;
- Forecast of flood wave evolution and propagation along the water courses;
- Warning of the authorities and the population about the extension, severity, and occurrence time of floods;
- Organization of the authorities and the population and response actions to emergencies;
- Assurance of (material, financial, and human) resources at the county level for effective intervention;
- Activation of the operational institutions, mobilization of resources, etc. [Government Decision, 2005]

After the flood, the following measures should be taken:

- Help the disaster-affected population, allowing satisfaction of their immediate necessities and their return to normal life;

- Reconstruction of the distressed buildings, infrastructure, and those parts of the protection systems against floods;
- Review of the flood management activities to improve the intervention planning process, to cope with future similar events in the affected areas, as well as other areas. [Government Decision, 2005].

Another way of classifying works to reduce the volume and effects of flash floods in the following categories of works:

1. Works on hillslopes (flow and erosion control works) - organizational works (restructuring land use categories: arable, vineyard, fruit trees, lawns etc., ensuring road networks through which to carry reducing transportation costs, internal organization of new categories of use, by creating territorial units of work, corresponding in shape and size, taking into account the achievement of maximum efficiency in agricultural mechanization and soil erosion control; creating the conditions for correct entering of soil erosion control systems and a corresponding agricultural technique that would ensure prevention and control of soil erosion and increase agricultural production); agro-phytotechnical works (soil preparation work, methods that do not change surface micro-relief, tillage methods that modify the surface micro-relief in order to disperse and slow leaks, special methods for retaining water in the fields, fertilization, sowing - on the contour line, crop rotation on hillslopes, crop systems after contour lines direction, culture into strips, grass strips, terraces, grassing); forestry works and measures (vineyards and orchards, reforestation on land eroded by water); hydro improvements and hydrotechnical works (horizontal or inclined earth waves and channels, terraces with horizontal or inclined platform, individual terraces with support wall, with picket fences, horizontal or inclined coastal channel outlets in natural or artificial valley).
2. Torrential formations arrangement. By ravines and torrent arrangements seeking to enforce a set of measures for regulating runoff from hillslopes and in drainage networks to restore the whole area of the catchment to corresponding economic and erosion control categories and to remove the damage caused by floods. After the application site of works, we worked in the peak zone and in the ravine origin zone (branch occlusion and modeling - leveling the area, retention of leakage through horizontal channels, interception and evacuating of leakage through inclined channels and outlets, forestation); arrangements along the drainage network (forestation, grassing, simple slopes, riprap of stone or concrete slabs, retaining walls, contra-banquette for banks); cross construction that can be filtered (wood works - fascine, layers of fascines, fill/packing, wattle work, crest and dams of stone and concrete - dry masonry, gabions, stone masonry cross works – with cement and concrete); longitudinal hydrotechnical works (works to strengthen the banks, river regularization works - to change direction or position of water current); arrangements in the natural area of deposit of silt (evacuation channels, setting deposited alluvial material, creating conditions for flood water evacuation).
3. Other non-construction measures: developing information systems for torrential flood warning and forecasting and decision systems for operative action; implementation and development of the monitoring system of triggers and favoring factors of dangerous phenomena of soil erosion; rethinking hydrotechnical arrangements operating rules; institutional reform, developing legal frameworks for coordination and institutions responsible for developing strategies and operational decisions; developing or updating of regulations, operational action plans and intervention models

for various manufacturing scenarios of dangerous erosion (flash floods, mud flow, landslides); developing risk maps (especially for landslides); placing restrictions on building permits for hazardous areas; activities awareness of decision makers at different levels and population located in areas with dangerous erosion phenomena's risk; developing appropriate economic instruments: the material property insurance through insurance - reinsurance companies, compensation schemes etc.; relocation of settlements and economic objectives from areas with high risk of producing dangerous erosion phenomena's. [Bacov, 1996], [Popovici, 1991], [Giurma, 1997], [Di Silvio, 1998], [Stanescu et al, 2002]

CASE STUDIES

Application of DUFLOW Model [Beilicci & Beilicci, 2007], [Lazar, 2007]

The studied territory – Manastiur hydrographical basin – is situated in the depression between Zarandului and Poiana Ruscai Mountains, at the crossover line between Lipova Plateau and Occidental Field (Figure 1).

Figure 1. Studied territory

For the model studied the phenomena related to liquid runoff were cuisse the sub-basin of Topla River with a length of 11.56 km, a surface of 2357.40 ha; that contains 56 sectors of secondary valleys with a total length of 31.58 km; area plan with the separation on sub-basins corresponding to each valley is represented in Figure 2 and the calculus scheme (the network) in Figure 3.

Figure 2. Area plan

The case study is based on the next hypotheses:

- a constant intensity of the rain on all surfaces of the hydrographical sub-basins
- the same type of soil for all hydrographical sub-basins - Typic Hapludalfs (with more subtypes), having middle and heavy texture (reduced permeability), making part of hydrological group C, with initial humidity of 70%
- usage of the field is the same all over the sub-basin surface
- do not exist works for fighting soil erosion.

Figure 3. The calculus scheme

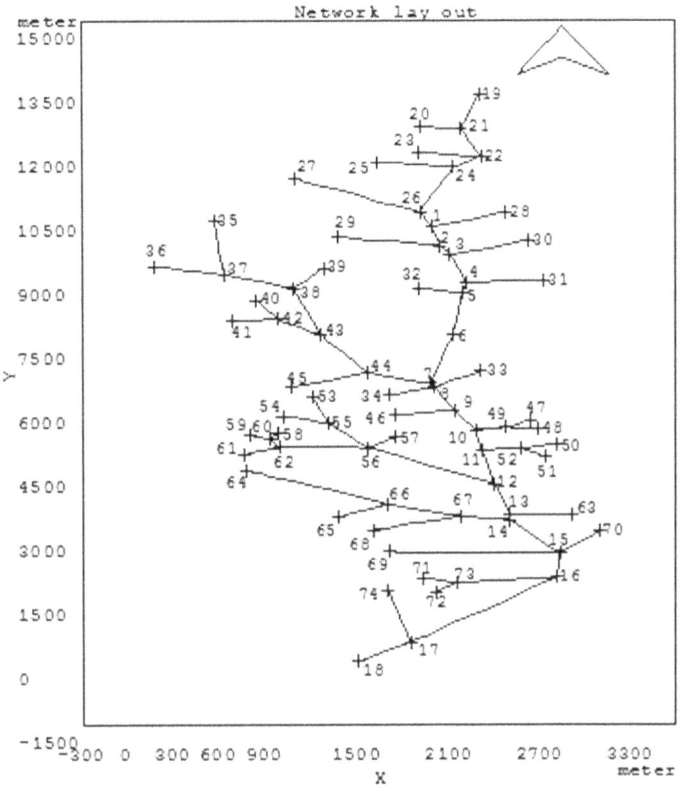

The hydrographical network is divided into flow sectors through 74 nodes. In each node is determined the cross sections and the water surface width for different water depths (from 0,2 to 0,2 m). In each node, the water discharges resulting from the overland flow on hillslopes (after establishing the catchments area regarding each node and the corresponding runoff coefficients, depending on soil hydrological type).

Through simulation with DUFLOW are determine the water discharge and water level in each node during the torrential rain, in four hypotheses (which have resulted from charts of torrential rain repartition with assurance of 10% and having the duration of 5, 15, 30, 60 minutes, obtained from measurements and statistical works of the meteorological data on the Bega hydrographical basin area):

1 - torrential rain with a duration of 5 minutes and calculus intensity i =3 mm/min
2 - torrential rain with a duration of 15 minutes and calculus intensity i =1,55 mm/min
3 - torrential rain with a duration of 30 minutes and calculus intensity i =0,96 mm/min
4 - torrential rain with a duration of 60 minutes and calculus intensity i =0,70 mm/min.

The maximum water discharges and the apparition time after the torrential rain begins in node 18 (outlet). Figures 4 and 5 are shown to illustrate the variation of water levels and discharges in hypothesis 4 (torrential rain with a duration of 60 minutes and calculus intensity i =0,70 mm/min).

Figure 4. The water level variations in nodes 16, 17, and 18

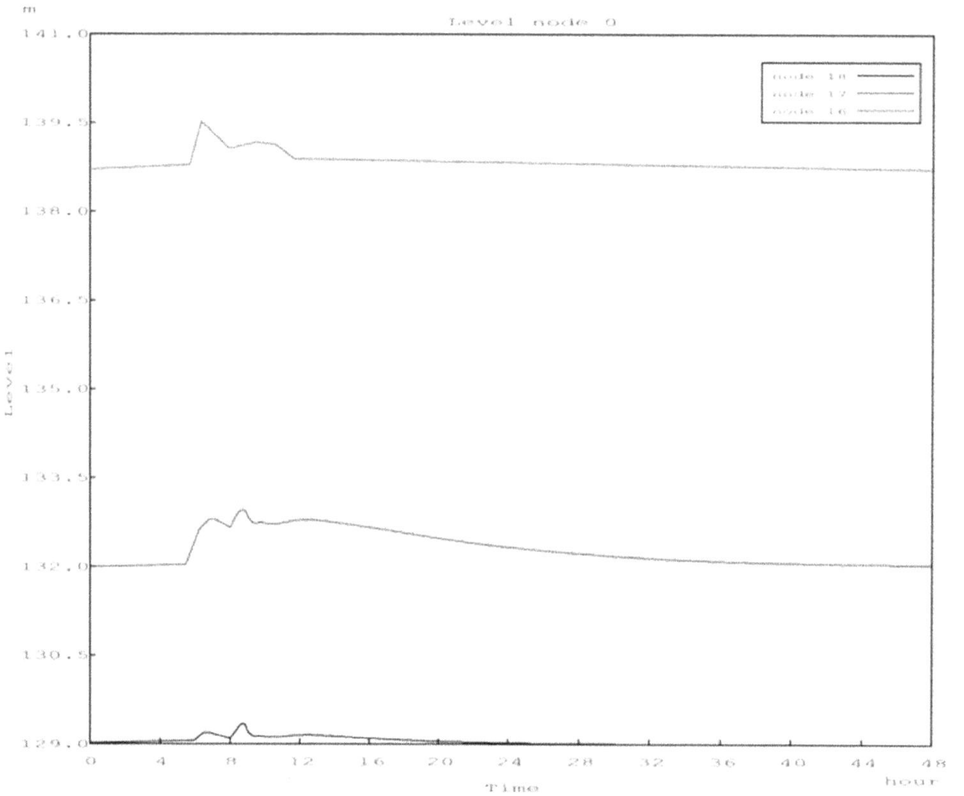

Figure 5. The water discharge variations in nodes 16, 17, and 18

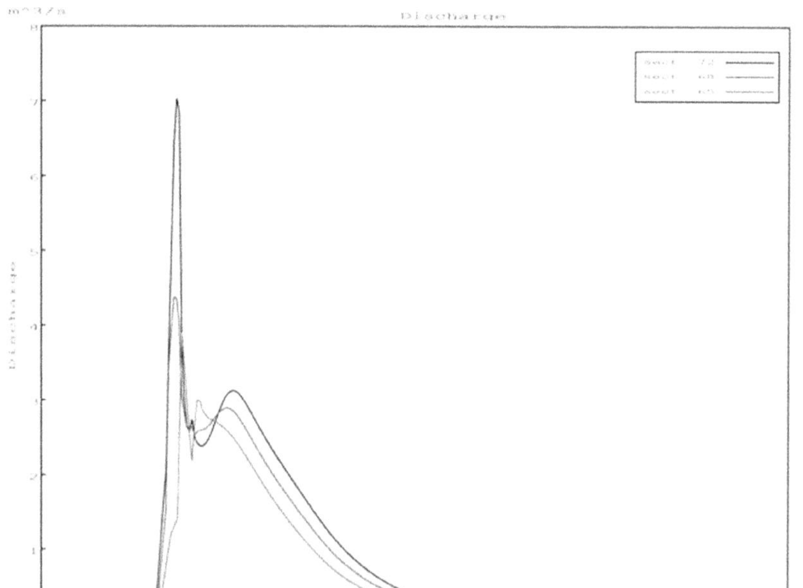

From the results analysis can be seen that in each hypothesis the water discharge hydrograph has two peaks, the value of the first peak is almost double than second peak. The duration of the flood is approximately 12 hours in each of 4 hypotheses. At the simulation start time, we considered that the water discharge in each node was zero (the riverbed is dry).

The values of water discharge are not so high, because the watershed is narrow and produces an important delay in flood wave propagation (we have considered the most unfavorable situation: that the rains has the same intensity on all area of the watershed and cover the whole watershed).

The water level in node 18 (outlet sections) does have no important variations because the cross-section of the riverbed is very wide.

Application of HEC-HMS Model [Beilicci & Beilicci, 2015b], [Fanta &Tadesse, 2022]

This case study simulated the rainfall-runoff phenomenon in the Valea Mare watershed, a component of Bega river catchments (Figure 6).

Figure 6. Studied territory

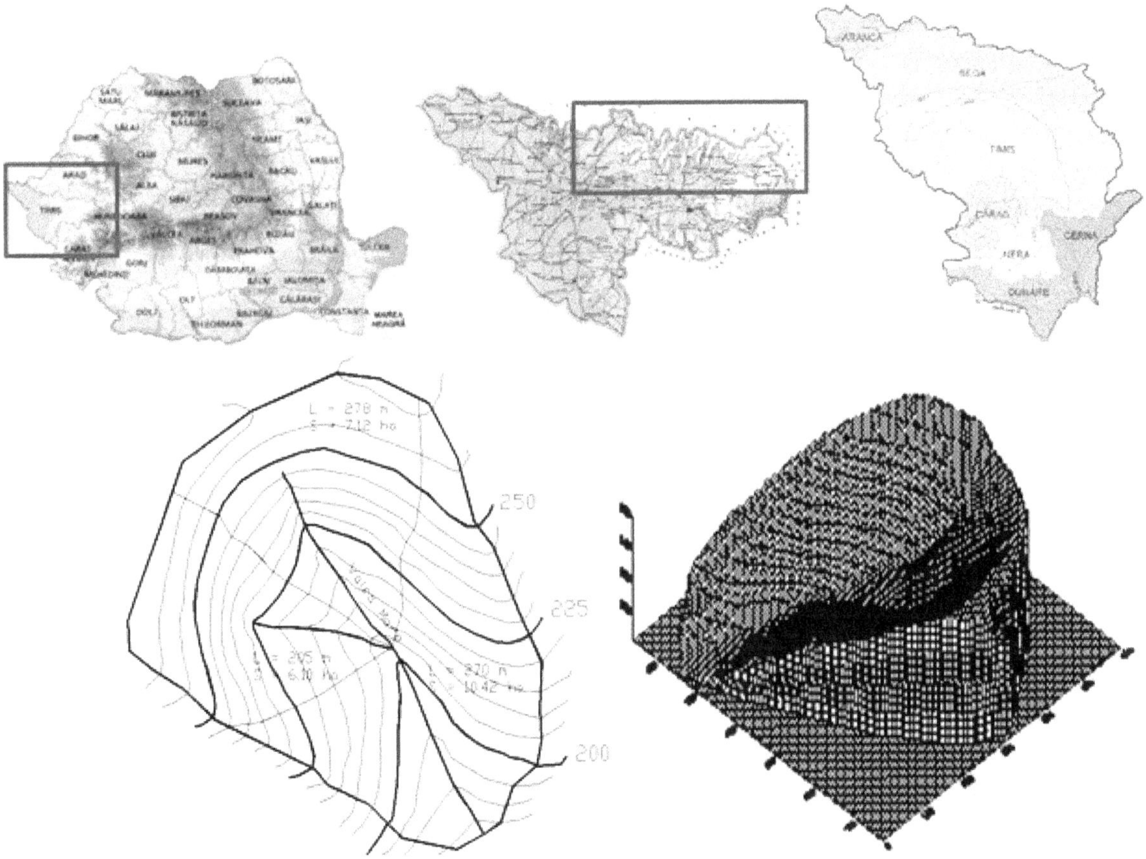

Hypotheses and input data were:

- Rainfall hydrograph – Figure 7
- SCS curve number - 79 (deciduous forests)
- Soil hidrological group - C
- Base flow - 0
- Initial abstraction - 0
- Catchment average slope - 20.64%
- Total area - 23.54 ha
- Simulation period – 06 May 2015, 12:00 AM – 09 May 2015, 11:00 PM
- Time step – 1 hour.

Figure 7. Rainfall

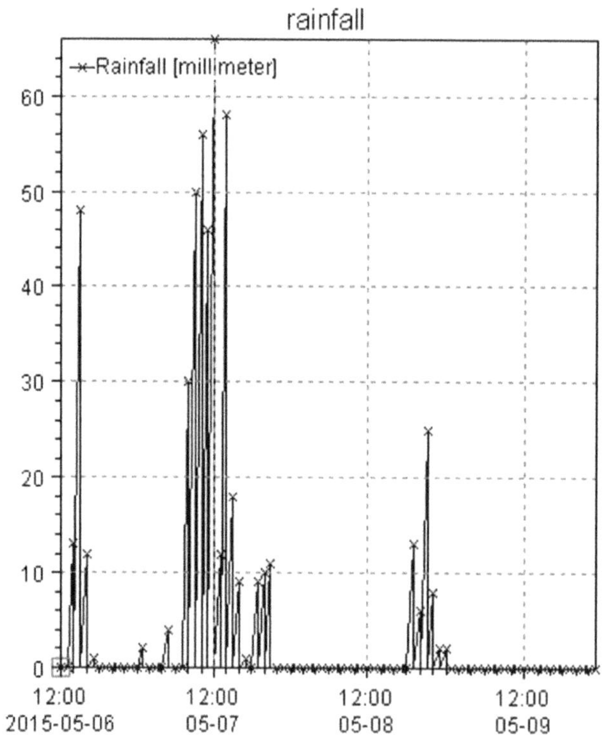

The results are shown in Figures 8, 9, and 10 for modeling with the HEC-HMS model.

Application of the WEPP Model [Beilicci & Beilicci, 2015a], [Saghafian et al, 2015]

The WEPP model was applied for the hydrographic basin presented in case study no. 1, with the related assumptions. Through simulation with the WEPP model, the water volume passing through in closing

Figure 8. Calculus scheme in HEC-HMS

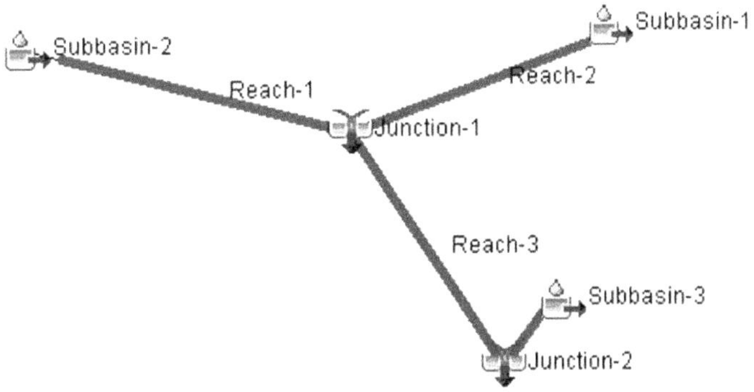

Figure 9. Detailed results from HEC-HMS

Project: buletin   Simulation Run: Runoff

Start of Run: 06May2015, 12:00   Basin Model: Basin
End of Run: 09May2015, 23:00   Meteorologic Model: ploaia
Compute Time: 11Dec2015, 14:43:33   Control Specifications: Control 1

Show Elements: All Elements   Volume Units: MM 1000 M3   Sorting: Hydrologic

Hydrologic Element	Drainage Area (KM2)	Peak Discharge (M3/S)	Time of Peak	Volume (MM)
Subbasin-1	0.071	0.8	07May2015, 14:00	452.35
Reach-2	0.071	0.8	07May2015, 14:00	452.39
Subbasin-2	0.061	0.7	07May2015, 14:00	452.35
Reach-1	0.061	0.7	07May2015, 14:00	452.38
Junction-1	0.132	1.6	07May2015, 14:00	452.38
Reach-3	0.132	1.6	07May2015, 14:00	452.43
Subbasin-3	0.104	1.2	07May2015, 14:00	452.35
Reach-4	0.104	1.2	07May2015, 14:00	452.36
Junction-2	0.236	2.8	07May2015, 14:00	452.40

Figure 10. Runoff hydrograph in catchment outlet section (HEC-HMS)

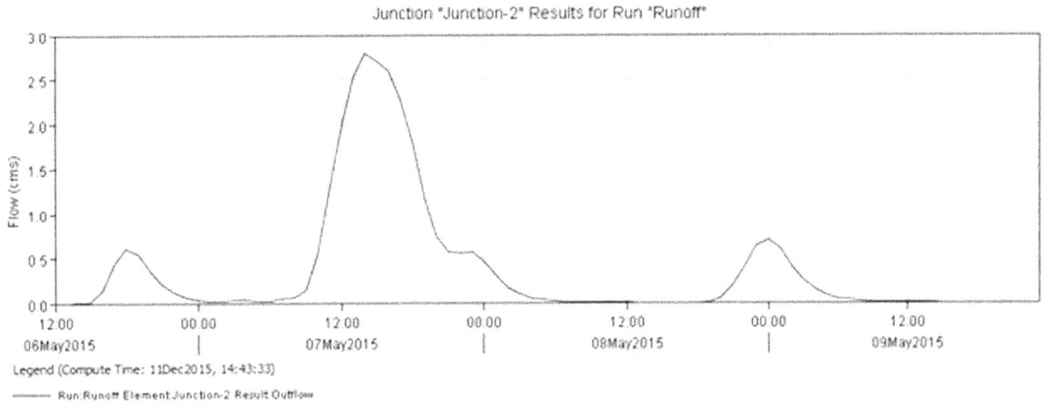

section 18, 340178 m^3, in hypothesis 4 (torrential rain with a duration of 60 minutes and calculus intensity i =0,70 mm/min).

Application of MIKE11byDHI model, UHM Method [Beilicci et al, 2013], [Kocsis et al, 2020]

For the rainfall-runoff modeling was chosen the representative hydrographical basin Sebes (Figure 11). The representative hydrographical basin Sebes is located in the northwest of the massive Mount Mic,

Figure 11. The location of the representative hydrographical basin Sebes

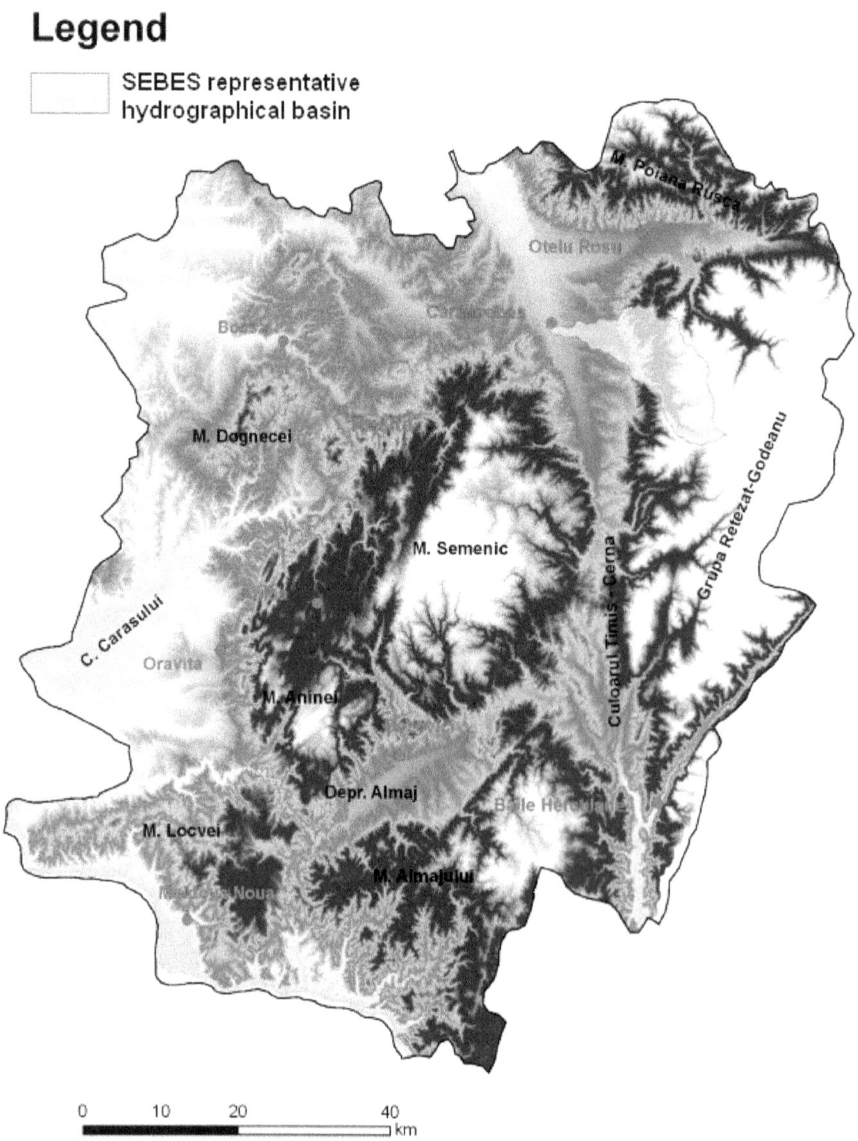

which is part of the Tarcu Mountains group (group Retezat-Godeanu, Carpathians). Sebes River, with a length of 20,1 km in the closing section, drains an area of 142 km^2 and is the right tributary of the river Timis, the confluence is situated in Caransebes town.

Because of the data they provide, the representative hydrographical basins complement the work undertaken at experimental stations and basins, an activity that is more accurate than in large river basins.

In developing the network of representative hydrographical basins was taken into account to obtain information about the average, maximum, and minimum flow, discharge of sediment transport, temperature of air and water, rainfall, and river frost phenomena evolution.

Considering the written above, was organized Sebes representative hydrographical basin.

The representative hydrographical basin location took into account several physical and geographical criteria, and not only (various physical and geographical conditions, different shapes of watersheds, unimportant changes of natural runoff, groundwater flow uniformity, the existence of meteorological stations, the existence of electrical power network up to high altitudes, the existence of human settlements closer the observation points) (Figure 12).

Figure 12. SEBES representative hydrographical basin

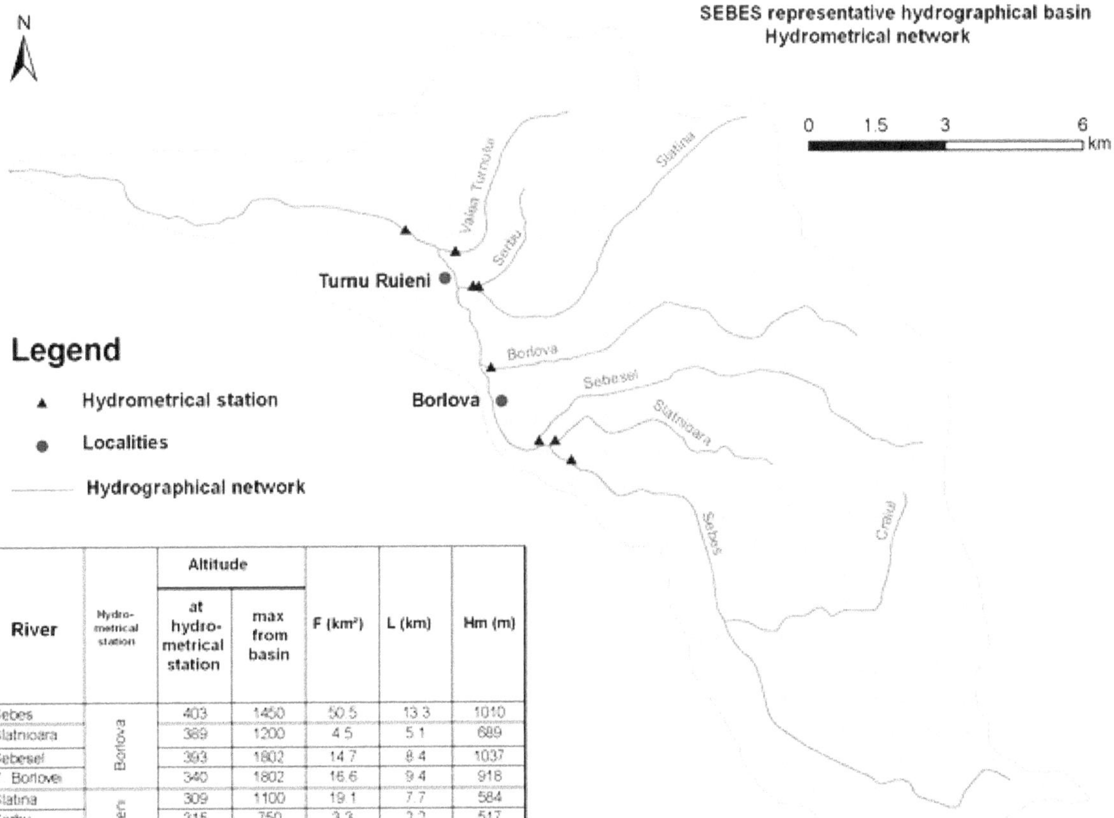

River	Hydro-metrical station	Altitude at hydro-metrical station	Altitude max from basin	F (km²)	L (km)	Hm (m)
Sebes	Borlova	403	1450	50.5	13.3	1010
Slatnioara	Borlova	389	1200	4.5	5.1	689
Sebesel	Borlova	393	1802	14.7	8.4	1037
V. Borlovei	Borlova	340	1802	16.6	9.4	918
Slatina	Tr Ruieni	309	1100	19.1	7.7	584
Serbu	Tr Ruieni	315	750	3.3	2.2	517
V. Turnului	Tr Ruieni	300	900	5.85	6.0	603
Sebes	Tr Ruieni	287	1802	124	20.3	815

Input data requirements:

- Area of the sub-catchment (GIS) – 50,50 km^2
- Hydraulic Length (GIS) – 13,30 km
- Land Use and percentage area of individual categories - mixed forest - coniferous, deciduous
- Soil Hydrologic Groups (A-D) and percentage area of individual categories – C
- Initial value of CN - 75
- Average Slope in the catchment (GIS) – 0,506
- Observed values: Rainfall – April 2005, Borlova hydrometrical station

Runoff (for calibration) - April 2005, Borlova hydrometrical station (Figure 13)

- Base flow – 0,267 m3/s
- Time series step – 1 day
- Time step for simulation – 30 seconds

Figure 13. Rainfall and observed discharge

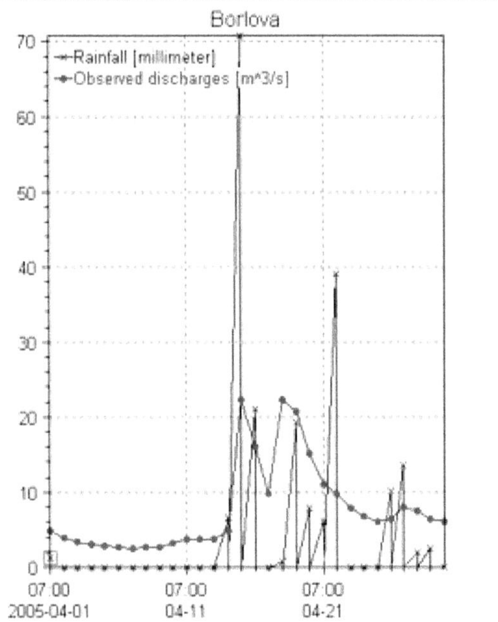

	Time	1:Rainfall [millimeter]	2:Observed discharges [m^3/s]
0	4/1/2005 7:00:00 AM	1.2	4.73
1	4/2/2005 7:00:00 AM	0	3.92
2	4/3/2005 7:00:00 AM	0	3.33
3	4/4/2005 7:00:00 AM	0	2.98
4	4/5/2005 7:00:00 AM	0	2.81
5	4/6/2005 7:00:00 AM	0	2.63
6	4/7/2005 7:00:00 AM	0	2.46
7	4/8/2005 7:00:00 AM	0	2.63
8	4/9/2005 7:00:00 AM	0	2.63
9	4/10/2005 7:00:00 AM	0	3.16
10	4/11/2005 7:00:00 AM	0	3.72
11	4/12/2005 7:00:00 AM	0	3.72
12	4/13/2005 7:00:00 AM	0	3.72
13	4/14/2005 7:00:00 AM	6.6	4.73
14	4/15/2005 7:00:00 AM	70.8	22.3
15	4/16/2005 7:00:00 AM	21.1	16
16	4/17/2005 7:00:00 AM	0	9.8
17	4/18/2005 7:00:00 AM	0.8	22.3
18	4/19/2005 7:00:00 AM	19.3	20.7
19	4/20/2005 7:00:00 AM	7.9	15.2
20	4/21/2005 7:00:00 AM	6	11.1
21	4/22/2005 7:00:00 AM	39.1	9.8
22	4/23/2005 7:00:00 AM	0	7.78
23	4/24/2005 7:00:00 AM	0	6.74
24	4/25/2005 7:00:00 AM	0	6
25	4/26/2005 7:00:00 AM	10.2	6.5
26	4/27/2005 7:00:00 AM	13.6	8.06
27	4/28/2005 7:00:00 AM	2	7.51
28	4/29/2005 7:00:00 AM	2.5	6.5
29	4/30/2005 7:00:00 AM	0	6

The results of the simulation are presented in Figure 14.

Model calibration consists of the comparison between the simulated hydrograph and observed flood event hydrograph (Figure 15); the goal is that both hydrographs should be similar and the purpose is to get a real model of rainfall-runoff relationship in an individual catchment based on real observed data.

For calibration can be used two methods for modifying uncertain parameters (CN curve numbers or Lag time). Modification of the CN curve consists of changing the hydrological group of soil or changing the CN curve number inside a default interval.

Figure 14. Net rainfall and simulated runoff

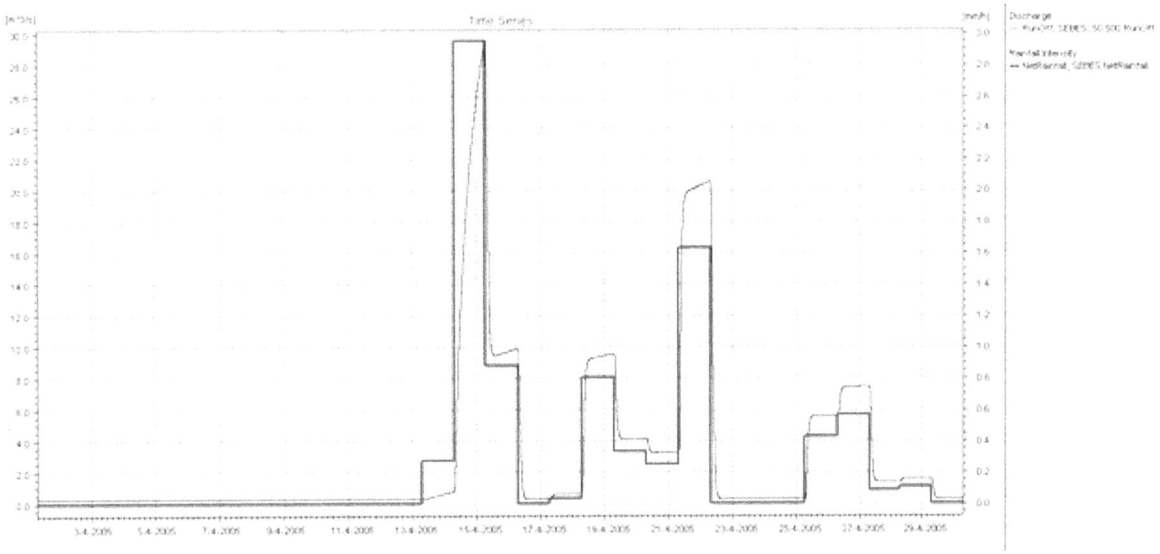

Figure 15. Comparison between the simulated hydrograph and observed flood event hydrograph

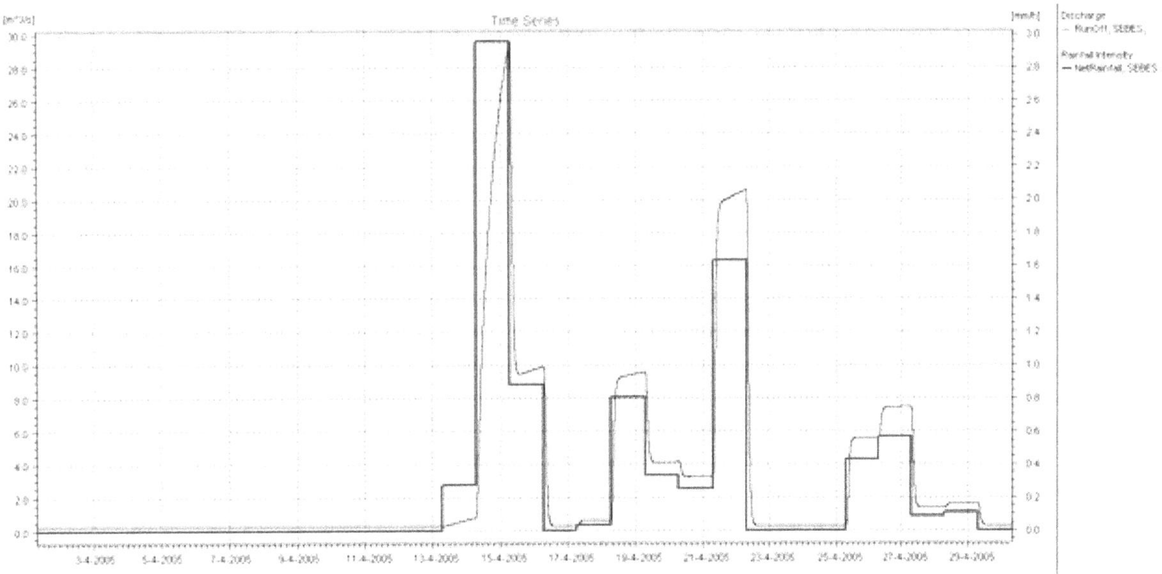

For example, mixed forest: minimum: 38 (HG A), respectively maximum: 81 (HG D). Do not go past the limits.

Modification of the Lag time is made according to the geomorphological characteristics of an individual sub-catchment.

For this model, calibration was modified CN curve number. For the value of CN 62, the maximum simulated discharge value result is approximately equal to the observed discharge (22,3 m3/s) (Figure 16).

Figure 16. Model calibration

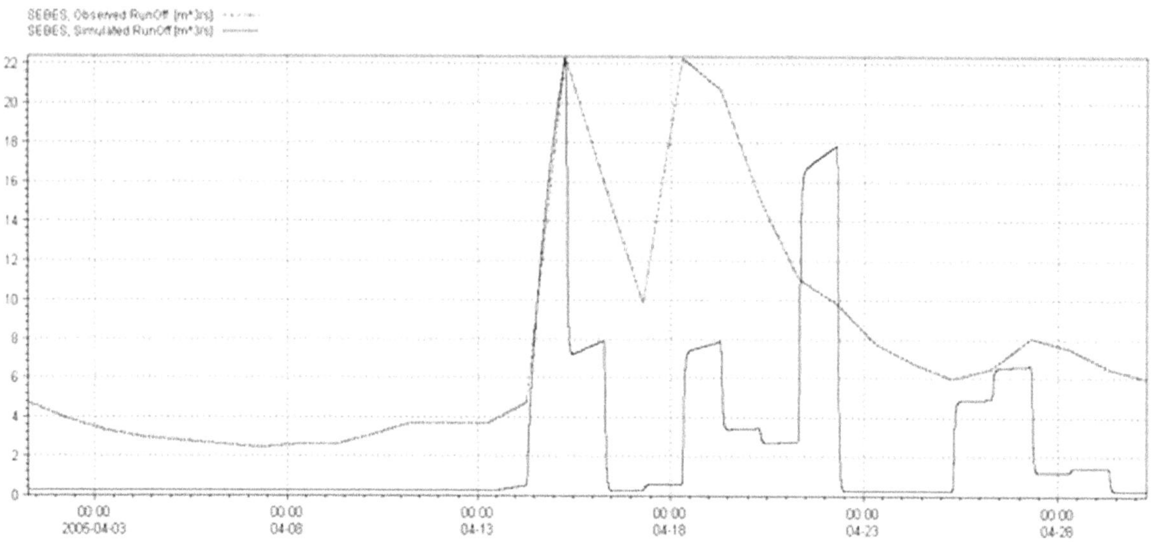

The differences between the observed and simulated hydrograph shape are due to the presence of small tributaries on the studied river. For a more accurate modeling should be known in detail the characteristics of these small tributaries a complex hydrodynamic model. Since it is a mountainous area may have difficulty in elevation data collection for longitudinal profiles and cross sections on these sectors.

Application of MIKE11 by DHI Model, NAM Method [Beilicci &Beilicci, 2016]

This case study simulated the runoff from the snowmelt phenomenon in the Valea Mare watershed, a component of Bega river catchments (Figure 6).

The snowmelt component of the runoff is incorporated as an integrated module within NAM, the temperature data is only required for simulation.

Hypotheses and input data were:

- Catchment average slope - 20.64%
- Total area - 23.54 ha
- Simulation period – 15 February 2016, 00:00 AM – 15 March 2016, 00:00 AM
- Time step – 30 minutes
- Rainfall is neglected
- Evaporation from the snow is neglected.

The used time series are illustrated in Figure 17.

The obtained results from the simulation are shown in Figure 18.

The variation of snow cover thickness in the simulation period is shown in Figure 19.

The discharges resulting from snow melting are not very high, because the average daily temperatures in the calculation period were within the specific climatological norms. If, however, the average daily temperatures were higher, then the resulting discharges would have higher values.

Figure 17. The time series

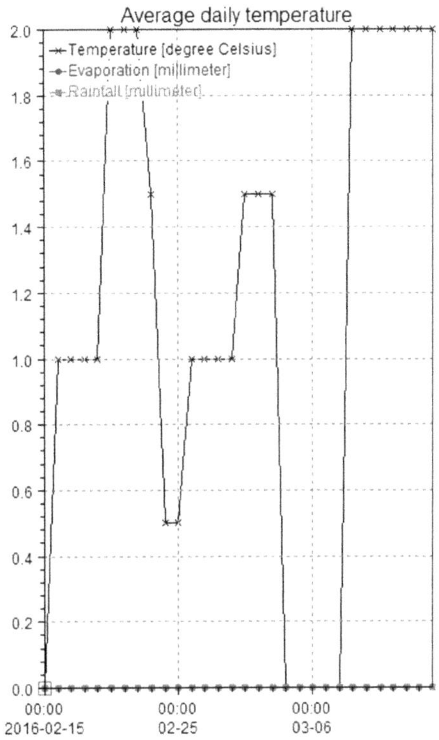

Figure 18. Time series runoff

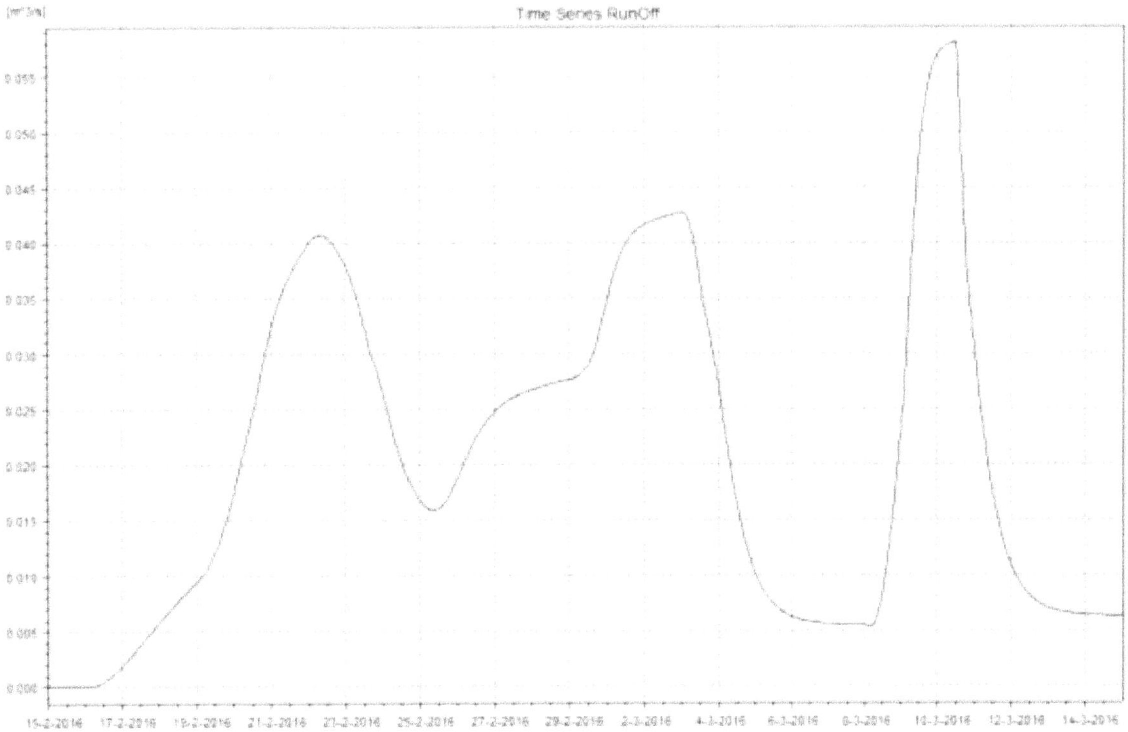

Figure 19. The variation of snow cover thickness

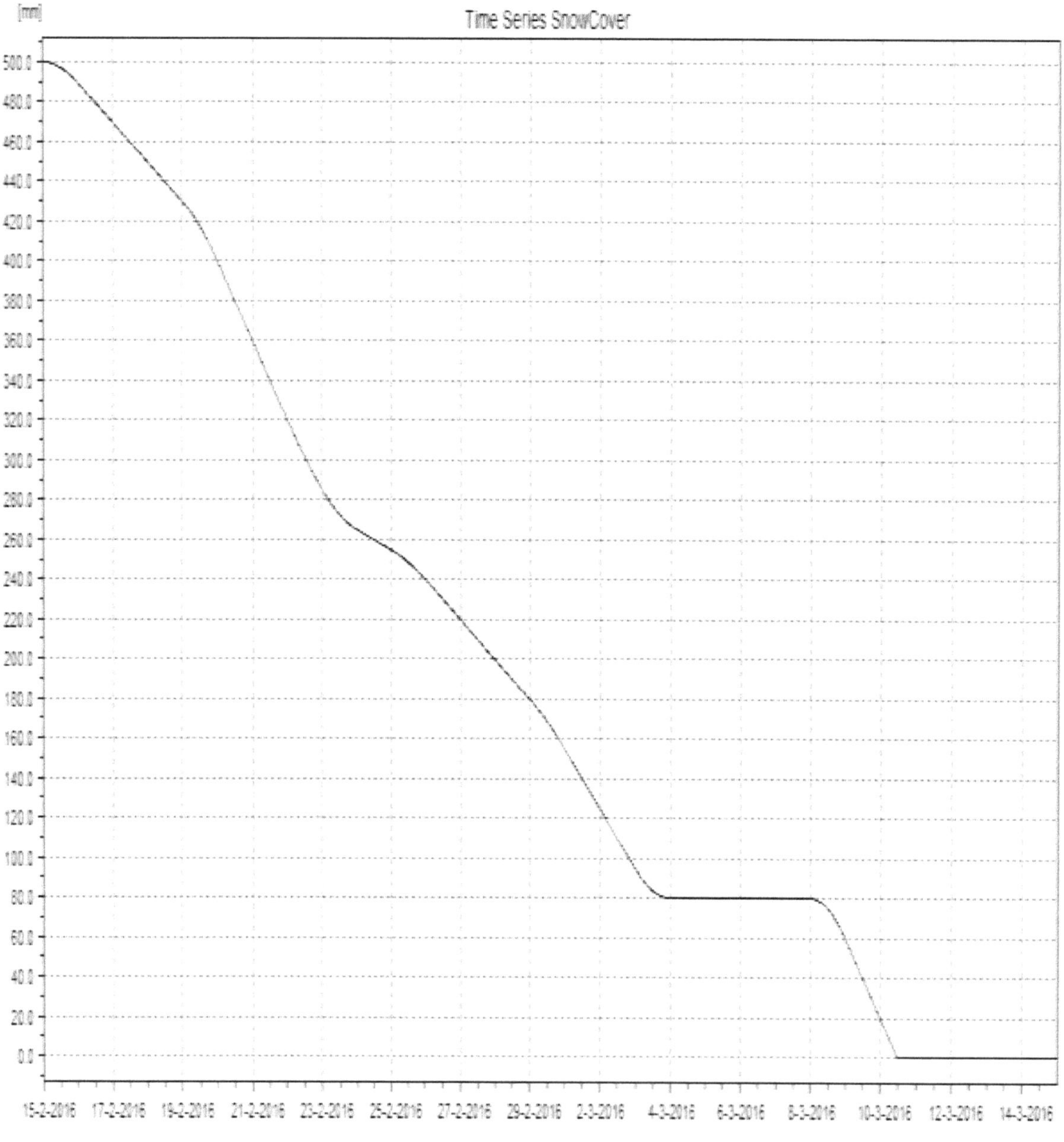

Comparisons on a Case Study Performed With the Same Input Data [Beilicci & Beilicci, 2015a], [Beilicci & Beilicci, 2015b]

For a comparison between the simulation results of the same phenomenon with the different presented software, the simulation of the rainfall-runoff phenomenon was performed with the WEPP, DUFLOW, and MIKE11 software for the hydrographic basin presented in case study no. 1, with the related assumptions and the same data series.

WEPP model gives the volume of water, DUFLOW gives the water discharge and water level and MIKE11 gives the water discharge in the closing section of the watershed. Through simulation with the WEPP model, the water volume passing through in closing section 18, 340178 m^3, in hypothesis 4 (torrential rain with a duration of 60 minutes and calculus intensity i =0,70 mm/min). The resulting maximum value of water discharge is 7.10 m^3/s with DUFLOW and 7.50 m^3/s with MIKE11. The resulting water level in the closing section of the watershed, given by DUFLOW, is 129.40 meters above sea level.

Another comparison was made for the hydrographic basin presented in case study no. 2, with the related assumptions and the same data series, with HEC-HMS and MIKE11 software.

The resulting maximum value of water discharge is different: 2.68 m^3/s for MIKE11 and 2.8 m^3/s for HEC-HMS, but the allure of runoff hydrographs was about the same.

REFERENCES

Bacov, A. (1996). *Hydrotechnical arrangements of hillslopes for soil erosion control.* Mirton Publishing.

Beilicci, E., & Beilicci, R. (2007). The water flow modellation through unpermanent hydrographic network. *Proceedings from: 32nd Internationally Attended Scientific Conference of the Military Technical Academy.* MTA.

Beilicci, E.& Beilicci, R. (2015a). Proceedings from: SGEM2015. *Flash floods - causes, effects, possibilities of modeling with hydroinformatic tools.*

Beilicci, E.& Beilicci, R. (2015b). A comparative analysis of two advanced hydroinformatic tools for rainfall–runoff phenomena modeling. *Scientific Bulletin of the Politehnica University of Timişoara, Transactions on Hydrotechnics, 60*(74), 55-58.

Beilicci, E., Beilicci, R., Girbaciu, A., & Girbaciu, C. (2013). Proceedings from: 6th International Conference on Environmental and Geological Science and Engineering. *Rainfall – Runoff modeling using advanced hydroinformatic tools.*

Danish Hydraulic Institute (DHI). (2011). MIKE 11 - A modeling system for rivers and channels, Short introduction and tutorial. Horsholm, Denmark.

Di Silvio, G. (1998). *Soil erosion and conservation - lecture notes.* International Institute for Hydraulic and Environmental Engineering.

Drobot, R. (2007). *Methodology to determine torrential hydrographical basins where exist localities exposed to the danger of flash floods.* Technical University of Civil Engineering.

Fanta, S. S., & Tadesse, S. T. (2022). Application of HEC–HMS for runoff simulation of Gojeb Watershed, Southwest Ethiopia. *Modeling Earth Systems and Environment, 8*(4), 4687–4705. doi:10.100740808-022-01397-4

Giurma, I. (1997). *Reservoirs silting.* Bucharest, Romania: H*G*A* Publishing House.

International Institute for Infrastructural, Hydraulic and Environmental Engineering (IHE). (1995). DUFLOW – Manual.

Lazar, Gh. I. (2007). Computer-assisted numerical modeling of currents with free level in the regulated regime, Timisoara, Romania: Politechnica Publishing House.

Ministry of Environment and Water Management. (2005). HG 1854/2005. Floods risk management national strategy. Bucharest, Romania.

National Oceanic and Atmospheric Administration. (2000). *What are flash floods?* NOAA.

National Oceanic and Atmospheric Administration, National Severe Storms Laboratory. (n.d.). *Flood Forecasting*. NOAA. https://www.nssl.noaa.gov/education/svrwx101/floods/forecasting/

Popovici, N. (1991). *Soil Erosion Control Practice*. Politechnical Institute Lithography.

Saghafian, B., Meghdadi, A. R., & Sima, S. (2015). Application of the WEPP model to determine sources of run-off and sediment in a forested watershed. *Hydrological Processes*, *29*(4), 481–497. doi:10.1002/hyp.10168

Stanescu, V.A., & Drobot, R. (2002). *Non-structural measures for flood management.* H*G*A* Publishing House.

US Army Corps of Engineers, Hydrologic Engineering Centers. (2000). *Hydrologic Modeling System HEC-HMS Technical Reference Manual*. US Army.

US Army Corps of Engineers, Hydrologic Engineering Centers. (2010). HEC-HMS Quick Start Guide version 3.5. US Army.

US Department of Agriculture, Agricultural Research Service (USDA-ARS). (1995). *WEPP User's Manual*. Purdue University.

KEY TERMS AND DEFINITIONS

Flash Flood: A flood caused by torrential rainfall, i.e. a large amount of water in a short time. Flash floods are usually characterized by high flows, and strong currents that produce significant erosion in their path.

Hydrograph: A graph that shows the rate of flow (discharge) or level over time, in a certain transversal section of the river, channel.

Hydroinformatics: A branch of informatics that concentrates on the application of information and communications technologies in addressing the increasingly serious problems of the equitable and efficient use of water for many different purposes; is the integration of computational fluid dynamics and of artificial intelligence. Computational hydraulics is the fusion of numerical methods of applied mathematics, hydrodynamics, and hydraulics.

Lag Time: The time interval between the moment of the peak of the precipitation isohyet and the moment of the peak of the runoff hydrograph in the outlet section of a hydrographic basin.

Numerical Model: A combination of a large number of mathematical equations describing a physical phenomenon, the solution of which can be done with the help of computers, to find an approximate solution to the underlying physical problem.

Numerical Simulation: A calculation performed on a computer by running a program that implements a mathematical model for a physical system. Numerical simulations are necessary to study the behavior of systems whose mathematical models are too complex to provide exact solutions, approximate solutions are needed, which are based on certain simplifying assumptions.

Runoff: Movement of water across the earth's surface owing to the force of gravity. Are divided into sheet flow and channel flow. Sheet flow is made up of rain and meltwater and moves down slopes, without fixed channels. Channel flow occurs in definite linear directions, in the channels of rivers and bottoms of ravines.

Time of Concentration of Runoff: The maximum time a raindrop falling in a hydrographical basin travels the distance required to reach the closure cross-section – outlet section.

Chapter 4
Flash Flood Early Warning Systems:
A Critical Review of Their Effectiveness in Reducing Disaster Risks

Anna Msigwa
https://orcid.org/0000-0002-4782-9145
The Nelson Mandela African Institution of Science and Technology, Tanzania & The University of Pretoria, South Africa

Ayodeji Samuel Makinde
https://orcid.org/0000-0002-6482-2949
Edo State University, Uzairue, Nigeria

Grite Nelson Mwaijengo
The Nelson Mandela African Institution of Science and Technology, Tanzania

ABSTRACT

Flash floods are natural disasters that significantly threaten communities and infrastructure worldwide. Flash flood early warning systems (FFEWS) have been developed to mitigate the negative impacts of flash floods. These systems rely on monitoring and forecasting weather conditions and providing timely warnings to potentially affected communities. However, their effectiveness in reducing disaster risks needs to be better established. This chapter aims to critically review flash flood early warning systems, examining their strengths, limitations, and challenges associated with their implementation and proposing an improved framework and algorithm for an early warning system to mitigate some of the challenges encountered in previous systems. By delving into the current state of these systems, their components, and the factors influencing their effectiveness, this chapter seeks to enhance our understanding of EWS and contribute to developing more robust and efficient strategies for mitigating the impacts of flash floods.

DOI: 10.4018/978-1-6684-8771-6.ch004

INTRODUCTION

Flash floods are rapid-onset natural disasters caused by heavy rainfall, often leading to landslides, mudflows, and damage to infrastructure (Sene, 2012; Elkhrachy, 2015). Climate change projections suggest an increase in intense rainfall, potentially exacerbating flash flooding (Zhang et al., 2021). Additionally, urbanization has expanded the vulnerable population. Effective flash flood forecasting requires accurate rainfall forecasts and an understanding of the hydrological processes in catchments. Small catchments are prone to flash floods due to steep slopes, saturated soils, and impermeable surfaces. However, many of these catchments lack monitoring data, posing challenges for hydrological modeling (Hapuarachchi & Wang, 2008).

In March 2023, heavy rains brought relief and devastation to the drought-stricken Horn of Africa (Emily, 2023). While parts of Ethiopia received higher-than-usual rainfall, bringing much-needed respite from the prolonged period of extreme drought, the torrential downpours led to flash flooding in Ethiopia and Somalia. Tragically, several lives were lost, and around 300,000 people were affected by the sudden deluge (Emily, 2023; Crop Monitor GEOGLAM, 2023, June). The juxtaposition of severe drought and destructive floods underscores the unpredictable and extreme nature of weather events in the region.

It is said that Water-related disasters account for a significant 90% of all-natural disasters worldwide, with floods being the most prevalent type of water-related disaster (Perera et al., 2019). Africa bears number two after Asia. Structural and non-structural flood mitigation measures have been implemented to minimize these disasters, but the economic damages caused by floods have continued to increase (Khan et al., 2021).

The situation in the Horn of Africa is not an isolated incident. Eastern Africa has been grappling with the impact of floods, exacerbating vulnerable communities' challenges. Since mid-2022, over 2.25 million individuals in the region have been affected by destructive floods and landslides, further compounding their hardships (Emily, 2023). With the increasing occurrence of flash floods attributed to climate change, anticipating and preparing for such events is crucial to minimize their impacts (Borga et al., 2008; Gaume et al., 2009; Marchi et al., 2010). Forecasting flash floods poses significant challenges due to the explosive nature of rainfall events and the lack of measurement and forecasting in poorly gauged areas. However, radar rainfall observations offer high-resolution data that can improve flash flood early warning systems (FFEWS) by providing quantitative precipitation estimates and forecasts (Hapuarachchi & Wang, 2008).

FFEWS can be categorized as flow-based or rainfall-based. Flow-based systems rely on real-time flow information from river sections, while rainfall-based systems use spatially averaged basin rainfall as the driving variable. The latter method is simpler to implement and has shown the capacity to detect flood events using rainfall threshold exceedances (Corral et al., 2019).

On the other hand, Henao & Zambrano (2022) conducted a systematic review of 19 case studies to compare and analyze methods used for rainfall threshold (RT) estimation in flash flood prediction, focusing on the El Guamo stream basin in Colombia and other regions. The review identifies four types of methods: empirical, hydrological/hydrodynamic, probabilistic, and compound, and discusses their indicators and predicting variables. The study emphasizes the importance of considering local factors when selecting an appropriate method for early warning systems. It also highlights the significance of hydrodynamic models and uncertainty analysis in improving RT-based systems. Challenges in the work include the complexities of predicting flash floods in regions with specific characteristics, limited data availability and quality, and difficulties in real-time hydrological forecasting. The paper calls for data collection and research improvements to enhance flash flood forecasting and response capabilities.

Several operational flash flood early warning systems have been implemented at the regional scale, such as the Flash Flood Guidance (FFG) concept, the European Flood Awareness System (EFAS), the FF-EWS, and AIGA system in France, among others (Clark et al., 2014; Thielen et al., 2009; Georgakakos, 2005; Corral et al., 2019; Alfieri et al., 2011; Park et al., 2017).

Flash floods have been a recurrent issue in various regions, causing significant loss of life and property. The importance of flash flood early warning systems cannot be overstated, as they play a crucial role in mitigating the devastating impacts of flash floods on communities worldwide. These systems provide timely and accurate information about impending flood events, empowering individuals, communities, and authorities to take proactive measures to protect lives and minimize property damage.

However, the effectiveness of early warning systems in reducing disaster risks is an area that requires comprehensive evaluation and analysis. This chapter aims to critically review flash flood early warning systems, examining their strengths, limitations, and challenges associated with their implementation. It proposes an improved framework and algorithm for an early warning system to mitigate some of the challenges encountered in previous systems. By delving into the current state of these systems, their components, and the factors influencing their effectiveness, this chapter seeks to enhance our understanding of EWS and contribute to developing more robust and efficient strategies for mitigating the impacts of flash floods.

FLASH FLOOD EARLY WARMING SYSTEMS

Overview of Flash Flood Early Warning Systems in Operational

Flash flood early warning systems are crucial tools for mitigating the impact of flash floods and saving lives. These systems employ various technologies to detect and monitor rainfall patterns, river levels, and other relevant parameters to warn at-risk communities promptly. Here are some types of early warning systems that have been applied in different regions (see Figure 1):

1. Rainfall-based Early Warning Systems: These systems focus on monitoring rainfall patterns and predicting potential flood events based on precipitation data. They utilize rain gauges, weather radar, and satellite imagery to estimate rainfall intensity and duration. By analyzing this information, they can issue warnings based on predefined thresholds. For example, the European Flood Awareness System (EFAS) uses rainfall forecasts and hydrological models to provide early warnings for flash floods in Europe [https://www.efas.eu/].
2. River Level-based Early Warning Systems: River level-based systems monitor water levels in rivers and streams to assess the risk of flash floods. They employ sensors and gauges installed at key locations along water bodies to measure water levels in real-time. By comparing the observed levels with predefined thresholds, these systems can trigger warnings when the water level exceeds a critical point. One example is the Flood Early Warning System (FEWS) implemented in Bangladesh, which uses river-level data to issue warnings to vulnerable communities (Reliefweb, 2022).

Figure 1. Types of early warning systems

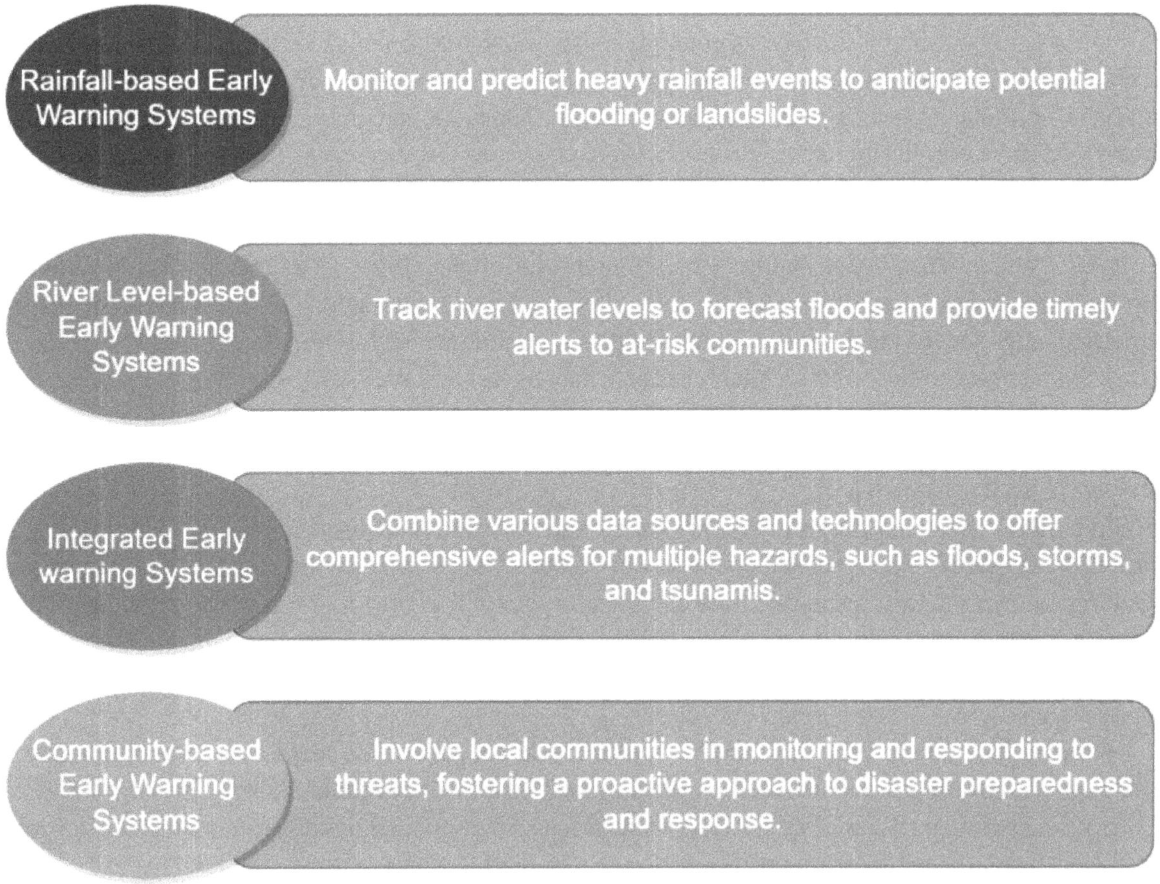

3. Integrated Early Warning Systems: Integrated early warning systems combine multiple data sources and technologies to provide comprehensive flash flood warnings. They integrate rainfall data, river levels, weather forecasts, and other relevant information into a unified system. These systems employ sophisticated modeling and prediction algorithms to generate accurate and timely warnings. The Integrated Flood Management System (IFMS) implemented in Japan is an example of an integrated early warning system that combines rainfall, river level, and weather data for flood forecasting (ICHARM, 2023).
4. Community-based Early Warning Systems: Community-based early warning systems involve local communities in the monitoring and dissemination of flood warnings. They empower communities by providing them with tools and training to recognize flood risk indicators and take appropriate actions. These systems often incorporate local knowledge and traditional forecasting methods. The Red Cross Red Crescent Climate Centre has implemented community-based early warning systems in various countries, including Uganda and Nepal, to enhance community resilience (Climate Centre, 2015).

Data Requirements for Robust Flash Flood Early Warning Systems

Flash flood early warning systems rely on comprehensive data from various sources to predict and respond to potential flash flood events accurately and timely. These systems are crucial in mitigating the impact of flash floods and protecting vulnerable communities. The key data components required for effective flash flood early warning systems include rainfall data, streamflow data, weather data, hydrological and topographical data, remote sensing data, historical data, early warning monitoring systems, communication, and information systems, historical flash flood data, and models and algorithms.

Rainfall data is of paramount importance as rainfall is the primary driver of flash floods. High-resolution and real-time rainfall data is essential for forecasting and risk assessment. This data is obtained from various sources such as rain gauges, weather radar, and satellite-based precipitation products. While rain gauges provide accurate point data, their coverage may be limited. Weather radar helps monitor rainfall over a broader area, while satellite-based products offer global rainfall estimation. Blending data from multiple sources improves the accuracy of Quantitative Precipitation Estimates (QPEs) and enhances hydrometeorological modeling.

Streamflow data from gauging stations along rivers and watercourses is crucial for detecting rapid rises in water levels, which indicate potential flash floods. Monitoring streamflow also helps in understanding the hydrological response of the catchment to rainfall events. Real-time meteorological data, including temperature, humidity, wind speed, and atmospheric pressure, provides valuable information about weather patterns and triggers for flash floods. Meteorological data is typically obtained from weather stations and meteorological agencies.

Hydrological and topographical data are essential for accurate flood modeling. Information about catchment characteristics, such as land use, soil properties, topography, and drainage networks, plays a vital role in hydrological modeling and flood simulations. High-quality digital elevation models (DEMs) provide elevation data necessary for precise flood modeling. Remote sensing technologies have vastly improved catchment characterization, especially in poorly gauged areas. Microwave satellite observations enable the estimation of surface soil moisture, land surface temperature, snow cover, and vegetation cover, contributing to hydrological modeling and flood forecasting.

Historical data on rainfall, streamflow, and past flood events are critical for calibrating and validating flood models. Analyzing historical data helps establish flood frequency analyses and understand the recurrence intervals of flash floods. Early warning monitoring systems, equipped with real-time sensors like water level sensors, rain gauges, and weather stations, continuously monitor meteorological and hydrological conditions, providing essential input for timely warnings.

Communication and information systems are vital components of flash flood early warning systems. Effective communication networks are required to relay warnings to the public, emergency responders, and relevant authorities. Disseminating information through various channels such as mobile phones, sirens, social media, and community engagement is essential for public awareness and preparedness.

Access to historical flash flood data provides valuable insights into the vulnerabilities of specific areas and helps improve forecasting and response strategies. Mathematical models, including hydrological models and flood inundation models, are fundamental for flash flood early warning systems. These models use input data to simulate flood events and predict their behavior, enabling timely warnings to be issued.

By integrating and analyzing data from these diverse sources, flash flood early warning systems can provide accurate and timely warnings to at-risk communities. These warnings empower communities to take proactive measures and minimize the devastating impact of flash floods, saving lives and protecting property. Continuous advancements in data collection, processing techniques, and modeling will further enhance the effectiveness of these warning systems in the face of flash flood hazards.

OPERATIONAL COUNTRY FLASH FLOOD EARLY WARNING SYSTEMS

The section highlights various operational country-specific flash flood forecasting systems used worldwide. One such system is Automated Local Evaluation in Real Time (ALERT) in Australia, operated by the Bureau of Meteorology (Perera et al., 2019). This real-time flood forecasting and warning system provides forecasts on a regional basis through flood warning centers in each State and the Northern Territory. Flash flood warnings are issued by state and local authorities using local systems. ALERT employs various methods, including empirical correlation of rainfall and peak flows, unit hydrograph modeling, and hydrological modeling. Additionally, the system utilizes an automated flash flood warning system called ALERT, which utilizes radio telemetry to transfer real-time rainfall and water level data to forecasting centers, enabling timely and accurate flood forecasting and warning dissemination.

In Northern America, the National Weather Service (NWS) operates the Gridded Flash Flood Guidance (GFFG) system (Bournas & Baltas, 2022). This system allows for generating county and catchment average flash flood guidance values based on hydrologic models. It uses multi-sensor precipitation estimates and forecasts from NEXRAD radar, rain gauges, and Numerical Weather Prediction (NWP) model outputs. However, it's worth noting that the general FFG method does not account for the threat of local flooding inside catchments due to heavy rainfall covering only a part of the catchment.

Central America benefits from the Central American Flash Flood Guidance (CAFFG) system, which has been operational since 2004. CAFFG provides flash flood guidance and warning for all seven Central American countries. It utilizes gauged rainfall and satellite-based rainfall derived by the Hydro-Estimator, along with short-range quantitative precipitation forecasts produced by the WS-ETa model. This system has been crucial in helping the region respond to flash flood events and mitigate their impacts.

In Northern Austria, a spatially distributed flash flood forecasting model has been operational since 2006. This model is grid-based and capable of simulating snow processes, soil moisture processes, and hill-slope scale routing for each grid cell. It integrates data from rain gauges, radar information, and deterministic precipitation forecasts from the Austrian Meteorological Office to provide accurate and location-specific flash flood forecasts.

For a broader scale, the European Flood/Flash Flood Forecasting System (EFFS) provides daily information on potential floods for large rivers and flash floods in small catchments across multiple European regions. The system uses global numerical weather prediction models, optional downscaling of global precipitation using regional NWP models, a catchment hydrology model, and a high-resolution flood inundation model to provide comprehensive and timely flood forecasts.

In Asia, the Decision Support System for Flash Flood Warning in Thailand is an adaptive mechanism that employs an artificial neural network model to forecast flash floods based on data from Automatic Weather Stations, meteorological satellite images, and NWP model outputs. This system helps provide timely flash flood warnings to vulnerable areas in Thailand, enabling communities to take appropriate precautions and actions.

In Malaysia, the GEOREX FLOOD system is operational in the state of Kelantan. It integrates remote sensing, Geographic Information System (GIS), hydrological modeling, and telecommunications technologies. The system uses the Stream-flow Synthesis and Reservoir Regulation (SSARR) model for flood forecasting, helping the authorities respond effectively to flash floods and manage water resources in the region.

Finally, in Israel, the Flash Flood Forecasting System of the Ayalon Stream uses an adaptive mechanism and a combination of sub-models to forecast flash floods in the Ayalon Stream. The system provides simultaneous forecasts for multiple lead times, enhancing the preparedness and response of the local authorities to flash flood events in the area.

These operational flash flood forecasting systems demonstrate the importance of having region-specific, technology-driven solutions to address the unique challenges posed by flash floods in different parts of the world. Such systems play a crucial role in minimizing the impacts of flash floods and safeguarding communities and infrastructure from the destructive forces of these natural disasters.

Case Studies: Developed Flash Flood Early Warning Systems

Early warning systems (EWS) for flash floods play a crucial role in disaster risk reduction and climate change resilience, safeguarding communities from the impacts of extreme weather events (Yasmin et al., 2023). The studies by Fernández-Nóvoa et al. (2020); Tu et al. (2022); and Nguyen et al. (2019), each present valuable insights into the development and implementation of EWS, focusing on different aspects of EWS design and performance as compared in Table 1.

Spain

Fernández-Nóvoa et al. (2020), introduce an innovative EWS that efficiently combines hydrological and hydraulic components, enabling the use of the Soil Conservation Service Curve Number (SCS-CN) method for continuous modeling (Fernández-Nóvoa et al., 2020). The integration of continuous and single-event modeling approaches provides the system with the main advantages of both methods, enhancing the accuracy of flash flood predictions. The hydrological model showcased a "very good" performance, accurately predicting the annual cycle of river flow, which is crucial for early flood warning (Fernández-Nóvoa et al., 2020). The statistical parameters, such as the Nash-Sutcliffe Efficiency (NSE) and Percent Bias (PBIAS), supported the robustness of the methodology. However, the model exhibited a slight overestimation of predicted river flows, necessitating further investigation into the reasons behind this overestimation and fine-tuning the model to reduce discrepancies (Fernández-Nóvoa et al., 2020). Addressing this overestimation is vital as it could lead to unnecessary alerts and resource allocation for flood management, affecting the system's overall effectiveness.

Moreover, the EWS presented by Fernández-Nóvoa et al., (2020), demonstrated the capability to predict 76% of alerts for hazardous flood situations, with a 28% false positive rate. While the system offers valuable information for decision-makers through detailed hazard maps, it also acknowledges the uncertainties in precipitation forecasts and their impact on flood predictions (Fernández-Nóvoa et al., 2020). To enhance the accuracy of the EWS, incorporating ensemble forecasting techniques or using probabilistic precipitation forecasts is needed to understand potential variations in predicting better outcomes. Furthermore, considering future climate change scenarios could significantly impact flood patterns and intensities, making it crucial to integrate climate change projections into the EWS to enhance its effectiveness in forecasting and managing potential changes in flood risk.

China

In contrast, Tu et al. (2020), focused on establishing early warning indicators for flash floods based on the variation in the rising rate of flood hydrographs in the rising limb (Tu et al., 2020). The study utilized the HEC-HMS model to obtain flood hydrographs under different rainfall events, emphasizing the importance of data integration from various sources, including land use, soil type, and slope information extracted using GIS technology (Tu et al., 2020). While the Baisha River flood early warning system demonstrated promising results, the authors identified certain limitations. The reliance on historical flood events and meteorological data for calibration and validation of the HEC-HMS model necessitates evaluating the current model performance using more recent data from 2015 to 2019 (Tu et al., 2020). Moreover, the lack of real-time monitoring is a concern, as it may limit the system's ability to respond to rapidly evolving flood events. Therefore, it is recommended that the integration of real-time monitoring and advanced forecasting technologies will enhance the system's effectiveness by providing timely and accurate warnings to local communities (Tu et al., 2020).

Vietnam

In their study, Nguyen et al. (2019), discussed the establishment of an EWS for flash floods in a specific region in Vietnam, highlighting the importance of considering various factors such as rainfall, basin characteristics, geomorphology, land cover, and risk factors for landslides in the flash flood forecasting process (Nguyen et al., 2019). The Multi-Criteria Analysis (MCA) method was utilized to build flash flood risk maps, taking into account the weight assigned to each layer of information (Nguyen et al., 2019). The study showcased the effectiveness of the system, but some areas for improvement were identified. Notably, the article lacked a clear introduction that sets the context and significance of establishing the early warning system for flash floods in the mentioned region. Additional details about the data collection process and specific methods used for data analysis and processing would strengthen the study's findings (Nguyen et al., 2019). Furthermore, the discussion on the system's accuracy and effectiveness could be more comprehensive, including real-world performance and user feedback. Additionally, incorporating examples of practical application through case studies of actual flash flood events and the system's response and impact would provide valuable insights (Nguyen et al., 2019).

Italy

Corral et al. (2019) compare two early warning systems (EWS) for real-time flash flood hazard forecasting at a regional scale. The systems are based on a gridded drainage network and use weather radar precipitation inputs to assess flash flood hazard levels using the return period as an indicator. The two systems under comparison are ERICHA, a rainfall-based system using upstream basin-aggregated rainfall, and Flood-PROOFS, a system based on a distributed rainfall-runoff model. The evaluation was conducted for three severe rainfall events in the Liguria region of Italy in 2014. The study contributes to understanding EWS performance by comparing the two approaches and revealing their strengths and limitations. It shows that ERICHA performs well in reproducing observed flows, even for extreme flood events, demonstrating the consistency of its flood hazard estimates. However, both systems are influenced by the quality of rainfall inputs, and limited observational data, uncertainty in rainfall inputs, concentration-time parameters, and the lack of sufficient streamflow observations pose challenges in conducting a comprehensive assessment. Additionally, the study points out that radar quantitative precipitation estimates (QPE) would have been preferable for more consistent rainfall thresholds in ERICHA. Despite these challenges, the research suggests that rainfall-based EWS like ERICHA can be as reliable as flow-based systems, particularly for large catchments and high-return period flood events.

Colombia

Acosta-Coll et al. (2018) address the rising occurrence of pluvial flash floods in urban areas due to climate change and human actions, leading to adverse effects on communities and infrastructure. To mitigate these impacts, the study emphasizes the importance of Early Warning Systems (EWS) to minimize casualties and damages. Through a systematic review, the study defines a basic structure for an effective EWS for rain flash floods. It covers factors influencing flash flood intensity, key components of an EWS, and various EWS architectures implemented worldwide. The review identifies shortcomings in existing EWS designs, prompting the proposal of a people-centered EWS architecture with four structures: Disaster Risk Knowledge, Forecasting, Dissemination and Communication of information, and Preparedness and Response. Real-time monitoring of rainfall and water levels using suitable sensor technology, such as rain gauges, weather radars, ultrasonic, or radar sensors, is highlighted as crucial for accurate forecasting and timely alerts. However, challenges in predicting flash floods due to their rapid onset, technical issues in data monitoring, resource constraints for implementation, and non-compliance with suggested norms in some projects are identified. The study provides valuable insights for designing effective EWS for pluvial flash floods in urban areas and calls for adherence to guidelines and best practices in future EWS development and implementation.

Mexico

Ibarreche et al. (2020) introduce the Emergency Water Information Network (EWIN), an early warning system (EWS) designed to predict flash floods and provide critical real-time information for timely evacuations. The system integrates various sensors, including hydrological and meteorological monitoring stations, "drifters," and a data mulling sniffer. EWIN successfully collects and disseminates information to decision-makers, government officials, and the public in the Colima and Villa de Alvarez metropolitan areas in Mexico. The main contribution of the study is the development and implementation of the EWS, offering a comprehensive solution to address the dangers of flash floods and potentially save lives. However, the system is still under development, and integrating weather data into the warning model remains challenging. Future improvements aim to enhance precision, user-friendliness, and scalability for wider applications. Continuous coordination with local authorities and maintenance support is crucial for the long-term success and impact of EWIN in flood-prone regions. Overall, the paper emphasizes the significance of ongoing enhancements to make the system more robust and effective in mitigating flash flood risks.

Choosing the "best" methodology requires considering accuracy, scalability, adaptability to different regions, real-time capabilities, and reliability. No single methodology might be universally applicable or superior. Each country's system is tailored to its specific needs, available technology, and environmental

Table 1. Flash flood early warning systems in different countries

Country	Authors/Study	Methodology	Strengths	Limitations	Recommendations
Spain	Fernández-Nóvoa et al. (2020)	Combined hydrological and hydraulic components, SCS-CN method, ensemble forecasting	Accurate prediction of river flow, detailed hazard maps	Slight overestimation of river flows, uncertainty in precipitation forecasts	Fine-tuning models, incorporate ensemble forecasting, integrate climate change projections
China	Tu et al. (2020)	HEC-HMS model, GIS integration, historical data	Promising results in warning indicators	Reliance on historical data, lack of real-time monitoring	Use real-time monitoring, integrate advanced forecasting, validate with recent data
Vietnam	Nguyen et al. (2019)	MCA method, various factors consideration	Effectiveness in flash flood risk mapping	Lack of context, insufficient detail on methodology, limited discussion on accuracy	Provide context, elaborate on methodology, include real-world performance and case studies
Italy	Corral et al. (2019)	ERICHA & Flood-PROOFS comparison, weather radar inputs	ERICHA's good performance, consistency in flood hazard estimates	Quality of rainfall inputs, limited observational data, uncertainty	Prefer radar QPE, enhance observational data, improve rainfall inputs
Colombia	Acosta-Coll et al. (2018)	Review of EWS structures, sensor technology importance	Emphasis on people-centered EWS architecture, key components identified	Challenges in prediction, technical issues, resource constraints	Focus on real-time monitoring, adherence to best practices, address technical issues
Mexico	Ibarreche et al. (2020)	EWIN integration, various sensors, real-time information	Comprehensive solution, real-time information dissemination	Challenges in weather data integration, system under development	Improve weather data integration, continuous coordination with authorities

conditions. The "best" methodology would likely involve an amalgamation of approaches, incorporating the strengths of different systems while addressing their limitations, to create a more comprehensive and robust early warning system for flash floods.

NUMERICAL MODELING SOFTWARE FOR DEVELOPING FLASH FLOOD EARLY WARNING SYSTEMS

Table 2 provides an overview of various numerical modeling software applications to develop flash flood early warning systems. The table includes information on the authorship, the specific numerical modeling software employed, and brief descriptions of each software's primary capabilities. The table aims to guide researchers and practitioners in selecting appropriate modeling tools based on the specific aspects of flash flood early warning systems they are addressing.

Table 2. Numerical modeling software for developing flash flood early warning systems

Author	Numerical Modeling Software	Description
Brunner, (2016)	HEC-RAS (Hydrologic Engineering Center's River Analysis System)	Specifically designed for river hydraulics and floodplain analysis. It's widely used for simulating and predicting riverine flooding.
Behrouz et al. (2020)	SWMM (Storm Water Management Model)	Primarily used for urban stormwater modeling, SWMM can simulate rainfall runoff, drainage systems, and their effects on floods.
Patro et al. (2009); Tansar et al. (2020)	MIKE Flood	Part of the MIKE software suite, MIKE Flood specializes in flood modeling, considering rivers, urban areas, and coastal zones. It's robust for comprehensive flood analysis.
Erena et al. (2018)	FLO-2D	This software focuses on floodplain and floodway modeling, offering detailed hydraulic and hydrologic analysis for both riverine and overland flow.
Phillips et al. (2005)	TUFLOW	A powerful hydrodynamic modeling software for simulating flooding, it's known for its ability to handle complex hydraulic and hydrological conditions.
Van Der Knijff et al. (2010)	LISFLOOD	A distributed hydrological model used for simulating floodplain inundation and river flow processes, often applied in large-scale flood assessments.
Cydzik and Hogue, (2009)	HEC-HMS (Hydrologic Engineering Center's Hydrologic Modeling System)	Primarily used for rainfall-runoff modeling, it helps in understanding the response of watersheds to precipitation and potential flooding.

CHALLENGES IN FLASH FLOOD EARLY WARNING SYSTEMS

Warning Dissemination and Communication Challenges

Perera et al. (2020) discuss flaws in warning dissemination and communication that can occur during flood events. The provided examples highlight the consequences of ineffective communication during critical moments. In the flash flood disaster in Malaysia, the misinterpretation of a siren activated during a water release led to casualties and property loss. Similarly, during Cyclone Eline in Southern Africa, a small percentage of respondents received warnings before, during, or after the floods. Vulnerable groups, such as those in remote or coastal areas, illiterate or impoverished individuals, and women, children, and the elderly in developing countries, often lack access to warning messages due to power outages, limited access to media, and language barriers. To address these challenges, targeted warnings, enhanced communication, and improved forecast accuracy are proposed to increase the credibility of early warnings and ensure effective dissemination to vulnerable communities (Perera et al., 2020).

Preparedness Challenges

Perera et al. (2020) also discuss the challenges in community preparedness for flood events. Limited public interest and risk awareness, lack of participation in decision-making, inadequate inclusion of vulnerable groups, and insufficient implementation of contingency plans hinder effective preparedness. Moreover, the lack of political commitment and reliance on post-disaster relief rather than risk mitigation can further hinder preparedness efforts. In regions with transboundary river basins, coordination and cooperation between countries become crucial for effective preparedness and response (Perera et al., 2020). The study by Yasmin et al. (2023) proposes a participatory approach known as "SMART convergence participatory research," which involves a shared understanding of risks, monitoring of risks, awareness-building, and pre-planning response actions on time. This approach seeks to develop an early warning system that is more resilient, context-specific, and inclusive, benefiting vulnerable communities and enhancing disaster preparedness and response efforts on multiple scales.

Forecasting Techniques and Operational Challenges

Hofmann and Schüttrumpf (2019) and Zanchetta and Coulibaly (2020) provide valuable insights into developing and implementing operational flood forecasting systems. The former discusses the development of a risk-based EWS for urban flooding in Aachen, Germany, utilizing hydro-numerical modeling, GIS mapping, and risk classification. The EWS integrates rainfall forecasting, a hydro-numeric model for flow simulation, and a GIS system to predict pluvial flooding and its impacts, issuing object-precise warnings for forecasted heavy rainfall events in specific urban areas. It is part of a larger multi-functional pluvial flood information system (MPFIS) that provides a holistic approach to flood management, analyzing historic, potential, and forecasted flood events (Hofmann and Schüttrumpf, 2019).

Zanchetta and Coulibaly (2020) highlight the opportunities and challenges in forecasting pluvial flash floods through EWS. The choice of forecasting criteria depends on the area covered and available resources, with hydrological models based on mixed conceptual-physical representations being preferred for their fewer parameters to calibrate. Improving monitoring capabilities, such as multi-constellation approaches and higher-resolution radar stations, can enhance flash flood forecasting accuracy. However, the forecasted extreme precipitation products remain the primary source of uncertainties in flash flood forecasting. Addressing this issue requires exploring multi-model approaches and deep learning-based techniques for more effective uncertainty estimations. The study also mentions the potential of data-driven surrogate modeling using machine learning and cellular automata methods to represent the influence of high momentum flows in flash floods. However, further validation is needed for these techniques (Zanchetta and Coulibaly, 2020).

Challenges in Designing and Implementing Early Warning Systems

Designing and implementing early warning systems for flash floods is challenging and involves overcoming various obstacles. One of the primary challenges is the availability and quality of data. Reliable and up-to-date data is crucial for accurate flood forecasting, but in some regions, data may be scarce, especially in developing areas or remote locations. Additionally, ensuring data quality and consistency can be difficult due to sensor malfunctions, data transmission errors, and limited data validation and quality control resources.

The technological infrastructure required for early warning systems poses another significant challenge. Establishing and maintaining monitoring networks, data processing systems, and communication channels demand a robust technological setup. However, this can be problematic in areas with limited resources, poor connectivity, or challenging terrain. Adequate funding and technical expertise are essential for successfully establishing and sustaining the necessary infrastructure.

Forecasting accuracy and lead time are critical factors for effective early warning systems. Achieving high accuracy in flood forecasting is a complex task that depends on multiple variables and factors. Sophisticated modeling techniques, reliable input data, and adequate computational resources are necessary for accurate predictions. Striking a balance between false alarms and missed warnings is crucial to provide meaningful and timely alerts to at-risk communities.

Community engagement and awareness are key to the success of early warning systems. However, it can be challenging to engage communities and ensure their understanding of warning messages and response actions. Cultural factors, language barriers, and limited education and literacy levels may hinder effective communication and response. Tailoring warnings to address different communities' specific needs and characteristics is vital to ensure their cooperation and active participation.

Institutional coordination and governance present additional challenges in early warning system implementation. Establishing clear roles, responsibilities, and coordination among various stakeholders is essential. However, bureaucratic processes, fragmented responsibilities, and conflicting mandates can impede effective institutional coordination. These challenges require strong governance structures and coordination mechanisms involving relevant authorities and agencies.

The sustainability and maintenance of early warning systems are ongoing challenges. Regular maintenance and updates are necessary to keep the systems effective. Ensuring sustained financial resources, the availability of skilled personnel, and addressing technological obsolescence are critical aspects of maintaining the functionality of early warning systems over time.

Social and cultural factors can significantly influence the acceptance and effectiveness of early warning systems. Community beliefs, trust in authorities, and risk perception can impact the response to warnings. Overcoming social and cultural barriers requires targeted social engagement efforts, involvement of local leaders, and consideration of traditional knowledge and practices in designing and implementing early warning systems.

PROPOSED METHODOLOGY FOR AN IMPROVED FFEWS

Proposed Conceptual Framework for an Improved FFEWS

The proposed conceptual framework for an improved FFEWS is presented in Figure 2, showing the implementation process for the FFEWS. Each of the steps is explained as follows:

a.) Sensor Deployment: the IoT sensors are installed strategically in flood-prone areas to collect relevant data. These sensors include rain gauges, water level sensors, streamflow sensors, soil moisture sensors, and weather stations. Each sensor has the capability of transmitting data wirelessly.

b.) Data Transmission: LoRaWAN IoT communication device was utilized to establish a network for transmitting sensor data. This allows the collected data to be sent to a central server or cloud platform in real-time.

Figure 2. Proposed conceptual framework for an improved FFEWS

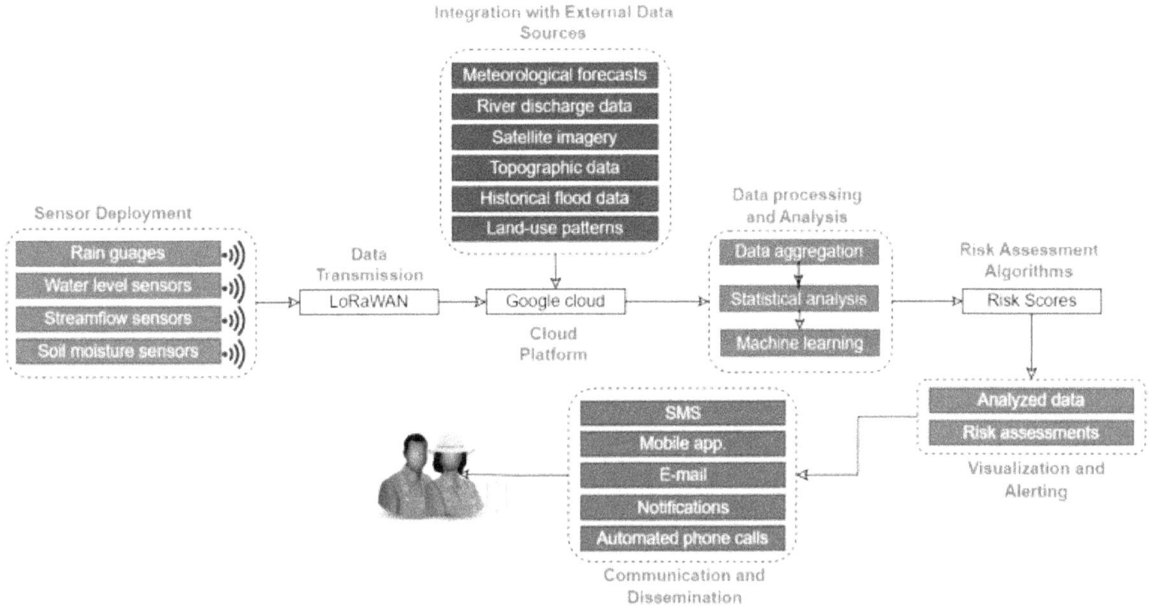

c.) Cloud Platform: Google Cloud platform, which provides the infrastructure for scalable and secure data storage, was used to receive, store, and process the sensor data.

d.) Data Processing and Analysis: Data analytics techniques were employed on the cloud platform to process and analyze the incoming sensor data. This involves real-time data aggregation, statistical analysis, and machine learning algorithms to identify patterns, thresholds, and anomalies.

e.) Risk Assessment Algorithms: The risk assessment algorithms integrate the collected sensor data with other relevant information, such as topographic data, historical flood data, and land-use patterns. These algorithms calculate risk scores based on predefined parameters and thresholds.

f.) Visualization and Alerting: A user-friendly dashboard to visualize the analyzed data and risk assessments. This enables users to monitor the current situation and historical trends. Establish alerting mechanisms to trigger notifications or warnings when the risk assessment algorithms detect potentially dangerous conditions.

g.) Integration with External Data Sources: Integrate IoT data with external data sources like meteorological forecasts, river discharge data, or satellite imagery. This provides additional context and enhances the accuracy of risk assessments.

h.) Stakeholder Communication: Establish communication channels to disseminate warnings and alerts to relevant stakeholders. This is done through SMS, mobile apps, email notifications, or automated phone calls.

Proposed Algorithm for an Improved FFEWS

Algorithm 1 outlines the steps for a flash flood early warning system that utilizes IoT sensor data and weather forecast information to generate a flash flood warning.

a.) Initialize Variables: In this step, the algorithm sets up the necessary thresholds and risk levels for triggering warnings. The rainfall intensity (rainfall_threshold), water level (water_level_threshold), and soil moisture (soil_moisture_threshold) are predefined values that determine when the corresponding parameters exceed a certain limit, indicating a potential risk of flash flooding. The alert level scale (low, moderate, high) and the corresponding risk thresholds (low_risk_threshold, moderate_risk_threshold) are used to classify the severity of the flash flood warning.

b.) Collect IoT Sensor Data: The algorithm reads real-time data from various IoT sensors, including rain gauges, water level sensors, and soil moisture sensors. These sensors are strategically deployed in flood-prone areas to collect data on rainfall intensity (rain_gauge_data), water levels in rivers and streams (water_level_data), and soil moisture levels (soil_moisture_data).

c.) Preprocess Sensor Data: Before using the sensor data for analysis, the algorithm cleans the data by handling any missing values and removing any outliers or errors. Preprocessing techniques are applied to normalize the data, bringing all sensor values on a similar scale to ensure fair comparisons during risk assessment.

d.) Collect Weather Forecast Data: The algorithm accesses weather forecast data, which includes predicted rainfall intensity (rainfall_intensity_forecast) and weather conditions (weather_condition_forecast). The forecast data provides additional context and helps anticipate potential flash flood events.

e.) Feature Engineering: In this step, the algorithm combines the real-time IoT sensor data and the weather forecast data to create a feature matrix. The feature matrix contains relevant attributes that will be input for the machine learning model. Feature engineering is crucial for generating meaningful predictions and assessing the risk accurately.

f.) Machine Learning Prediction: Using the feature matrix, the algorithm employs a trained machine learning model to predict the likelihood of a flash flood event. The model has learned patterns and relationships from historical data and can make real-time predictions based on the feature matrix.

g.) Flash Flood Risk Assessment: The algorithm compares the predicted flash flood probability with the predefined thresholds in the initialization step. It evaluates the risk based on the individual sensor readings and assigns risk levels (high, moderate, or low) to each parameter (rain_risk, water_risk, soil_risk) based on whether they exceed their respective thresholds.

h.) Aggregate Risk Levels: After assessing the risk levels for each parameter, the algorithm calculates an overall risk level based on the highest risk among the three (high_risk). This ensures that the flash flood warning reflects the most severe risk detected across all factors.

i.) Generate Flash Flood Warning: Using the aggregated risk level (high_risk), the algorithm generates a flash flood warning message. The warning message corresponds to the severity of the risk:

If the high_risk is above the moderate_risk_threshold, the warning is set to "High Risk - Take Immediate Action".

If the high_risk is between the low_risk_threshold and moderate_risk_threshold, the warning is set to "Moderate Risk - Stay Alert".

If the high_risk is below the low_risk_threshold, the warning is set to "Low Risk - Exercise Caution".

j.) Output: The final output of the algorithm is the flash_flood_warning, which represents the warning message generated based on the current sensor readings and the predicted risk of flash flooding.

Algorithm 1. Flash flood early warning system with IoT and deep learning

Input: Rainfall intensity (rain_gauge_data), Water level in rivers and streams (water_level_data), Soil moisture (soil_moisture_data), Predicted rainfall intensity, Predicted weather conditions
Output: Flash flood warning (flash_flood_warning)

Step 1: *#Initialize Variables*
- Set a threshold for rainfall intensity, water level, and soil moisture as rainfall_threshold, water_level_threshold, and soil_moisture_threshold *#to trigger a warning*
- Set an alert level scale (low, moderate, high) and the corresponding risk thresholds (low_risk_threshold, moderate_risk_threshold) *#to classify the warning severity*

Step 2: *#Collect IoT Sensor Data*
Read real-time data from IoT sensors (rain_gauge_data, water_level_data, soil_moisture_data)
Step 3: *#Pre-process Sensor Data*
Clean and handle missing values in the sensor data if necessary.
Normalize the data to bring all sensor values on a similar scale.
Step 4: *#Collect Weather Forecast Data*
Read the weather forecast data (rainfall_intensity_forecast, weather_condition_forecast).
Step 5: *#Feature Engineering*
Create a feature matrix by combining the sensor data and weather forecast data.
Step 6: *#Machine Learning Prediction*
Use the trained machine learning model to predict the likelihood of a flash flood based on the feature matrix.
Step 7: *#Flash Flood Risk Assessment*
#Compare the predicted flash flood probability with predefined thresholds
  If the rainfall intensity is above the rainfall_threshold, then
    set rain_risk = high.
  If the water level is above the water_level_threshold, then
    set water_risk = high.
  If the soil moisture is above the soil_moisture_threshold, then
    set soil_risk = high.
Step 8: *#Aggregate Risk Levels*
Calculate an overall risk level based on the individual risk levels (rain_risk, water_risk, soil_risk).
Determine the highest risk level among the three (high_risk).
Step 9: *#Generate Flash Flood Warning*
  If the high_risk is above the moderate_risk_threshold, then
    set the flash_flood_warning to "High Risk - Take Immediate Action."
  If the high_risk is between the low_risk_threshold and moderate_risk_threshold, then
    set the flash_flood_warning to "Moderate Risk - Stay Alert."
  If the high_risk is below the low_risk_threshold, then
    set the flash_flood_warning to "Low Risk - Exercise Caution."
Step 10: *#Output*
The algorithm should return the flash_flood_warning as the final output.

This warning can be disseminated to relevant stakeholders (e.g. farmers) to take appropriate actions and precautions to mitigate potential flash flood risks.

CONCLUSION

Flash flood early warning systems (FFEWS) are indispensable tools in mitigating the devastating impacts of flash floods and protecting vulnerable communities. This chapter has critically reviewed the key components, challenges, and proposed methodologies associated with FFEWS. The increasing occurrence of flash floods, often exacerbated by climate change, underscores the urgency of implementing effective early warning systems. Flash floods pose a significant risk, particularly in regions with vulnerable populations and poorly gauged areas. The integration of advanced technologies, such as high-resolution radar rainfall observations, remote sensing, and data-driven models, has shown promise in improving the accuracy of flash flood forecasting.

One of the significant challenges FFEWS faces is the dissemination and communication of warnings to at-risk communities. Ineffective communication during flood events can lead to casualties and property loss. To address this, targeted warnings and enhanced communication methods must ensure that warning messages reach vulnerable populations promptly, as shown in the proposed framework in Figure 2. Community engagement and preparedness are vital components of successful early warning systems. Empowering communities with knowledge about flood risks and involving them in decision-making processes can enhance the effectiveness of FFEWS. Community-based early warning systems incorporating local knowledge and traditional forecasting methods have successfully enhanced community resilience.

Designing and implementing FFEWS is not without its challenges. Data availability and quality, technological infrastructure, forecasting accuracy, institutional coordination, sustainability, and social and cultural factors must be carefully considered and addressed. Thus, flash flood early warning systems play a critical role in minimizing the impact of flash floods on communities worldwide. By integrating advanced technologies, effective communication strategies, and community engagement, FFEWS can provide timely and accurate warnings, empowering individuals and authorities to take proactive measures and save lives. As climate change intensifies the frequency and severity of extreme weather events, developing and enhancing robust FFEWS is more crucial than ever in protecting vulnerable populations and reducing disaster risks.

REFERENCES

Acosta-Coll, M., Ballester-Merelo, F., Martinez-Peiró, M., & De la Hoz-Franco, E. (2018). Real-time early warning system design for pluvial flash floods—A review. *Sensors (Basel)*, *18*(7), 2255. doi:10.339018072255 PMID:30002358

Alfieri, L., Velasco, D., & Thielen, J. (2011). Flash flood detection through a multi-stage probabilistic warning system for heavy precipitation events. *Advances in Geosciences*, *29*, 69–75. doi:10.5194/adgeo-29-69-2011

Behrouz, M. S., Zhu, Z., Matott, L. S., & Rabideau, A. J. (2020). A new tool for automatic calibration of the Storm Water Management Model (SWMM). *Journal of Hydrology (Amsterdam)*, *581*, 124436. doi:10.1016/j.jhydrol.2019.124436

Borga, M., Gaume, E., Creutin, J. D., & Marchi, L. (2008). Surveying flash flood response: Bournas, A., & Baltas, E. (2022). Investigation of the gridded flash flood Guidance in a Peri-Urban basin in the greater Athens area, Greece. *Journal of Hydrology (Amsterdam)*, *610*, 127820.

Brunner, G.W. (2016). *HEC-RAS River Analysis System: Hydraulic Reference Manual, Version 5.0.* US Army Corps of Engineers–Hydrologic Engineering Center.

Clark, R. A., Gourley, J. J., Flamig, Z. L., Hong, Y., & Clark, E. (2014). CONUS-wide evaluation of National Weather Service flash flood guidance products. *Weather and Forecasting*, *29*(2), 377–392. doi:10.1175/WAF-D-12-00124.1

Climate Centre. (2015). *Floods and mudslides trigger renewed concern over glacier-melt hazard across Central Asia*. Climate Centre. https://www.climatecentre.org/1801/floods-and-mudslides-trigger-renewed-concern-over-glacier-melt-hazard-across-central-asia/

Corral, C., Berenguer, M., Sempere-Torres, D., Poletti, L., Silvestro, F., & Rebora, N. (2019). Comparison of two early warning systems for regional flash flood hazard forecasting. *Journal of Hydrology (Amsterdam)*, *572*, 603–619. doi:10.1016/j.jhydrol.2019.03.026

Cydzik, K., & Hogue, T. S. (2009). Modeling postfire response and recovery using the hydrologic engineering center hydrologic modeling system (HEC-HMS) 1. *Journal of the American Water Resources Association*, *45*(3), 702–714. doi:10.1111/j.1752-1688.2009.00317.x

Elkhrachy, I. (2015). Flash flood hazard mapping using satellite images and GIS tools: A case study of Najran City, Kingdom of Saudi Arabia (KSA). *The Egyptian Journal of Remote Sensing and Space Sciences*, *18*(2), 261–278. doi:10.1016/j.ejrs.2015.06.007

Emily Cassidy. (2023). *Heavy Rains Hit Drought-Stricken Horn of Africa*. NASA, Earth Observatory. https://earthobservatory.nasa.gov/images/151208/heavy-rains-hit-drought-stricken-horn-of-africa

Erena, S. H., Worku, H., & De Paola, F. (2018). Flood hazard mapping using FLO-2D and local management strategies of Dire Dawa city, Ethiopia. *Journal of Hydrology. Regional Studies*, *19*, 224–239. doi:10.1016/j.ejrh.2018.09.005

Fernández-Nóvoa, D., García-Feal, O., González-Cao, J., de Gonzalo, C., Rodríguez-Suárez, J. A., Ruiz del Portal, C., & Gómez-Gesteira, M. (2020). MIDAS: A new integrated flood early warning system for the Miño River. *Water (Basel)*, *12*(9), 2319. doi:10.3390/w12092319

Gaume, E., Bain, V., Bernardara, P., Newinger, O., Barbuc, M., Bateman, A., Blaškovicová, L., Blöschl, G., Borga, M., Dumitrescu, A., Daliakopoulos, I., Garcia, J., Irimescu, A., Kohnova, S., Koutroulis, A., Marchi, L., Matreata, S., Medina, V., Preciso, E., & Viglione, A. (2009). A collation of data on European flash floods. *Journal of Hydrology (Amsterdam)*, *367*(1-2), 70–78. doi:10.1016/j.jhydrol.2008.12.028

Hapuarachchi, H. P., & Wang, Q. J. (2008). A review of methods and systems available for flash flood forecasting. *Report for the Bureau of Meteorology*. Bureau of Meteorology.

Henao Salgado, M. J., & Zambrano Nájera, J. (2022). Assessing flood early warning systems for flash floods. *Frontiers in Climate*, *4*, 787042. doi:10.3389/fclim.2022.787042

Hofmann, J., & Schüttrumpf, H. (2019). Risk-based early warning system for pluvial flash floods: Approaches and foundations. *Geosciences*, *9*(3), 127. doi:10.3390/geosciences9030127

Ibarreche, J., Aquino, R., Edwards, R. M., Rangel, V., Pérez, I., Martínez, M., Castellanos, E., Álvarez, E., Jimenez, S., Rentería, R., Edwards, A., & Álvarez, O. (2020). Flash flood early warning system in Colima, Mexico. *Sensors (Basel)*, *20*(18), 5231. doi:10.339020185231 PMID:32937798

ICHARM. (2023). Integrated Flood Analysis System (IFAS). *International Centre for Water Hazard and Risk Management under the auspices of UNESCO*. PWRI. https://www.pwri.go.jp/icharm/research/ifas/ifas.html

Khan, I., Lei, H., Shah, A. A., Khan, I., & Muhammad, I. (2021). Climate change impact assessment, flood management, and mitigation strategies in Pakistan for a sustainable future. *Environmental Science and Pollution Research International, 28*(23), 29720–29731. doi:10.100711356-021-12801-4 PMID:33566292

Marchi, L., Borga, M., Preciso, E., & Gaume, E. (2010). Characterization of selected extreme flash floods in Europe and implications for flood risk management. *Journal of Hydrology (Amsterdam), 394*(1–2), 118–133. doi:10.1016/j.jhydrol.2010.07.017

Nguyen, T. N., Pham, C. X., Nguyen, H. Q., & Dang, T. N. B. (2019). Establishing an early warning system for flash floods in Hoang Su Phi District, Ha Giang Province, Vietnam. *Singapore Journal of Tropical Geography, 40*(2), 312–333. doi:10.1111jtg.12276

Park, S., Berenguer, M., Sempere-Torres, D., Baugh, C., & Smith, P. (2017). Toward seamless high-resolution flash flood forecasting over Europe based on radar nowcasting and NWP: An evaluation with case studies. *Geophysical Research Abstracts, 19*, EGU2017–EGU12158.

Patro, S., Chatterjee, C., Mohanty, S., Singh, R., & Raghuwanshi, N. S. (2009). Flood inundation modeling using MIKE FLOOD and remote sensing data. *Photonirvachak (Dehra Dun), 37*(1), 107–118. doi:10.100712524-009-0002-1

Perera, D., Agnihotri, J., Seidou, O., & Djalante, R. (2020). Identifying societal challenges in flood early warning systems. *International Journal of Disaster Risk Reduction, 51*, 101794. doi:10.1016/j.ijdrr.2020.101794

Phillips, B. C., Yu, S., Thompson, G. R., & De Silva, N. (2005). *1D and 2D modeling of urban drainage systems using XP-SWMM and TUFLOW*. In 10th international conference on urban drainage (pp. 21-26). Copenhagen, Denmark.

Reliefweb. (2022). *Practical Action's approach in disseminating localized flood early warning in Bangladesh*. ReliefWeb. https://reliefweb.int/report/bangladesh/practical-actions-approach-disseminating-localized-flood-early-warning-bangladesh

Sene, K. (2012). *Flash floods: forecasting and warning*. Springer Science & Business Media.

Tansar, H., Babur, M., & Karnchanapaiboon, S. L. (2020). Flood inundation modeling and hazard assessment in Lower Ping River Basin using MIKE FLOOD. *Arabian Journal of Geosciences, 13*(18), 1–16. doi:10.100712517-020-05891-w

Tu, H., Wang, X., Zhang, W., Peng, H., Ke, Q., & Chen, X. (2020). Flash flood early warning coupled with hydrological simulation and the rising rate of the flood stage in a mountainous small watershed in Sichuan Province, China. *Water (Basel), 12*(1), 255. doi:10.3390/w12010255

Van Der Knijff, J. M., Younis, J., & De Roo, A. P. J. (2010). LISFLOOD: A GIS-based distributed model for river basin scale water balance and flood simulation. *International Journal of Geographical Information Science, 24*(2), 189–212. doi:10.1080/13658810802549154

Yasmin, T., Khamis, K., Ross, A., Sen, S., Sharma, A., Sen, D., Sen, S., Buytaert, W., & Hannah, D. M. (2023). Brief communication: Inclusiveness in designing an early warning system for flood resilience. *Natural Hazards and Earth System Sciences*, *23*(2), 667–674. doi:10.5194/nhess-23-667-2023

Zanchetta, A. D., & Coulibaly, P. (2020). Recent advances in real-time pluvial flash flood forecasting. *Water (Basel)*, *12*(2), 570. doi:10.3390/w12020570

Zhang, Y., Wang, Y., Chen, Y., Xu, Y., Zhang, G., Lin, Q., & Luo, R. (2021). Projection of changes in flash flood occurrence under climate change at tourist attractions. *Journal of Hydrology (Amsterdam)*, *595*, 126039. doi:10.1016/j.jhydrol.2021.126039

Chapter 5
Analysis Methods for Hydrological Drought

Carmen Maftei
https://orcid.org/0000-0002-7150-608X
Transilvania University of Brasov, Romania

Gabriel Dobrica
https://orcid.org/0000-0003-1260-7021
Ovidius University of Constanta, Romania

Alina Barbulescu
Transilvania University of Brasov, Romania

Cristina Serban
Ovidius University of Constanta, Romania

ABSTRACT

According to the World Meteorological Organization, the total economic impact of drought in Europe in the past 30 years was close to €100 billion. A recent publication by the European Environment Agency states that meteorological and hydrological droughts have become more frequent and severe in most of Europe, "being the highest in southern Europe." Significant changes in worldwide hydrography have occurred recently, as evidenced by the drying up of some lakes and a decrease in river flows. In a handbook published by WMO and Global Water Partnership, a set of indices and indicators were proposed to investigate different drought types. The data needed to calculate these indices refers to different hydro-meteorological variables, and their analysis must be carried out. The indices based on remote sensing data could offer a viable alternative to improve the observation. Based on different applications realized by the authors, we propose a review of the methods used to identify hydrological drought. It suggests a methodology appropriate to the Dobrogea region in southeastern Romania.

DOI: 10.4018/978-1-6684-8771-6.ch005

INTRODUCTION

World Meteorological Organization reported (World Meteorological Organization -WMO, 2021) that between 1970 and 2019, in Europe, out of all natural disasters, 3% were droughts and 5% were heat-waves. The five hazards that led to the most economic losses were river floods (44%), windstorms (36%), and droughts (10%). The dangers that caused the most significant loss of human life in the top ten natural catastrophes are (World Meteorological Organization -WMO, 2021), in descending order, droughts (650,000 deaths), storms (577,232 deaths), floods (58,700 deaths), and excessive temperatures (55,736). A recent publication offered by the European Environment Agency (European Environment Agency, 2021) states that "the frequency and severity of meteorological and hydrological droughts have increased in most parts of Europe, being the highest in southern Europe." A recent publication by the World Bank experts (World Bank, 2018) referring to drought shows that droughts already affect 48% of Romania's surface, especially the territories located in the southern part of the country (Romanian Plain and Getic Plateau), southeast (Dobrogea), and east (Moldova). Over the last 60 years, drought has caused substantial losses in agriculture, but the most severe drought was the one recorded in 2007, followed by the one in 2011–2012 when there was a total of 8 months of drought. The drying of natural lakes represents one of the main effects of prolonged droughts and anthropogenic influence; possibly the most well-known example is the Aral Sea, classified as a closed sea. The examples could continue with Lake Urmia, an endorheic lake in Iran that was classified as the sixth largest Salt Lake in the world and which, starting from the end of 2007, has lost 10% of its surface area of 5200 km^2. In 2021, the surface area has been reduced by half. In recent years, there have been numerous ecological catastrophes of this type in Romania. In October 2020, Iezer Lake in Călăraşi County, with an area of approximately 400 hectares, completely dried up. The same event occurred in August 2020 at Lake Nuntaşi-Tuzla (Figure 1(a)) in the Dobrogea region. Lake Amara (Figure 1(b)), located in Buzau County, with an area of over 800 hectares, completely dried up in June 2022.

Figure 1. Photos from (a) Nuntasi Lake and (b) Amara Lake
photo a) National Water Agency; photo b) author

(a) (b)

In the context of climate change, the increasing frequency of drought occurrence, to support the efforts of all stakeholders involved in drought management, WMO and GWP (Global Water Partnership) published a handbook (World Meteorological Organization -WMO & Global Water Partnership - GWP,

2016) in which a set of indices and indicators is used to investigate different types of drought. The data needed to calculate these indices refers to different hydrometeorological parameters such as rainfall, temperature, discharge, stream/lake level, etc. The handbook proposed a set of indices based on the type of drought. A set of indices based on remote sensing data, such as the Normalized Difference Water Index (NDWI), MNDWI (Modified Normalized Difference Water Index), Water Ratio Index (WRI), Weighted Normalized Difference Water Index (WNDWI), could offer a viable alternative to improve the observation (Şerban et al., 2022).

Based on different applications realized by the authors, this paper proposes a review of the methods used to identify hydrological drought. It suggests a methodology appropriate to the Dobrogea region, recognized as Romania's driest area.

HYDROLOGICAL DROUGHT

Drought. Concept, Terminology, Type of Drought

There are many drought definitions. National Integrated Drought Information System (NIDIS) mentions on its website (https://www.drought.gov/what-is-drought/drought-basics) that from 1980 more than 150 definitions related to drought have been published. Drought has been categorized into conceptual and operational definition categories starting with Wilhite & Glantz (Wilhite & Glantz, 1985). American Meteorological Society provides a classical conceptual definition of drought: "a period of abnormally dry weather sufficiently long enough to cause a serious hydrological imbalance" (*Drought - Glossary of Meteorology*, n.d.). Whereas the conceptual definition refers more to the concept itself and can be used in establishing policies and strategies to combat droughts, the operational definition tries to establish when drought events start and end and the severity of the dry period. Yevjevich pointed out in 1969 the importance of defining and describing drought and the wide variety of terminologies used in various study fields to describe droughts (Yevjevich, 1969). From a disciplinary perspective, Wilhite and Glantz (Wilhite & Glantz, 1985) established four types of drought: meteorological, hydrological, agricultural, and socioeconomic. A list of definitions is presented in the following.

Box 1. Definitions

Meteorological drought	represents a "degree of dryness (in comparison to some normal or average amount) and the duration of the dry period." Some meteorological definitions include different thresholds as (i) there has been no "appreciable" amount of precipitation recorded for 14 days, meaning one-tenth of a millimeter or more. (Hepites, 1906) (ii) less than 2.5 mm of rainfall in forty-eight hours (Blumenstock, 1942).
Hydrological drought	refers to a "lack of water" in the streamflow network due to "a lack of precipitation over a large area and period" a long-term and extensive lack of naturally available water supplies or "period during which stream flows are inadequate to supply established uses under a given water management system" (Dracup et al., 1980; Linsley et al., 1975) or "period of below normal conditions for one or a combination of factors, such as streamflow, reservoir storage, and groundwater" (Dalezios et al., 2017)
Agricultural drought	occurs when "soil moisture is insufficient to support corps" (Tallaksen & Lanen, 2004) or "plant water availability falls below that required by a plant community during a critical growth stage" (Dalezios et al., 2017)
Socio-economic drought	defined "in terms of loss from an average or expected return" (Dalezios et al., 2017)

The last period a new drought type was introduced was the ecological drought, which refers to a long-term and extensive lack of naturally available water supplies "causing stress on ecosystems" (Tallaksen & Lanen, 2004). What is the physical relevance/significance of these definitions? The responses to these concerns are essential because they help us assess a definition's accuracy and applicability. In this context, the scientists propose a series of indices that could characterize different types of droughts.

Drought. Indicators and Indices

Determining parameters that describe drought is of great importance for water resource management. Regarding the parameters that can describe drought, there are two different types: indicators and indices: (i) indicators refer to those variables used to describe drought conditions. Examples include precipitation, temperature, streamflow, groundwater and lake levels, soil moisture, and snowpack; (ii) indices are numerical representations of drought severity and duration of drought events, evaluated using climatic or hydrometeorological inputs. The multitude of numerical indices developed to monitor drought (there are more than 100 indices) shows that a single method for describing and evaluating drought severity could not be defined (Lloyd-Hughes, 2014) and used in all circumstances.

A series of indices was developed for each drought mentioned above. For example, the most known index for meteorological drought is the Palmer Drought Severity Index (PDSI), introduced by Palmer in 1965 (Wilhite & Glantz, 1985). It makes a quantitative comparison between the quantity of precipitation that fell in a given location during a specific period and the average or typical amount that was anticipated during that period (Palmer, 1965). The supply-and-demand theory behind the water balance equation is the foundation of PDSI. Because this index provides a poor indication of short-term changes in moisture status influencing crops and farming activities (Alley, 1984; Kogan, 1995; Smakhtin & Hughes, 2004) stated that PDSI had little value for nations outside of the USA. This indicator is inappropriate for addressing drought issues in semiarid and subhumid areas. To give a solution to this issue, McKee et al. (1993) developed the Standard Precipitation Index (SPI) based on the probability of precipitation for different time scales. The SPI calculation for every location (e.g., meteorological stations) is based on the long-term rainfall record fitted to a gamma probability distribution, which is normalized (McKee et al., 1993). At the Inter-Regional Workshop on Drought Early Warning Systems in Lincoln, Nebraska (USA) in 2009, the participants recognized that no single drought index suits all situations. They agreed that SPI should be used to characterize meteorological drought worldwide (Hayes et al., 2009) due to its simplicity.

As specified in the Introduction, in the WMO and GWP handbook (World Meteorological Organization -WMO & Global Water Partnership - GWP, 2016), the drought indexes are classified according to ease of use into green - easy, orange/yellow - moderate, and red - difficult. This grouping was based on the following criteria.

Box 2.

EASY	There is software available online. Daily data is not required. Missing data is allowed. It is widely used.
MODERATE	Multiple parameters are required for calculations. The mathematical apparatus is minimal in complexity. There is no software available online.
DIFFICULT	There is no developed calculation software. Inputs can be parameters modeled using other models. The index is not available.

According to the above criteria, some information about the most known and used indices is presented in the following paragraphs.

Box 3.

Palmer Hydrological Drought Index (PHDI)	is a modification of the meteorological drought index- the Palmer Drought Severity Index (PDSI), developed by Palmer (Palmer, 1965).	Evaluate the water budget using as input: precipitation, temperature, available water content
SRSI - Standardized Reservoir Supply Index	is similar to the SPI (McKee et al., 1993) but applied to reservoirs. It was developed by Gusvey (Gusyev et al., 2015).	The input is the monthly discharge and average reservoir storage volume.
SSFI - Standardized Streamflow Index	is similar to the SPI. It was developed by Modarres (Modarres, 2007) and investigated by (Telesca et al., 2012).	It is used as input discharge data at monthly or daily timescale. It was implemented in SPI software.
SDI - Streamflow Drought Index	is similar to the SPI developed by (Nalbantis and Tsakiris, 2009).	The input is formed by the monthly streamflow volumes. It accepts gaps in the data series, but long time series are required.
ADI – Aggregate Dryness Index	It was developed by a consortium of U.S. University led by Dracup (Keyantash & Dracup, 2004)	The input is formed by precipitation, temperature, streamflow, reservoir storage, soil moisture content, and snow water content. Software does not exist but in the scientific literature, there are sufficient explanations for its use.
SMRI- Standardized Snow Melt and Rain Index	Is similar to SPI developed by Staudinger (Staudniger et al.,2014).	The input is the daily sum of snow melting and precipitation.

All these indices are based on the SPI mathematics and use different meteorological and/or hydrological variables as input.

Hydrological Drought Characteristics

Using operational definitions requires data (hydro-meteorological data series) on different time steps (daily, monthly, etc.). Their analysis must be carried out first. The necessary steps in data analysis are: (i) Obtaining and preparing data; (ii) Exploratory Data Analysis; (iii) Application of Statistical Tests; and (iv) Interpretation of Results. After applying this methodology, a set of drought indices could be calculated.

Based on Tallaksen's work (Tallaksen & Lanen, 2004), World Meteorological Organization (WMO, 2008) proposed a methodology for characterizing hydrological drought that can be divided into two categories: (i) determining the minimum flow based on the flow frequency curve (FDC), the minimum annual flow (MAM), and the base flow index (BFI); and (ii) determining the deficit characteristics using the Threshold Level Method (TLM) and the Sequential Peak Algorithm (SPA). Indices in category (i) do not provide information about the duration of the drought period, the beginning of the drought, or its end, and methods in category (ii) provide information on drought duration, severity, event intensity, and the minimum value of each event.

Data Analysis Methodology

According to Dobrica (Dobrica, 2023), combining the two above-mentioned methods is suitable for a good representation of hydrological drought. The proposed methodology is presented in Figure 2.

Figure 2. Methodology to evaluate hydrological drought

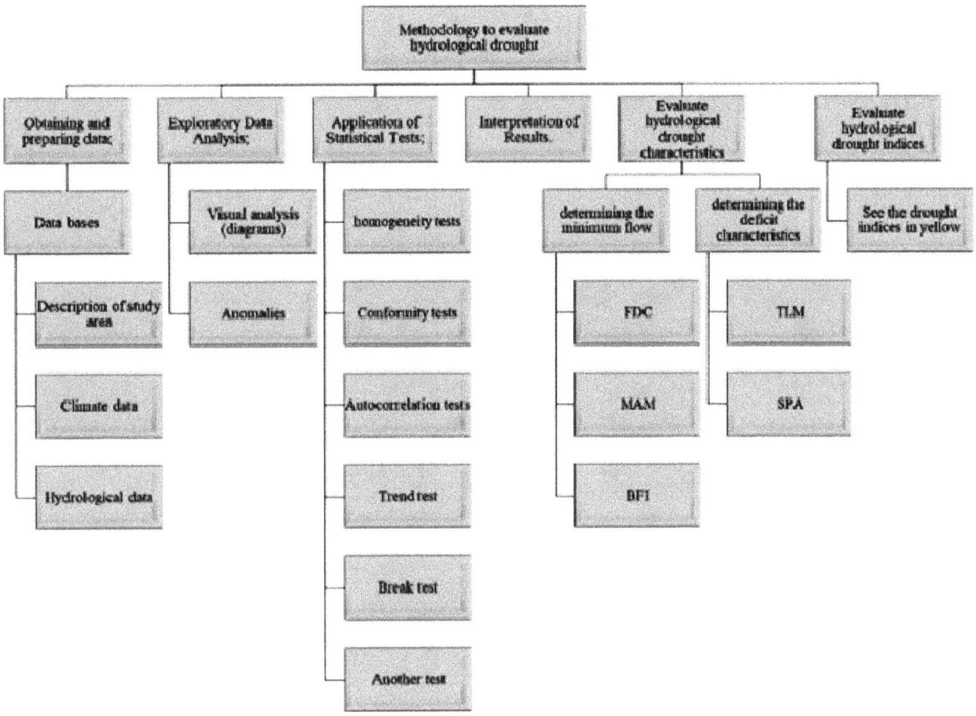

Obtaining and Preparing Datasets

The first step in data analysis is to obtain and prepare the data. Figure 3 lists the most usual data necessary for hydrological drought analysis. In any hydrological study, the characterization of the watershed from a climatic and hydrological perspective should not be overlooked. The topography/relief, geology, soil, land cover, and hydrographic network are mainly necessary information to describe the basin behavior. The essential information is related to the relief because some basin psychographics characteristics (surface, shape, elevation, slope, etc.) could influence the hydrological response Precipitation and temperature significantly influence the hydrological regime. The hydrological regime is also influenced by other factors such as anthropogenic influences.

Different sources could be used to obtain this information. According to the EU 2007/2/EC Directive (The European Parliament and the Council of the European Union, 2007), the European countries make their official map series searchable, viewable, and downloadable to the public. The information provided by the ANCPI geoportal (ANCPI, n.d.). could be used for Romania. U.S. Geological Survey (USGS) provides topographic maps at different scales (https://apps.nationalmap.gov/downloader/#/).

ESRI Topographic Map (ESRI, 2017) offers a product that could be utilized as a base map. To transform a topographic map into a digital map, some specific procedures are necessary: (i) scanning (if the map is a "hard-copy" paper) or downloading from the above-specified portal; (ii) georeferencing (the maps downloaded using different geoportals are in "*.geoTIFF" format and can store georeferencing information; for this kind of maps, this stage it is not necessary); (iii) digitizing. An interpolation step is necessary to obtain a Digital Elevation Model (DEM) after digitizing the contour lines. For these steps, software is necessary. ArcGIS from ESRI or QGIS (free of charge) may be used. This procedure is time-consuming.

Figure 3. Chart flow of data processing
(Dobrica, 2023)

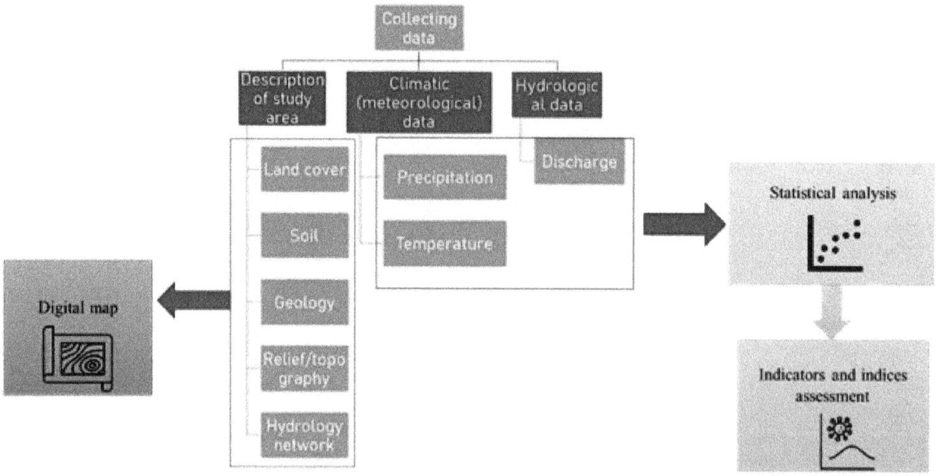

Alternatively, Digital Elevation Model could be obtained from different sources: (i) SRTM (Shuttle Radar Topographic Mission) with 30m or 90m resolution from USGS Earth Explorer website (https://earthexplorer.usgs.gov/); (ii) ASTER Global Digital Elevation Model from USGS Earth Explorer website; (iii) ALOS PALSAR product from Japan Exploration Aerospace Agency (JAXA) with 30m resolution. For flat regions, the 12.5m resolution is recommended; (iv) LiDAR (Light Detection and Ranging) solutions, some of them available on the USGS Earth Explorer website.

To obtain the land cover map, the CORINE Land Cover (CLC) product may be utilized. It has been produced since 1990 by the "Eionet National Reference Centers Land Cover" based on the interpretation of high-resolution satellite images. The satellite data used to create CLC-2000 was Landsat 7, and for the 2018 version, Sentinel-2 and Landsat-8 satellites (EEA & JRC, 2011).

Regarding the hydrographic network, the following solutions may be employed: (i) using a topographic map and digitizing the river network; (ii) through a DEM, using a watershed delineation solution or (iii) using an existing river networks file provided by different government agencies. Soil and geologic maps are available through the national agencies at different scales.

As results from the brief description of the necessary products, they are of various types, including scanned maps, digital maps in Geo Tiff format, raster images, polygon vectors, and georeferenced data in different systems. Before processing them, they must be brought into the same system to provide a unified product.

Concerning climate and hydrologic datasets for a hydrological study, at least precipitation, temperature, and discharge datasets are necessary. A long data time series (more than 30 years) is needed for both statistical analysis and hydrological drought evaluation. Generally, the data time series are provided by the national agencies. For Romania, for example, the monthly data series of precipitation and temperature (in Excel format) are free of charge and offered by the government agency, but only for the principal meteorological station. ROCADA product (Dumitrescu & Birsan, 2015) can also be used. It offers homogenized daily values of precipitation and temperature for the period 1961-2013, covering Romania at a spatial resolution of 0.1°. Daily temperature and precipitation datasets are available on the NASA website (https://data.giss.nasa.gov/) or, for Europe, on the ECAD website (https://www.ecad.eu/). Temperature and precipitation data could be extracted from the remote sensing product. A good review related to the main remote sensing product is provided by Jiao (Jiao et al., 2021). Some information and methods are provided by (Serban et al., 2010; Şerban et al., 2022; Serban & Maftei, 2011).

Regardless of how the data series are obtained, they are processed to remove errors and organized on databases. Old measurements are tabulated on specific forms and stored in databases. If the databases are not in digital format, they must be transformed. Excel is suitable for this kind of operation due to its simplicity. Once the digital database is created, it can be processed, and the following types of datasets can be extracted for different periods: daily, monthly, and annual precipitation/temperature/discharge, isohyets, and isotherms maps (if it is necessary). The records could be presented as diagrams or tables depending on the study research nature.

As an example, we presented in the following paragraphs a description of the studied watershed. Dobrogea region is situated between the lower Danube and the Black Sea (Figure 4). Located on the Black Sea littoral at 35km Nord of Constanta, Nuntasi River is a tributary to Nuntasi-Tuzla Lake which is part of the Razim-Sinoe lagoon complex.

This area is a mixed natural reserve (botanical and faunal) with significant archaeological value (the ancient Greek acropolis Histria is located a few kilometers away, on the shore of Lake Istria) – Figure 4. The studied region is part of the Central Dobrogea Plateau, specifically the Istria Plateau subunit, and has a coastal plain aspect (Hamangia Meadow) developed between the Casimcea Plateau and the Black Sea, being a combination of marine plains (Săcele and Chituc), sandy barriers, and shallow lakes (Sinoe, Histria, and Nuntasi) - figure 4 (Maftei et al., 2021; Vespremeanu-Stroe et al., 2013).

The GeoTIFF images provided by ALOS PALSAR were processed in ArcGIS 10.8, and a DEM (Digital Elevation Model) was obtained for the study area. Based on the DEM, we obtained a slope map which shows that the average slope of the terrain in the BH Nuntaşi area is 3.7%. According to the hypsometric curve, the elevation varies between 218m (in the northern part of the basin) and close to 0m along the tributary valleys of the lake (figure no. 5). 34% of the basin's surface is located between elevations of 20 and 40m and 30% between elevations of 160 and 180m.

From a geological point of view, the entire area is composed of limestone and calcareous sandstones, calcareous shales, and calcareous marls that are covered by 2 to 30 meters of loess-like deposits. Green schists predominate in the valleys (Figure 6).

The climate is temperate continental with marine influences. The temperature in this area varies from 10.6°C to 11.6°C. The Dobrogea area is recognized as the driest area. The basin is situated approximately between the isohyets of 410mm and 430mm (Figure 8). Dobrogea hydrographic area is very poor in water resources; it consists of several rivers and lakes situated on the littoral and the Danube Riverbanks.

Figure 4. Dobrogea and Nuntasi location
(Maftei et al., 2021)

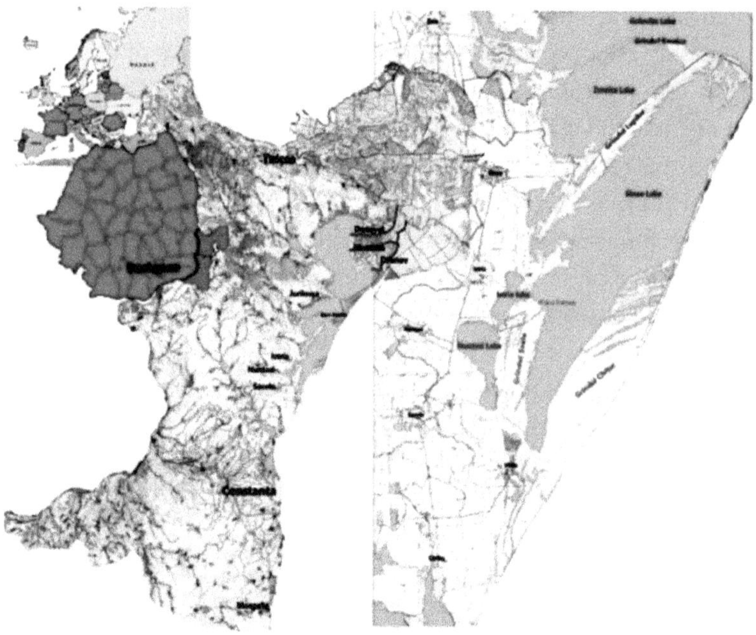

The Nuntaşi-Tuzla Lake Basin is located on the Dobrogea Littoral Water Basin and is situated at the confluence of two rivers: Nuntaşi with a length of 22 km – (according to GIS, the length is 24.96 km) and Săcele with a length of 10 km (according to GIS, the length is 11.8 km).

The Nuntaşi River receives a right tributary, Cogealac, with a length of 15.5 km. Groundwater is present in alluvial deposits (in the western part of the region), sand or gravel deposits (along the coastal strip), and above the impermeable or relatively impermeable layer (the suspended water). The groundwater level varies between 2 meters (around the lakes) and over 15 meters within the region (Grumezea et al., 1990).

Pedological studies are based on the map of soil provided by the local administrative agency and show that the dominant soils are various types of chernozems. The soils located in the Nuntaşi-Tuzla Lake Water Basin, being formed on loess deposits, have predominantly clayey textures (41.5%) or clayey-silty textures (51.55%). Considering the type of texture and the fact that the groundwater is not at a shallow depth to significantly influence the runoff regime, the soils have a moderate drainage potential (Dobrica, 2023).

For the study area, the vegetation map was compiled based on the Corine Land Cover database (*CORINE Land Cover — Copernicus Land Monitoring Service*, n.d.) from which we extracted the main types of cultivated vegetation found in the basin today. Arable lands cover 82% of the territory.

This area has undergone significant transformation over the past 120 years. According to (Pons, 1992) in the mid-19th century, attempts were made to create many polders for agriculture, but they were doomed to failure due to poor soil drainage. During the same period, hydraulic work was carried out to regulate the flow of the Danube's branches and to dredge and deepen two small watercourses that connected the St. Gheorghe Danube branch to the Razim-Sinoe Lake to make it usable for fishing. The most significant human intervention took place in the mid-20th century, when a large irrigation system was implemented, transforming Razim-Sinoe Lake into a large freshwater reservoir. To do this, the channels connecting

Figure 5. DEM map of Nuntasi -Tuzla Lake Basin
(Dobrica, 2023)

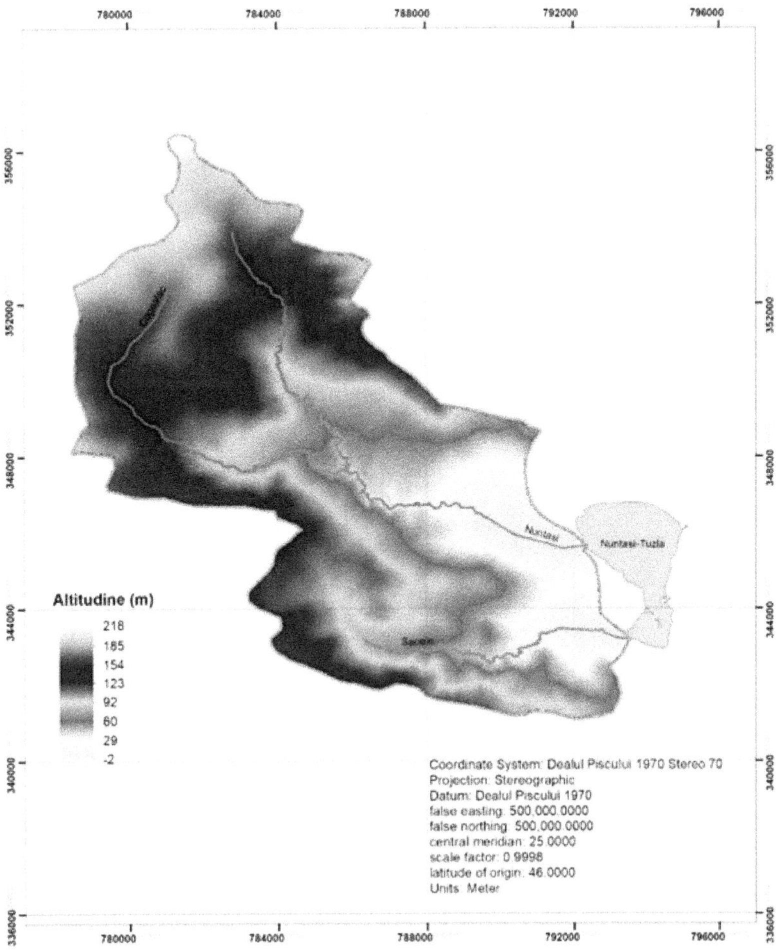

the Sfântu Gheorghe branch to the Razim-Sinoe Lake were enlarged, and the old connections between the lake and the Black Sea were closed off (Dobrica, 2023; Maftei et al., 2021; Şerban et al., 2022). The Nuntaşi-Tuzla Lake Water Basin is located within the Sinoe irrigation system (57,162 hectares). The water source is a pumping station (SPA Sinoe) designed for a flow rate of 46.1 cubic meters per second and a pumping height of 55.7 meters. The water is lifted through nine pumping stations up to an elevation of 200 meters. The irrigation system was operated from 1975 to 1989 and partially (20%) to 1996 (Dobrica, 2023). All these anthropological interventions have altered the hydrological dynamics of the area, resulting in increasing river discharge and lake water levels.

Data Analysis; Exploratory and Statistical Analysis

Exploratory data analysis (EDA) involves a visual analysis of a data series (using graphs or charts) to summarize the series' main characteristics. The most used graphs are time series plots (Figure 10), his-

Figure 6. Geological map of Nuntasi-Tuzla Lake Basin
(Dobrica, 2023)

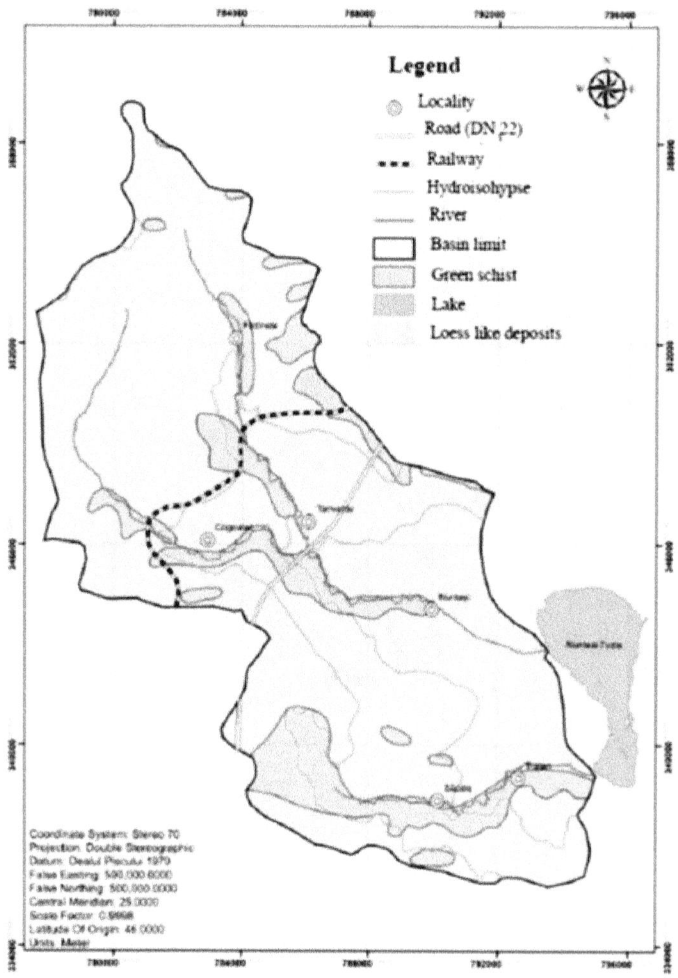

tograms, step diagrams, and anomaly plots (Figure 11). The term "anomaly" refers to a deviation from a reference value or long-term average value. A positive anomaly indicates that the observed value was higher than the reference value, while a negative anomaly indicates that the observed value was lower than the reference value.

Statistical analysis refers to the application of statistical tests, like normality, homoscedasticity, autocorrelation, and change points. Kolmogorov – Smirnov, Anderson-Darling, Jarque - Bera, or Shapiro-Wilk tests can be used to test the null hypothesis that the data series is Gaussian against the hypothesis that it is not normally distributed. The Q - Q plots can also be employed to visually analyze the normality against the hypothesis that the series is not Gaussian. In the case when the normality hypothesis was not rejected, the Bartlett test can be used to test the homoscedasticity hypothesis against the heteroskedasticity. If the normality hypothesis was rejected, the Levene test must be utilized instead of the Bartlett test. The series autocorrelation is generally tested by building the autocorrelation function and the correlogram. Visual inspection of the correlogram emphasizes the autocorrelation order. For specific problems,

Figure 7. Isotherms map for Dobrogea and Nuntasi-Tuzla Lake Basin (in the left corner)
(Dobrica, 2023)

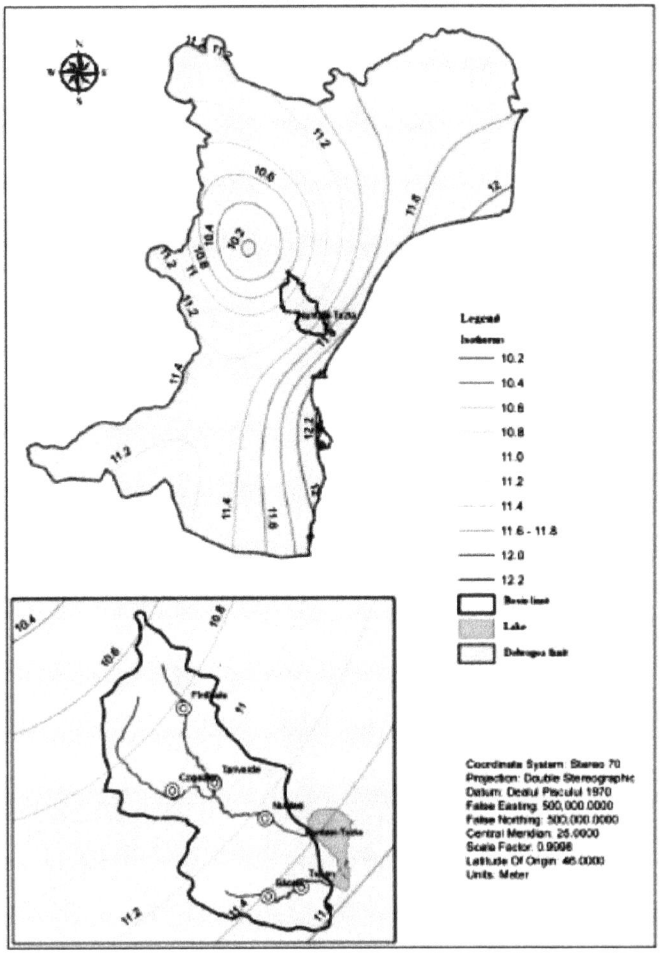

different statistical tests can be used to test the autocorrelation of different orders. For details on these tests the reader may see (Bărbulescu, 2016; Sheskin, 2000). Kundewicz (Kundzewicz & Robson, 2000) considers that a "change" in a time series could be given by a trend (increasing or decreasing) or an abrupt change i.e., a break.

The Mann-Kendall (MK) test is a non-parametric test that tests the null hypothesis H_0 that "there is no change in the mean" versus the alternative hypothesis H_1: "the mean of the series is increasing (or decreasing)." Many software's could be used to test the trend existence in a time series. MAKESENS (Salmi et al., 2002) software was developed in Excel by the Finnish Meteorological Institute in 2002. It is a user-friendly tool, free of charge, that implements the Mann-Kendal test. The slope estimation uses the non-parametric Sen's method (Salmi et al., 2002). The slope magnitude per year corresponds

$$Q_i = \frac{x_j - x_k}{j - k}, j > k$$

Figure 8. Isohyets map for Dobrogea and Nuntasi-Tuzla Lake Basin (in the left corner)
(Dobrica, 2023)

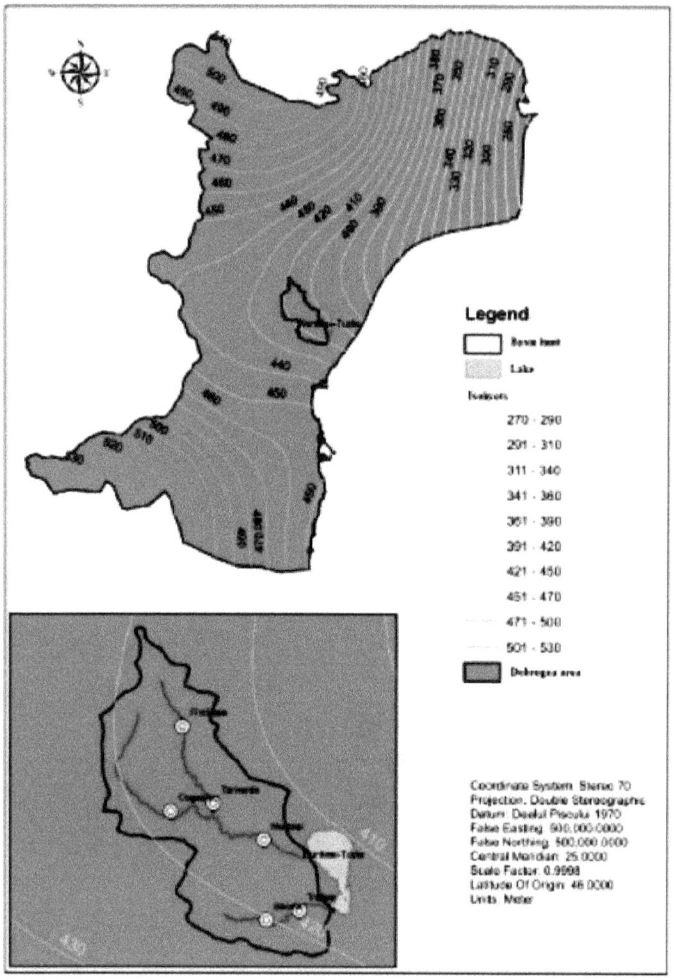

Figure 9. Discharge annual evolution at Nuntași Săcele River station
(Dobrica, 2023)

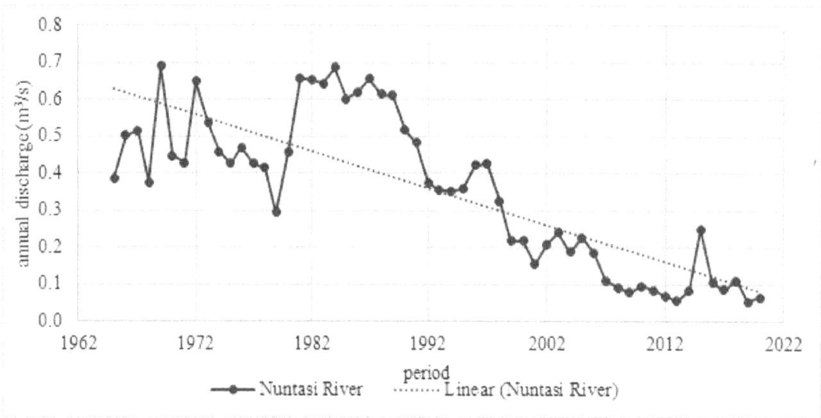

Figure 10. Precipitation anomalies at Nuntasi Meteorological Station
(Dobrica, 2023)

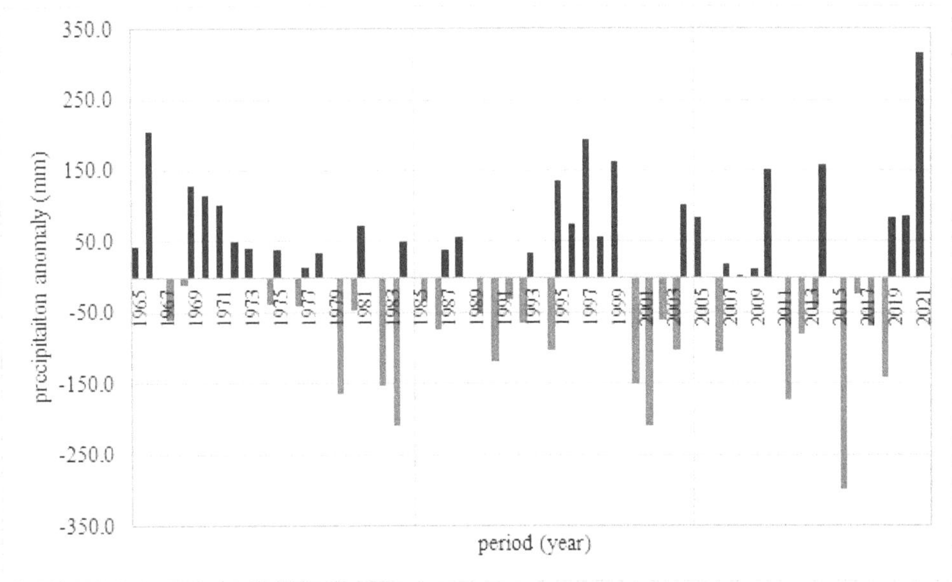

Figure 11. Results of MK test for Nuntași discharge dataset
(Dobrica, 2023)

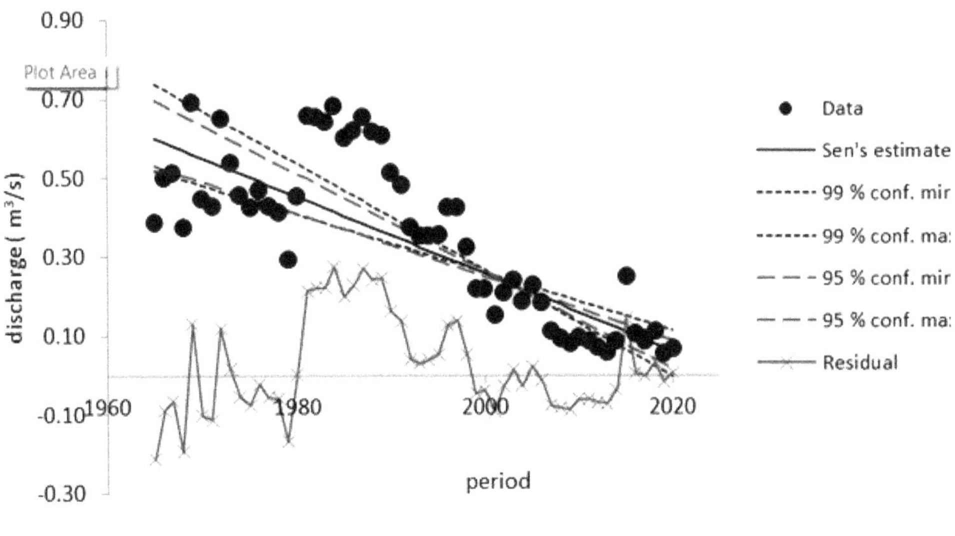

where x_j, x_k are values of the studied time series (j, k =1,2, …, n), n being the number of the series elements. For n value of $x_{j,}$ $N = n(n\text{-}1)/2$ slope estimations will be obtained. The values of the slope, Q_i, are ranked from smallest to largest, and the median, Q_{med}, of the slope or Sen's slope estimator is calculated. The sign of Q_{med} reflects the trend direction (positive or negative), while its magnitude indicates the slope of the trend. The Q_{med} slope is statistically different from zero if the two limits (Q_{min} and Q_{max}) have the same sign.

MAKESENS also produces table results and graphs. As an example, Figure 12 presents the trend statistic of the Nuntasi discharge time series and the computed residuals.

Breakpoint tests are used to detect "breaks" in a time series. Pettitt (Pettitt, 1979), Buishand (Buishand, 1984), and Lee & Heghinian tests (Lee & Heghinian, 1977)detect a single break. The Hubert segmentation procedure (Hubert, 2000) can detect multiple breaks. The above-mentioned tests are implemented in the Khronostat software (Boyer, 2002) and presented in the articles (Maftei et al., 2007; Maftei & Barbulescu, 2008, 2012), where they were used for breakpoints detection of the series from the Dobrogea region. In all these tests, the null hypothesis is that there is no break in the series and the alternative is that there is at least one breakpoint. The Buishand and Lee & Heghinian tests are based on the hypothesis that the time series data is Gaussian. Khronostat software searches for some transformations (e.g., Box-Cox) to normalize the series, if possible. When the transformation fails, the two tests are not performed. For Nuntasi's discharge annual time series data (Figure 10) the results are: (i) the Pettitt test rejected the null hypothesis and admitted a break in 1996; (ii) the Hubert segmentation procedure detects 4 break points in 1980, 1989, 1996 and 2006 (Dobrica, 2023).

To estimate the probability of change points based on the confidence level, the CPA procedure calculated a cumulative sum (CUSUM) chart together with a bootstrap rank statistic technique (Taylor, 2000). The CUSUM chart represents the total sum of differences between observations and the average (Taylor, 2000). This technique was used to detect point change for the discharge data of the Taita River (Maftei & Barbulescu, 2008) and some climate and hydrological parameters of Techirghiol Lake (Maftei et al., 2019). The results obtained with this procedure are like those provided by the above-mentioned statistical tests. In the article (Maftei & Barbulescu, 2008), the authors detect a breakpoint in the temperature series registered at the main meteorological stations in the Dobrogea region, in 1997. The same year was obtained (as a change point) with the Khronostat software. We recommend using this procedure in combination with statistical tests.

Their work (Pelletier & Turcotte, 1997) shows that "*Persistence for a temperature time series means that warm years (or days or weeks) are, more often than not, followed by warm years and cold years by cold years*." Comparable behavior can be observed for other time series, such as precipitation series and river discharge. Based on the work of Pelletier and his colleague, we applied The Hurst exponent procedure for precipitation and temperature time series recorded at 10 meteorological stations situated in the Dobrogea region. The results are presented in (Bărbulescu et al., 2010). The same procedure was applied to the discharge datasets of the Taita River (Barbulescu & Maftei, 2021; Maftei et al., 2016). The results obtained demonstrate this procedure could be applied to drought hazard assessment.

Interpretation of Results

After the introduction of relevant results, a detailed discussion is necessary. The principal steps to follow are related to the significance, importance, and applicability of your findings. Depending on the nature of the research, the interpretations will take many forms, but the following are some requirements information: (i) identified correlations and/or trends among the data; (ii) setting the results in the context of existing research; (iii) describing the unexpected results and determining their significance (if it is the case); (iv) recognize the study limitation; (v) some recommendations could be offered. A brief example is presented below.

Figure 10 shows the evolution of annual discharge at the Nuntasi River hydrological station (situated in the Dobrogea region, Romania) for the investigated period (1965-2021). Based on the daily series, the monthly and annual datasets were built. An evident negative trend of the flow is remarked. The multiannual mean of Nuntasi's discharge is 0,348 m3/s. Starting with 1992, the annual discharge at this station is greater than the multiannual value, with one exception (1996-1997). The negative trend is also confirmed by the results of the Mann-Kendall (MK) test (Figure 12). The statistics value related to this test is Z=-6.71, showing a pronounced negative trend in this series. The hypothesis that the residual series is random was also rejected. In line with the results obtained with Khronostat software, Dobrica stipulated in his work (Dobrica, 2023) that each period detected via the Hubert procedure is related to the irrigation operation's system or/and an increasing trend of temperature in the region (Dobrica, 2023; Maftei & Barbulescu, 2008). Table 1 presents the results obtained with Khronostat software.

Table 1. Discharge average at the Nuntași hydrological station and the subseries detected with Khronostat software (adapted on Dobrica, 2023)

Subseries	Multiannual average (m^3/s)	Observation
1965-1980	0.467	Before the irrigation system (SI) enters into operation and in the first years of operations
1981-1989	0.638	(SI) operating
1990-1996	0.409	Partial disruption of SI
1997-2006	0.239	Total disruption of SI and partially increasing temperature
2007-2020	0.092	increasing of temperature

Estimation of Hydrological Drought Characteristics

In this section, the low flow and deficit characteristics are presented. Low flow index refers to a percentile from the Flow Duration Curve (FDC), annual minimum flow (MAM), and Base flow index (BFI).

The FDC plots river flow's empirical cumulative frequency (EFQ) against the exceedance frequency. Based on the work of (Tallaksen & and Lanen, 2004), we have adapted the worked example to our datasets and estimated FDC. This procedure was used by the authors for the Taita and Nuntasi catchments. The results are published in the articles (Dobrica, 2023; Maftei, 2013; Maftei et al., 2009). The FDC graph construction is based on the daily streamflow data recorded at Nuntasi station (located on the Nuntasi River) for 1965-2021. The difference between Q95%, Q80%, and Q75% is small (Table 2). In this case, the FDC is flat. This result is typical for a river with low variability in the discharge (Figure 7). The obtained values (Table 2) were used as a threshold to detect the hydrological droughts via the TLM method.

Hydrological droughts were determined using the TLM method (Gregor, 2010), implemented in the software with the same name, a HydroOffice component (https://hydrooffice.org/Tool/TLM). The software offers several separation methods, including constant and variable threshold values. Using a constant threshold value of Q90%, 61 drought periods were obtained, spanning from one day to 225 days (2021). The inconvenience of this type of calculation lies in the fact that it identifies minor droughts of just one day, and there is no possibility to configure the software to eliminate deficits below a certain threshold, thereby excluding these minor droughts from the calculations.

Figure 12. FDC for Nuntasi hydrological station
(Dobrica, 2023)

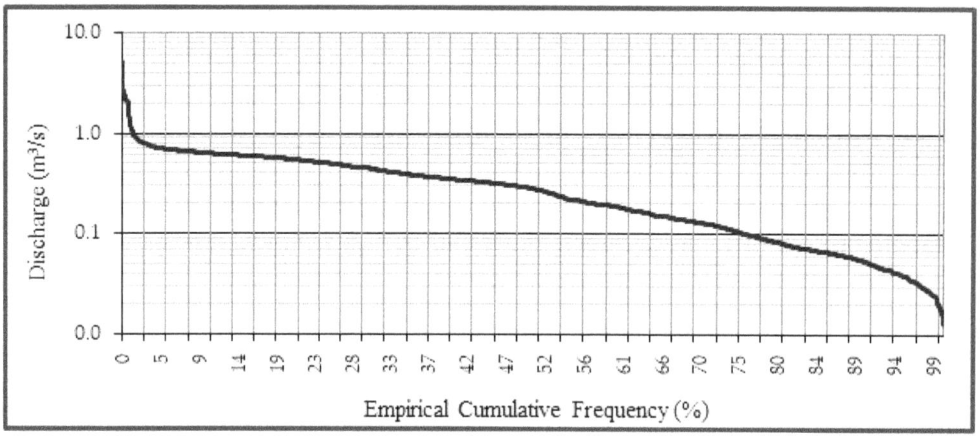

Table 2. Selected Exceedance Values

P_Q (%)	Q ($m^3 s^{-1}$) - Nuntași
90	0.056
80	0.083
75	0.108

Estimation of Hydrological Drought Index

SSFI - Standardized Streamflow Index was used to identify the hydrological drought on the Nuntasi River. The computation is based on a technique developed by McKee et al. (McKee et al., 1993), which applies a Gamma distribution function to a long time series of precipitation or flow rates and then transforms the results into a normal distribution (Cismaru et al., 2004) facilitating the classification of drought in seven classes from dry to wet. An SSFII value under -1 indicates a flow deficit, while a value above +1,0 means an excess. For accuracy, some authors like McKee (McKee et al., 1993) and Giddings (Giddings & Soto, 2005) divide the values of the normal index, ranging from (-1,+1), into two categories: (i) the (-1,0) interval represents weak-dry conditions (mild drought), and (ii) the (0,1) interval represents slightly wet conditions. In this context, Table 3 contains the SPI classification. A color code was awarded from dark blue to dark orange.

The SPI generator software (https://drought.unl.edu/monitoring/SPI/SPIProgram.aspx), developed by the National Drought Mitigation Center, could be used to compute this index. SSFI can be calculated over different time intervals: 1, 3, 6, 9, 12, 24, and 48 months. An SSFI for 12 months (SSFI-12) was chosen to evaluate drought for a long accumulation period.

The variation of SSFI is presented in Figure 13. It could be observed that from 1979 to 1999 (21 years), the SSFI for the Nuntași River had values up to 1, except for the period from 1981 to 1984 when the values exceeded 1, even reaching above 2 (in 1981 and 1982). From 1990 to 1999, the SSFI-12 index consistently decreased. From 2000 onwards, the Nuntași River entered a prolonged drought, with SSFI-12

Table 3. SPI classification

Interval	Regimen	Color Code
SSFI>2	Extremely wet	EW
1.5<=SSFI<=2	Very wet	VW
1.0<=SSFI<=1.5	Moderately wet	MW
1<=SSFI<=0	Slightly wet	SW
-1<=SSFI<=0	Weak-dry	WD
-1.5<=SSFI<=-1	Moderately dry	MD
-2<= SSFI <=-1.5	Very dry	VD
S SSFI >-2	Extremely dry	ED

Figure 13. Variation of SSFI values for Nuntasi River
(Dobrica, 2023)

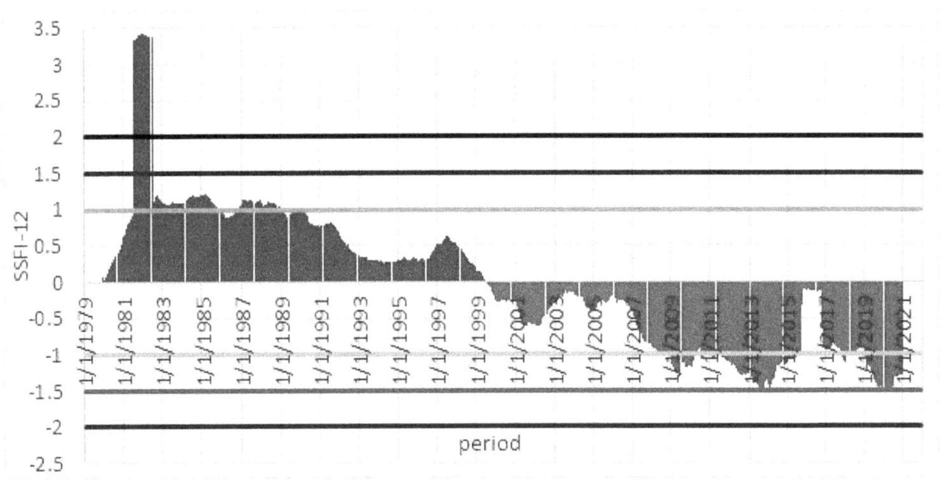

values continuously decreasing (except for the period from 2004 to 2007 when the values varied around -0.2). In 2018, the SSFI-12 index reached its lowest value, from -1.07 to -2.81. Based on the frequency distribution (Figure 14) 30% are mild droughts, 22% are moderate droughts, 1% are severe droughts, and there are no extreme droughts.

CONCLUSION

This document is not a manual, it represents part of the author's works related to the investigation of Dobrogea climate and hydrological conditions in this area. We have thus attempted to share with readers our experience regarding the drought, especially hydrological drought. We recommend the methodology proposed as the free software because we strongly believe that they have been useful in the development of the Drought Management Plan (DMP) related to the hydrological basin.

Figure 14. Histogram variation for the Nuntasi River
(Dobrica, 2023)

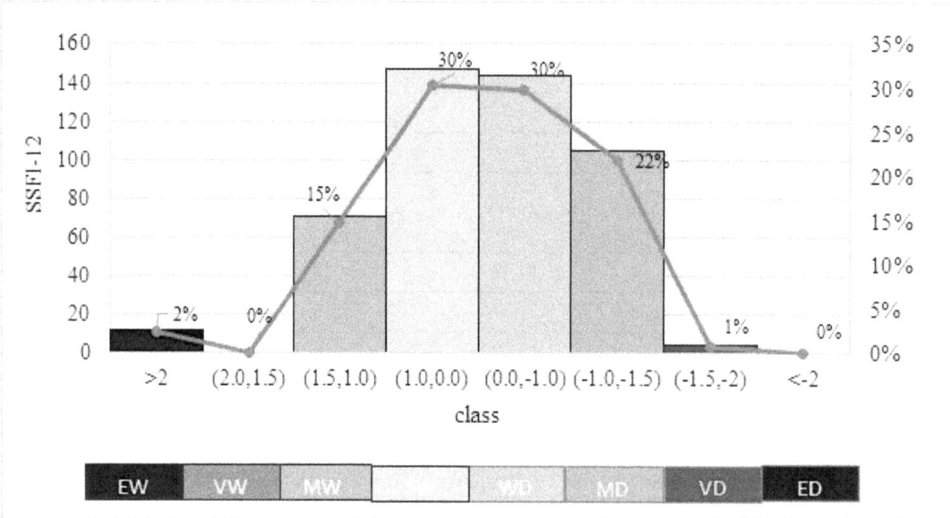

Even though the characterization of the watershed is a large time consumer its analysis is very important to understand its behavior from a hydrological point of view considering all the factors which could influence the hydrological response. An additional reason is the recommendation of the Water Framework Directive -WFD (2000/60/EC) related to the integration of DMP in the Basin Management Plan (BMPs). It is furthermore a good source for the ungauged basin to evaluate the discharge using different hydrological models.

Data collection is essential to the success of such a project. Climate and hydrological measures from credible sources, such as national or international agencies, are preferable since they are free from gross errors. A minimum of 30 years of daily/monthly data are necessary. The old data are generally presented as tables with specially designed forms and must be transformed into digital data using Microsoft Excel or other products. In this case, a minimum inspection is necessary (e.g., misplaced decimal points, copying errors, or mathematical mistakes). Given the quantitative and qualitative importance of the data, it is important to organize them.

The method used by SPI Generator software has several advantages but also disadvantages. The main advantage is that it uses only one input parameter (precipitation, and respective streamflow). This advantage also becomes the main disadvantage because drought is caused by a combination of factors such as evapotranspiration, soil moisture, wind, etc. The main disadvantage is that the parameters of the Gamma function used to determine SPI/SSFI require a historical series of cumulative non-zero precipitation. The more zero values the series has, the more difficult it is to calibrate the Gamma distribution. Another solution is to use DrinC software (Tigkas et al., 2013) which provides the Pearson distribution. The software includes other modules to calculate evapotranspiration, and hydrological drought as the Reconnaissance Drought Index (RDI) and the Streamflow Drought Index (SDI). This software is also free of charge.

REFERENCES

Alley, W. M. (1984). The Palmer Drought Severity Index: Limitations and Assumptions. *Journal of Climate and Applied Meteorology*, *23*(7), 1100–1109. doi:10.1175/1520-0450(1984)023<1100:TPDSIL>2.0.CO;2

ANCPI. (n.d.). *Geoportal ANCPI*. ANCPI. https://geoportal.ancpi.ro/portal/home/

Bărbulescu, A. (2016). *Studies on Time Series Applications in Environmental Sciences* (*Vol. 103*). Springer International Publishing. doi:10.1007/978-3-319-30436-6

Bărbulescu, A., Gherghina, C., & Maftei, C. (2010). Evaluation of Hurst exponent for precipitation time series. *LATEST TRENDS on COMPUTERS*, *II*, 590–595.

Barbulescu, A., & Maftei, C. (2021). *STATISTICAL APPROACH OF THE BEHAVIOR OF HAMCEARCA RIVER (ROMANIA). 703.*

Blumenstock, G. (1942). *Drought in the United States Analyzed using the Theory of Probability* (Technical Bulletin No. 819). USDA.

Boyer. (2002). *Khronostat – HydroSciences Montpellier*. Hydro Sciences. http://www.hydrosciences.org/index.php/2020/09/04/khronostat/

Buishand, T. A. (1984). Tests for detecting a shift in the mean of hydrological time series. *Journal of Hydrology (Amsterdam)*, *73*(1), 51–69. doi:10.1016/0022-1694(84)90032-5

Cismaru, C., Gabor, V., Bartha, I., & Scripcariu, D. (2004). *Gestiunea Secetelor | PDF*. Performantica. https://ro.scribd.com/document/450217576/Gestiunea-secetelor

CORINE Land Cover—Copernicus Land Monitoring Service. (n.d.). [Land Section]. Copernicus Land Monitoring. https://land.copernicus.eu/pan-european/corine-land-cover

Dalezios, N. R., Dunkel, Z., & Eslamian, S. (2017). Meteorological Drought Indices: Definitions. In Handbook of Drought and Water Scarcity. Principles of Drought and Water Scarcity. Routledge Handbooks Online. doi:10.1201/9781315404219-3

Dobrica, G. (2023). *Evaluarea secetei hidrologice in bazinul hidrografic al lacului Nuntasi-Tuzla* [PhD thesis, Ovidius University of Constanta, Romania].

Dracup, J. A., Lee, K. S., & Paulson, E. G. Jr. (1980). On the definition of droughts. *Water Resources Research*, *16*(2), 297–302. doi:10.1029/WR016i002p00297

Dumitrescu, A., & Birsan, M.-V. (2015). ROCADA: A gridded daily climatic dataset over Romania (1961–2013) for nine meteorological variables. *Natural Hazards*, *78*(2), 1045–1063. doi:10.100711069-015-1757-z

EEA. E. E. A., & JRC, J. R. C. (2011). *CLC 2018—Copernicus Land Monitoring Service* [Land item]. Copernicus Land Monitoring. https://land.copernicus.eu/pan-european/corine-land-cover/clc2018

ESRI. (2017). *World Topographic Map—Overview*. ESRI. https://www.arcgis.com/home/item.html?id=7dc6cea0b1764a1f9af2e679f642f0f5

European Environment Agency. (2021). *Meteorological and hydrological droughts in Europe* [Indicator Assessment]. EEA. https://www.eea.europa.eu/data-and-maps/indicators/river-flow-drought-3/assessment

Giddings, L., & Soto, M. (2005). Standardized Precipitation Index Zones for México. *Atmosfera6, 18*, 33–56.

Gregor, M. (2010). *TLM 2.1 User's Manual* [User Manual]. Department of Hydrogeology, Faculty of Natural Science, Comenius University. https://hydrooffice.org/Files/UM%20TLM.pdf

Grumezea, N., Kleps, C., & Tusa, C. (1990). *Evolutia Nivelului Si Chimismului Apei Freatice Din Amenajarile De Irigatii In Interrelatie Cu Mediul Inconjurator*. Intreprinderea poligrafica Oltenia.

GusyevM.HasegawaA.MagomeJ.KuribayashiD.SawanoH.LeeS. (2015, December 4). *Drought assessment in the Pampanga River basin, the Philippines – Part 1: Characterizing the role of dams in historical droughts with standardized indices*. doi:10.13140/RG.2.1.4931.4321

Hayes, M., Svoboda, M., Wall, N., & Widhalm, M. (2009). *THE LINCOLN DECLARATION ON DROUGHT INDICES: Universal Meteorological Drought Index Recommended*. CORE. https://core.ac.uk/reader/20047170

Hepites, S. (1906). Secetele în România. *Buletinul Societăţei Geografice Române, XXVII*. dSpace. http://dspace.bcu-iasi.ro/static/web/viewer.html?file=http://dspace.bcu-iasi.ro/bitstream/handle/123456789/54680/BCUIASI_CARTE_IIIB-3.944_1906.pdf?sequence=2&isAllowed=y

Hubert, P. (2000). The segmentation procedure as a tool for discrete modeling of hydrometeorological regimes. *Stochastic Environmental Research and Risk Assessment*, *14*(4), 297–304. doi:10.1007/PL00013450

Jiao, W., Wang, L., & McCabe, M. F. (2021). Multi-sensor remote sensing for drought characterization: Current status, opportunities and a roadmap for the future. *Remote Sensing of Environment*, *256*, 112313. doi:10.1016/j.rse.2021.112313

Keyantash, J. A., & Dracup, J. A. (2004). An aggregate drought index: Assessing drought severity based on fluctuations in the hydrologic cycle and surface water storage. *Water Resources Research*, *40*(9), 2003WR002610. doi:10.1029/2003WR002610

Kogan, F. N. (1995). Droughts of the Late 1980s in the United States as Derived from NOAA Polar-Orbiting Satellite Data. *Bulletin of the American Meteorological Society*, *76*(5), 655–668. doi:10.1175/1520-0477(1995)076<0655:DOTLIT>2.0.CO;2

Kundzewicz, Z. W., & Robson, A. (2000). Detecting Trend and Other Changes in Hydrological Data: Vol. WMO/TD-No. 1013.

Lee, A. F. S., & Heghinian, S. M. (1977). A Shift Of The Mean Level In A Sequence Of Independent Normal Random Variables—A Bayesian Approach—. *Technometrics*, *19*(4), 503–506. doi:10.1080/00401706.1977.10489592

Linsley, R. K. Jr, Kohler, M. A., & Paulhus, J. L. H. (1975). *Hydrology for Engineers* (2nd ed.). McGraw-Hill. http://www.esalq.usp.br/lepse/imgs/conteudo_thumb/Understanding-the-Drought-Phenomenon-The-Role-of-Definitions.pdf

Lloyd-Hughes, B. (2014). The impracticality of a universal drought definition. *Theoretical and Applied Climatology*, *117*(3), 607–611. doi:10.100700704-013-1025-7

Maftei, C. (2013, October 17). *Hydrological Drought Assessment in the Dobrogea Region*. Kos 2013-Facets of Uncertainty" Statistical Hydrology (STAHY '13), Kos Island, Greece.

Maftei, C., & Barbulescu, A. (2008). Statistical analysis of climate evolution in Dobrudja region. *Proceedings of the World Congress on Engineering*. Research Gate.

Maftei, C., & Barbulescu, A. (2012). Statistical analysis of precipitation time series in the Dobrudja region. *Mausam (New Delhi)*, *63*(4), 553–564. doi:10.54302/mausam.v63i4.416

Maftei, C., Barbulescu, A., Buta, C., & Serban, C. (2007). Change points detection and variability analysis of some precipitation series. *Recent Researches in Computational Techniques, Non-Linear Systems and Control*, (pp. 978–1).

Maftei, C., Barbulescu, A., & Carsteanu, A. A. (2016). Long-range dependence in the time series of Taiţa River discharges. *Hydrological Sciences Journal*, *61*(9), 1740–1747. doi:10.1080/02626667.2016.1171869

Maftei, C., Buta, C., Draghici, G., & Filip, C. (2019). Analysis of changes in hydrometeorological variables of Techirghiol Lake. *IOP Conference Series. Earth and Environmental Science*, *344*(1), 012025. doi:10.1088/1755-1315/344/1/012025

Maftei, C., Dobrica, G., Cerneaga, C., & Buzgaru, N. (2021). Drought Land Degradation and Desertification—Case Study of Nuntasi-Tuzla Lake in Romania. In A. Vaseashta & C. Maftei (Eds.), *Water Safety, Security and Sustainability: Threat Detection and Mitigation* (pp. 583–597). Springer International Publishing., doi:10.1007/978-3-030-76008-3_25

Maftei, C., Roşu, L., & Teodorescu, D. (2009). The influence of water works on the flow regime of Taita catchment. *Lucrări Ştiinţifice, Universitatea de Ştiinţe Agricole Şi Medicină Veterinară "Ion Ionescu de La Brad" Iaşi. Seria Horticultură*, *52*, 909–924.

McKee, T. B., Doesken, N. J., & Kleist, J. (1993). *The Relationship of Drought Frequency and Duration to Time Scales*. 179–184.

Modarres, R. (2007). Streamflow drought time series forecasting. *Stochastic Environmental Research and Risk Assessment*, *21*(3), 223–233. doi:10.100700477-006-0058-1

Nalbantis, I., & Tsakiris, G. (2009). Assessment of Hydrological Drought Revisited. *Water Resources Management*, *23*(5), 881–897. doi:10.100711269-008-9305-1

Palmer, W. C. (1965). Meteorological Drought. *Weather Bureau Research Paper, 45*, 1–58.

Pelletier, J. D., & Turcotte, D. L. (1997). Long-range persistence in climatological and hydrological time series: Analysis, modeling, and application to drought hazard assessment. *Journal of Hydrology (Amsterdam)*, *203*(1), 198–208. doi:10.1016/S0022-1694(97)00102-9

Pettitt, A. N. (1979). A Non-Parametric Approach to the Change-Point Problem. *Applied Statistics*, *28*(2), 126–135. doi:10.2307/2346729

Pons, L. J. (1992). Natural resources. In *Conservation Status of the Danube Delta* (Vol. 4, pp. 23–36). IUCN.

Salmi, T., Määttä, A., Pia, A., Ruoho-Airola, T., & Amnell, T. (2002). Detecting Trends of Annual Values of Atmospheric Pollutants by the Mann-Kendall Test and Sen's Solpe Estimates the Excel Template Application MAKESENS. *Publications on Air Quality, 31.*

Serban, C., & Maftei, C. (2011). Thermal analysis of climate regions using remote sensing and grid computing. *ArXiv Preprint ArXiv:1101.4474.*

Serban, C., Maftei, C., & Barbulescu, A. (2010). Assessment of evapotranspiration using remote sensing data and grid computing and application. *WSEAS Transactions on Computers*, *9*(11), 1245–1254.

Şerban, C., Maftei, C., & Dobrică, G. (2022). Surface Water Change Detection via Water Indices and Predictive Modeling Using Remote Sensing Imagery: A Case Study of Nuntasi-Tuzla Lake, Romania. *Water (Basel)*, *14*(4), 4. doi:10.3390/w14040556

Sheskin, D. J. (2000). *Parametric and Nonparametric Statistical Procedures* (2nd ed.). Chapman & Hall.

Smakhtin, V. U., & Hughes, D. A. (2004). *Review, Automated Estimation and Analyses of Drought Indices in South Asia* (No. 83; Working Paper 83). International Water Management Institute.

Staudniger, M., Stahl, K., & Seibert, J. (2014). A drought index accounting for snow. *Water Resources Research*, *50*(10), 7861–7872. doi:10.1002/2013WR015143

Tallaksen, L., & van Lanen, H. A. J. (2004). *Hydrological drought. Processes and estimation methods for streamflow and groundwater*. Elsevier. https://research.wur.nl/en/publications/hydrological-drought-processes-and-estimation-methods-for-streamf

Taylor, W. (2000). *Change-Point Analysis: A Powerful New Tool For Detecting Changes*. Variation. https://www.variation.com/cpa

Telesca, L., Lovallo, M., Lopez-Moreno, I., & Vicente-Serrano, S. (2012). Investigation of scaling properties in monthly streamflow and Standardized Streamflow Index (SSI) time series in the Ebro basin (Spain). *Physica A*, *391*(4), 1662–1678. doi:10.1016/j.physa.2011.10.023

Tigkas, D., Vangelis, H., & Tsakiris, G. (2013). *The Drought Indices Calculator*. DrinC.

Vespremeanu-Stroe, A., Preoteasa, L., Hanganu, D., Brown, A., Bîrzescu, I., Toms, P., & Timar-Gabor, A. (2013). The impact of the Late Holocene coastal changes on the rise and decay of the ancient city of Histria (southern Danube delta). *Quaternary International*, *293*, 245–256. doi:10.1016/j.quaint.2012.11.039

Wilhite, D. A., & Glantz, M. H. (1985). Understanding the Drought Phenomenon: The Role of Definitions. *Water International*, *10*(3), 111–120. doi:10.1080/02508068508686328

World Bank (Ed.). (2018). *From uneven growth to inclusive development: Romania's path to shared prosperity*. World Bank Group. doi:10.1596/9781464813177

World Meteorological Organization (WMO). (2008). Guide to Hydrological Practices (WMO-No. 168), Volume I: Hydrology – From Measurement to Hydrological Information (Edition 2008, updated in 2020). WMO.

World Meteorological Organization -WMO. (2021). *WMO Atlas of Mortality and Economic Losses from Weather, Climate and Water Extremes (1970–2019) (WMO-No. 1267)*. WMO.

World Meteorological Organization -WMO, & Global Water Partnership - GWP. (2016). *Handbook of Drought Indicators and Indices* (M. Svoboda & B. Fuchs, Eds.). WMO. https://public.wmo.int/en/resources/library/handbook-of-drought-indicators-and-indices

Yevjevich, V. M. (1969). An objective approach to definitions and investigations of continental hydrologic droughts. *Journal of Hydrology (Amsterdam)*, *7*(3), 353. doi:10.1016/0022-1694(69)90110-3

Chapter 6
Google Earth Engine (GEE) for Modeling and Monitoring Hydrometeorological Events Using Remote Sensing Data

Khaled Hazaymeh
https://orcid.org/0000-0002-2162-7385
Yarmouk University, Jordan

Mohammad Zeitoun
Yarmouk University, Jordan

ABSTRACT

Google Earth Engine (GEE) has emerged as a powerful platform for modeling and monitoring extreme hydrometeorological events. In recent years, GEE has been used extensively for studying floods, droughts, and other natural disasters. It offers a comprehensive suite of tools that can help researchers and practitioners better understand the complex interactions between weather, climate, and water resources. By providing access to a wealth of satellite imagery, climate data, and geospatial datasets, GEE enables users to model and monitor these events with unprecedented accuracy and efficiency. This book chapter explores the various ways in which GEE can be used for modeling and monitoring extreme hydrometeorological events, understanding hydrometeorological events and their monitoring needs, including case studies and practical examples. It's worth noting that this chapter mainly focuses on using GEE for remote sensing and geospatial data analysis into hydrometeorological modeling and monitoring.

INTRODUCTION

GEE is a cloud-based platform that provides access to a vast collection of global and regional remote sensing datasets for analysis and visualization (Amani et al., 2020). It allows for the analysis of large-scale geospatial data using Google's computing infrastructure. With over 40 years of satellite imagery

DOI: 10.4018/978-1-6684-8771-6.ch006

and other geospatial data available, GEE has become an important tool for the analysis of earth systems (Tamiminia et al., 2020), including hydrometeorological modeling and monitoring. This platform is particularly useful in this area, as it provides access to large amounts of historical and real-time data, making it possible to generate models and predictions of hydrometeorological processes.

Hydrometeorology is the study of the interaction between the atmosphere and water on the Earth's surface. This includes processes such as precipitation, evapotranspiration, runoff, and groundwater recharge. Hydrometeorological models are used to simulate and predict these processes, providing important information for water resources management, flood and drought prediction, and climate change adaptation. In recent years, there has been a growing interest in the use of GEE and deep learning for hydrometeorological modeling and monitoring, due to its ability to handle large amounts of data and provide near-real-time analysis (Lee et al., 2021).

In this book chapter, we will review the use of GEE for hydrometeorological extreme events modeling and monitoring. We will begin by discussing the hydrometeorological events and their monitoring needs using GEE data sources available in GEE, including satellite imagery, weather data, and other geospatial data. We will then describe the preprocessing of remotely sensed data for hydrometeorological modeling within GEE. Followed by providing a discussion to understand the time series analysis and machine learning techniques and the integration between ground-based measurements with remotely sensed data for modeling hydrometeorological events using GEE. Then, we will discuss the potential of GEE in improving early warning systems and disaster response strategies for extreme hydrometeorological events. Finally, we provide example case studies for using GEE in studying hydrometeorological extreme events.

UNDERSTANDING HYDROMETEOROLOGICAL EVENTS AND THEIR MONITORING NEEDS

Hydrometeorological events, such as floods, droughts, and extreme weather events, can have significant impacts on water resources, agriculture, infrastructure, and human lives (Matthies-Wiesler and Quevauviller, 2022). Understanding these events and monitoring their occurrence is essential for developing effective management strategies and mitigating their impacts.

One of the primary challenges in understanding hydrometeorological events is the availability of accurate and reliable long-term data. Also, Traditional monitoring methods, such as ground-based observations and manual data collection, can be expensive, time-consuming, and limited in spatial coverage. However, advancements in technology have led to the development of remote sensing tools and data platforms that can provide more comprehensive and accurate data (Hazaymeh and Hassan 2016). Satellite-based remote sensing systems, such as NASA's Earth Observing System (EOS), National Oceanic and Atmospheric Administration's (NOAA) and the European Space Agency's (ESA) Sentinel satellites, provide a wealth of data on various hydrometeorological variables, including precipitation, soil moisture, evapotranspiration, and land surface temperature. These systems use a variety of sensors, such as optical, thermal, radar, lidar, and passive microwave sensors, to measure these variables at different spatial and temporal scales (Petropoulos and Islam, 2017). One of the significant advantages of satellite-based remote sensing systems is their ability to provide data over large areas and in real-time. This capability is particularly important for monitoring hydrometeorological events, such as floods, droughts and dust storms, which can occur over extensive areas and require timely responses (Lopez et al., 2020). Additionally, these systems can provide data on hard-to-reach areas, such as remote or inaccessible regions, where traditional monitoring methods may be challenging or impossible.

Another critical aspect is the use of modeling tools to simulate and predict their occurrence (Brunner et al., 2021). Hydrometeorological models use various data sources, including remote sensing data, to simulate and predict the behavior of water resources (Valipour et al., 2020). These models can provide valuable insights into the spatial and temporal distribution of water resources, including precipitation, soil moisture, and streamflow. In this context, GEE is one such platform that allows researchers to develop and apply customized algorithms for hydrometeorological modeling and monitoring (Yousefi et al., 2022). The platform's cloud-based system provides easy and quick access to large amounts of data, making it an ideal tool for time-sensitive projects. Additionally, the platform's ability to integrate various data sources, including remote sensing data, allows for more comprehensive and accurate modeling.

Furthermore, the monitoring needs for hydrometeorological events depend on the specific characteristics of the event and its potential impacts. For instance, monitoring droughts requires a different approach than monitoring floods or extreme weather events. Droughts are typically monitored through indicators such as soil moisture, precipitation deficits, and vegetation health (Hazaymeh and Hassan 2016), while floods are monitored through streamflow measurements and satellite observations of inundation extent (Schumann et al., 2022). The monitoring needs for extreme weather events, such as hurricanes, typhoons, and tornadoes, require a more complex approach. These events often have significant impacts on coastal regions, including flooding, storm surges, and high winds. Therefore, monitoring systems must provide accurate and timely information on the event's track, intensity, and potential impacts. This information is critical for developing emergency response plans and evacuating vulnerable populations (Erickson et al., 2019).

Another critical aspect of hydrometeorological monitoring is the integration of data from multiple sources and at different spatial and temporal scales. For example, drought measurements from ground-based stations can provide valuable information on local water resources, while satellite-based remote sensing data can provide a more comprehensive view of drought variables at a regional or global scale. Integrating data from multiple sources can help to reduce uncertainties and improve the accuracy of monitoring systems (Alkaraki and Hazaymeh 2023). Moreover, the use of hydrometeorological monitoring systems is not limited to predicting and managing the impacts of extreme events. Hydrometeorological monitoring systems can also play a vital role in supporting agriculture, particularly in regions that depend on rain-fed agriculture (Alkaraki and Hazaymeh 2023). Continuous monitoring of precipitation, soil moisture, and other relevant variables can help farmers make informed decisions about planting, irrigation, and crop management. This information can also support early warning systems for potential droughts or floods, allowing farmers to take necessary measures to protect their crops and livelihoods. The use of hydrometeorological monitoring systems can also support the management of water infrastructure, such as dams, reservoirs, and irrigation systems (de Lima Moraes et al., 2020). Monitoring the inflow, outflow, and water levels of these structures can provide critical information for decision-making, including water allocation and release strategies during periods of high demand or emergency situations. In addition, hydrometeorological monitoring systems can also support the management of water quality (de Lima Moraes et al., 2020). Water quality monitoring is critical for ensuring the safety of drinking water supplies and maintaining ecosystem health. Continuous monitoring of water quality parameters such as temperature, pH, dissolved oxygen, and nutrient concentrations can help in identifying potential contamination sources, assessing the effectiveness of water treatment systems, and tracking the health of aquatic ecosystems (Goyal et al., 2022).

In conclusion, the use of hydrometeorological monitoring systems is critical for understanding, predicting, and managing the impacts of extreme events such as floods, droughts, and extreme weather events. It is also essential for regular monitoring and management of water resources, supporting agriculture, and managing water infrastructure. Advances in technology and modeling tools, such as remote sensing systems and GEE, have improved our ability to monitor hydrometeorological events and water resources. Continued research and development in this field are necessary to improve the accuracy and reliability of monitoring systems and support sustainable water management practices.

PREPROCESSING REMOTELY SENSED DATA FOR HYDROMETEOROLOGICAL MODELING

Remote sensing data plays an important role in hydrometeorological modeling, providing information on key variables such as precipitation, soil moisture, and vegetation cover. However, before these data can be used in modeling, they must undergo preprocessing to ensure they are of sufficient quality and in the correct format. This preprocessing includes several steps, including calibration, correction for atmospheric effects, and spatial and temporal averaging.

Calibration is the process of converting raw sensor measurements to physical units such as watts per square meter, millimeters of precipitation, degrees of land surface temperature, or percentage of spectral reflectance. This is necessary because raw sensor measurements are often in arbitrary units and vary depending on factors such as sensor type, gain settings, and temperature. Calibration is typically performed using field measurements or reference data to establish a relationship between sensor measurements and physical units (Cheema and Bastiaanssen, 2012). Correction for atmospheric effects and gap filling are necessary because the atmosphere features can absorb, scatter, and reflect radiation, affecting the accuracy of remote sensing measurements. This correction involves the use of atmospheric models and algorithms to estimate and remove the effects of the atmosphere and to fill the gaps in the remote sensing data. For instance, filling gaps in remote sensing data caused by clouds and other atmospheric effects can be addressed using machine learning algorithms (Wang et al., 2022) and spatio-temporal statistics (Shew and Ghosh, 2017). These approaches can help researchers gain a better understanding of changes in the Earth's surface over time and provide valuable insights into the impacts of natural and human-induced changes on the environment. For example, the widely used Landsat and MODIS land surface temperature product uses a physical model to correct atmospheric effects and provide accurate temperature measurements.

Spatial and temporal averaging involves aggregating remote sensing data to a consistent spatial and temporal resolution (Bradley et al., 2007). This is necessary because different sensors have different spatial and temporal resolutions, and modeling requires data at a consistent resolution. Spatial averaging involves aggregating data to a coarser resolution, while temporal averaging involves aggregating data over a specific period. The appropriate spatial and temporal resolution for a given hydrometeorological modeling application depends on the scale of the process being modeled and the data availability, therefore, it is important to have the necessary dataset to perform the modelling accurately.

In addition, there are other important procedures in the preprocessing of remotely sensed data for hydrometeorological modeling. One of these procedures is data fusion, which involves combining data from multiple sensors to produce a more accurate and complete dataset (Hazaymeh and Hassan 2015). Data fusion can be particularly useful in situations where data from a single sensor are incomplete or insufficient. Another important procedure is data quality control, which involves identifying and correcting errors or anomalies in the data. It Is also important to consider the effects of topography and land cover on remote sensing data, particularly for variables such as precipitation and soil moisture (Temimi et al., 2010).

Other few more important procedures in the preprocessing of remotely sensed data for hydrometeorological modeling involve:

- Data interpolation: Remotely sensed data may have gaps or missing values due to cloud cover or other factors. Data interpolation techniques can be used to estimate values for these missing pixels based on surrounding data. Common interpolation methods are the geostatistical methods (Tayebi et al., 2023) such as nearest neighbor, inverse distance weighting, and kriging.
- Image registration: When using data from multiple sensors or time periods, it is important to ensure that the images are registered, or aligned, with each other. This involves correcting differences in sensor geometry, such as sensor position and orientation, coordinate transformation and projection techniques, as well as differences in image scale and resolution.
- Data transformation: Depending on the specific application, it may be necessary to transform the raw data into a different format or coordinate system. For example, some models may require data in a specific projection or units or may require the data to be resampled to a coarser or finer resolution.
- Noise reduction: Remotely sensed data can be subject to various types of noise, including random noise, systematic noise, and speckle noise. Techniques such as filtering and smoothing can be used to reduce these types of noise and improve the accuracy of the data (Fu et al., 2022).
- Data assimilation: In addition to preprocessing the remote sensing data, it is often necessary to assimilate this data into hydrologic or atmospheric models (Duan et al., 2021). This involves integrating the remote sensing data with other data sources, such as gauge observations or weather forecasts, and using the combined dataset to drive the model.

In addressing all the above preprocessing procedures of remotely sensing data, several tools and software packages are available, including the GEE platform which provides a range of preprocessing functions for remote sensing data, including atmospheric correction and spatial and temporal aggregation. These functions can be used to preprocess data from a range of sensors, including MODIS, Landsat, and Sentinel-2, for use in hydrometeorological modeling applications (Roy et al., 2014).

TIME SERIES ANALYSIS AND MACHINE LEARNING TECHNIQUES FOR MODELING HYDROMETEOROLOGICAL EVENTS USING GEE

Hydrometeorological extreme events pose significant challenges to societies, ecosystems, and infrastructure. Understanding the patterns, trends, and drivers of these events is crucial for developing effective mitigation and adaptation strategies. Time series analysis techniques provide powerful tools for modeling and analyzing hydrometeorological extreme events over time (Valipour et al., 2020). GEE provides a powerful platform for performing advanced time series analysis on a global and regional scale, enabling researchers to extract valuable insights into the complex processes underlying these events. GEE offers access to a wide range of remote sensing and climate data, including precipitation, temperature, soil moisture, and streamflow data. These data sources can be combined and analyzed using a range of time series analysis techniques to model hydrometeorological extreme events.

Various time series analysis and machine learning techniques have been used in modelling hydrometeorological extreme events using GEE such as:

Trend Analysis

Trend analysis in GEE is a powerful technique for identifying long-term patterns and trends in hydrometeorological data. This method involves fitting a linear or nonlinear model to a time series and testing for significant trends over time. Trend analysis can be used to identify changes in the frequency, intensity, and duration of extreme events, such as floods and droughts (Weilnhammer et al., 2021). It can lead to an accurate and timely forecast of extreme events that can mitigate negative impacts and enhance preparedness.

Seasonal Decomposition

Seasonal decomposition is a technique for separating a time series into its seasonal, trend, and residual components. This method can be used in GEE to identify the seasonal patterns underlying extreme events, such as the timing and duration of floods and droughts. Seasonal decomposition can also help identify long-term trends in extreme events that are masked by seasonal variability (Tylkowski and Hojan, 2019).

Machine Learning Techniques

In addition to traditional time series analysis techniques, machine learning techniques can also be applied to hydrometeorological data in GEE to model extreme events. Machine learning algorithms such as random forests and neural networks can be trained to identify complex patterns in the data that may not be captured by traditional techniques (Farhadi and Najafzadeh, 2021). For example, in a recent study, researchers used a random forest algorithm to model drought risk in the semiarid environment, achieving high accuracy in predicting drought occurrence (Alkaraki, and Hazaymeh, 2023).

Integration With GIS

GEE provides integration with Geographic Information System (GIS) tools, allowing researchers to visualize and analyze hydrometeorological data in a spatial context (Teodoro and Duarte, 2022). By combining time series analysis techniques with GIS, researchers can identify hotspots of extreme events, explore the spatial distribution of drivers, and develop targeted mitigation and adaptation strategies (Kumar et al., 2022).

Citizen Science and Community Engagement

GEE also provides opportunities for citizen science and community engagement in modeling hydrometeorological extreme events. It enables users to access and analyze data from anywhere in the world, making it a valuable tool for engaging communities in data-driven decision-making processes and develop flood early warning systems (Perera et al., 2020). For example, in a recent study, researchers used GEE to develop coastal management in Guyana, enabling community members to contribute local knowledge to mapping efforts and coastal management (Oyedotun et al., 2022).

Ensemble Modeling

Ensemble modeling involves combining multiple models or approaches to better capture the complexity of hydrometeorological extreme events. In GEE, ensemble modeling can be used to combine traditional time series analysis techniques with machine learning algorithms, or to combine multiple data sources to improve model accuracy (Hosseini et al., 2020). Ensemble modeling can help overcome the limitations of individual models and provide more robust and accurate predictions of extreme events. For example, researchers have developed an integrated approach for the assessment of climate change impacts on the hydro-meteorological characteristics of the Mahaweli river basin, Sri Lanka (Selvarajah et al., 2021) and drought monitoring (Alkaraki, and Hazaymeh, 2023)

Forecasting

Forecasting involves predicting the behavior of hydrometeorological events in the future, based on historical data and other relevant variables. Forecasting can be based on traditional time series analysis techniques, such as autoregressive models, or on machine learning algorithms, such as random forests or neural networks (Alsumaiei and Alrashidi, 2020). GEE provides a range of tools for forecasting, including time series analysis, machine learning algorithms, and model validation techniques.

Decision Support Systems

Time series analysis in GEE can also be used to develop decision support systems (DSS) for managing hydrometeorological extreme events. DSS provide decision-makers with real-time information and predictive models to inform decision-making processes, such as flood forecasting and drought monitoring. By combining time series analysis with machine learning and GIS tools, DSS can provide a comprehensive and user-friendly platform for managing extreme events (Al-Waili et al., 2023). In addition to the above-discussed techniques, GEE offers a range of other advanced time series analysis methods, such as wavelet analysis, autoregressive models, which can be used to model and predict complex events, such as droughts, floods, and land cover changes.

INTEGRATING GROUND-BASED MEASUREMENTS WITH REMOTELY SENSED DATA HYDROMETEOROLOGICAL MODELING AND MONITORING IN GEE

Hydrometeorological modeling and monitoring are crucial for managing natural resources and mitigating the impact of extreme hydrometeorological events such as floods and droughts. In recent years, there has been a growing trend towards integrating ground-based measurements with remotely sensed data to improve the accuracy of hydrometeorological modeling and monitoring. GEE is a platform that can facilitate this integration by allowing users to access and analyze large amounts of remote sensing data and ground data in a single environment. In this section, we will discuss the pros and cons of the integration between remote sensing data and ground data for extreme hydrometeorological events using GEE.

Remote sensing data, such as satellite imagery and radar data, can provide a wide range of information on hydrometeorological events. For example, satellite imagery can be used to estimate the extent and severity of floods by detecting changes in surface water extent (DeVries et al., 2020), while radar data can be used to estimate the depth and velocity of floodwaters (Hamidi et al., 2023). However, remote sensing data alone may not provide the level of detail required for accurate modeling and monitoring of hydrometeorological events. Ground data, such as rainfall measurements, streamflow measurements, and soil moisture measurements, can provide detailed information on local conditions that is not captured by remote sensing data (Galkate et al., 2023). GEE can be used to integrate remote sensing data and ground data to improve the accuracy of hydrometeorological modeling and monitoring.

GEE provides access to a wide range of remote sensing data, including Landsat imagery, MODIS data, and radar data, as well as ground data, such as streamflow measurements and weather station data. By combining these two types of data, GEE users can develop more accurate models of extreme hydrometeorological events and monitor these events in real-time (Kumar et al., 2022). For example, GEE can be used to develop flood models by combining remote sensing data on water extent and depth with ground data on rainfall and streamflow (Vanama et al., 2020). GEE can also be used to monitor drought conditions by combining remote sensing data on vegetation health with ground data on soil moisture and rainfall (Alkaraki and Hazaymeh, 2023). In both cases, GEE allows users to analyze large amounts of data quickly and efficiently, providing real-time monitoring and modeling capabilities.

Cons and Pros of Integrating Within GEE

One of the advantages of using GEE for integrating remote sensing data and ground data is the ability to perform analysis at different spatial and temporal scales. GEE allows users to analyze data at a global scale or zoom in to a specific region or watershed, providing a flexible tool for monitoring and modeling hydrometeorological events. Furthermore, GEE is a cloud-based platform, which means that users can access and analyze data from anywhere in the world with an internet connection, making it an accessible tool for researchers and practitioners (Tamiminia et al., 2020).

Another benefit of integrating remote sensing data and ground data using GEE is the ability to create custom workflows that can be tailored to specific research questions or management objectives. For example, GEE can be used to develop early warning systems for floods by integrating remote sensing data on precipitation, surface water, and soil moisture with ground data on streamflow and river levels (Perera et al., 2020). This type of workflow can be adapted to different locations and river systems, allowing for the development of site-specific early warning systems that can improve preparedness and response to flood events.

Furthermore, GEE can incorporate machine learning and artificial intelligence algorithms into data analysis. GEE provides a platform for developing and testing machine learning models that can be used to predict future hydrometeorological events based on historical data (Tamiminia et al., 2020). For example, machine learning algorithms can be trained to predict the likelihood of drought conditions based on a combination of remote sensing data and ground data on soil moisture and rainfall (Alkaraki and Hazaymeh, 2023).

Moreover, integrating ground-based measurements with remotely sensed data in GEE provide a more comprehensive understanding of hydrometeorological processes by filling gaps in knowledge and improve the accuracy of models. It provides the capability for real-time monitoring on local conditions, while remotely sensed data can provide information on larger spatial scales. Integrating these two types of data can provide a more comprehensive picture of hydrometeorological conditions in real-time which can be critical for decision-making and emergency response.

The cost effectiveness is another advantage of the integration of ground-based measurements and remote sensing. For instance, ground-based data can be expensive and time-consuming, while remotely sensed data can provide information at a lower cost and with greater efficiency. Integrating these two types of data can provide a cost-effective approach to hydrometeorological modeling and monitoring using GEE. GEE provides flexible and accessibility for integrating and analyzing data at different scales, from global to local, allowing for customized workflows to be developed to address specific research questions or management objectives. It is a cloud-based platform, which means that users can access and analyze data and share their results and codes from anywhere in the world with an internet connection, making it an accessible tool for researchers and practitioners.

Despite the benefits of using GEE for integrating remote sensing data and ground data, there are also limitations that should be considered. One of the main limitations is the availability and quality of remote sensing data, which can be affected by cloud cover, sensor limitations, and atmospheric conditions. In addition, the accuracy and resolution of ground data can vary depending on the location and the type of data being collected, which can limit the accuracy of models and analyses. Some other challenges associated with integrating remote sensing data and ground data using GEE is the differences in the spatial and temporal resolution of the remote sensing data and the ground data, which can make it difficult to combine the two datasets. In addition to data quality, the data availability may possess some limitations which can impact the feasibility of integrating the two datasets. Another limitation could be related to technical expertise as GEE requires technical expertise in remote sensing and geospatial analysis, which can limit its accessibility for some users.

Overall, integrating remote sensing and ground-based data in GEE can provide significant benefits for hydrometeorological monitoring and modeling. However, careful consideration should be given to the quality and availability of data, as well as technical expertise required to effectively use the platform.

POTENTIAL OF GEE IN IMPROVING EARLY WARNING SYSTEMS AND DISASTER RESPONSE STRATEGIES FOR EXTREME HYDROMETEOROLOGICAL EVENTS

Before discussing the potential of GEE in early warning systems (EWS) and disaster response strategies for extreme hydrometeorological events, we should provide a general overview of the EWS for hydrometeorological extreme events and their benefits. EWS are essential tools that help to reduce the risk of disasters and save lives and enable effective disaster response efforts (Glago, 2021). These systems are designed to monitor and assess weather and hydrological conditions and provide timely information to decision-makers, communities, and individuals to take appropriate action to protect themselves and their property. The benefits of early warning systems for hydrometeorological extreme events monitoring and assessment and modeling are numerous, and some of the key benefits include:

- Early Identification of Hazardous Conditions: EWS can help to identify hazardous conditions such as heavy rainfall, floods, drought, and heat waves. This information can be used as a warning and to trigger the appropriate response, such as evacuations or the deployment of emergency supplies.

- Improved Planning and Preparation: EWS can provide critical information that can be used to plan and prepare for impending weather and hydrological conditions. For example, emergency responders can pre-position equipment and supplies, and individuals can take measures to secure their property and protect themselves (Sarker et al., 2020).
- Reduced Risk of Loss of Life and Property: EWS can reduce the risk of loss of life and property by providing advanced notice of impending disasters. This allows individuals and communities to take appropriate actions to protect themselves and their property, such as evacuating or moving valuables to higher ground.
- Enhanced Disaster Response Efforts: EWS can enhance disaster response efforts by providing real-time data and information that can be used to coordinate response efforts. For example, emergency responders can use this information to prioritize response efforts and allocate resources where they are most needed through using internet of things-based interconnected devices (Sharma et al., 2021).
- Increased Community Awareness and Participation: EWS can increase community awareness and participation in disaster preparedness efforts. By providing information on impending weather and hydrological conditions, individuals and communities can take an active role in preparing for disasters and reducing their vulnerabilities (Bajracharya et al., 2020).
- Cost Savings: EWS can help to reduce the cost of disasters by providing advanced notice of impending disasters. This allows communities and individuals to take preventative measures, reducing the need for emergency response efforts and the associated costs.
- Improved Decision Making: EWS can provide critical information to decision-makers at all levels, enabling them to make informed decisions about resource allocation and emergency response efforts.
- Increased Awareness: EWS can help to increase awareness of the risks associated with hydrometeorological extreme events (Potter et al., 2021). By raising awareness, individuals and communities can take a more proactive approach to disaster preparedness and response.
- Mitigating the impact of shock: EWS can improve methods of dealing with extreme weather conditions, as early warning to the community reduces the impact of shock.

In this context, GEE has great potential in improving EWS and disaster response strategies for extreme hydrometeorological events. As discussed earlier in this chapter, GEE is a cloud-based platform that provides access to a vast archive of remote sensing data and computing resources, making it a powerful tool for analyzing and visualizing large amounts of data.

One of the key advantages of GEE is its ability to process and analyze data in near real-time (Pandey etal., 2022). This means that data from satellite imagery, weather stations, and other sources can be processed and analyzed quickly, allowing for the timely detection of extreme hydrometeorological events such as floods (Trambadia et al., 2023), droughts (Veneros and García, 2021), and landslides (Subiyantoro et al., 2022). Early warning systems can be developed using GEE that can alert communities and authorities of potential disasters, allowing them to take preemptive measures to mitigate the impact (Perera et al., 2020). GEE can also be used to develop disaster response strategies. By analyzing data from satellite imagery, GEE can help identify areas that have been affected by extreme hydrometeorological events and provide a detailed view of the damage caused. This information can be used to guide relief efforts and prioritize response efforts, allowing for a more efficient and effective response (Alam and Masum, 2021).

In addition to the advantages, GEE has several other features that make it a valuable tool for improving early warning systems and disaster response strategies for extreme hydrometeorological events. Some of these features include:

- Machine learning algorithms: GEE has a range of machine learning algorithms that can be used to analyze remote sensing data and identify patterns and anomalies. For example, machine learning algorithms can be used to identify areas that are at high risk of flooding (Saleh et al., 2020).
- Collaborative platform: GEE is a collaborative platform that allows users to share data, code, and analysis tools. This means that experts from different fields and locations can work together to develop and refine early warning systems and disaster response strategies.
- Historical data analysis: GEE provides access to historical data that can be used to analyze trends and patterns in extreme hydrometeorological events. This can be used to develop predictive models that can help anticipate future events and develop more effective response strategies (Zhang et al., 2020).
- Interactive visualization tools: GEE has a range of interactive visualization tools that can be used to communicate complex data and analysis results to a wide audience. This can help raise awareness of the risks associated with extreme hydrometeorological events and encourage communities to take action to reduce their vulnerability.

Overall, GEE has tremendous potential in improving early warning systems and disaster response strategies for extreme hydrometeorological events. Its ability to process and analyze large amounts of data in near real-time, access to remote sensing data and machine learning algorithms, collaborative platform, historical data analysis, and interactive visualization tools can provide timely and accurate information to authorities and communities, allowing for a more proactive and effective response strategies and ultimately save lives and reduce the impact of these events.

EXAMPLE CASE STUDIES FOR USING GEE IN STUDYING HYDROMETEOROLOGICAL EXTREME EVENTS

GEE has been used for studying of hydrometeorological extreme events at various local, regional, and global scales. In this section, we will provide some example case studies on the leverage of GEE at the global and regional scales in studying some hydrometeorological extreme events.

Drought

Drought is one of the most devastating natural disasters that can have serious environmental, social, and economic consequences. In recent years, the severity and frequency of drought events have increased due to climate change, population growth, decreasing vegetation cover and increasing demand for water resources. Extreme drought events can have significant impacts on food security, water resources, and natural ecosystems. With the advanced processing capabilities and vast amounts of data, GEE can be a valuable tool for monitoring and detecting extreme drought conditions at the global scale due to ability to process and analyze different types of data, including satellite imagery, climate data, and ground-based measurements. For example, Khan and Gilani (2021a and b) used GEE and big geospatial datasets to monitor global drought conditions. In their first work, they describe how they utilized GEE to calculate the drought severity index (DSI), which is a widely accepted indicator of drought severity. The DSI considers various factors such as precipitation, temperature, and soil moisture to provide a holistic view of drought conditions. The authors then used this index to analyze drought trends over a 20-year period, from 2000 to 2020, and identified areas that experienced frequent droughts. In the second paper, they also utilized GEE, but focused on leveraging big geospatial datasets to monitor global drought conditions. The authors used various datasets such as precipitation, temperature, and soil moisture, and combined them with machine learning techniques to create a drought monitoring system. The system was able to provide near real-time monitoring of drought conditions globally, allowing for early detection and response. Overall, both papers demonstrate the potential of GEE to effectively monitor and analyze drought conditions on a global scale.

Floods

Floods are among the most devastating natural disasters, causing loss of life, damage to property, and disruption of economic activities. Accurate and timely information on the extent and severity of floods is critical for effective disaster management, including early warning systems, response planning, and relief efforts. Traditionally, flood mapping and modeling have relied on ground-based observations, satellite imagery, and hydrological models, which are time-consuming and resource-intensive. With the availability of cloud-based platforms like GEE, it is now possible to analyze and visualize large volumes of earth observation data quickly and efficiently. One example of the use of GEE in global flood mapping and modeling is the Global Flood Mapper (Tripathy and Malladi, 2022) as a new application developed in GEE that leverages Sentinel-1 SAR data to quickly detect and map floods worldwide. The system uses a combination of pre- and post-flood SAR images to identify changes in the landscape, such as flood water extent and inundation depth. The authors demonstrate the effectiveness of the Global Flood Mapper by comparing its results to existing flood maps and field observations, showing good agreement. Another study (Tellman et al., 2016) emphasized the importance of historical flood data for developing machine learning models to predict future floods. The authors describe how they compiled a global database of historic flood events, spanning several decades, and used it to train a machine learning algorithm to predict future floods. They also utilized GEE to process and analyze the large amounts of data involved in this project. The authors demonstrate the effectiveness of their approach by comparing their model's predictions to observed flood events, showing good agreement. Overall, both works demonstrated the potential of GEE to improve flood monitoring and prediction by utilizing cutting-edge technology and data analysis techniques.

Landslides

Landslides are one of the most common geomorphological hazards that can cause severe environmental, social, and economic impacts. Landslides can lead to property damage, displacement of people, and loss of life. In recent years, the severity and frequency of landslide events have increased due to climate change, urbanization, and land-use changes. GEE can be a valuable tool for monitoring, modeling, and forecasting extreme landslide conditions through analyzing different types of data, including satellite imagery, topographic data, and ground-based measurements. One of the example-based GEE tools is the HazMapper (Scheip and Wegmann,2021) which is an open-access application developed using

GEE that is designed to identify and analyze natural disasters quickly and accurately. It is meant to be used by both the scientific and emergency management communities. HazMapper is especially helpful in tracking changes to the landscape caused by natural disasters or human activities, particularly those that result in the removal or recovery of vegetation. However, the platform is not currently appropriate for analyzing non-vegetated environments like polar, high-altitude, or desert regions. An example case study on using HazMapper was performed in regions in eastern Africa experienced many days of intense rainfall. West Pokot County, Kenya, located in the rugged terrain of the East African Rift valley in western Kenya after an extreme rainfall event with more than 400 mm of rainfall within the period 23–25 November 2019. The heavy rainfall triggered landslides on the steep hillslopes in the western rim of the East African Rift valley, leading to debris flows that caused significant damage to homes, agricultural fields, and infrastructure. HazMapper was used to identify the location of the debris flows. It also detected the loss of riparian vegetation and sedimentation along the banks of lower-gradient rivers that drained the mountainous terrain where the mass wasting occurred. Benz and Blum (2019) conducted a worldwide study to identify clusters of landslides triggered by the same rainfall event. They developed an algorithm that looks for a sequence of landslide events that (a) occur no more than 2 days apart and (b) have precipitation levels at the event location that are correlated with at least one other event. To carry out their analysis, they applied the algorithm to the Global Landslide Catalog (GLC) maintained by NASA. They used the Climate Hazards Group InfraRed Precipitation with Station data (CHIRPS) for their rainfall analysis, which have a resolution of 0.05 x 0.05 degrees and provide daily measurements. For each location of a landslide event, they downloaded the CHIRPS data using GEE for the 10 years leading up to the event and up to 2 days after the event. In conclusion, GEE can be a valuable tool for monitoring, modeling, and forecasting extreme landslide conditions. The platform's advanced processing capabilities and vast amounts of data enable users to analyze and predict landslide occurrence, volume, and movement. The Example Case Studies discussed here highlight the importance of GEE for effective landslide monitoring and management strategies, especially in regions prone to frequent landslide events. As climate change continues to exacerbate landslide conditions, it is crucial to develop effective and timely landslide monitoring and warning systems to mitigate the impacts of landslides on communities and ecosystems.

Hurricanes, Tornados, and Storms

Hurricanes, tornados, and storms are among the most powerful and destructive weather phenomena on Earth. Their monitoring and modeling are critical for disaster preparedness and response. In recent years, satellite-based remote sensing has emerged as a valuable tool for post-disaster monitoring. GEE can be used to analyze and process remote sensing data, making it a valuable tool for mapping and monitoring post-disaster damages in the affected areas. With its advanced processing capabilities and vast amounts of available data, GEE can provide a comprehensive understanding of several large-scale and zoomed scenes for visual interpretation and the corresponding assessment analysis. Example case studies from different regions of the world were performed for damage assessment of these destructive weather phenomena. Kakooei et al., (2022) assessed various weather disasters, including the Joplin MO Tornado in 2011, Hurricane Harvey in 2017, and Hurricane Michael in 2018, using remote sensing techniques. The study utilized GEE to obtain and process several satellite sensors such as Landsat 5 and Sentinel 2, as well as airborne imagery obtained through the National Agriculture Imagery Program (NAIP) and the National Oceanic and Atmospheric Administration (NOAA). They compared different remote sensing methods, including pixel- and object-based classification techniques, and spectral/spatial feature analysis, to determine the potential of vertical and oblique images in generating regional- and building-level damage maps. The study presented several large-scale and zoomed scenes for visual interpretation, along with the corresponding assessment analysis. Kakooei and Baleghi (2020) used High Resolution (HR) and Very High Resolution (VHR) remote sensing images for building damage assessment after Hurricane Michael 2018 in Florida. They developed a method that uses a hierarchical approach of unsupervised and supervised methods to detect damaged and undamaged areas within each cluster of buildings, implemented on the GEE platform. The images were used to evaluate the results, with damage information presented at the building-level as a fractional number between 0 and 1, the R-squared (R2) value between actual and predicted damage scores was 0.9688. These studies among others found that satellite-based damage assessment within GEE platform can provide accurate and timely information to support disaster response efforts.

Heat Wave

A heat wave (HW) is a prolonged period of excessively high temperature which is accompanied by high humidity with a significant decrease in wind speed, that is significantly above the average temperature for a particular region and time of year. HW can be dangerous, particularly for vulnerable populations such as the elderly, young children, and people with pre-existing health conditions, as they can lead to heat exhaustion, heat stroke, and other heat-related illnesses. HW can also have significant impacts on the environment, including drought, wildfires, and damage to crops and ecosystems. GEE provides access to a wide range of geospatial data, including temperature data, land cover data, and satellite imagery. These datasets can be used to develop models for heatwave forecasting and monitoring. For example, by analyzing historical temperature data and land cover data, users can develop models that predict heatwave intensity and duration. The platform's advanced processing capabilities allow users to analyze and process large datasets quickly, making it an ideal tool for heatwave forecasting and monitoring. Zhang et al., (2022) developed the Heat Wave Tracker, a multi-method, multi-source toolkit for measuring HW using GEE. It incorporates satellite data, meteorological data, and land surface temperature data, to provide a comprehensive assessment of heat wave intensity, frequency, and duration. The toolkit is built using GEE, which allows for the efficient processing and analysis of large-scale satellite and geospatial data. They demonstrate the application of the Heat Wave Tracker in Australia. The results show that the toolkit is effective in measuring HW across different regions and can provide valuable information for HW mitigation and adaptation efforts. The authors conclude that the Heat Wave Tracker can contribute to a better understanding of HW and their impacts on human health and the environment. The toolkit can also be used by policymakers, urban planners, and public health officials to inform decision-making and develop effective heat wave mitigation and adaptation strategies. Rao et al., (2021) conducted a spatio-temporal analysis of land surface temperature for identification of heat wave risk and vulnerability hotspots in Indo-Gangetic Plains of India. They used the freely available Moderate Resolution Imaging Spectrometer (MODIS) Land Surface Temperature (LST) data on the GEE platform. HW indices were identified and computed on a 10 km * 10 km grid size, focusing on the duration, frequency, and intensity of HW. To identify HW risk and vulnerability hotspots, spatial modeling was conducted using HW indices, land use and land cover (LULC) changes, and population density. The results showed that the National Capital Territory (NCT) Delhi and its surrounding region had the highest risk of HW with high vulnerability. The analysis revealed a strong positive correlation between variability of HW indicators and increasing built-up areas, indicating that built-up surfaces have a significant impact on HW conditions.

Sand and Dust Storms

Sand and dust storms (SDS) also known are hydrometrical-related hazards that can cause significant damage to infrastructure, health, and the environment. They are common in arid and semi-arid regions and can be triggered by natural and human activities. Dust storm monitoring and forecasting are crucial for early detection and effective management of dust storms. GEE provides access to a wide range of geospatial data, including satellite imagery, weather data, and topography data, which can be used for dust storm monitoring and forecasting. By analyzing historical dust storm events and geospatial data, users can develop models that identify areas at risk of dust storms and predict their behavior. The platform's advanced processing capabilities allow users to analyze and process large datasets quickly, making it an ideal tool for dust storm monitoring and forecasting. Different methods were proposed for studying SDS in various regions. Wang et al., (2022) proposed a method for detecting SDS through utilizing the support vector machine (SVM) algorithm on the GEE cloud computing platform and a spectral index to label training samples automatically. The study evaluated the effectiveness and accuracy of the proposed method using eight SDS events captured by MODIS over Arid Central Asia (ACA) and compared it with traditional approaches. The results showed that the proposed method can distinguish between mixed pixels, such as thin cloud and land surface, and SDS pixels more effectively than traditional methods. The proposed method achieved over 98% in SDS detection. In another study Wang et al., (2023) used GEE platform to predict SDS sources in ACA using four machine learning (ML) methods. Fourteen factors, both meteorological and terrestrial, were chosen as controlling factors for SDS source susceptibility and applied in the modeling process. Results showed that the random forest (RF) algorithm performed best, followed by the gradient boosting tree (GBT), maximum entropy (MaxEnt) model, and support vector machine (SVM). The RF model revealed that wind speed played the most significant role in SDS source prediction, followed by the normalized difference vegetation index (NDVI). This study could assist in the development of programs aimed at reducing SDS risks in arid and semi-arid regions, particularly in ACA. Alshammari et al., (2022) conducted a meta-analysis review to investigate the application of machine learning including GEE to dust storm prediction. They provided a comparison between different models and types of data used in the literature and discusses the properties of SDS in relation to the region to predict the location of SDS events. The review summarized the output classes and various performance metrics observed and highlighted the capabilities of machine learning models in predicting dust storms. The analysis identifies two main categories: early detection and dust storm prediction. Most models used for dust storm early detection from satellite images are support vector machines (SVM), while SVM and random forests are the most used models for dust storm prediction from meteorological data. They concluded the challenges and future trends in the field and the potential of applying deep learning algorithms for long-range predictions of SDS duration and intensity. The example case studies highlighted the importance of effective SDS monitoring and management strategies, particularly in areas facing significant dust storm risks. As climate change continues to exacerbate natural hazards globally, it is crucial to develop effective and timely SDS monitoring and management systems to protect lives and property.

CONCLUSION

This chapter has investigated the capabilities of Google Earth Engine (GEE) in the feild of modeling and monitoring hydrometeorological processes. The integration of satellite imagery, geospatial data, and powerful computational resources within the GEE platform has opened up new frontiers in understanding and managing the complex dynamics of Earth's hydrological and meteorological systems. We have explored how GEE facilitates the combine the multi-source geospatial data, enabling researchers and practitioners to create comprehensive models for hydrological and meteorological analysis. The ability to access a vast repository of historical and near-real-time data has empowered users to develop accurate and timely assessments of hydrometeorological phenomena, ranging from precipitation patterns to land cover changes. Furthermore, the efficiency of GEE has been highlighted, emphasizing its suitability for large-scale studies and real-world applications. The platform's cloud-based infrastructure not only accelerates data processing but also encourages collaboration by providing a centralized environment for researchers worldwide.

The chapter has also touched the significance of the functions of GEE, such as time-series analysis, machine learning algorithms, and data visualization techniques. These functions contribute to the development of robust models and enhance our ability to monitor and predict hydrometeorological events with greater precision. As we navigate an era marked by increasing environmental challenges, the role of advanced technologies like GEE becomes dominant. Its integration into hydrometeorological research equips us with the tools needed to address critical issues such as water resource management, flood monitoring and forecasting, and climate change impacts.

In essence, the exploration of GEE in this chapter underscores its potential as a powerful ally in the ongoing mission to model, monitor, and manage hydrometeorological events. As we look forward, the continued refinement and utilization of such technologies promise to push our capacity for informed decision-making and sustainable resource management.

REFERENCES

Al-Waili, S. M., Abd Latif, Z., & Salleh, S. A. (2023). GIS-Based Decision Support System and Analytical Hierrachical Process for Integrated Flood Management. *International Journal of Intelligent Systems and Applications in Engineering*, *11*(4s), 392–399.

Alam, S., & Masum, F. (2021). Rapid mapping of flood inundation using Google Earth Engine: A case study of Bangladesh. *Journal of Water and Climate Change*, *12*(1), 238–252. doi:10.2166/wcc.2020.190

Alkaraki, K. F., & Hazaymeh, K. (2023). A comprehensive remote sensing-based Agriculture Drought Condition Indicator (CADCI) using machine learning. *Environmental Challenges*, *11*, 100699. doi:10.1016/j.envc.2023.100699

Alshammari, R. K., Alrwais, O., & Aksoy, M. S. (2022). Machine Learning Applications to Dust Storms: A Meta-Analysis. *Aerosol and Air Quality Research*, *22*(12), 220183. doi:10.4209/aaqr.220183

Alsumaiei, A. A., & Alrashidi, M. S. (2020). Hydrometeorological drought forecasting in hyper-arid climates using nonlinear autoregressive neural networks. *Water (Basel)*, *12*(9), 2611. doi:10.3390/w12092611

Amani, M., Ghorbanian, A., Ahmadi, S. A., Kakooei, M., Moghimi, A., Mirmazloumi, S. M., Moghaddam, S. H. A., Mahdavi, S., Ghahremanloo, M., Parsian, S., Wu, Q., & Brisco, B. (2020). Google earth engine cloud computing platform for remote sensing big data applications: A comprehensive review. *IEEE Journal of Selected Topics in Applied Earth Observations and Remote Sensing*, *13*, 5326–5350. doi:10.1109/JSTARS.2020.3021052

Bajracharya, S. R., Khanal, N. R., Nepal, P., Rai, S. K., Ghimire, P. K., & Pradhan, N. S. (2021). Community assessment of flood risks and early warning system in Ratu Watershed, Koshi Basin, Nepal. *Sustainability (Basel)*, *13*(6), 3577. doi:10.3390u13063577

Benz, S. A., & Blum, P. (2019). Global detection of rainfall-triggered landslide clusters. *Natural Hazards and Earth System Sciences*, *19*(7), 1433–1444. doi:10.5194/nhess-19-1433-2019

Bradley, B. A., Jacob, R. W., Hermance, J. F., & Mustard, J. F. (2007). A curve fitting procedure to derive inter-annual phenologies from time series of noisy satellite NDVI data. *Remote Sensing of Environment*, *106*(2), 137–145. doi:10.1016/j.rse.2006.08.002

Brunner, M. I., Slater, L., Tallaksen, L. M., & Clark, M. (2021). Challenges in modeling and predicting floods and droughts: A review. *WIREs. Water*, *8*(3), e1520. doi:10.1002/wat2.1520

Cheema, M. J. M., & Bastiaanssen, W. G. (2012). Local calibration of remotely sensed rainfall from the TRMM satellite for different periods and spatial scales in the Indus Basin. *International Journal of Remote Sensing*, *33*(8), 2603–2627. doi:10.1080/01431161.2011.617397

de Lima Moraes, A. G., Bocardo-Delgado, E., Bowling, L. C., Daneshvar, F., Pinto, J., Watkins, A. H., & Cherkauer, K. A. (2020). Assessment of Arequipa's hydrometeorological monitoring infrastructure to support water management decisions. *Journal of Contemporary Water Research & Education*, *171*(1), 27–48. doi:10.1111/j.1936-704X.2020.3343.x PMID:34790284

DeVries, B., Huang, C., Armston, J., Huang, W., Jones, J. W., & Lang, M. W. (2020). Rapid and robust monitoring of flood events using Sentinel-1 and Landsat data on the Google Earth Engine. *Remote Sensing of Environment*, *240*, 111664. doi:10.1016/j.rse.2020.111664

Duan, W., Maskey, S., Chaffe, P. L., Luo, P., He, B., Wu, Y., & Hou, J. (2021). Recent advancement in remote sensing technology for hydrology analysis and water resources management. *Remote Sensing (Basel)*, *13*(6), 1097. doi:10.3390/rs13061097

Erickson, T. B., Brooks, J., Nilles, E. J., Pham, P. N., & Vinck, P. (2019). Environmental health effects attributed to toxic and infectious agents following hurricanes, cyclones, flash floods and major hydrometeorological events. *Journal of Toxicology and Environmental Health. Part B, Critical Reviews*, *22*(5-6), 157–171. doi:10.1080/10937404.2019.1654422 PMID:31437111

Farhadi, H., & Najafzadeh, M. (2021). Flood risk mapping by remote sensing data and random forest technique. *Water (Basel)*, *13*(21), 3115. doi:10.3390/w13213115

Fu, B., Lan, F., Yao, H., Qin, J., He, H., Liu, L., Huang, L., Fan, D., & Gao, E. (2022). Spatio-temporal monitoring of marsh vegetation phenology and its response to hydro-meteorological factors using CCDC algorithm with optical and SAR images: In case of Honghe National Nature Reserve, China. *The Science of the Total Environment*, *843*, 156990. doi:10.1016/j.scitotenv.2022.156990 PMID:35764147

Galkate, R. V., Jain, S., Jaiswal, R. K., Pandey, R. P., Lohani, A. K., Yadav, S., & Yadava, R. N. (2023). Application of GIS and Remote Sensing Tools in Assessment of Drought Using Satellite and Ground-Based Data. In *Application of Remote Sensing and GIS in Natural Resources and Built Infrastructure Management* (pp. 105–121). Springer International Publishing.

Glago, F. J. (2021). Flood disaster hazards; causes, impacts and management: a state-of-the-art review. *Natural hazards-impacts, adjustments and resilience*, 29-37.

Goyal, M. K., Gupta, A. K., & Gupta, A. (Eds.). (2022). *Hydro-Meteorological Extremes and Disasters*. Springer Nature. doi:10.1007/978-981-19-0725-8

Hamidi, E., Peter, B. G., Muñoz, D. F., Moftakhari, H., & Moradkhani, H. (2023). Fast Flood Extent Monitoring With SAR Change Detection Using Google Earth Engine. *IEEE Transactions on Geoscience and Remote Sensing*, *61*, 1–19. doi:10.1109/TGRS.2023.3240097

Hazaymeh, K., & Hassan, Q. K. (2015). Fusion of MODIS and Landsat-8 surface temperature images: A new approach. *PLoS One*, *10*(3), e0117755. doi:10.1371/journal.pone.0117755 PMID:25730279

Hazaymeh, K., & Hassan, Q. K. (2016). Remote sensing of agricultural drought monitoring: A state of art review. *AIMS Environmental Science*, *3*(4), 604–630. doi:10.3934/environsci.2016.4.604

Hosseini, F. S., Choubin, B., Mosavi, A., Nabipour, N., Shamshirband, S., Darabi, H., & Haghighi, A. T. (2020). Flash-flood hazard assessment using ensembles and Bayesian-based machine learning models: Application of the simulated annealing feature selection method. *The Science of the Total Environment*, *711*, 135161. doi:10.1016/j.scitotenv.2019.135161 PMID:31818576

Kakooei, M., & Baleghi, Y. (2020). A two-level fusion for building irregularity detection in post-disaster VHR oblique images. *Earth Science Informatics*, *13*(2), 459–477. doi:10.100712145-020-00449-6

Kakooei, M., Ghorbanian, A., Baleghi, Y., Amani, M., & Nascetti, A. (2022). Remote sensing technology for post-disaster building damage assessment. In *Computers in Earth and Environmental Sciences* (pp. 509–521). Elsevier. doi:10.1016/B978-0-323-89861-4.00047-6

Khan, R., & Gilani, H. (2021a). Global drought monitoring with drought severity index (DSI) using Google Earth Engine. *Theoretical and Applied Climatology*, *146*(1-2), 411–427. doi:10.100700704-021-03715-9

Khan, R., & Gilani, H. (2021b). Global drought monitoring with big geospatial datasets using Google Earth Engine. *Environmental Science and Pollution Research International*, *28*(14), 17244–17264. doi:10.100711356-020-12023-0 PMID:33394397

Kumar, R., Devrani, R., Dangwal, A., Deshmukh, B., & Dutt, S. (2022). Extreme Hydrological Event-Induced Temporal Variation in Soil Erosion of the Assiganga River Basin, NW Himalaya. *Advances in Remote Sensing Technology and the Three Poles*, 230-246.

Lee, T., Singh, V. P., & Cho, K. H. (2021). *Deep Learning for Hydrometeorology and Environmental Science*. Springer. doi:10.1007/978-3-030-64777-3

Llasat-Botija, M., & Llasat, M. C. (2022). 6. FLOODUP, a citizen science project to increase flood risk awareness and collective knowledge. *Information Technologies and Social Media: New Scientific Methods for the Anthropocene*, 106.

Lopez, T., Al Bitar, A., Biancamaria, S., Güntner, A., & Jäggi, A. (2020). On the use of satellite remote sensing to detect floods and droughts at large scales. *Surveys in Geophysics*, *41*(6), 1461–1487. doi:10.100710712-020-09618-0

Matthies-Wiesler, F., & Quevauviller, P. (Eds.). (2022). *Hydrometeorological Extreme Events and Public Health*. John Wiley & Sons. doi:10.1002/9781119259350

Oyedotun, T. D. T., Moonsammy, S., Renn-Moonsammy, J., Renn-Moonsammy, D. M., Nedd, G. A., Bishop, J., & Burningham, H. (2022). Data consolidation for coastal management: The development of the Guyana coastal database platform. *Interdisciplinary Environmental Review*, *22*(2), 143–162. doi:10.1504/IER.2022.125019

Pandey, A. C., Kaushik, K., & Parida, B. R. (2022). Google Earth Engine for large-scale flood mapping using SAR data and impact assessment on agriculture and population of Ganga-Brahmaputra basin. *Sustainability (Basel)*, *14*(7), 4210. doi:10.3390u14074210

Perera, D., Seidou, O., Agnihotri, J., Mehmood, H., & Rasmy, M. (2020). Challenges and technical advances in flood early warning systems (FEWSs). *Flood impact mitigation and resilience enhancement*.

Petropoulos, G. P., & Islam, T. (Eds.). (2017). *Remote sensing of hydrometeorological hazards*. CRC Press. doi:10.1201/9781315154947

Potter, S., Harrison, S., & Kreft, P. (2021). The benefits and challenges of implementing impact-based severe weather warning systems: Perspectives of weather, flood, and emergency management personnel. *Weather, Climate, and Society*, *13*(2), 303–314. doi:10.1175/WCAS-D-20-0110.1

Pradhan, B., Moneir, A. A. A., & Jena, R. (2018). Sand dune risk assessment in Sabha region, Libya using Landsat 8, MODIS, and Google Earth Engine images. *Geomatics, Natural Hazards & Risk*, *9*(1), 1280–1305. doi:10.1080/19475705.2018.1518880

Rao, P., Gupta, K., Roy, A., & Balan, R. (2021). Spatio-temporal analysis of land surface temperature for identification of heat wave risk and vulnerability hotspots in Indo-Gangetic Plains of India. *Theoretical and Applied Climatology*, *146*(1-2), 567–582. doi:10.100700704-021-03756-0

Roy, D. P., Wulder, M. A., Loveland, T. R., Woodcock, C. E., Allen, R. G., Anderson, M. C., & Zhu, Z. (2014). Landsat-8: Science and product vision for terrestrial global change research. *Remote Sensing of Environment*, *145*, 154–172. doi:10.1016/j.rse.2014.02.001

Saleh, A., Salama, M. S., Elhag, M., & Elagib, N. (2020). Assessing the accuracy of machine learning algorithms in mapping flood inundation using Sentinel-1 SAR imagery and Google Earth Engine platform. *Water (Basel)*, *12*(5), 1405. doi:10.3390/w12051405

Sarker, M. N. I., Peng, Y., Yiran, C., & Shouse, R. C. (2020). Disaster resilience through big data: Way to environmental sustainability. *International Journal of Disaster Risk Reduction*, *51*, 101769. doi:10.1016/j.ijdrr.2020.101769

Scheip, C. M., & Wegmann, K. W. (2021). HazMapper: A global open-source natural hazard mapping application in Google Earth Engine. *Natural Hazards and Earth System Sciences*, *21*(5), 1495–1511. doi:10.5194/nhess-21-1495-2021

Schumann, G., Giustarini, L., Tarpanelli, A., Jarihani, B., & Martinis, S. (2022). Flood Modeling and Prediction Using Earth Observation Data. *Surveys in Geophysics*, 1–26.

Selvarajah, H., Koike, T., Rasmy, M., Tamakawa, K., Yamamoto, A., Kitsuregawa, M., & Zhou, L. (2021). Development of an integrated approach for the assessment of climate change impacts on the hydro-meteorological characteristics of the Mahaweli river basin, Sri Lanka. *Water (Basel)*, *13*(9), 1218. doi:10.3390/w13091218

Sharma, K., Anand, D., Sabharwal, M., Tiwari, P. K., Cheikhrouhou, O., & Frikha, T. (2021). A disaster management framework using internet of things-based interconnected devices. *Mathematical Problems in Engineering*, *2021*, 1–21. doi:10.1155/2021/9916440

Shew, A. M., & Ghosh, A. (2017). USING MULTI-TEMPORAL REMOTE SENSING DATA TO ANALYZE THE SPATIO-TEMPORAL PATTERNS OF DRY SEASON RICE PRODUCTION IN BANGLADESH. *ISPRS Annals of the Photogrammetry, Remote Sensing and Spatial Information Sciences*, *IV-4*(W2), 4. doi:10.5194/isprs-annals-IV-4-W2-61-2017

Subiyantoro, A., Van Westen, C. J., Den Bout, B. V., Yuniawan, R. A., & Mulyana, A. R. (2022, November). Semi-automatic Landslide Detection Using Google Earth Engine, a Case Study in Poi Village, Central Sulawesi. In *2022 IEEE International Conference on Aerospace Electronics and Remote Sensing Technology (ICARES)* (pp. 1-4). IEEE. 10.1109/ICARES56907.2022.9993507

Tamiminia, H., Salehi, B., Mahdianpari, M., Quackenbush, L., Adeli, S., & Brisco, B. (2020). Google Earth Engine for geo-big data applications: A meta-analysis and systematic review. *ISPRS Journal of Photogrammetry and Remote Sensing*, *164*, 152–170. doi:10.1016/j.isprsjprs.2020.04.001

Tayebi, A., Kasmaeeyazdi, S., Tinti, F., & Bruno, R. (2023). Contributions from experimental geostatistical analyses for solving the cloud-cover problem in remote sensing data. *International Journal of Applied Earth Observation and Geoinformation*, *118*, 103236. doi:10.1016/j.jag.2023.103236

Tellman, B., Sullivan, J., Kettner, A., Brakenridge, G. R., Slayback, D. A., Kuhn, C., & Doyle, C. (2016, December). Developing a global database of historic flood events to support machine learning flood prediction in Google Earth Engine. In AGU Fall Meeting Abstracts (Vol. 2016, pp. H13Q-06). Research Gate.

Temimi, M., Leconte, R., Chaouch, N., Sukumal, P., Khanbilvardi, R., & Brissette, F. (2010). A combination of remote sensing data and topographic attributes for the spatial and temporal monitoring of soil wetness. *Journal of Hydrology (Amsterdam)*, *388*(1-2), 28–40. doi:10.1016/j.jhydrol.2010.04.021

Teodoro, A. C., & Duarte, L. (2022). The synergy of remote sensing and geographical information systems in the management of natural disasters. In *Nanotechnology-Based Smart Remote Sensing Networks for Disaster Prevention* (pp. 217–230). Elsevier. doi:10.1016/B978-0-323-91166-5.00023-9

Trambadia, N. K., Patel, D. P., Patel, V. M., & Gundalia, M. J. (2023). Flood inundation assessment in the data-scarce region using hydrodynamic models and google earth engine-A case of Ghed region, Ozat River basin, Gujarat, India. Research Gate.

Tripathy, P., & Malladi, T. (2022). Global Flood Mapper: A novel Google Earth Engine application for rapid flood mapping using Sentinel-1 SAR. *Natural Hazards*, *114*(2), 1341–1363. doi:10.100711069-022-05428-2

Tylkowski, J., & Hojan, M. (2019). Time decomposition and short-term forecasting of hydrometeorological conditions in the south Baltic Coastal zone of Poland. *Geosciences*, *9*(2), 68. doi:10.3390/geosciences9020068

Valipour, M., Bateni, S. M., Dalezios, N. R., Almazroui, M., Heggy, E., Şen, Z., & Angelakis, A. N. (2020). Hydrometeorology: Review of Past, Present and Future Observation Methods. *Hydrology*.

Vanama, V. S. K., Mandal, D., & Rao, Y. S. (2020). GEE4FLOOD: Rapid mapping of flood areas using temporal Sentinel-1 SAR images with Google Earth Engine cloud platform. *Journal of Applied Remote Sensing*, *14*(3), 034505–034505. doi:10.1117/1.JRS.14.034505

Veneros, J., & García, L. (2021). Application of the Standardized Vegetation Index (SVI) and Google Earth Engine (GEE) For Drought Management In Peru. *Tropical and Subtropical Agroecosystems*, *25*(1). Advance online publication. doi:10.56369/tsaes.3784

Wang, W., Samat, A., Abuduwaili, J., De Maeyer, P., & Van de Voorde, T. (2023). Machine learning-based prediction of sand and dust storm sources in arid Central Asia. *International Journal of Digital Earth*, *16*(1), 1530–1550. doi:10.1080/17538947.2023.2202421

Wang, W., Samat, A., Abuduwaili, J., Ge, Y., De Maeyer, P., & Van de Voorde, T. (2022). A novel hybrid sand and dust storm detection method using MODIS data on GEE platform. *European Journal of Remote Sensing*, *55*(1), 420–428. doi:10.1080/22797254.2022.2093278

Wang, Y., Zhou, X., Ao, Z., Xiao, K., Yan, C., & Xin, Q. (2022). Gap-Filling and Missing Information Recovery for Time Series of MODIS Data Using Deep Learning-Based Methods. *Remote Sensing (Basel)*, *14*(19), 4692. doi:10.3390/rs14194692

Weilnhammer, V., Schmid, J., Mittermeier, I., Schreiber, F., Jiang, L., Pastuhovic, V., Herr, C., & Heinze, S. (2021). Extreme weather events in europe and their health consequences–A systematic review. *International Journal of Hygiene and Environmental Health*, *233*, 113688. doi:10.1016/j.ijheh.2021.113688 PMID:33530011

Yousefi, E., Sayadi, M. H., & Chamenhpour, E. (2022). Google Earth Engine platform to calculate the hydrometeorology and hydrological water balance of wetlands in arid areas and predict future changes. *Journal of Applied Research in Water and Wastewater*, *9*(1), 52–68.

Zhang, M., Yang, X., Cleverly, J., Huete, A., Zhang, H., & Yu, Q. (2022). Heat wave tracker: A multi-method, multi-source heat wave measurement toolkit based on Google Earth Engine. *Environmental Modelling & Software*, *147*, 105255. doi:10.1016/j.envsoft.2021.105255

Zhang, X., Shi, P., Liu, X., Liu, J., & Lu, Y. (2020). Analyzing the spatiotemporal pattern of extreme precipitation using Google Earth Engine: A case study in China. *Journal of Hydrology (Amsterdam)*, *590*, 125384. doi:10.1016/j.jhydrol.2020.125384

Chapter 7
Application of Machine Learning Techniques in Hydrometeorological Event Prediction

Anna Msigwa
https://orcid.org/0000-0002-4782-9145
The Nelson Mandela African Institution of Science and Technology, Tanzania & The University of Pretoria, South Africa

Ayodeji Samuel Makinde
https://orcid.org/0000-0002-6482-2949
Edo State University, Uzairue, Nigeria

ABSTRACT

Hydrometeorological events, such as floods and droughts, pose significant challenges to societies worldwide, causing loss of life and economic damage. Traditional methods of predicting such events often rely on statistical and physical models that are limited by their assumptions, uncertainties, and computational requirements. Machine learning (ML) techniques, with their ability to extract knowledge and insights from data, have shown great potential for improving the accuracy and lead time of hydrometeorological event prediction. This chapter reviews the use of ML for predicting hydrometeorological events focusing on flood and drought events. The chapter provides an overview of the application of ML techniques or algorithms for hydrometeorological events prediction. The chapter discusses data type, collection, and analysis for ML applications for predicting hydrometeorological events. The chapter presents case studies from different regions and highlights the benefits of ML-based approaches and the challenges. Finally, the chapter identifies future research directions.

DOI: 10.4018/978-1-6684-8771-6.ch007

INTRODUCTION

Hydrometeorological events are natural phenomena due to the complex interplay between atmospheric, hydrological, and meteorological processes. These events can profoundly and often devastate human lives, economies, and the environment. Among the most common types of hydrometeorological events are floods, droughts, storms, landslides, and heat waves, each with unique characteristics and consequences.

Floods, one of the most destructive hydrometeorological events, result from heavy rainfall, snowmelt, or a combination of both, leading to an overflow of water in rivers, streams, and other water bodies. The inundation of land causes widespread damage to property, infrastructure, and agricultural fields and poses a significant threat to human safety. The destruction caused by floods often requires extensive recovery and rehabilitation efforts (Morán-Tejeda et al., 2019; Quesada-Román & Villalobos-Chacón, 2020).

Conversely, **drought** occurs when a region experiences an extended period of inadequate precipitation, resulting in a water deficit. These dry spells can have severe economic and social impacts, particularly in agriculture-dependent regions. Crop failure, water scarcity, and food insecurity are expected consequences of prolonged droughts, which can trigger economic losses and social upheavals (Haile et al., 2020).

Storms encompass a variety of intense weather phenomena, such as hurricanes, tornadoes, and thunderstorms. Strong winds, heavy rainfall, lightning, and thunder characterize these events. Storms can cause significant damage to infrastructure, homes, and natural ecosystems. Hurricanes, in particular, are notorious for their destructive potential, causing widespread devastation in coastal regions (Yang et al., 2023).

Landslides are another hazardous hydrometeorological event occurring when soil, rock, or other materials move down a slope. Heavy rainfall, earthquakes, or human activities like construction or mining can trigger landslides. The impact of landslides can be catastrophic, resulting in the destruction of homes, roads, and other infrastructure and posing a severe risk to human life (Lacroix et al., 2020).

Heatwaves are prolonged periods of excessively high temperatures and humidity. These extreme weather events can have severe health implications, particularly for vulnerable populations such as the elderly, young children, and individuals with pre-existing medical conditions. Heat-related illnesses and deaths can significantly increase during heat waves, making them a pressing public health concern (Zeppetello et al., 2022).

This chapter primarily focuses on predicting drought and flood events using machine learning. Understanding and predicting drought and flood events are crucial for effective disaster preparedness, risk mitigation, and climate adaptation. These events can have far-reaching consequences, and their increasing frequency and intensity due to climate change necessitate advanced prediction and early warning systems. The application of machine learning techniques in hydrometeorological event prediction holds immense promise in improving the accuracy and effectiveness of forecasting models. By harnessing the power of data-driven approaches and innovative algorithms, we can enhance our ability to forecast floods, and drought events, empowering communities to take proactive measures and build resilience against these natural disasters.

Overview of Machine Learning and Its Potential for Hydrometeorological Event Prediction

In recent years, machine learning techniques have emerged as a powerful and promising approach to enhance the prediction of drought and flood events. The Role of Machine learning, a subfield of artificial intelligence, focuses on developing algorithms that enable computers to learn patterns and relationships from data and make predictions or decisions without explicit programming. Applying machine learning in hydrometeorological event prediction offers several advantages over traditional modeling approaches.

One key advantage is the data-driven approach of machine learning techniques. These models leverage historical hydrometeorological data, including rainfall patterns, river discharge, soil moisture, and other relevant variables, to build predictive models. By analyzing large datasets and learning from past events, machine learning models can identify complex patterns and non-linear relationships that may not be apparent through conventional methods. Another significant benefit is the enhanced prediction accuracy achieved by machine learning models. Their ability to capture intricate patterns and relationships in data leads to more accurate predictions of hydrometeorological events. Improved prediction accuracy can enable early warning systems and timely interventions, which are crucial in mitigating the impact of disasters and supporting effective decision-making (Vaseashta, 2022).

Machine learning techniques are highly flexible and adaptable to different types of hydrometeorological data and geographic regions. They can be applied to various spatial and temporal scales, allowing for localized predictions and the development of region-specific models. This adaptability is especially valuable in addressing the unique challenges of different climatic and geographical conditions.

Current research in machine learning for hydrometeorological event prediction is witnessing significant progress. Researchers have been actively exploring the application of various machine learning algorithms, such as Support Vector Machines (SVM), Random Forests, Artificial Neural Networks (ANN), and Long Short-Term Memory (LSTM), to develop robust prediction models.

For flood prediction, machine learning models have shown promise in accurately forecasting flood events by analyzing rainfall data, river gauge readings, topographic features, and historical flood records. These models can aid in flood risk assessment, flood mapping, and early warning systems, enabling communities to prepare and respond proactively to flood events (Bayat & Tavakkoli, 2020; Mahdizadeh Gharakhanlou & Perez, 2022; Razali et al., 2020; Tayfur et al., 2018).

Similarly, machine learning techniques have been employed for drought prediction to analyze meteorological variables, satellite data, and hydrological indicators. These models can help in identifying drought-prone regions, assessing agricultural impacts, and implementing drought management strategies to mitigate water scarcity challenges (Chen et al., 2023; Kafy et al., 2023; Khan et al., 2020; Mohamadi et al., 2020; Piri et al., 2023). For example, the study conducted by Hassan and Azelan (2019) focuses on flood prediction using machine learning algorithms, specifically comparing the performance of Artificial Neural Network (ANN), Support Vector Machine (SVM), and Decision Tree (DT). The research emphasizes the significance of data-driven models like machine learning in flood prediction, as they offer faster analysis and high accuracy based on historical data.

In a similar vein, Herath et al. (2023) conducted a study that compared two types of neural network models, Feedforward Neural Network (FFNN) and Long Short-Term Memory (LSTM), for predicting water levels in various wetlands. The research demonstrated that LSTM outperformed FFNN regarding predictive accuracy, suggesting the importance of capturing temporal relationships in daily water levels. The results of this study shed light on the potential of LSTM in accurately predicting average and extreme water levels in wetlands, which can aid in effective water level management and resource planning. Khan et al. (2020) compared the performance of the k-Nearest Neighbours (KNN) algorithm, one of the oldest machine learning algorithms, with the latest machine learning algorithms (Support Vector Machine - SVM and Artificial Neural Network - ANN) in developing drought prediction models. The results showed that KNN-based drought models had limited performance compared to SVM and ANN-based models during validation. SVM-based models outperformed ANN-based models in capturing the temporal and spatial characteristics of droughts in Pakistan, consistent with previous studies. However, all ML-based models showed limited ability to capture extreme droughts during the Rabi season, as documented in other research. The superior generalization skills of SVM, ensuring a global optimum solution, might contribute to its better performance than ANN, which can be trapped at local optima.

Regarding model predictor selection, the study identified important indicators of droughts during both the Rabi and Kharif seasons. Relative humidity, atmospheric temperature, and wind speed were significant indicators of droughts in Pakistan. The study recommends considering these indicators in developing drought prediction models for the region that encompasses the Mediterranean Sea, the region north of the Caspian Sea, the Indian Ocean, and the Arabian Sea. Acharki et al. (2023) examined agricultural drought using three indices (VCI, TCI, and VHI) based on NDVI and LST data. The results showed regions affected by vegetation stress due to water availability deficits, and higher temperatures in some areas indicated dry conditions. Moderate drought was reported in certain locations during specific years, but extreme and severe droughts were not observed throughout the studied period. The research also explored using machine learning algorithms (RSS, RF, M5P, and REPTree) to predict SPI on two-time scales. The REPTree model demonstrated better predictive performance than other models and could be applied for drought forecasting and management in Mediterranean regions.

The study presented here provides valuable insights into historical drought patterns in the Loukkos basin and offers a practical approach for drought prediction and management using machine learning algorithms. Further research with alternative ensembles or hybrid ML algorithms is proposed to improve prediction capabilities in the future. Figure 1 shows the general overview of the conceptual schematic of the machine learning process for hydrometeorological event prediction.

Figure 1. Conceptual schematic of ML process for hydrometeorological event prediction

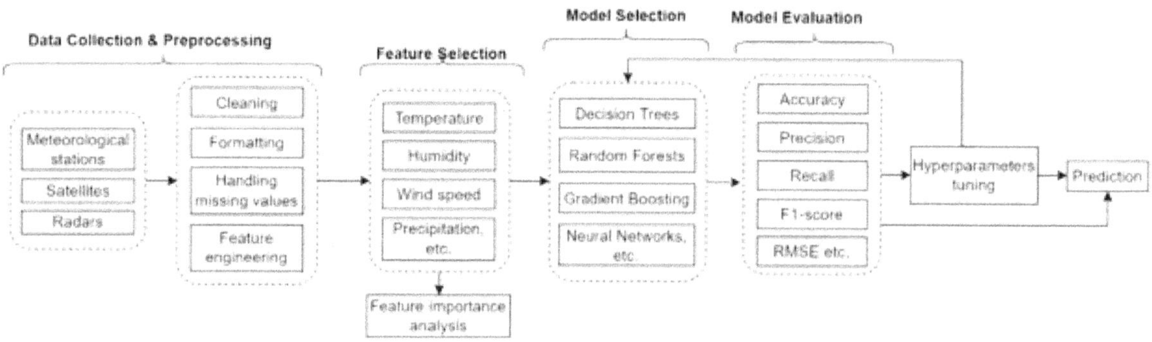

DATA COLLECTION AND PREPROCESSING

Types of Data Used for Prediction

Machine learning models for hydrometeorological events utilize various data types to make accurate predictions. When creating machine learning (ML) prediction models for hydrometeorological events, historical hydrometeorological events records such as flood, and ground-based weather stations that collect data on temperature, humidity, wind speed, and other atmospheric conditions at specific locations and real-time cumulative data from rain gauges or other sensing devices are commonly used (Seo & Breidenbach, 2002). The dataset sources typically include rainfall and water level measurements from ground rain gauges or remote-sensing technologies such as satellites, multisensor systems, and radars. Remote sensing, especially weather radar observations, provides higher-resolution and more reliable data than rain gauges (Grecu & Krajewski,2000). Hence, using radar-based datasets for building prediction models has been shown to offer higher accuracy (Maddox et al., 2002). Regardless of whether radar-based or ground gauge data is used, the historical dataset is divided into individual sets for training, validation, verification, and testing of the learning models.

Challenges in Data Collection and Preprocessing

Collecting and preprocessing hydrometeorological data presents several challenges to researchers and practitioners. One significant challenge is the availability of real-time data, especially for sources like ground station datasets. The ground station data are typically space with a lot of missing data. However, also, satellite data come with difficulties in accessing as they may require advanced processing techniques. Additionally, the quality of data from different sources can vary, leading to meticulous data cleaning and validation to ensure accurate and reliable predictions (Kia et al., 2012).

Techniques for Cleaning and Filtering Data

Several techniques are employed in the data cleaning and filtering process to ensure the quality and reliability of data for ML prediction. Outlier detection helps identify and handle data points that deviate significantly from most of the data, reducing the impact of erroneous readings on the model's performance. Noise reduction techniques filter out unwanted variations in the data, improving the signal-to-noise ratio and enhancing the model's ability to discern meaningful patterns. Smoothing methods are applied to remove short-term fluctuations and focus on the long-term trends within the data, providing a more robust foundation for the predictive models.

Strategies for Dealing With Missing Data

Dealing with missing data is crucial to ensure the accuracy of the ML prediction model is maintained. Imputation is a common strategy involving statistical techniques to fill in missing values based on the available data. Methods like mean, median, or interpolation can be applied to replace missing values, maintaining the integrity of the dataset. In cases where the amount of missing data is relatively small, deletion of the affected samples might be an option. For time-series data, forecasting models can be employed to predict missing values and maintain temporal continuity in the dataset. By implementing these strategies, researchers can handle missing data effectively and produce reliable predictions in hydrometeorological event forecasting (Li et al., 2015).

FEATURE SELECTION

Overview of Feature Selection Techniques

Feature selection is choosing relevant and important predictors (also known as features or variables) from the available data to build a predictive model. The aim is to reduce the dimensionality of the data, eliminate irrelevant or redundant information, and improve model performance and interpretability.

Nearest-Sample Choosing (NSC) in Hydrometeorological Event Prediction

In the domain of hydrometeorological event prediction, particularly in the context of heavy precipitation events, such events are often rare compared to non-extreme precipitation clusters (NEPCs). This class imbalance in the dataset can lead to biased results in predictive modeling, with traditional classifiers favoring the majority class (NEPCs) and performing poorly in predicting the rare events of interest (heavy precipitation events). To address this issue, Nearest-Sample Choosing (NSC) is introduced as a feature selection technique with a focus on building a more balanced and informative training dataset for classification. NSC leverages data mining algorithms to identify NEPCs closest to the decision boundary between the two classes. These selected NEPCs are crucial as they contain essential information about the regions where the rare heavy precipitation events and NEPCs intersect.

By including the most relevant NEPCs close to the boundary, NSC aims to provide a more representative training set that captures the complexities and variabilities of the data, leading to improved predictive performance for rare events. Moreover, NSC does not require additional data labeling or synthetic data generation, making it practical and cost-effective. In a hydrometeorological context, NSC can be implemented as follows: First, clustering techniques or density-based algorithms are applied to identify NEPCs in the dataset. Next, proximity measures are used to determine the NEPCs that are nearest to the boundary. The samples selected through NSC are then incorporated into the training dataset for classification models, helping to mitigate the class imbalance problem and enhancing the overall accuracy and robustness of hydrometeorological event prediction.

Online Fast Streaming Feature Selection (Fast-OSFS) in Hydrometeorological Event Prediction

Hydrometeorological datasets often present challenges in the form of high dimensionality, where a large number of features are available for modeling. This can lead to increased computational complexity and potential overfitting, hampering the accuracy and efficiency of predictive models. In the context of hydrometeorological event prediction, Online Fast Streaming Feature Selection (Fast-OSFS) is introduced as a powerful technique to tackle the high-dimensional data problem. Fast-OSFS operates in a streaming manner, meaning it processes features one by one as they flow in over time. This approach significantly reduces computational overhead and memory requirements, making it well-suited for handling vast and continuously evolving datasets.

Fast-OSFS aims to identify the most relevant and non-redundant features in the data that contribute significantly to the classification task, such as distinguishing between different types of precipitation events. The algorithm adopts a Bayesian-based approach to model the dependencies and relationships between features and the target variable, allowing it to make informed decisions about feature relevance and redundancy. In hydrometeorological event prediction, Fast-OSFS is particularly beneficial because it allows the algorithm to adapt dynamically to changes in the data distribution, a common occurrence in weather-related datasets. As precipitation patterns and climatic conditions shift, new relevant features may emerge, and irrelevant ones may become obsolete. By processing features in a streaming manner, Fast-OSFS can continuously update its feature subset, ensuring that the model remains up-to-date and accurate in predicting extreme weather events.

Discriminative Feature Selection (DFS) with Discriminative Locality Alignment (DLA) in Hydrometeorological Event Prediction

In hydrometeorological event prediction, identifying and selecting relevant features from high-dimensional data plays a crucial role in building accurate and interpretable predictive models. Discriminative Feature Selection (DFS) is a technique designed to address this challenge by identifying a subset of discriminative features that have the most significant impact on distinguishing between different types of precipitation events. Traditional linear dimension reduction techniques like Principal Component Analysis (PCA) and Fisher's Linear Discriminant Analysis (LDA) are not always suitable for hydrometeorological datasets. These methods may assume data distributions, such as Gaussian distributions in the case of PCA, that may not hold true for weather-related variables. Additionally, they may overlook local relationships between data points, which are essential in understanding the underlying patterns of hydrometeorological events.

Discriminative Locality Alignment (DLA) is employed as the specific method for DFS in the context of hydrometeorological event prediction. DLA belongs to the family of subspace learning algorithms, which consider both local and global characteristics of the data. In hydrometeorology, local learning is essential, as each data point's impact on its neighboring points can significantly influence weather patterns. DLA aims to create a low-dimensional feature representation that preserves the discriminative information while aligning the local neighborhoods of data points in the transformed space. By doing so, DLA ensures that samples from the same class are grouped together, enhancing class separability and capturing the intrinsic structure of the data.

DFS with DLA can help researchers and meteorologists uncover the critical variables and relationships that govern extreme weather events in hydrometeorological event prediction. By reducing the data's dimensionality and focusing on discriminative features, DFS with DLA enables more accurate and interpretable predictive models, leading to better-informed decision-making in weather-related applications.

Physical Element-Based Feature Selection

Zhang & Ye (2021) use physical element-based feature selection whereby predictors are selected based on physical elements related to weather phenomena. The three categories of physical elements mentioned are a.) Direct Outputs of Numerical Weather Predictions (NWPs): These are the raw outputs of weather models, such as precipitation predictions, temperature, and wind speed. b.) Diagnostic Physical Elements: These are calculated from essential factors and may include quantities like water vapor flux. c.) Geographical Attributes: These are features related to the location, such as latitude, longitude, and elevation.

Spatial Feature Selection

Spatial features are related to the geographical distribution of data. In the study conducted by Zhang & Ye (2021), all data within the watershed are used as spatial information to reduce modeling effort. However, this strategy may become less desirable as the study area increases in scale. Previous approaches involved building postprocessing models for each grid cell or site separately to ensure local accuracy. However, more recent studies have introduced a hyperparameter radius, where predictors within a specific radius are selected for each grid cell. The value of the hyperparameter radius is selected by adaptive algorithms and may vary depending on topography, climate type, and study area.

Time-Dimensional Feature Selection

Also, in the same study conducted by Zhang & Ye (2021), a forward rolling time window strategy is used, which is different from the traditional rolling time window strategy. The traditional method selects several periods before and after the forecast day as predictors to consider regional climatology information and increase the sample size. However, the correlation skill analysis suggests that forecasts with larger time windows may not provide much additional useful information. The forward rolling time window is more effective in capturing uncertainty in the weather state.

Case Studies Illustrating Successful Feature Selection Approaches

Huang et al. (2023) employed manual feature elimination techniques to enhance the performance and interpretability of their prediction model. Initially, a manual feature elimination process was conducted to eliminate irrelevant or redundant features, resulting in 116 remaining features. Subsequently, the Sequential Backward Selection (SBS) technique was utilized to refine the feature set further. During SBS, the researchers monitored the changes in the coefficient of determination (R2) with the number of features, aiming to identify the optimal subset that would maximize the model's performance. The final outcome of the feature selection process was the selection of the top 16 features along with precipitation, which were used as inputs for their Random Forest (RF) prediction model.

To gain insights into the behavior of the black-box RF model and to understand the relationships between predicted soil moisture (SM) and its influencing factors, an integrated Explainable Artificial Intelligence (XAI) method was employed (Huang et al., 2023). The XAI framework included several post-hoc interpretable methods, namely Mean Feature Importance (MFI), Sensitivity Value (SV), One-Dimensional Accumulated Local Effects (1D ALE), Individual Conditional Expectation (ICE), Derivative of Individual Conditional Expectation (DICE), and Two-Dimensional Accumulated Local Effects (2D ALE). These methods provided various perspectives on the model's behavior and highlighted the impact of specific features on SM predictions. Through MFI and SV, the researchers ranked features based on their importance and contribution to the model's precision. At the same time, 1D ALE allowed them to explore the average SM evolution with variations in individual features. Additionally, ICE and DICE enabled the visualization of the SM evolution for each instance with one feature variation, while 2D ALE provided insights into the interaction effects of two features on predicted SM. This comprehensive XAI tool played a crucial role in overcoming the black-box nature of AI models, enabling the researchers to interpret the RF model effectively and achieve a more intelligent application in hydrometeorological forecasting.

Di et al. (2015) compared the efficacy of the Nearest-Sample Choosing (NSC) method for feature selection with the Down-sampling (DS) method. Both methods aim to reduce the number of non-extreme precipitation clusters (NEPCs) in the training data set to form a balanced dataset, enabling accurate prediction of extreme precipitation clusters (EPCs). The study uses a dataset containing records of weather events from previous years as the training set and data from 2010 as the testing set. The DS method reduces the number of NEPCs using an integer factor of 11, resulting in a balanced training set. Meanwhile, the NSC method selects NEPCs that are closest to the decision boundary, thereby maintaining a balanced dataset while targeting the grey boundaries between EPCs and NEPCs. Results show that the NSC method outperforms the DS method in terms of recall. The NSC method achieves a recall of 39.47%, indicating that it correctly classifies a considerable proportion of EPCs, whereas the DS method has a null recall, misclassifying all EPCs as NEPCs. The superiority of the NSC method in capturing rare EPCs suggests that it effectively preserves critical boundary information, leading to more accurate and reliable predictions of extreme weather events in hydrometeorological forecasting.

Also, Di et al. (2015) considered using Fast Online Streaming Feature Selection (Fast-OSFS) and Discriminative Locality Alignment (DLA) algorithms on dimensionality reduction and feature selection. The dataset contains a large number of spatio-temporal features (466,650) for each sample, most of which are redundant and not strongly relevant to weather prediction in Iowa. The study employs Fast-OSFS to select the most relevant and non-redundant features. As a result, only 15 features are identified as highly influential in predicting EPCs in Iowa for the year 2010. Additionally, the Fast-OSFS algorithm prioritizes features with varying lead times, highlighting the most influential features at different time intervals. Next, the study utilizes the DLA algorithm for further feature selection and dimensionality reduction. DLA aligns the local neighborhoods of data points in the transformed feature space, ensuring that samples from the same class are grouped together while those from different classes are well separated. This step refines the feature space and removes additional redundancies. The combination of Fast-OSFS and DLA significantly reduces the feature space to only the most discriminative features, improving predictive accuracy and interpretability in hydrometeorological event prediction.

MACHINE LEARNING ALGORITHMS

Overview of Different Types of Machine Learning Algorithms

Long-term Recurrent Convolutional Network (LRCN)

Guastavino et al. (2022) proposed the application of the Long-term Recurrent Convolutional Network (LRCN), a machine learning algorithm designed for short-term extreme event prediction using radar image video classification. This approach combines the power of two deep learning architectures, namely the Convolutional Neural Network (CNN) and the Long Short-Term Memory (LSTM) network, creating an integrated spatio-temporal model. To feed the LRCN, the input consists of a sequence of ten radar reflectivity images representing a 1.5-hour video. These images are captured at three different levels: CAPPI 2, CAPPI 3, and CAPPI 5, and are resized to 128x256 pixels for computational efficiency. The CNN component is responsible for automatically extracting spatial features from this image set, which are then passed sequentially into the LSTM network for further analysis. The output of the LSTM layer is then fed into a fully connected layer with a sigmoid activation function, which generates the probability distribution for the positive class, enabling the prediction of extreme event occurrences.

The CNN architecture within the LRCN model is structured into three blocks, each featuring a convolutional layer with a stride of (2, 2), followed by a batch normalization layer for increased stability. These convolutional layers use ReLU as the activation function and are followed by max-pooling operations with size (4, 4) and stride (2, 2). The number of kernels in the convolutional layers is 8, 16, and 32, with corresponding sizes of (5, 5), (3, 3), and (3, 3). The input to the CNN is a sequence of size (T, 128, 256, 3), where T represents the number of frames in each video, and 128 and 256 indicate the image dimensions, while 3 denotes the three levels of CAPPI data. The utilization of the "TimeDistributed" layer in the Keras library allows for parallel training of the T convolutional flows. After obtaining the CNN output, the feature vectors are flattened and fed into the LSTM layer, which comprises 50 hidden neurons. To prevent overfitting, a dropout layer with a dropout rate of 0.5 is employed during training to randomly deactivate 50% of neurons in each iteration. The hyperparameters of the neural network are determined through an empirical trial-and-error optimization process in multiple experiments.

Multi-Phase Maximal Overlap Discrete Wavelet Transform (MODWT) Combined With the Multivariate Adaptive Regression Splines (MARS) Model

Ahmed et al. (2023) combined two machine learning models, Multi-Phase Maximal Overlap Discrete Wavelet Transform (MODWT) with the Multivariate Adaptive Regression Splines (MARS) model. The MODWT is utilized for wavelet transformation, and it involves identifying wavelet-scaling filter types and decomposition levels to create an effective wavelet transformation model. The NCA approach is employed for feature selection, and it filters the predictor variables before combining them with the wavelet transformation to create the MODWT-MARS model. The authors have employed a trial and error strategy to choose the optimal filter, as there is no one approach to make this decision.

The proposed MODWT-MARS model aims to forecast dissolved oxygen (DO) levels in a river system. To achieve this, the authors addressed issues related to the wavelet decomposition process, which can lead to inaccurate forecasting inputs. The study identified three problems: (1) improper use of future data, (2) unsuitable selection of decomposition levels and filters, and (3) incorrect division of validation and calibration data. Additionally, the study used various strategies for selecting the time series of lagged memories of DO and predictors, but the literature did not specify a definitive method. The autocorrelation function (ACF), partial autocorrelation function (PACF), and cross-correlation function (CCF) approaches were considered. The authors found a significant antecedent behavior in terms of the lag of DO from the predictors, with PACF being used as the predictor. The cross-correlation function was crucial in determining the statistical similarity between the predictors and the target variable (DO). It helped establish the significant input combinations by assessing the cross-correlation of each predictor with DO. The cross-correlation function was also used for the decomposed predictor variables. The predictor data sets were normalized between 0 and 1 to prevent the overestimation of one variable.

Exponential Smoothing-Long Short-Term Memory (ES-LSTM) Structure, Recurrent Neural Networks (RNNs), Artificial Neural Network (ANN) model and Decision Tree (DT) Algorithm

Hayder et al. (2023) employed machine learning algorithms to develop a comprehensive flood forecasting and prediction model aimed at minimizing damages and casualties caused by flood disasters. The main focus was on accurately predicting hourly precipitation patterns and classifying precipitation as rain or not rain. The machine learning algorithms used in this content are the Exponential Smoothing-Long Short-Term Memory (ES-LSTM) structure and Recurrent Neural Networks (RNNs) for forecasting and the Artificial Neural Network (ANN) model and Decision Tree (DT) algorithm for precipitation classification. The ES-LSTM and RNN models play a critical role in forecasting by leveraging historical precipitation data to predict future patterns. These models excel at capturing sequential dependencies and trends, making them suitable for time series forecasting. The ANN model and DT algorithm contribute to precipitation classification, providing a mechanism to distinguish between rain and non-rain events based on historical data.

The implementation of an early stopping mechanism and the use of the Adam optimization method showcase efforts to improve training efficiency and prevent overfitting. Including decision tree-based classification adds interpretability to the model's predictions, which can enhance its real-world applicability. In the final decision stage, the machine learning models contribute to the decision-making process by analyzing sensor data and historical patterns. This integration allows for informed responses, such as sending notifications to activate rain detector sensors based on predicted rain occurrence probabilities. The system's ability to adapt to different rain percentage thresholds and its power-saving measures demonstrate practical considerations in applying the developed model.

Artificial Neural Networks (ANNs)

Artificial Neural Networks (ANNs) are efficient mathematical models that imitate the structure and functioning of biological neural networks through interconnected neuron units. They have gained popularity as versatile and efficient learning algorithms for modeling complex flood processes with high fault tolerance and accurate approximation (Abbot and Marohasy, 2014; Kim et al., 2016). Compared to traditional statistical models, ANNs have shown greater accuracy in flood prediction (Li et al., 2010). Since their first use in the 1990s, ANNs have become the most popular method for flood prediction due to their reliance on historical data rather than catchment physical characteristics (Wu et al., 2010]). ANNs are reliable data-driven tools for constructing black-box models of complex and nonlinear relationships, including rainfall-flood and river flow-discharge forecasting (Sulaiman and Wahab, 2018; Kar et al., 2010). Despite their advantages, ANNs have some drawbacks related to network architecture, data handling, and the interpretability of the modeled system (Deo and Sahin,2015).

Multilayer Perceptron (MLP)

Multilayer Perceptron (MLP) is a class of ANNs characterized by its feed-forward structure, where information flows from input nodes through hidden layers to output nodes. MLP has gained popularity among hydrologists and flood modelers due to its nonlinear activation function and multiple layers (Kim and Singh, 2013). It has been found to be more efficient with better generalization ability than other ANN variations, although it may be more challenging to optimize (Senthil et., 2005).

Adaptive Neuro-Fuzzy Inference System (ANFIS)

The Adaptive Neuro-Fuzzy Inference System (ANFIS) is an advanced form of fuzzy-rule-based modeling that combines neural networks and fuzzy logic. ANFIS uses the Takagi-Sugeno fuzzy modeling technique and has proven reliable for complex systems, including flood modeling (Shu et al., 2008). By applying neural learning rules, ANFIS can identify and tune the parameters and structure of a fuzzy inference system, making it highly effective in flood modeling (Lafdani, et al., 2013).

Wavelet Neural Network (WNN)

Wavelet Neural Network (WNN) is based on wavelet transform, which is a mathematical tool used for data analysis. In flood modeling, discrete wavelet transform (DWT) is often employed to decompose time series data into different bands, improving data quality and prediction accuracy. WNN has gained popularity in flood modeling, especially for short-term predictions of streamflow, rainfall-runoff, water level, and flash floods (Shoaib et al., 2014; Shafaei et al., 2017; Kumar et al., 2015; Seo et al., 2015; Sudhishri et al., 2016).

Support Vector Machine (SVM)

Support Vector Machine (SVM) is a powerful supervised learning algorithm known for its robustness and efficiency in flood prediction. It works based on statistical learning theory and structural risk minimization, providing good generalization and superior efficiency compared to other methods (Dibike et al., 2001; Bao et al., 2014). SVMs have been applied successfully in various flood prediction applications, such as extreme rainfall, streamflow, and soil moisture (Bao et al., 2014; Nayak and Ghosh, 2013; Jajarmizadeh et al., 2015).

Decision Tree

Decision Tree is a widely used algorithm for predictive modeling in flood simulation. It uses a tree structure with branches representing decision rules and leaves representing target values. DTs are fast algorithms and are often used in ensemble forms for flood modeling (Tehrany et al., 2013). Random Forests (RF) is a popular DT method used for flood prediction (Wang et al., 2015), and the M5 decision-tree algorithm is another significant DT approach (Etemad-Shahidi and Mahjoobi, 2009).

Advantages and Disadvantages of Each Type of Algorithm for Hydrometeorological Event Prediction

Long-Term Recurrent Convolutional Network (LRCN) (Guastavino et al., 2022)

Advantages

The Long-term Recurrent Convolutional Network (LRCN) offers several advantages for spatio-temporal learning tasks. Firstly, it effectively captures both spatial and temporal patterns in radar image videos, making it well-suited for tasks involving the analysis of image sequences over time, such as predicting extreme events. Additionally, the combination of Convolutional Neural Network (CNN) and Long Short-Term Memory (LSTM) allows for automatic feature extraction from the images by the CNN, followed by the LSTM's analysis of temporal dependencies, reducing the need for manual feature engineering. LRCN has demonstrated promising results in various spatio-temporal tasks, even achieving state-of-the-art performance in certain cases. Furthermore, the use of the "TimeDistributed" layer enables parallel training of the CNN, significantly speeding up the training process for large sequences.

Disadvantages

Despite its strengths, the Long-term Recurrent Convolutional Network (LRCN) also comes with several disadvantages. One major concern is its computational complexity, particularly for long video sequences or large input image sizes. This complexity demands substantial computing resources and may lead to longer training times. Moreover, LRCN's performance can be sensitive to the choice of hyperparameters, making the optimization process more challenging and time-consuming. Overfitting is another issue to consider, as with many deep learning models. LRCN can be susceptible to overfitting, especially when working with limited datasets. While dropout and other regularization techniques can help alleviate this problem, careful tuning and monitoring are necessary. Additionally, the interpretability of LRCN poses challenges due to its complex architecture, often rendering it as a "black box." Understanding the decision-making process and the model's reasoning for predictions can be difficult, limiting its interpretability in certain scenarios.

Multi-Phase Maximal Overlap Discrete Wavelet Transform (MODWT) Combined With the Multivariate Adaptive Regression Splines (MARS) Model (Ahmed et al., 2023)

Advantages

The NCA approach used in this model helps in selecting relevant predictor variables, ensuring that only significant features are considered, leading to a more robust and interpretable model. MARS is a non-linear regression technique that can capture complex relationships between predictors and the target variable, making it suitable for modeling intricate and nonlinear patterns in the data. The MODWT allows for the decomposition of time-series data into different scales, which can help capture both low-frequency and high-frequency patterns in the data, improving the model's forecasting accuracy.

Disadvantages

The trial and error strategy used for selecting the optimal filter in the MODWT can be time-consuming and may not always guarantee the best filter choice, potentially affecting the model's performance. The accuracy of the wavelet-based forecasting model can be affected by the selection of boundary conditions during the wavelet decomposition process. This sensitivity might introduce errors in the predictions, particularly at the edges of the time series. Although the study considers various approaches for input selection, there is no standard method mentioned in the literature, leaving room for ambiguity in choosing the most appropriate approach for a given dataset. The use of wavelet transformation and non-linear regression techniques can increase the computational complexity of the model, especially for large datasets, potentially requiring more computational resources and time for model training and evaluation.

Exponential Smoothing-Long Short-Term Memory (ES-LSTM) Structure, Recurrent Neural Networks (RNNs), Artificial Neural Network (ANN) Model, and Decision Tree (DT) Algorithm (Hayder et al., 2023)

Advantages

The advantages of the machine learning algorithms used in this study include their ability to handle complex and non-linear relationships in the data, making them suitable for modeling and predicting natural phenomena like rainfall. The ES-LSTM and RNN models are well-suited for time series prediction tasks, allowing them to capture temporal patterns and dependencies in precipitation data. The ANN model and DT algorithm are effective tools for classification tasks, enabling the accurate categorization of precipitation as rain or not rain.

Disadvantages

However, these machine learning algorithms also come with certain disadvantages. The complexity of the models, especially the ES-LSTM and RNN, may lead to longer training times and increased computational resources. Additionally, overfitting is a concern, where the models may perform well on the training data but fail to generalize to new, unseen data. Hyperparameter tuning is crucial to optimize the performance of these models, which can be a time-consuming process. The ANN model's architecture, such as the number of layers and neurons, requires careful design to achieve optimal results. The decision tree algorithm, while interpretable and simple, may struggle to capture complex relationships in the data without extensive tuning.

MODEL EVALUATION

Common Metrics Used for Evaluating Machine Learning Models for Hydrometeorological Event Prediction

Hydrometeorological event prediction involves forecasting weather-related phenomena such as rainfall, floods, hurricanes, and other hydrological events. Evaluating the performance of machine learning models for hydrometeorological event prediction is crucial to ensure accurate and reliable forecasts. Here are some common metrics used for evaluating such models, along with their mathematical equations:

Accuracy (ACC)

Accuracy measures the proportion of correctly predicted instances among the total instances in the dataset (Yacouby & Axman, 2020).

$$Accuracy = \frac{Number\ of\ Correct\ Predictions}{Total\ Number\ of\ Predictions} \quad (1)$$

Precision

Precision measures the proportion of true positive predictions among all positive predictions. It indicates how well the model identifies true positives and avoids false positives (Yacouby & Axman, 2020).

$$Precision = \frac{True\ Positives}{True\ Positives + False\ Positives} \quad (2)$$

Recall (Sensitivity or True Positive Rate)

Recall measures the proportion of true positive predictions among all actual positive instances. It indicates the model's ability to capture positive instances (Yacouby & Axman, 2020).

$$Recall = \frac{True\ Positives}{True\ Positives + False\ Negatives} \quad (3)$$

F1-Score

The F1-Score is the harmonic mean of precision and recall. It provides a balance between precision and recall and is useful when dealing with imbalanced datasets (Yacouby & Axman, 2020).

$$F1\_Score = \frac{2 \times Precision \times Recall}{Precision + Recall} \quad (4)$$

Mean Absolute Error (MAE)

MAE measures the average absolute difference between predicted and actual values. It gives an idea of the model's prediction error magnitude (Chicco et al., 2021).

$$MAE = \frac{1}{n}\sum_{i=1}^{n}\left|Actual_i - Predicted_i\right| \quad (5)$$

Mean Squared Error (MSE)

MSE measures the average of the squared differences between predicted and actual values. It emphasizes larger errors more than MAE (Chicco et al., 2021).

$$MSE = \frac{1}{n}\sum_{i=1}^{n}\left(Actual_i - Predicted_i\right)^2 \quad (6)$$

Root Mean Squared Error (RMSE)

RMSE is the square root of the MSE and provides an interpretable measure of the model's prediction error (Chicco et al., 2021).

$$RMSE = \sqrt{\frac{1}{n}\sum_{i=1}^{n}\left(Actual_i - Predicted_i\right)^2} \tag{7}$$

Coefficient of Determination (R-squared)

R-squared measures the proportion of the variance in the dependent variable that is predictable from the independent variables. It indicates how well the model's predictions match the actual variability (Chicco et al., 2021).

$$R^2 = 1 - \frac{\sum_{i=1}^{n}\left(Actual_i - Predicted_i\right)^2}{\sum_{i=1}^{n}\left(Actual_i - Mean\ Actual\right)^2} \tag{8}$$

The metrics in equations 1 - 8 provide valuable insights into different aspects of a machine learning model's performance for hydrometeorological event prediction. Depending on the specific application and requirements, certain metrics may be more relevant than others. It's important to choose metrics that align with the goals of the prediction task and consider the trade-offs between precision, recall, and prediction errors.

Techniques for Cross-Validation

Cross-validation and model selection are critical steps in the development of machine-learning models for hydrometeorological event prediction. These techniques help assess the model's performance, generalize its capabilities, and select the best model among various alternatives.

Cross-Validation Techniques

Holdout Validation (Zhang et al., 2019)

1. Split the dataset into training (X_{train}, y_{train}) and validation (X_{vol}, y_{vol}) sets.
2. Train the model on the training set and evaluate it on the validation set.
3. Useful when you have a large dataset, but may lead to high variance in performance estimation due to the random split.

Model= Train Model(X_{train}, y_{train})

Validation Score= Evaluate(Model, X_{vol}, y_{vol}) (9)

K-Fold Cross-Validation (Marzban et al., 2022)

1. Divide the dataset into K folds: (X_1,y_1), (X_2,y_2), …, (X_k,y_k).
2. Train the model on K-1 folds and validate it on the remaining fold iteratively, such that each fold is used as a validation set once.
3. Compute the average performance across all iterations.
4. Reduces variance in performance estimation compared to holdout validation.
5. 5. Commonly used values for K are 5 or 10.

Total Score= 0

for i in 1 to K:

Model= Train Model(X_{train}, y_{train})

Validation Score$_i$= Evaluate(Model, $X_{vol,i}$, $y_{vol,i}$)

Total Score= Total Score + Validation Score$_i$

$$AverageScore = \frac{TotalScore}{K} \tag{10}$$

Stratified K-Fold Cross-Validation (Zhang et al., 2022)

1. Similar to K-Fold, but ensures that each fold maintains the same class distribution as the original dataset.
2. Particularly useful when dealing with imbalanced classes in hydrometeorological event prediction.

Total Score= 0

for i in 1 to K:

Stratified Sampling(X_{train},y_{train}, $X_{train,i}$,$y_{train,i}$)

Model= Train Model($X_{train,i}$,$y_{train,i}$)

Validation Score$_i$= Evaluate(Model, $X_{vol,i}$,$y_{vol,i}$)

Total Score= Total Score + Validation Score$_i$

$$AverageScore = \frac{TotalScore}{K} \tag{11}$$

Leave-One-Out Cross-Validation (LOOCV) (Fu et al., 2022)

1. A special case of K-Fold where K is set to the number of data points.
2. Train and validate the model K times, each time leaving out one data point for validation.
3. Can provide a more accurate estimate of model performance but can be computationally expensive for large datasets.

Let denote:

K as the total number of data points (samples)
X_i as the feature vector for the i-th data point.
y_i as the corresponding true target value for the i-th data point.
M_{-i} as the model trained on all data points except the i-th data point.
$\hat{y}_{-i}$ as the predicted value for the i-th data point using the model M_{-i}.

The LOOCV process can be expressed as follows:

For i= 1 to K:

a. Exclude the i-th data point: $X_{-i}= [X_1,X_2,\ldots,X_{i-1}, X_{i+1},\ldots,X_K]$
b. Exclude the corresponding target: $y_{-i}= [y_1,y_2,\ldots,y_{i-1}, y_{i+1},\ldots,y_K]$
c. Train a model M_{-i} on the dataset (X_{-1},y_{-1}).
d. Predict the i-th data point: $\hat{y}_{-i} = Predict\left(M_{-i}, X_i\right)$.
e. Calculate the prediction error for the i-th data point: $E_i = y_i - \hat{y}_{-i}$.
f. Calculate the Mean Squared Error (MSE) for LOOCV:

$$MSE_{LOOCV} = \frac{1}{K}\sum_{i=1}^{K} E_i^2 \tag{12}$$

Strategies for Interpreting Model Results and Identifying Areas for Improvement

Interpreting model results and identifying areas for improvement in hydrometeorological event prediction is crucial for ensuring the reliability and accuracy of the predictions. Here are some best strategies for effectively interpreting model results and pinpointing areas that need improvement.

Feature Importance Analysis (Papacharalampous & Tyralis, 2022)

This involves the use of techniques like feature importance (FI_i) scores from tree-based models (e.g., Random Forest, Gradient Boosting) to identify which features have the most influence on predictions. This helps in understanding which meteorological variables or factors contribute significantly to the prediction of the target event. Finally, consider domain knowledge to validate the importance of these features (X_i).

$$Feature\ Importance = \frac{\sum_{i=1}^{n} FI_i}{n} \tag{13}$$

Partial Dependence Plots (PDPs) and Individual Conditional Expectation (ICE) Plots (Huang et al., 2023)

PDPs show how the predicted outcome changes with variations in a single feature (X_k) while keeping other features constant. ICE plots provide a more granular view, showing the impact of a single feature (X_k) on predictions for individual instances (i). Analyzing these plots can reveal nonlinear relationships between features and predictions.

$$PDP\left(X_k\right) = \frac{1}{n}\sum_{i=1}^{n} Model\left(X_1,\ldots,X_{k-1},X_k,X_{k+1},\ldots,X_p\right) \tag{14}$$

$$ICE(X_k,i) = \text{Model}(X_1,\ldots,X_{k-1}, X_k, X_{k+1}, \ldots,X_p) \tag{15}$$

SHAP (SHapley Additive exPlanations) Values (Fan et al., 2023)

SHAP values provide (SV_i) provide a unified measure of feature importance for each prediction. They explain the contribution of each feature (X_i) to the difference between the model's prediction and the average prediction. SHAP summary plots can help visualize the impact of features on individual predictions and the overall model behavior.

$$SHAP\ Value = \frac{1}{n}\sum_{i=1}^{n} SV_i \tag{16}$$

Residual Analysis (Meis et al., 2022)

Analyze the residuals (R_i) (the differences between predicted $\left(\hat{y}_i\right)$ and actual (y_i) values) to understand how well the model captures the underlying patterns in the data. Patterns in residuals can indicate areas where the model consistently overestimates or underestimates predictions. Look for patterns in the residuals over time, space, or other relevant dimensions.

$$R_i = y_i - \hat{y}_i \tag{17}$$

Temporal Analysis (Aristizábal et al., 2020)

For time series data, analyze the model's performance across different time periods (T_j) (e.g., seasons, years) using evaluation metric (E_j) to identify temporal patterns and variations in prediction accuracy. Examine how well the model captures short-term fluctuations and long-term trends.

$$Temporal\ Performance = \frac{\sum_{j=1}^{m} E_j}{m} \tag{18}$$

Spatial Analysis (Mavhura, 2019)

For spatial data, assess the model's performance across different geographic locations (L_k) using evaluation metric (E_k) to identify spatial patterns and variations in prediction accuracy. Identify areas where the model performs well and areas where it struggles to make accurate predictions.

$$Spatial\ Performance = \frac{\sum_{k=1}^{q} E_k}{q} \tag{19}$$

Error Analysis

Classify prediction errors into different categories, such as false positives (*FP*), false negatives (*FN*), true positives (*TP*), and true negatives (*TN*) as shown in equation 2 and 3. Understand the consequences of different types of errors in the context of hydrometeorological events (e.g., missing a flood prediction vs. issuing a false alarm). Investigate specific instances where the model's predictions were particularly accurate or inaccurate.

Model Comparison and Ensembling

Compare the performance of different models (M_i) using evaluation metrics (E_i). Ensemble models (combining predictions from multiple models) can often lead to improved performance and more robust predictions.

$$\text{Best Model} = \arg\max_i E_i \tag{20}$$

Feedback From Domain Experts

Collaborate with meteorologists and hydrologists to gain insights into model behavior and potential areas for improvement. Experts' feedback ($F_{expert)}$ can provide valuable context and domain-specific knowledge to validate and interpret model results.

$$\text{Adjusted Model} = \text{Model} + F_{expert} \tag{21}$$

FUTURE DIRECTIONS

The field of hydrometeorological event prediction stands at the threshold of exciting advancements and transformative possibilities, fueled by the rapid progress in machine learning and data science. As we look ahead, several key directions and opportunities emerge that have the potential to reshape the landscape of hydrometeorological prediction and its applications.

One of the most promising trajectories involves the integration of multiple data sources. The fusion of data from diverse origins, including remote sensing platforms, weather stations, social media, and crowdsourced citizen observations, can amplify the accuracy and precision of predictive models. This holistic approach enriches the data pool and empowers early warning systems with comprehensive and real-time insights, ultimately enhancing our ability to anticipate and respond to hydrometeorological events.

Improved uncertainty estimation techniques represent another vital avenue for exploration. Hydrometeorological predictions inherently carry uncertainty, and developing robust methods to quantify and communicate these uncertainties is crucial for effective decision-making. By enhancing our capacity to assess and convey prediction uncertainties, we can empower decision-makers and stakeholders with a clearer understanding of the risks involved, enabling more informed and adaptive strategies.

The realm of online learning and adaptive models holds tremendous potential for creating prediction systems that are dynamic and responsive. Real-time updates and continuous learning mechanisms that adapt to changing data patterns and evolving environmental conditions can significantly enhance the accuracy and relevance of predictions. This adaptive paradigm ensures that models remain attuned to the latest developments, enabling timely interventions and accurate forecasts.

Ensemble learning, where diverse models collaborate to produce collective predictions, offers an exciting opportunity for enhancing predictive accuracy. Combining the strengths of different algorithms and techniques can mitigate individual model limitations and provide more reliable and robust predictions. Furthermore, this collaborative approach opens doors to exploring novel ensemble methods and model stacking strategies that can push the boundaries of prediction performance.

The spirit of interdisciplinary collaboration is poised to play a pivotal role in shaping the future of hydrometeorological prediction. Bridging the expertise of meteorologists, hydrologists, data scientists, and domain specialists will yield a holistic and comprehensive understanding of complex systems. This synergy will not only refine model development but also enrich the interpretation and application of predictions, fostering a deeper connection between scientific advancements and real-world impact.

Advancements in explainable AI are essential for building trust and transparency in predictive models. As these models become more intricate, the ability to provide meaningful explanations for predictions becomes increasingly critical. Techniques like SHAP values, partial dependence plots, and local interpretable model-agnostic explanations (LIME) will be key in ensuring that stakeholders can comprehend and trust the reasoning behind predictions. Finally, the automation of feature engineering holds the promise of unlocking the potential of complex datasets. As data volumes grow, the extraction of meaningful features becomes a challenge. Automated methods that identify and engineer relevant features can amplify prediction accuracy and enable models to capture subtle relationships within the data.

CONCLUSION

In conclusion, the future of hydrometeorological event prediction is illuminated by the convergence of cutting-edge technologies and interdisciplinary collaboration. As machine learning algorithms continue to evolve and data availability expands, our capacity to anticipate and mitigate the impacts of hydrometeorological events will reach unprecedented heights. By embracing challenges, harnessing opportunities, and nurturing collaborative innovation, we are poised to usher in an era where accurate and actionable predictions guide our response to the complex interplay of weather and water systems. The journey ahead holds the promise of safer communities, informed decision-making, and a resilient future in the face of hydrometeorological challenges.

REFERENCES

Abbot, J., & Marohasy, J. (2014). Input selection and optimization for monthly rainfall forecasting in Queensland, Australia, using artificial neural networks. *Atmospheric Research*, *2014*(138), 166–178. doi:10.1016/j.atmosres.2013.11.002

Acharki, S., Singh, S. K., do Couto, E. V., Arjdal, Y., & Elbeltagi, A. (2023). Spatio-temporal distribution and prediction of agricultural and meteorological drought in a Mediterranean coastal watershed via GIS and machine learning. *Physics and Chemistry of the Earth Parts A/B/C*, *131*, 103425. doi:10.1016/j.pce.2023.103425

Ahmed, A. A. M., Jui, S. J. J., Chowdhury, M. A. I., Ahmed, O., & Sutradha, A. (2023). The development of a dissolved oxygen forecast model using a hybrid machine learning algorithm with hydro-meteorological variables. *Environmental Science and Pollution Research International*, *30*(3), 7851–7873. doi:10.100711356-022-22601-z PMID:36045185

Al Kafy, A., Bakshi, A., Saha, M., Al Faisal, A., Almulhim, A. I., Rahaman, Z. A., & Mohammad, P. (2023). Assessment and prediction of index-based agricultural drought vulnerability using machine learning algorithms. *The Science of the Total Environment*, *867*, 161394. doi:10.1016/j.scitotenv.2023.161394 PMID:36634773

Aristizábal, E., Arango Carmona, M. I., Gómez, F. J., López Castro, S. M., De Villeros Severiche, A., & Riaño Quintanilla, A. F. (2020). Hazard analysis of hydrometeorological concatenated processes in the Colombian Andes. In *Advances in Natural Hazards and Hydrological Risks: Meeting the Challenge: Proceedings of the 2nd International Workshop on Natural Hazards (NATHAZ'19),* (pp. 7-10). Springer International Publishing. 10.1007/978-3-030-34397-2_2

Bao, Y., Xiong, T., & Hu, Z. (2014). Multi-step-ahead time series prediction using multiple-output support vector regression. *Neurocomputing*, *2014*(129), 482–493. doi:10.1016/j.neucom.2013.09.010

Bayat, M., & Tavakkoli, O. (2020). Application of machine learning in flood forecasting. *Future Technology*, *1*(1), 1–6. doi:10.55670/fpll.futech.1.1.1

Chen, X., Li, X., Jiang, B., Su, J., Zheng, X., & Wang, G. (2023). Prediction of spring agricultural drought using machine learning algorithms in the southern Songnen Plain, China. *Land Degradation & Development*, *34*(13), 3836–3849. doi:10.1002/ldr.4720

Chicco, D., Warrens, M. J., & Jurman, G. (2021). The coefficient of determination R-squared is more informative than SMAPE, MAE, MAPE, MSE and RMSE in regression analysis evaluation. *PeerJ. Computer Science*, *7*, e623. doi:10.7717/peerj-cs.623 PMID:34307865

Deo, R. C., & Şahin, M. (2015). ¸Sahin, M. (2015). Application of the artificial neural network model for prediction of monthly standardized precipitation and evapotranspiration index using hydrometeorological parameters and climate indices in Eastern Australia. *Atmospheric Research*, *161*, 65–81. doi:10.1016/j.atmosres.2015.03.018

Di, Y., Ding, W., Mu, Y., Small, D. L., Islam, S., & Chang, N. B. (2015, April). Developing machine learning tools for long-lead heavy precipitation prediction with multi-sensor data. In *2015 IEEE 12th International Conference on Networking, Sensing and Control* (pp. 63-68). IEEE. 10.1109/ICNSC.2015.7116011

Dibike, Y. B., Velickov, S., Solomatine, D., & Abbott, M. B. (2001). Model induction with support vector machines: Introduction and applications. *Journal of Computing in Civil Engineering*, *2001*(15), 208–216. doi:10.1061/(ASCE)0887-3801(2001)15:3(208)

Etemad-Shahidi, A., & Mahjoobi, J. (2009). Comparison between m50 model tree and neural networks for prediction of significant wave height in lake superior. *Ocean Engineering*, *2009*(36), 1175–1181. doi:10.1016/j.oceaneng.2009.08.008

Fan, M., Zhang, L., Liu, S., Yang, T., & Lu, D. (2023). Investigation of hydrometeorological influences on reservoir releases using explainable machine learning methods. *Frontiers in Water*, *5*, 1112970. doi:10.3389/frwa.2023.1112970

Fu, B., Lao, Z., Liang, Y., Sun, J., He, X., Deng, T., He, W., Fan, D., Gao, E., & Hou, Q. (2022). Evaluating optically and non-optically active water quality and its response relationship to hydro-meteorology using multi-source data in Poyang Lake, China. *Ecological Indicators*, *145*, 109675. doi:10.1016/j.ecolind.2022.109675

Grecu, M., & Krajewski, W. (2000). A large-sample investigation of statistical procedures for radar-based short-term quantitative precipitation forecasting. *Journal of Hydrology (Amsterdam)*, *2000*(239), 69–84. doi:10.1016/S0022-1694(00)00360-7

Guastavino, S., Piana, M., Tizzi, M., Cassola, F., Iengo, A., Sacchetti, D., Solazzo, E., & Benvenuto, F. (2022). Prediction of severe thunderstorm events with ensemble deep learning and radar data. *Scientific Reports*, *12*(1), 20049. doi:10.103841598-022-23306-6 PMID:36414648

Haile, G. G., Tang, Q., Li, W., Liu, X., & Zhang, X. (2020). Drought: Progress in broadening its understanding. *WIREs. Water*, *7*(2), e1407. doi:10.1002/wat2.1407

Hassan, F. H., & Azelan, N. A. (2019). Comparing the performance of machine learning algorithms in a flood prediction model with real data sets. *International Journal of Advanced Trends in Computer Science and Engineering, 8*(1.4 S1), 152–157. doi:10.30534/ijatcse/2019/2381.42019

Hayder, I. M., Al-Amiedy, T. A., Ghaban, W., Saeed, F., Nasser, M., Al-Ali, G. A., & Younis, H. A. (2023). An Intelligent Early Flood Forecasting and Prediction Leveraging Machine and Deep Learning Algorithms with Advanced Alert System. *Processes (Basel, Switzerland)*, *11*(2), 481. doi:10.3390/pr11020481

He, G., Chen, Y., Fang, G., & Li, Z. (2022). Hydrometeorological Forecast of a Typical Watershed in an Arid Area Using Ensemble Kalman Filter. *Water (Basel)*, *14*(23), 3970. doi:10.3390/w14233970

Herath, M., Jayathilaka, T., Hoshino, Y., & Rathnayake, U. (2023). Deep Machine Learning-Based Water Level Prediction Model for Colombo Flood Detention Area. *Applied Sciences (Basel, Switzerland), 13*(4), 2194. doi:10.3390/app13042194

Huang, F., Shangguan, W., Li, Q., Li, L., & Zhang, Y. (2023). Beyond prediction: An integrated post-hoc approach to interpret complex model in hydrometeorology. *Environmental Modelling & Software, 167*, 105762. doi:10.1016/j.envsoft.2023.105762

Jajarmizadeh, M., Lafdani, E. K., Harun, S., & Ahmadi, A. (2015). Application of SVM and swat models for monthly streamflow prediction, a case study in South of Iran. *KSCE Journal of Civil Engineering, 2015*(19), 345–357. doi:10.100712205-014-0060-y

Kar, A. K., Lohani, A. K., Goel, N. K., & Roy, G. P. (2010). Development of flood forecasting system using statistical and ANN techniques in the downstream catchment of Mahanadi basin, India. *Journal of Water Resource and Protection, 2010*(2), 880.

Khan, N., Sachindra, D. A., Shahid, S., Ahmed, K., Shiru, M. S., & Nawaz, N. (2020). Prediction of droughts over Pakistan using machine learning algorithms. *Advances in Water Resources, 139*, 103562. Advance online publication. doi:10.1016/j.advwatres.2020.103562

Kia, M. B., Pirasteh, S., Pradhan, B., Mahmud, A. R., Sulaiman, W. N. A., & Moradi, A. (2012). An artificial neural network model for flood simulation using GIS: Johor River Basin, Malaysia. *Environmental Earth Sciences, 67*(1), 251–264. doi:10.100712665-011-1504-z

Kim, S., Matsumi, Y., Pan, S., & Mase, H. (2016). A real-time forecast model using an artificial neural network for after-runner storm surges on the Tottori Coast, Japan. *Ocean Engineering, 2016*(122), 44–53. doi:10.1016/j.oceaneng.2016.06.017

Kim, S., & Singh, V. P. (2013). Flood forecasting using neural computing techniques and conceptual class segregation. *Journal of the American Water Resources Association, 2013*(49), 1421–1435. doi:10.1111/jawr.12093

Kumar, S., Tiwari, M. K., Chatterjee, C., & Mishra, A. (2015). Reservoir inflow forecasting using ensemble models based on neural networks, wavelet analysis, and bootstrap method. *Water Resources Management, 2015*(29), 4863–4883. doi:10.100711269-015-1095-7

Lacroix, P., Handwerger, A. L., & Bièvre, G. (2020). Life and death of slow-moving landslides. *Nature Reviews. Earth & Environment, 1*(8), 404–419. doi:10.103843017-020-0072-8

Lafdani, E. K., Nia, A. M., Pahlavanravi, A., Ahmadi, A., & Jajarmizadeh, M. (2013). Research article daily rainfall-runoff prediction and simulation using ANN, ANFIS, and conceptual hydrological MIKE11/NAM models. Int. J.nEng. *Technol., 2013*(1), 32–50.

Li, L., Xu, H., Chen, X., & Simonovic, S. (2010). Streamflow forecast and reservoir operation performance assessment under climate change. *Water Resources Management, 2010*(24), 83–104. doi:10.100711269-009-9438-x

Li, P., Stuart, E. A., & Allison, D. B. (2015, November 10). Multiple Imputation: A Flexible Tool for Handling Missing Data. *Journal of the American Medical Association*, *314*(18), 1966–1967. doi:10.1001/jama.2015.15281 PMID:26547468

Maddox, R. A., Zhang, J., Gourley, J. J., & Howard, K. W. (2002). Weather radar coverage over the contiguous United States. *Weather and Forecasting*, *2002*(17), 927–934. doi:10.1175/1520-0434(2002)017<0927:WRCOTC>2.0.CO;2

Mahdizadeh Gharakhanlou, N., & Perez, L. (2022). Spatial Prediction of Current and Future Flood Susceptibility: Examining the Implications of Changing Climates on Flood Susceptibility Using Machine Learning Models. *Entropy (Basel, Switzerland)*, *24*(11), 1630. doi:10.3390/e24111630 PMID:36359720

Marzban, C., Liu, J., & Tissot, P. (2022). On Variability due to Local Minima and K-Fold Cross Validation. *Artificial Intelligence for the Earth Systems*, *1*(4), e210004. doi:10.1175/AIES-D-21-0004.1

Mavhura, E. (2019). Systems analysis of vulnerability to hydrometeorological threats: An exploratory study of vulnerability drivers in Northern Zimbabwe. *International Journal of Disaster Risk Science*, *10*(2), 204–219. doi:10.100713753-019-0217-x

Meis, M., Llano, M. P., & Rodriguez, D. (2022). A statistical tool for a hydrometeorological forecast in the lower La Plata Basin. *International Journal of River Basin Management*, 1–12.

Mohamadi, S., Sammen, S. S., Panahi, F., Ehteram, M., Kisi, O., Mosavi, A., Ahmed, A. N., El-Shafie, A., & Al-Ansari, N. (2020). Zoning map for drought prediction using integrated machine learning models with a nomadic people optimization algorithm. *Natural Hazards*, *104*(1), 537–579. doi:10.100711069-020-04180-9

Morán-Tejeda, E., Fassnacht, S. R., Lorenzo-Lacruz, J., López-Moreno, J. I., García, C., Alonso-González, E., & Collados-Lara, A. J. (2019). Hydro-meteorological characterization of major floods in Spanish mountain rivers. *Water (Basel)*, *11*(12), 2641. doi:10.3390/w11122641

Nayak, M. A., & Ghosh, S. (2013). Prediction of extreme rainfall events using weather pattern recognition and support vector machine classifier. *Theoretical and Applied Climatology*, *2013*(114), 583–603. doi:10.100700704-013-0867-3

Papacharalampous, G., & Tyralis, H. (2022). Time series features for supporting hydrometeorological explorations and predictions in ungauged locations using large datasets. *Water (Basel)*, *14*(10), 1657. doi:10.3390/w14101657

Piri, J., Abdolahipour, M., & Keshtegar, B. (2023). Advanced Machine Learning Model for Prediction of Drought Indices using Hybrid SVR-RSM. *Water Resources Management*, *37*(2), 683–712. doi:10.100711269-022-03395-8

Quesada-Román, A., & Villalobos-Chacón, A. (2020). Flash flood impacts of Hurricane Otto and hydrometeorological risk mapping in Costa Rica. *Geografisk Tidskrift*, *120*(2), 142–155. doi:10.1080/00167223.2020.1822195

Razali, N., Ismail, S., & Mustapha, A. (2020). Machine learning approach for flood risk prediction. *IAES International Journal of Artificial Intelligence*, *9*(1), 73–80. doi:10.11591/ijai.v9.i1.pp73-80

Senthil Kumar, A., Sudheer, K., Jain, S., & Agarwal, P. (2005). Rainfall-runoff modeling using artificial neural networks: Comparison of network types. *Hydrological Processes*, *2005*(19), 1277–1291. doi:10.1002/hyp.5581

Seo, D. J., & Breidenbach, J. (2002). Real-time correction of spatially nonuniform bias in radar rainfall data using rain gauge measurements. *Journal of Hydrometeorology*, *2002*(3), 93–111. doi:10.1175/1525-7541(2002)003<0093:RTCOSN>2.0.CO;2

Seo, Y., Kim, S., Kisi, O., & Singh, V. P. (2015). Daily water level forecasting using wavelet decomposition and artificial intelligence techniques. *Journal of Hydrology (Amsterdam)*, *2015*(520), 224–243. doi:10.1016/j.jhydrol.2014.11.050

Shafaei, M., & Kisi, O. (2017). Predicting river daily flow using wavelet-artificial neural networks based on regression analyses in comparison with artificial neural networks and support vector machine models. *Neural Computing & Applications*, *2017*(28), 15–28. doi:10.100700521-016-2293-9

Shoaib, M., Shamseldin, A. Y., & Melville, B. W. (2014). Comparative study of different wavelet-based neural network models for rainfall–runoff modeling. *Journal of Hydrology (Amsterdam)*, *2014*(515), 47–58. doi:10.1016/j.jhydrol.2014.04.055

Shu, C., & Ouarda, T. (2008). Regional flood frequency analysis at ungauged sites using the adaptive neuro-fuzzy inference system. *Journal of Hydrology (Amsterdam)*, *2008*(349), 31–43. doi:10.1016/j.jhydrol.2007.10.050

Sudhishri, S., Kumar, A., & Singh, J. K. (2016). Comparative evaluation of neural network and regression-based models to simulate runoff and sediment yield in an outer Himalayan watershed. *J. Agric. Sci. Technol.*, *2016*(18), 681–694.

Sulaiman, J., & Wahab, S. H. (2017). Heavy rainfall forecasting model using artificial neural network for flood-prone areas. In It Convergence and Security 2017. Springer: Singapore.

Tayfur, G., Singh, V. P., Moramarco, T., & Barbetta, S. (2018). Flood hydrograph prediction using machine learning methods. *Water (Basel)*, *10*(8), 968. doi:10.3390/w10080968

Tehrany, M. S., Pradhan, B., & Jebur, M. N. (2013). Spatial prediction of flood susceptible areas using rule-based decision tree (DT) and a novel ensemble bivariate and multivariate statistical models in GIS. *Journal of Hydrology (Amsterdam)*, *2013*, 504. doi:10.1016/j.jhydrol.2013.09.034

Vaseashta, A. (2022). Applying Resilience to Hybrid Threats in Infrastructure, Digital, and Social Domains Using Multisectoral, Multidisciplinary, and Whole-of-Government Approach. NATO Science for Peace and Security Series - D. *Information and Communication Security.*, *61*, 42–59. doi:10.3233/NICSP220017

Wang, Z., Lai, C., Chen, X., Yang, B., Zhao, S., & Bai, X. (2015). Flood hazard risk assessment model based on random forest. *Journal of Hydrology (Amsterdam)*, *2015*(527), 1130–1141. doi:10.1016/j.jhydrol.2015.06.008

Wu, C., & Chau, K. W. (2010). Data-driven models for monthly streamflow time series prediction. *Engineering Applications of Artificial Intelligence*, *2010*(23), 1350–1367. doi:10.1016/j.engappai.2010.04.003

Yacouby, R., & Axman, D. (2020, November). Probabilistic extension of precision, recall, and f1 score for more thorough evaluation of classification models. In *Proceedings of the first workshop on evaluation and comparison of NLP systems* (pp. 79-91). IEEE. 10.18653/v1/2020.eval4nlp-1.9

Yang, F., Koukoula, M., Emmanouil, S., Cerrai, D., & Anagnostou, E. N. (2023). Assessing the power grid vulnerability to extreme weather events based on long-term atmospheric reanalysis. *Stochastic Environmental Research and Risk Assessment*, *37*(11), 1–16. doi:10.100700477-023-02508-y

Zeppetello, L. R. V., Battisti, D. S., & Baker, M. B. (2022). The physics of heat waves: What causes extremely high summertime temperatures? *Journal of Climate*, *35*(7), 2231–2251. doi:10.1175/JCLI-D-21-0236.1

Zhang, B., Salem, F. K. A., Hayes, M. J., & Tadesse, T. (2022). *Quantitative Assessment of Drought Impacts Using XGBoost based on the Drought Impact Reporter.* arXiv preprint arXiv:2211.02768.

Zhang, K., Wang, Q., Chao, L., Ye, J., Li, Z., Yu, Z., Yang, T., & Ju, Q. (2019). Ground observation-based analysis of soil moisture spatiotemporal variability across a humid to semi-humid transitional zone in China. *Journal of Hydrology (Amsterdam)*, *574*, 903–914. doi:10.1016/j.jhydrol.2019.04.087

Zhang, Y., & Ye, A. (2021). Machine learning for precipitation forecasts postprocessing: Multimodel comparison and experimental investigation. *Journal of Hydrometeorology*, *22*(11), 3065–3085.

Chapter 8

Spatial Trend Analysis of Vegetation Dynamics and Their Responses to Climate Change on Black Sea Coasts, Romania From 2000 to 2021

Vahid Nasiri
University of Agriculture, Krakow, Poland

Soroor Rahmanian
https://orcid.org/0009-0003-1752-7490
German Centre for Integrative Biodiversity Research (iDiv) Halle-Jena-Leipzig, Germany

Zohreh Hashemi Aslani
University of Tehran, Iran

Carmen Maftei
https://orcid.org/0000-0002-7150-608X
Transilvania University of Brasov, Romania

Ashok Vaseashta
https://orcid.org/0000-0002-5649-0067
International Clean Water Institute, USA

ABSTRACT

Based on the normalized difference vegetation index (NDVI) time series of moderate-resolution imaging spectroradiometer products, meteorological observations, and spatial trend analysis, the authors analyze how climate change and vegetation cover have changed in the Black Sea coasts over the past 22 years. Moreover, spatial correlation analysis was used to examine the correlation between climatic factors and NDVI as a proxy of vegetation productivity. With growth rates of 0.003, 0.5°C, and 5.5 mm per year, respectively, the inter-annual variation of NDVI, temperature, and precipitation revealed a noticeably growing trend. At the seasonal time scale, a similar trend was observed, which was statistically significant in some cases. The spatial-temporal trend of NDVI showed a greening trend in the inter-annual and seasonal time scales (except autumn). The greening trend in vegetation refers to the increase in vegetation productivity and density over time. The authors conclude that the most effective factor in the vegetation greening process is an increasing trend in temperature and precipitation.

DOI: 10.4018/978-1-6684-8771-6.ch008

INTRODUCTION

Climate change refers to the long-term changes in the Earth's climate patterns, and temperatures (Thuiller 2007; Loarie et al. 2009), primarily caused by human activities (Jiang et al. 2017; Huang et al. 2020; Gao et al. 2022) that increase the concentration of greenhouse gases in the atmosphere, leading to the enhanced greenhouse effect and warming of the planet (Zheng et al. 2019; Eskander and Fankhauser 2020). Climate change effects are already being noticed worldwide, including rising sea levels (Mukul et al. 2019), melting glaciers (Nabi et al. 2018), more frequent and severe heatwaves (Lhotka et al. 2018), droughts (Holden et al. 2022), floods (Liu et al. 2023), and storms (Leal et al. 2022). These impacts are projected to worsen in the future and have significant economic, social, and environmental consequences (Hoffmann et al. 2022; Abbas et al. 2023).

Vegetation plays a critical role in providing ecosystem services such as carbon sequestration, water regulation, and habitat for wildlife (Heilmayr et al. 2020; Di Sacco et al. 2021). One of the largest climate shifts is anticipated to have a big influence on vegetation and the ecosystems that depend on it (Afuye et al. 2021; Cui et al. 2022). Previous research discovered that climate change was having a significant impact on vegetation, such as reducing plant growth and productivity (Pouyan et al. 2022; Wang et al. 2023), changing plant phenology (Praveen and Sharma 2019; Chaudhry and Sidhu 2022), shifting species diversity and distribution (Richardson et al. 2013; Piao et al. 2019), increasing the risk of wildfire (Williams et al. 2019; Nasiri et al. 2022), increasing pest and disease pressure (Jiranek et al. 2023), changing carbon storage capacity (Gatti et al. 2021), and so on. Also, vegetation serves as an important indicator of climate change, because changes in the vegetation phenology, structure, diversity, distribution and productivity can all signal shifts in climate patterns (Morecroft and Keith 2009; Shen et al. 2023). Therefore, monitoring the effect of climate change on vegetation is critical for understanding the impacts of climate change on ecosystems, biodiversity, and human well-being, and for developing effective strategies to mitigate and adapt to these impacts.

The effects of climate change on vegetation have generally been observed at the landscape level, needing a long-term and extensive monitoring procedure that can detect changes in both climatic factors and vegetation dynamics (Bowman et al. 2015; Piao et al. 2019). However, such monitoring procedure at the landscape level using ground-based observations is challenging because of the large area, spatial heterogeneity, and accessibility of observation facilities. In the recent decades, with the development of satellite remote sensing, spatial statistical approaches, and computing platforms, large-scale monitoring and investigation of vegetation dynamics and its interactions with climatic conditions are now possible (Chen et al. 2023; Kolarik et al. 2023).

Remote sensing time series are the first component of any large-scale monitoring procedure. In this regard, vegetation indices (VIs) are widely used in monitoring vegetation dynamics (Wang et al. 2020; Chen et al. 2022). VIs are mathematical combinations of spectral reflectance values from different wavelengths of light, typically in the visible and near-infrared regions of the electromagnetic spectrum, which provide information about vegetation biomass, vigor, and health (da Silva et al. 2020; Khunrattanasiri 2023; Wu et al. 2023). Monitoring vegetation dynamics using vegetation indices involves comparing vegetation index values over time to detect their spatial and temporal variations. After constructing the vegetation index time series, spatial statistical methods such as linear regression or trend analysis can be used to detect changes and trends in vegetation over time (Zhou et al. 2021). In addition, the intermingling of these methods with climatic factors can be used to determine what type of variation occurred as a result of climate change (Wang et al. 2020). In this regard, some previous studies have used the tools mentioned above to examine the spatial and temporal trends of vegetation indices and climatic factors and infer their relationship in different biomes (Baniya et al. 2019; Wang et al. 2020; Wei et al. 2022). In the Loess Plateau region (a Chinese landform), for instance, Li et al. (2021) intergraded NDVI and meteorological time series and reported an upward trend in the vegetation cover under the influence of rising air temperature and precipitation. According to the literature that is currently available, changes in vegetation cover and climatic factors varied between different ecosystems, vegetation zones and elevation gradients. Also, clearly various vegetation types and land use/cover classes respond to climate change in different patterns. For example, Ghebrezgabher et al. (2020) observed that the NDVI of grassland, savanna, and shrub land is more susceptible to climate change than the NDVI of forest and woodland.

The climate in Romania has changed. According with a recent study (Maftei and Barbulescu 2012; Prăvălie et al. 2019) throughout the previous 50 years, the territory of Romania had generally been drier. It has been demonstrated that recent climatic water deficits have caused reduced yields of agricultural products in some regions of the Romanian Extra-Carpathians (Prăvălie et al. 2016). Additional studies also indicated decreasing streamflow rates particularly in the summer (Birsan et al. 2014; Şerban et al. 2022). These examples demonstrate the importance of main climatic factors' influence on vegetation coverage as a controlling factor in the dynamics of both natural and human-made processes.

The main objective of this study was to identify the effects of key climate drivers on vegetation coverage (NDVI) across Southeastern Romania over the past 22 years (2000–2021). To do this, we integrated freely available satellite time series, meteorological observations, and spatial analysis tools to, 1) identify the spatial and temporal variation of vegetation dynamics, 2) detect the spatial and temporal trend of climatic factors including temperature and precipitation; and 3) analyze the relationship between NDVI and climatic factors at different time scales.

STUDY AREA

This study was performed in Dobrogea region, located in Southeastern part of Romania between lower Danube (West and Nord), Black Sea (East) and Bulgaria (South) (Fig. 1). This area is located at 43°73′ to 45°46′ N and 27°27′ to 29°72′ E and the elevation ranges between -27 and 463 m.a.s.l. The Dobrogea Plateau, the Danube floodplain, and the Danube Delta are the three main geomorphological divisions of the Dobrogea terrain. Dobrogea's climate is divided into two units, one on the littoral (covering 20-50km from the littoral to interior) and the other on the western part of the region (Maftei and Barbulescu 2008). Meteorological observations from 1965 to 2021 (based on 12 meteorological stations) characterized mean annual precipitation and temperature to be around 450mm mm 11.8°C. At regional level February-April period and August are the driest months and May is the wettest month, respectively.

DATASETS

NDVI Time Series

The NDVI time series was constructed based on the Moderate Resolution Imaging Spectroradiometer (MODIS) sensor products. The MODIS sensor is on board of two satellites, Terra (https://terra.nasa.gov/) and Aqua (https://aqua.nasa.gov/), provide time-series Global vegetation indices with a revisit time of 16

Figure 1. Geographical location of the study area

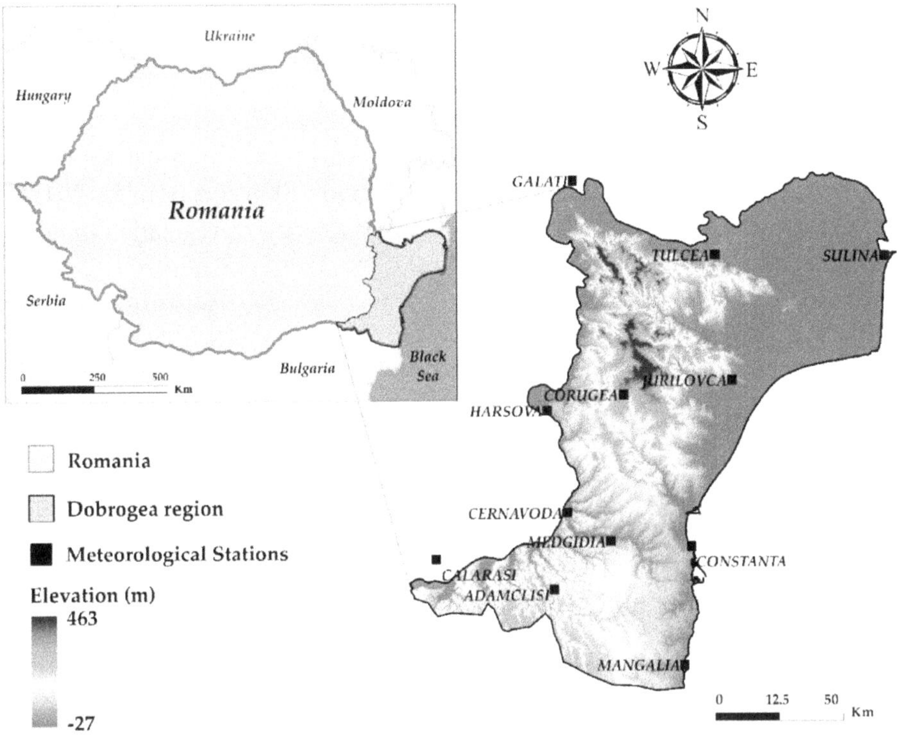

days and a spatial resolution of 250 m. In this study, the Google Earth Engine (GEE) cloud computing platform was used to integrate the MOD13Q1.061 Terra ("MODIS/061/MOD13Q1") and MYD13Q1.061 Aqua ("MODIS/061/MYD13Q1") vegetation indices products and generate a dense NDVI time series with a temporal resolution of 8 days for the period of 2000–2021. Based on the produced image collection, the annual and seasonal NDVI layers were generated using the maximum value synthesis method. In this regard, initially the monthly maximum-value composite (MVC) was obtained for each month, and then using the monthly MVC layers, the annual and seasonal mean NDVI max layers were computed. This method was widely used in previous studies because of its ability to reduce the effects of clouds, the atmosphere, and solar altitude angles on reflectance values (Li et al. 2021; Qian et al. 2021; Fengjin and Qiufeng 2023). The seasonal NDVI time series were produced based on meteorological standards including winter (from December to February of the following year), spring (from March to May), summer (from June to August), and autumn (from September to November).

Climate Datasets

The climate datasets, including daily precipitation and temperature from 12 meteorological stations (see Fig. 1), were obtained from the Dobrogea Regional Meteorological Center (CMR) for 22 years (2000–2021). For each station, the seasonal and annual means of precipitation and temperature were computed before using the Kriging interpolation technique to generate the raster layers. The temperature and precipitation layers were interpolated with a spatial resolution of 250×250 m in order to coincide with the NDVI time series.

Land Use Land Cover Maps

The CORINE (Coordination of Information on the Environment) Land Cover dataset for two different time points (2000 and 2018) with a spatial resolution of 25 m for 2000 and 10 m for 2018 was used to assess land use and land cover (LULC) changes and the interpretation of vegetation trends (Fig. A1).

STATISTICAL ANALYSIS

Analysis of the Spatial and Temporal Trend of NDVI

Various parametric and non-parametric statistical techniques were used to examine the spatial and temporal variation of the NDVI from 2000 to 2021. In the first section, the linear regression method was used to detect the spatial-temporal variation (equation 1), and the ordinary least squares regression (OLS) was used to calculate the coefficients of linear regression, including slope and intercept.

$$NDVI_i = a + b + i + \varepsilon i \quad i=1,2,\ldots,n$$

Where $NDVI_i$ represents the mean NDVI in the time series, n stands for the number of years in time series (in this study=22), a and b are the intercept and slope of the linear regression, and εi denotes the error term. The slope parameter in OLS, measures the NDVI's inter-annual growth rate throughout the research period and can be calculated using the following formula:

$$Slope = \frac{n\sum_{i=1}^{n} i \times NDVI_i - \sum_{i=1}^{n} i \times \sum_{i=1}^{n} NDVI_i}{n\sum_{i=1}^{n} i^2 - (\sum_{i=1}^{n} i)^2}$$

The results depict the present trend of the time series (greater than zero = increasing trend, less than that = decreasing trend). The Pearson correlation coefficient (r) is also calculated to figure out whether the time series (NDVI and climatic factors) reveals any significant trend. As recommended by Wei et al. (2022) the unbiased correlation coefficient (equation 3) is more persuasive for correlation analysis when the sample size is less than 30.

$$r^* = r\left[1 + \frac{1-r^2}{2(n-4)}\right]$$

Where n is the number of years in the series and r is the Pearson correlation coefficient. The linear trend has passed the significance test with 95% confidence intervals when $|r| \geq 0.444$ ($P < 0.05$).

The linear regression, which is a parametric method, assumes that the data follow a multivariate normal distribution. Therefore, in order to better comprehend the existing trend, the pixel-based Theil-Sen median and Mann-Kendall test were also applied since typically there are no requirements regarding the distribution of the datasets (distribution-free test). Because of its resistance to outlier data, measurement errors, and noise, as well as its excellent computational efficiency, the Theil-Sen median method is a popular method for detecting trend in long-term time series and can be calculated as follows:

$$\beta = Median\left(\frac{NDVI_j - NDVI_i}{j-i}\right), \forall j > i$$

Where β represents the trend of NDVI ($\beta > 0$ = increasing trend, $\beta < 0$ = decreasing trend), and $NDVI_j$ and $NDVI_i$ denote the NDVI value at the time *j* and *i*. Also, the Mann-Kendall nonparametric test was used to statistically evaluate if there is a monotonic increasing or decreasing trend of the NDVI time series over time which can be explained as follows:

If we have a list of data in the order in which it was collected over time: x_1, x_2..., xn, where x represents the measurements taken at t = 1, t = 2..., t = n, respectively, the differences of all possible pairs are calculated as follows:

$$x_2 - x_1, x_3 - x_2, x_3 - x_n, \ldots, x_n - x_{n-1}, x_{n-2} - x_{n-1}$$

Then the following equation defines the test statistic S.

$$S = \sum_{i-1}^{n-1} \sum_{j=i+1}^{n} sgn\left(x_j - x_i\right); sgn(\theta) = \begin{Bmatrix} 1, if\ \theta > 0 \\ 0, if\ \theta = 0 \\ -1, if\ \theta < 0 \end{Bmatrix}$$

And the variance of S is calculated as:

$$Var(S) = \frac{n(n-1)(2n+5) - \sum_{t-1}^{m} t_i (t_i - 1)(2t_i - 5)}{18}$$

Where *n* is the number of years in the series, *m* is the number of repeated values (tides), and t_i is the number of values in the tied *i*. finally the Z statistic can be calculated based on the *S* and *Var (S)* by:

$$Z = \begin{cases} \dfrac{S-1}{\sqrt{Var(S)}} & S > 0 \\ 0 & S = 0 \\ \dfrac{S+1}{\sqrt{Var(S)}} & S < 0 \end{cases}$$

In this study, the significance level α = 0.05 (confidence interval of 95%) was used. The critical value in the normal distribution table for the significant level of 0.05 is 1.96, therefore the resulted raster layer was classified accordingly in to four classes: significant increasing trend (Z > 0 and pass the 0.05 test), insignificant increasing trend (Z > 0 and fail the 0.05 test), stable (Z = 0), obvious decreasing trend (Z < 0 and pass the 0.05 test) and slight decreasing trend (Z < 0 and fail the 0.05 test).

CORRELATION ANALYSIS

A raster-based correlation coefficient method based on the R programming language was utilized for assessing the relationship between the NDVI and climatic factors. This correlation was calculated as:

$$r = \frac{\sum_{i=1}^{n}(x_i - \overline{x})\left(NDVI_i - \overline{NDVI}\right)}{\sqrt{\sum_{i=1}^{n}(x_i - \overline{x})^2}\sqrt{\sum_{i=1}^{n}(NDVI_i - \overline{NDVI})^2}}$$

where n is the number of observations, i represents the time, x_i is the value of climatic factor and $NDVI_i$ is the value of NDVI. The three levels of correlation coefficient implemented in this study including; high correlation (±0.8 < r⩽±1), moderate correlation (±0.3 < r⩽±0.8), and negligible correlation (0 < r⩽±0.3).

RESULTS

Inter-Annual Variation of NDVI and Climatic Factors

The inter-annual variation of NDVI, precipitation, and temperature is represented in Fig. 2. Based on the results from 2000 to 2021, the inter-annual NDVI time series showed a significant increasing trend (r = 0.7015, P-value = 0.0002) with a growth rate of 0.003/year. The mean value of 22-year NDVI is ~ 0.50 and the lowest and highest values across the research site were recorded in 2003 and 2021, resulting in 0.4550 and 0.5462, respectively (Fig. 2, a). The precipitation time series showed an increasing trend with an inter-annual growth rate of 5.5 mm, even though this trend is not statistically significant (r = 0.2920 and P-value = 0.1872). The mean value of precipitation during the study period is 474.2 mm, the lowest measurement was 275.6 mm (2000), and the highest one was 692.7 mm (2005). The inter-annual mean temperature, on the other hand, showed a statistically significant upward trend (r = 0.5528 and P-value = 0.0076) with a growth rate of 0.05°C. The mean temperature from all observations was 12.4°C, and in 2003 (11°C) and 2019, respectively (13.6°C), the mean temperature was at its lowest and highest points ever (Fig. 2, c).

Seasonal Variation of NDVI and Climatic Factors

The findings of linear regression and the unbiased correlation coefficient, which depict seasonal variation in NDVI, precipitation, and temperature, are shown in Table 1 and, for better presentation, in Fig. 3 and

Figure 2. Inter-annual variation of NDVI, precipitation and temperature across the study site from 2000-2021

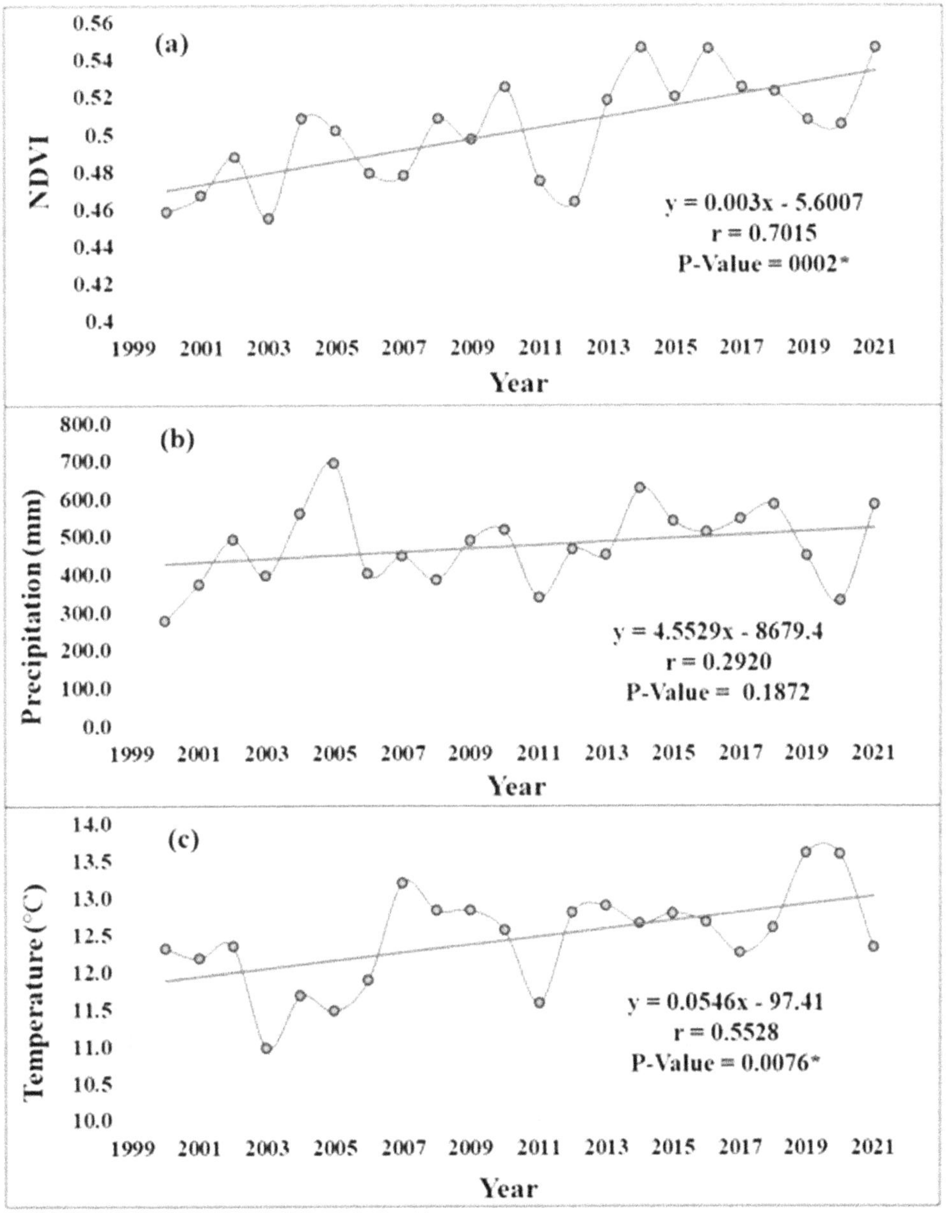

4. According to the findings, generally, the NDVI has an increasing trend in all seasons. The maximum value of NDVI was recorded in the summer of 2004 (0.64) and the minimum one in the winter of 2005 (0.25). The highest and lowest standard deviations were observed in the spring (SD = 0.06) and autumn (SD = 0.02) time series, meaning that these seasons had the highest and lowest inter-annual variations, respectively. The autumn NDVI time series had the lowest correlation coefficient (r = 0.047), which shows a mostly slight and insignificant trend (P-value = 0.8358), whereas the spring had the highest correlation coefficient (r = 0.700), showing a significantly increasing trend (P-value = 0.0002) with a

growth rate of 0.0066/year. In the period 2013–2021 of the spring NDVI time series, where the maximum value was 0.61 in 2016 and the minimum value was 0.36 in 2003, a dramatic upward trend was observed. Also, no remarkable trend was observed in the summertime series (see Table 1).

Based on the multi-seasonal the autumn and summer precipitation time series showed the highest (SD = 42.65) and lowest (SD = 17.79) variation (Table 1). Also, the mean precipitation showed an increasing trend in all seasons (Fig. 3). The results also revealed that the trend of precipitation in the spring was statistically significant at the 5% confidential level (r = 0.672, P = 0.0006) with a growth rate of 3.7 mm per year and ranging between 57.88 mm and 192.11 mm. Moreover, there was a remarkable upward trend in the winter precipitation with a growth rate of 2.15 mm per year (ranging between 61.93 and 183.24 mm) but this trend was not statistically significant (r = 0.392 and P-value = 0.0711).

In terms of the temperature time series, the maximum and minimum mean values were obtained in summer 2011 (23.83 °C) and winter 2003 (-0.66 °C), respectively. The seasonal mean temperature increased in all four seasons (Fig. 4); the winter season (r = 0.555, P-value = 0.0079) and the autumn season (r = 0.592, P-value = 0.0037) were significant at the 5% level (Table 1). Also, the winter season recorded the highest growth rate (0.0966/year) and largest standard division (1.12), which points to a clear increasing trend with considerable fluctuation. The summertime series also experienced a noticeable increasing trend with a growth rate of 0.0373 ranging between 21.37 and 23.83, but this trend was not statistically significant at the 5% level (r = 0.421, P-Value = 0.0509). Also based on the results, in all seasons from 2004 we observed a continuous increasing trend in mean temperature.

Spatial and Temporal Trend of NDVI

At the inter-annual time scale, the Mann-Kendal trend test results revealed an increasing trend in the vegetation cover of the study area. In terms of spatial distribution, in the most part of the Dobrogea region, we observed a statistically significant increasing trend, and the areas with a significant decreasing trend were mainly scattered in the northeast of it (the Danube Delta, Fig. 5, a). At the seasonal time scale, we

Table 1. Statistical trend analysis of NDVI, precipitation, and temperature at seasonal time scale

Trend analysis	Regression	R^2	r (unbiased)	P-Value	SD
Winter NDVI	y = 0.0027x + 5.204	0.164	0.390	0.0728	0.04
Spring NDVI	y = 0.0066x – 12.72	0.476	0.700	0.0002*	0.06
Summer NDVI	y = 0.0004x – 0.315	0.005	0.072	0.7511	0.03
Autumn NDVI	y = 0.0004x – 0.373	0.007	0.047	0.8358	0.02
Winter Precipitation	y = 2.1655x - 4244.5	0.170	0.392	0.0711	30.19
Spring Precipitation	y = 3.7008x - 7324.8	0.046	0.672	0.0006*	34.30
Summer Precipitation	y = 1.0251x - 1966.3	0.133	0.341	0.1204	17.79
Autumn Precipitation	y = 0.769x - 1410.5	0.002	0.115	0.6579	42.65
Winter Temperature	y = 0.0966x + 192.5	0.295	0.555	0.0079*	1.12
Spring Temperature	y = 0.0304x – 49.53	0.113	0.346	0.1143	0.57
Summer Temperature	y = 0.0373x – 51.81	0.170	0.421	0.0509	0.57
Autumn Temperature	y = 0.0577x – 102.7	0.337	0.592	0.0037*	0.62

Figure 3. The variation of NDVI and precipitation at seasonal time scale

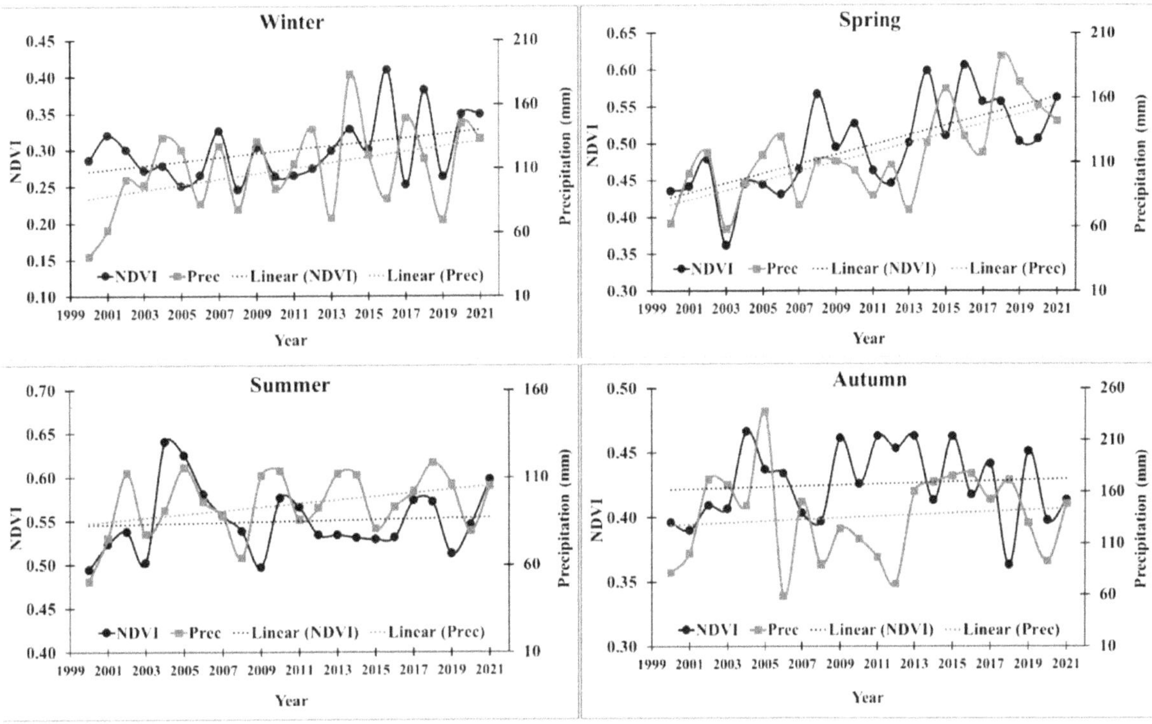

Figure 4. The variation of NDVI and precipitation at seasonal time scale

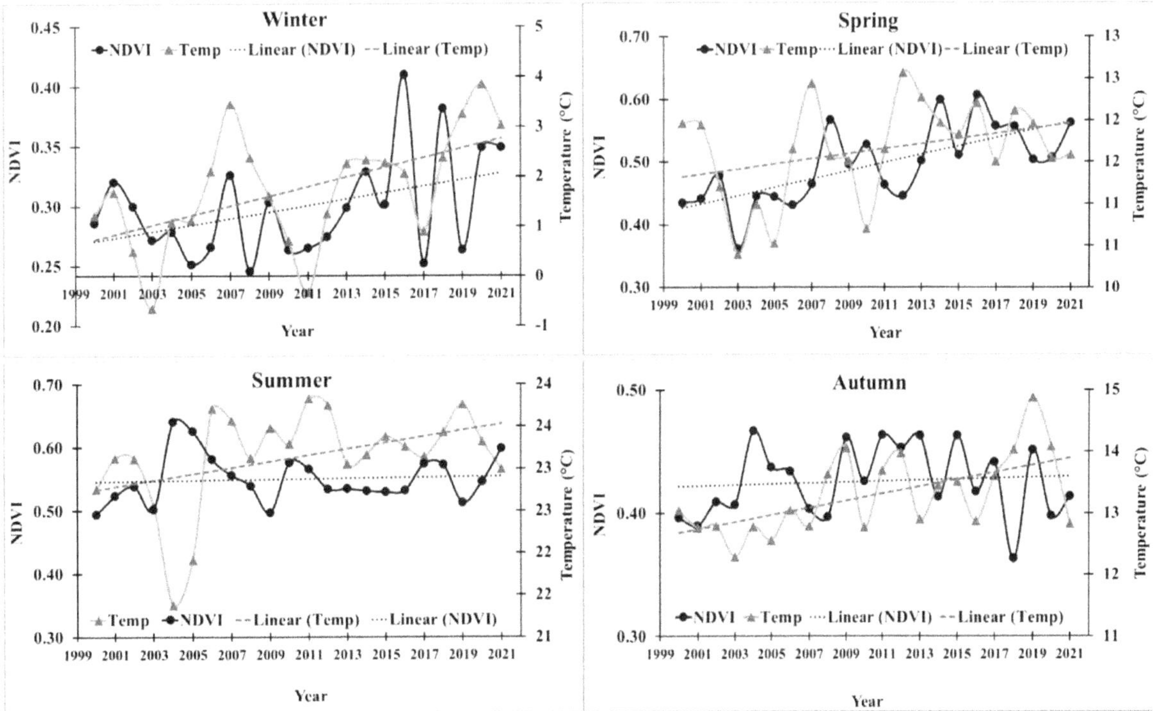

observed an increasing trend in both winter and spring, with the difference that the spring trend was statistically significant in the majority of the study area. On the summertime scale, a negligible growing trend (primarily in the far north) and a slight decline (in the interior regions) were observed, while the significant and insignificant decreasing trend appeared in the autumn (Fig. 5, e).

The Relationship Between NDVI and Climatic Factors

The spatial correlation analysis results between NDVI and climatic factors (precipitation and temperature) at inter-annual time scale are provided in Fig. 6. Based on the results, in most parts of the study area, mainly covered by croplands, we observed a moderately positive correlation between NDVI and precipitation. In some parts of the area covered by forests and wetlands, a negligible correlation between NDVI and precipitation was observed. Regarding the NDVI and temperature, a changeable correlation was detected, in that there is a high negative correlation in the north and southwest and the far north, which are mainly covered by croplands and grasslands, and a negligible correlation in the interior parts.

In winter and autumn, there was virtually no correlation between NDVI and precipitation on a seasonal time scale. We found a moderately positive correlation between NDVI and precipitation in the spring. Also, throughout the summer, a high positive correlation and a negligible correlation were found, primarily in regions with forests and croplands, respectively. Also, a changing correlation between the NDVI and temperature was discovered throughout the seasons. In the winter and autumn, we found no correlation between NDVI and temperature, despite some scattered moderate positive correlation in au-

Figure 5. Spatial and temporal trend of NDVI at different time scales: (a) inter-annual, (b) winter, (c) spring, (d) summer, and (e) autumn over the last 22 years (2000-2021)

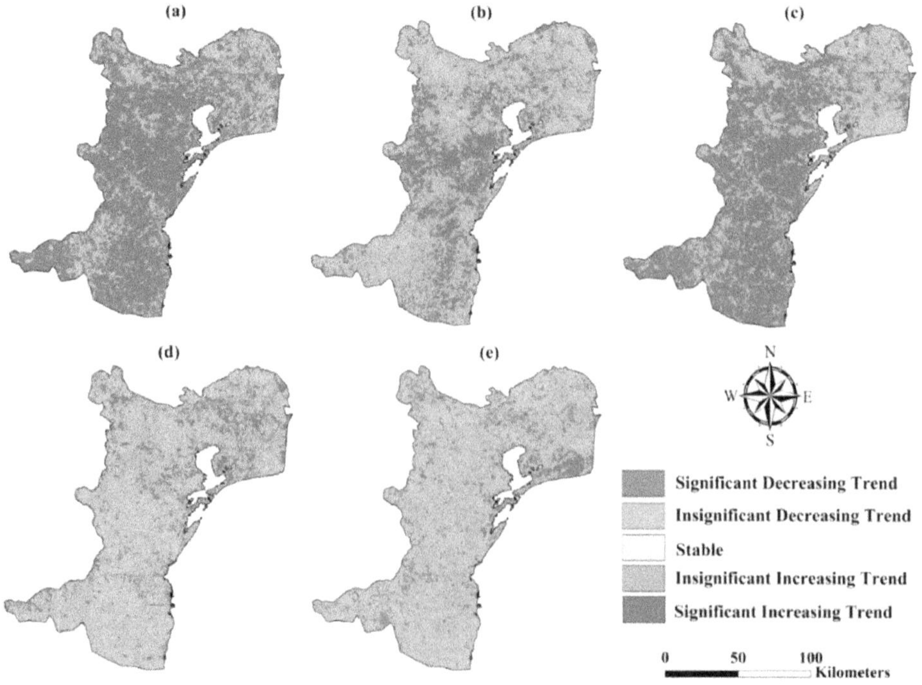

Figure 6. Pearson's correlation coefficient between NDVI and climatic factors at inter-annual time scale; a: precipitation and b: temperature

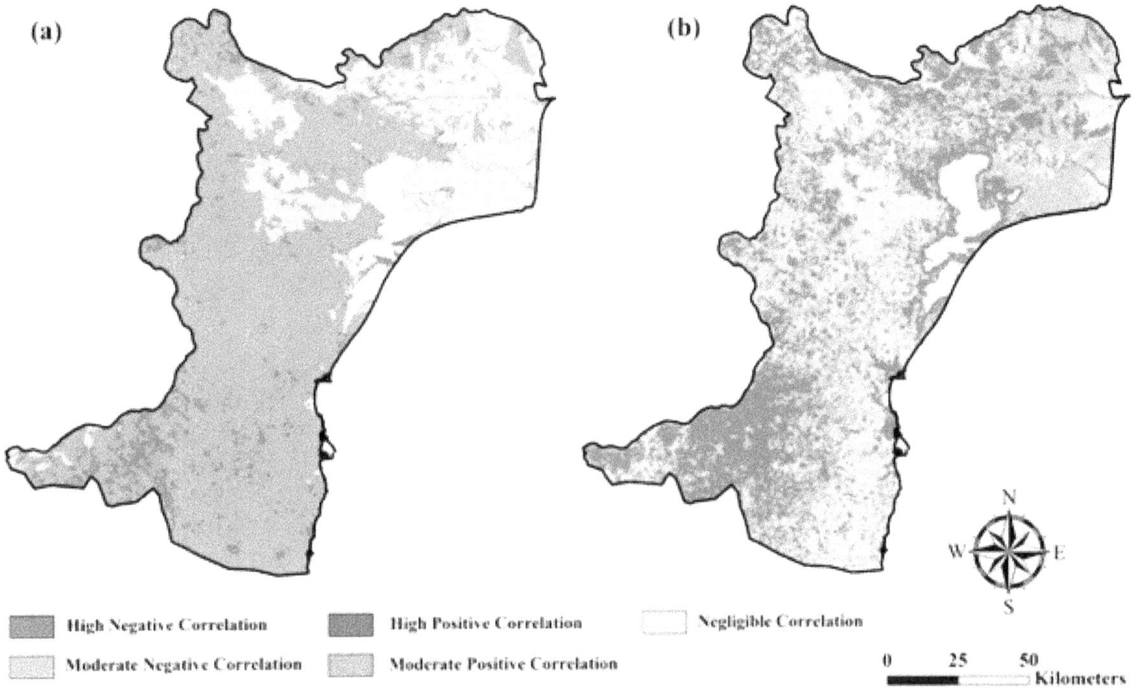

tumn. In the spring, a high positive correlation was identified in most parts of the study area; conversely, in the summer, a moderately negative correlation was observed.

DISCUSSION

Climate Change Trends and Impacts on Vegetation

Based on the results, we observed some of the key impacts of climate change in the study area. As mentioned in the result section, the internal-mean temperature experienced a significant increasing trend with a growth rate of 0.05°C. The same results were observed in the spring and autumn time series. These results are consistent with previous studies mentioning that climate change has been affecting European countries in various ways. According to the European Environment Agency (EEA), the average temperature in Europe has increased by 1.2°C above pre-industrial levels, and this warming trend is projected to continue in the coming decades (EEA, 2021). Several studies have investigated the causes and impacts of this warming trend. For example, Sippel et al., 2016 found that the increase in temperature in Europe between 1960 and 2014 was primarily due to human-caused greenhouse gas emissions. At regional scale, based on previous studies, Romania has been experiencing climate change in recent decades. Consistent with our results, for example, Micu et al. (2021) found that average temperatures in Romania have increased by 0.4°C per decade over the period 1961–2018. On the other hand, the annual

Figure 7. Pearson's correlation coefficient between NDVI and climatic factors at seasonal time scale

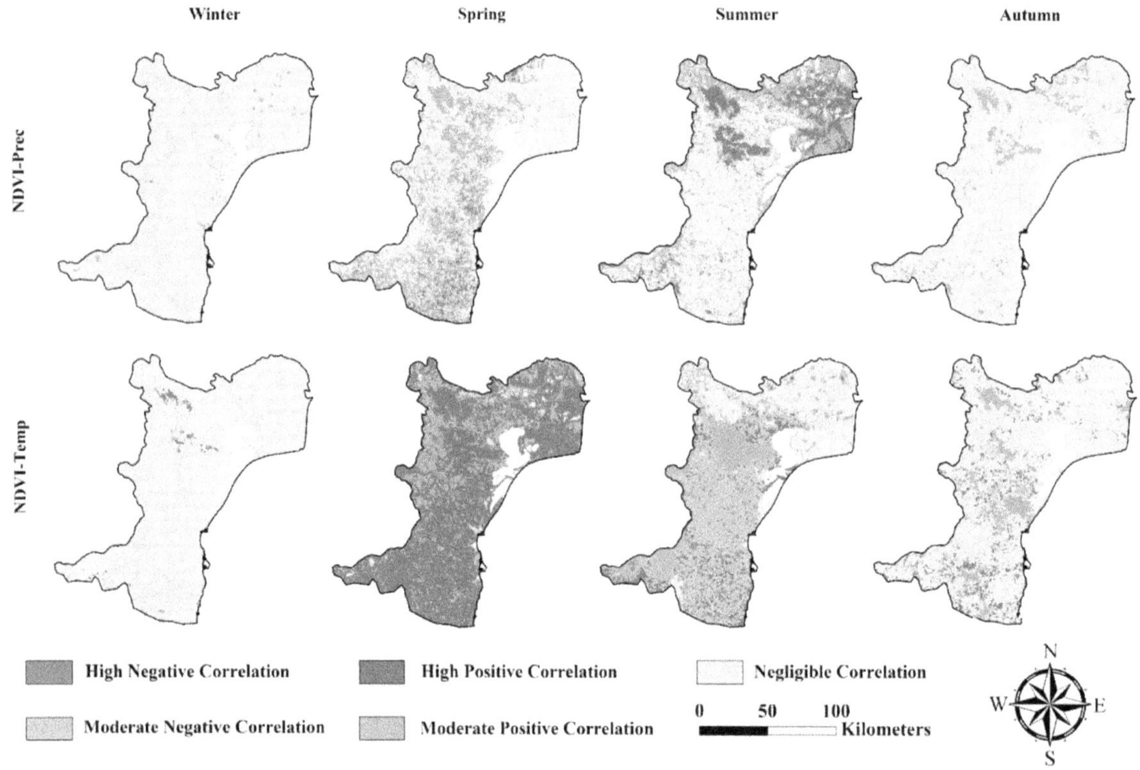

precipitation time series showed an increasing trend that was repeated at the seasonal time scale, especially in the spring when we observed a significant increasing trend. Also, climate change is expected to have significant impacts on precipitation patterns in Europe, with some areas becoming drier while others become wetter. In this regard, Kovats et al., 2014 found that climate change is expected to lead to changes in the seasonal distribution of precipitation in Europe, with winter precipitation increasing and summer precipitation decreasing in many parts of the continent. This finding partially replicates the results of our study because we observed an increasing trend in all seasons. In line with our findings, Birsan et al. (2019), indicated that total precipitation in Romania has increased slightly over the past century, but with significant spatial and temporal variability.

Increasing temperature and precipitation can have both positive and negative effects on vegetation. Precipitation is a key driver of ecosystem processes, and changes in precipitation patterns can have significant impacts on ecosystem functioning and biodiversity. For example, García-Palacios et al. (2018) reported that, increased precipitation in grassland ecosystems led to higher rates of primary productivity, as well as changes in plant species composition. While increased precipitation can have positive effects on some aspects of ecosystem functioning, it can also lead to increased soil erosion and flooding. Increased water availability due to precipitation and longer growing seasons due to warmer temperatures contributed to some positive effects such as enhancing the photosynthetic capacity and water use efficiency, which could facilitate the growth of vegetation and improve their carbon sequestration capacity (Chen et al. 2019). Regarding negative effects, some scientists noted that rising temperatures and precipitation may increase competition for resources and lead to alterations in the composition of the microbial community as a result of the changing climate. In a grassland environment in the Netherlands, a study by Oram et al. (2020) found that warming and more precipitation decreased the diversity of plant communities. Also, according to Halofsky et al. (2020), rising temperatures and precipitation may make some ecosystems more susceptible to drought and wildfire, both of which can be damaging to vegetation. For instance, additional vegetation growth driven by increased water availability from precipitation may enhance the fuel load and fire risk.

Spatial and Temporal Trend of NDVI and Vegetation Greening

The temporal and spatial trends of the inter-annual NDVI time series showed an increasing or greening trend in vegetation cover. Based on the results, at inter-annual time scale greening in vegetation cover happened in 87.22% of the study area. This growing trend was also observed at seasonal time scale, mainly in winter, spring, and autumn. The greening trend in vegetation refers to the increase in vegetation productivity and density over time. This trend has been documented in many regions around the world, particularly in the Northern Hemisphere. For example, (Vicca et al. 2012) analyzed the trends in the vegetation productivity of Europe between 1982 and 2006 using satellite data and found that, the vegetation productivity of Europe increased by 11% over the 25-year period. At regional scale, a study by Petrişor et al. (2020) analyzed the vegetation cover change in Romania between 1990 and 2012 using remote sensing data. The study found that the vegetation cover increased by 15.5% during this period, mainly due to the expansion of forests and grasslands. In another study, Prăvălie et al. (2022) analyzed the trend in vegetation greenness in Romania between 1987 and 2018 using satellite data and the Mann-Kendall test and reported a significant increasing trend in vegetation greenness during this period, with the strongest trend in spring. Vegetation greening and browning trends can be affected by various environmental factors, including climate change, land use change, and natural disturbances such as wildfires and insect outbreaks (Tucker et al. 2005; Lenton et al. 2008; Piao et al. 2010; Wei et al. 2011). IPCC (2013) reported that climate change increased vegetation productivity in Europe by 7.5% between 1982 and 2016, while land use change contributed to a 3.2% increase in vegetation productivity over the same period. In this regard, we also included LULC change detection in our investigation to assess how LULC change affected the vegetation's trend toward greening or browning. The results revealed unremarkable changes in LULC over the past 21 years. These findings support our earlier hypothesis that vegetation cover in Romania is undergoing a greening trend because of climate change.

The Relationship Between NDVI and Climatic Factors

The study found a moderately positive correlation between inter-annual NDVI and precipitation time series in most cropland and wetlands areas. However, in some areas covered by forests, a negligible correlation between NDVI and precipitation was observed. These results are not surprising because it is evident that water is an essential input for plant growth and development. Precipitation provides the necessary moisture for plants to carry out photosynthesis and grow. The negotiable correlation between precipitation and NDVI in forests may be related to other environmental factors such as soil moisture, temperature, nutrient availability, and disturbance regimes that also have a high impact on forests productivity. There are some previous studies that reported a negligible or weak correlation between NDVI and precipitation in forests. Anderegg et al. (2013) found that while precipitation was a significant driver of vegetation productivity in forest ecosystems, other factors such as species composition, biotic interactions, and disturbance regimes may also be important in determining vegetation productivity and could contribute to the weak correlation between NDVI and precipitation in some regions. Other studies have also suggested that the relationship between NDVI and precipitation can be influenced by local climate variability and other environmental factors. For example, a study by Tucker et al. (2005) found that the relationship between NDVI and precipitation varied across different land cover types, with a stronger correlation observed in grasslands compared to forests or shrublands. In most areas of the research area, we found that there was a strong negative correlation between NDVI and temperature. When the NDVI and temperature have a negative connection, it signifies that as the temperature increases, the NDVI values fall. This shows that increased temperatures may be affecting the growth and productivity of vegetation. This is due to the possibility that higher temperatures could lead to water stress, which would inhibit plant growth and photosynthesis. At the seasonal time scale, there was virtually no correlation between NDVI and climatic factors in winter and autumn, which can be related to low vegetation cover. In this time scale, the highest level of correlation was observed between NDVI and temperature in the spring. There is a significant body of research that has investigated the effect of increasing temperature on vegetation productivity in spring (Neumann et al. 2019; Zhou et al. 2020; Kelsey et al. 2021). For example, Chen et al. (2021) studied the impact of temperature on vegetation productivity in the Tibetan Plateau, a region with high sensitivity to climate change and found that an increase in temperature in spring led to a significant increase in vegetation productivity. Increasing temperature can enhance photosynthesis and respiration rates in plants, which in turn leads to increased growth and productivity. As temperature increases, plants can also have a higher rate of carbon dioxide uptake, which can further boost their growth and productivity. Also, with increasing temperatures, the spring season can start earlier and last longer, providing more time for plants to grow and produce biomass (Lehnebach et al. 2021). But this relationship changed in the summer when a negative correlation between temperature and NDVI was observed. High temperatures can increase evapotranspiration rates in plants, leading to water stress and decreased productivity. Also, high temperatures can cause heat stress and increase pest and disease pressure on vegetation, leading to reduced growth and productivity (Rivero et al. 2022).

CONCLUSION

The main aim of this study was to assess the effects of climate change on vegetation cover across Southeastern Romania over the past 22 years (2000–2021). To do this, different kinds of dense time series, including NDVI and climatic factors (precipitation and temperature), were analyzed based on statistical and trend analysis approaches. The methodology and findings provide the following conclusions: Temperature and precipitation experience an increasing trend with growth rates of 0.05 °C and 4.5 mm, respectively. The inter-annual NDVI time series, along with the Mann-Kendall trend analysis, showed an overall greening trend. At seasonal time scale, both temperature and precipitation showed an increasing trend in all seasons. Also, we observed an increasing trend in both the winter and spring NDVI time series. A negligible growing trend and a slight decreasing trend were also observed in the summer and autumn time series. The spatial correlation analysis between NDVI and climatic factors at inter-annual time scale revealed a moderately positive correlation between NDVI and precipitation and a high negative correlation between NDVI and temperature. Also, results revealed that vegetation cover in the study area is highly correlated with spring temperature and precipitation.

The proposed methodology, which is based on freely available satellite time series, meteorological datasets, and spatial statistical methods, could provide us with valuable information about the mechanism and intensity of climate change and its effect on vegetation cover. This mythology can be integrated into regional and global monitoring strategies to provide key information for mitigation and adaptation to climate change. As a novelty in this study, we also focused on seasonal time scales, which would provide a more precise understanding of the interaction between climate and vegetation growth.

Data Availability

The data supporting the findings of this study are available from the first author (V.N.) upon reasonable request.

Funding

The authors declare that no funds, grants, or other support were received during the preparation of this manuscript.

Conflict of Interest

The authors declare no conflict of interest – financial or otherwise.

Author Contributions

Writing - Original Draft, Formal Analysis, Visualization, Data Curation, Investigation, Methodology, Software, Validation, Supervision. V.N., Software, Visualization. S.R., Data Curation, Visualization. A.H., Supervision, C.M., Investigation, Writing - Review & Editing, Methodology. A.V.

REFERENCES

Abbas, A., Ekowati, D., Suhariadi, F., & Fenitra, R.M. (2023). Health implications, leaders societies, and climate change: A global review. *Ecological footprints of climate change: Adaptive approaches and sustainability*, 653-675.

Afuye, G. A., Kalumba, A. M., & Orimoloye, I. R. (2021). Characterisation of vegetation response to climate change: A review. *Sustainability (Basel)*, *13*(13), 7265. doi:10.3390u13137265

Anderegg, L. D., Anderegg, W. R., Abatzoglou, J., Hausladen, A. M., & Berry, J. A. (2013). Drought characteristics' role in widespread aspen forest mortality across Colorado, USA. *Global Change Biology*, *19*(5), 1526–1537. doi:10.1111/gcb.12146 PMID:23504823

Baniya, B., Tang, Q., Pokhrel, Y., & Xu, X. (2019). Vegetation dynamics and ecosystem service values changes at national and provincial scales in Nepal from 2000 to 2017. *Environmental Development*, *32*, 100464. doi:10.1016/j.envdev.2019.100464

Birsan, M.-V., Micu, D.-M., Nita, A., Mateescu, E., & Szep, R. (2019). Spatio-temporal changes in annual temperature extremes over Romania (1961–2013). *Romanian Journal of Physics*, *64*(7-8), 816.

Birsan, M. V., Zaharia, L., Chendes, V., & Branescu, E. (2014). Seasonal trends in Romanian streamflow. *Hydrological Processes*, *28*(15), 4496–4505. doi:10.1002/hyp.9961

Bowman, D. M., Perry, G. L., & Marston, J. (2015). Feedbacks and landscape-level vegetation dynamics. *Trends in Ecology & Evolution*, *30*(5), 255–260. doi:10.1016/j.tree.2015.03.005 PMID:25837918

Chaudhry, S., & Sidhu, G. P. S. (2022). Climate change regulated abiotic stress mechanisms in plants: A comprehensive review. *Plant Cell Reports*, *41*(1), 1–31. doi:10.100700299-021-02759-5 PMID:34351488

Chen, A., Huang, L., Liu, Q., & Piao, S. (2021). Optimal temperature of vegetation productivity and its linkage with climate and elevation on the Tibetan Plateau. *Global Change Biology*, *27*(9), 1942–1951. doi:10.1111/gcb.15542 PMID:33528057

Chen, J., Shao, C., Jiang, S., Qu, L., & Zhao, F. (2019). Effects of changes in precipitation on energy and water balance in a Eurasian meadow steppe. *Ecological Processes*, *8*, 1–15. doi:10.3390/pr8010001

Chen, X., Chen, W., & Xu, M. (2022). Remote-Sensing Monitoring of Postfire Vegetation Dynamics in the Greater Hinggan Mountain Range Based on Long Time-Series Data: Analysis of the Effects of Six Topographic and Climatic Factors. *Remote Sensing (Basel)*, *14*(13), 2958. doi:10.3390/rs14132958

Chen, Z., Liu, J., Qian, Z., Li, L., Zhang, Z., Feng, G., Ruan, S., & Sun, G. (2023). Monitoring Land Degradation through Vegetation Dynamics Mathematical Modeling: Case of Jornada Basin (in the US). *Remote Sensing (Basel)*, *15*(4), 978. doi:10.3390/rs15040978

Cui, L., Pang, B., Zhao, G., Ban, C., Ren, M., Peng, D., Zuo, D., & Zhu, Z. (2022). Assessing the Sensitivity of Vegetation Cover to Climate Change in the Yarlung Zangbo River Basin Using Machine Learning Algorithms. *Remote Sensing (Basel)*, *14*(7), 1556. doi:10.3390/rs14071556

da Silva, V. S., Salami, G., da Silva, M. I. O., Silva, E. A., & Monteiro, J. J. (2020). Methodological evaluation of vegetation indexes in land use and land cover (LULC) classification. *Geology, Ecology, and Landscapes*, *4*(2), 159–169. doi:10.1080/24749508.2019.1608409

Di Sacco, A., Hardwick, K. A., Blakesley, D., Brancalion, P. H., Breman, E., Cecilio Rebola, L., Chomba, S., Dixon, K., Elliott, S., Ruyonga, G., Shaw, K., Smith, P., Smith, R. J., & Antonelli, A. (2021). Ten golden rules for reforestation to optimize carbon sequestration, biodiversity recovery and livelihood benefits. *Global Change Biology*, *27*(7), 1328–1348. doi:10.1111/gcb.15498 PMID:33494123

Eskander, S. M. S. U., & Fankhauser, S. (2020). Reduction in greenhouse gas emissions from national climate legislation. *Nature Climate Change*, *10*(8), 750–756. doi:10.103841558-020-0831-z

Fengjin, X., & Qiufeng, L. (2023). An evaluation of vegetation loss due to the super typhoon Sarika in Hainan Island of China. *Natural Hazards*, *115*(2), 1677–1695. doi:10.100711069-022-05613-3

Gao, W., Zheng, C., Liu, X., Lu, Y., Chen, Y., Wei, Y., & Ma, Y. (2022). NDVI-based vegetation dynamics and their responses to climate change and human activities from 1982 to 2020: A case study in the Mu Us Sandy Land, China. *Ecological Indicators*, *137*, 108745. doi:10.1016/j.ecolind.2022.108745

García-Palacios, P., Gross, N., Gaitán, J., & Maestre, F. T. (2018). Climate mediates the biodiversity–ecosystem stability relationship globally. *Proceedings of the National Academy of Sciences of the United States of America*, *115*(33), 8400–8405. doi:10.1073/pnas.1800425115 PMID:30061405

Gatti, L. V., Basso, L. S., Miller, J. B., Gloor, M., Gatti Domingues, L., Cassol, H. L. G., Tejada, G., Aragão, L. E. O. C., Nobre, C., Peters, W., Marani, L., Arai, E., Sanches, A. H., Corrêa, S. M., Anderson, L., Von Randow, C., Correia, C. S. C., Crispim, S. P., & Neves, R. A. L. (2021). Amazonia as a carbon source linked to deforestation and climate change. *Nature*, *595*(7867), 388–393. doi:10.103841586-021-03629-6 PMID:34262208

Ghebrezgabher, M. G., Yang, T., Yang, X., & Sereke, T. E. (2020). Assessment of NDVI variations in responses to climate change in the Horn of Africa. *The Egyptian Journal of Remote Sensing and Space Sciences*, *23*(3), 249–261. doi:10.1016/j.ejrs.2020.08.003

Halofsky, J. E., Peterson, D. L., & Harvey, B. J. (2020). Changing wildfire, changing forests: The effects of climate change on fire regimes and vegetation in the Pacific Northwest, USA. *Fire Ecology*, *16*(1), 1–26. doi:10.118642408-019-0062-8

Heilmayr, R., Echeverría, C., & Lambin, E. F. (2020). Impacts of Chilean forest subsidies on forest cover, carbon and biodiversity. *Nature Sustainability*, *3*(9), 701–709. doi:10.103841893-020-0547-0

Hoffmann, R., Muttarak, R., Peisker, J., & Stanig, P. (2022). Climate change experiences raise environmental concerns and promote Green voting. *Nature Climate Change*, *12*(2), 148–155. doi:10.103841558-021-01263-8

Holden, P. B., Rebelo, A. J., Wolski, P., Odoulami, R. C., Lawal, K. A., Kimutai, J., Nkemelang, T., & New, M. G. (2022). Nature-based solutions in mountain catchments reduce impact of anthropogenic climate change on drought streamflow. *Communications Earth & Environment*, *3*(1), 51. doi:10.103843247-022-00379-9

Huang, J., Zhang, G., Zhang, Y., Guan, X., Wei, Y., & Guo, R. (2020). Global desertification vulnerability to climate change and human activities. *Land Degradation & Development*, *31*(11), 1380–1391. doi:10.1002/ldr.3556

IPCC. (2013). The Physical Science Basis. In T. F. Stocker, D. Qin, G.-K. Plattner, M. Tignor, S. K. Allen, J. Boschung, A. Nauels, Y. Xia, V. Bex, & P. M. Midgley (Eds.), *Climate Change 2013: Contribution of Working Group I to the Fifth Assessment Report of the Intergovernmental Panel on Climate Change*. Cambridge University Press.

Jiang, L., Bao, A., Guo, H., & Ndayisaba, F. (2017). Vegetation dynamics and responses to climate change and human activities in Central Asia. *The Science of the Total Environment*, *599*, 967–980. doi:10.1016/j.scitotenv.2017.05.012 PMID:28505889

Kelsey, K. C., Pedersen, S. H., Leffler, A. J., Sexton, J. O., Feng, M., & Welker, J. M. (2021). Winter snow and spring temperature have differential effects on vegetation phenology and productivity across Arctic plant communities. *Global Change Biology*, *27*(8), 1572–1586. doi:10.1111/gcb.15505 PMID:33372357

Khunrattanasiri, W. (2023). Application of Remote Sensing Vegetation Indices for Forest Cover Assessments. In *Concepts and Applications of Remote Sensing in Forestry* (pp. 153–166). Springer.

Kolarik, N., Roopsind, A., Pickens, A., & Brandt, J. (2023). A satellite-based monitoring system for quantifying surface water and mesic vegetation dynamics in a semi-arid region. *Ecological Indicators*, *147*, 109965. doi:10.1016/j.ecolind.2023.109965

Kovats, R. S., Valentini, R., Bouwer, L. M., Georgopoulou, E., Jacob, D., Martin, E., Rounsevell, M., & Soussana, J. F. (2014) Europe. In: Barros VR, Field CB, Dokken DJ, Mastrandrea MD, Mach KJ, Bilir TE, Chatterjee M, Ebi KL, Estrada YO, Genova RC, Girma B, Kissel ES, Levy AN, MacCracken, Mastrandrea PR, White LL (eds) Climate change 2014: impacts, adaptation, and vulnerability. Part B: regional aspects. Contribution of Working Group II to the Fifth Assessment Report of the Intergovernmental Panel on Climate Change

Leal, K. B., Robaina, L. E. S., & De Lima, A. S. (2022). Coastal impacts of storm surges on a changing climate: A global bibliometric analysis. *Natural Hazards*, *114*(2), 1455–1476. doi:10.100711069-022-05432-6

Lehnebach, R., Campioli, M., Gričar, J., Prislan, P., Mariën, B., Beeckman, H., & Van den Bulcke, J. (2021). High-resolution X-ray computed tomography: A new workflow for the analysis of xylogenesis and intra-seasonal wood biomass production. *Frontiers in Plant Science*, *12*, 698640. doi:10.3389/fpls.2021.698640 PMID:34421949

Lenton, T. M., Held, H., Kriegler, E., Hall, J. W., Lucht, W., Rahmstorf, S., & Schellnhuber, H. J. (2008). Tipping elements in the Earth's climate system. *Proceedings of the National Academy of Sciences of the United States of America*, *105*(6), 1786–1793. doi:10.1073/pnas.0705414105 PMID:18258748

Lhotka, O., Kyselý, J., & Farda, A. (2018). Climate change scenarios of heat waves in Central Europe and their uncertainties. *Theoretical and Applied Climatology*, *131*(3-4), 1043–1054. doi:10.100700704-016-2031-3

Li, P., Wang, J., Liu, M., Xue, Z., Bagherzadeh, A., & Liu, M. (2021). Spatio-temporal variation characteristics of NDVI and its response to climate on the Loess Plateau from 1985 to 2015. *Catena*, *203*, 105331. doi:10.1016/j.catena.2021.105331

Liu, W., Feng, Q., Engel, B. A., Yu, T., Zhang, X., & Qian, Y. (2023). A probabilistic assessment of urban flood risk and impacts of future climate change. *Journal of Hydrology (Amsterdam)*, *618*, 129267. doi:10.1016/j.jhydrol.2023.129267

Loarie, S. R., Duffy, P. B., Hamilton, H., Asner, G. P., Field, C. B., & Ackerly, D. D. (2009). The velocity of climate change. *Nature*, *462*(7276), 1052–1055. doi:10.1038/nature08649 PMID:20033047

Maftei, C., & Barbulescu, A. (2008). Statistical analysis of climate evolution in Dobrudja region. Proceedings of the World Congress on Engineering, Maftei C, Barbulescu A (2012). Statistical analysis of precipitation time series in Dobrudja region. *Mausam (New Delhi)*, *63*(4), 553–564. doi:10.54302/mausam.v63i4.416

Micu, D. M., Amihaesei, V. A., Milian, N., & Cheval, S. (2021). Recent changes in temperature and precipitation indices in the Southern Carpathians, Romania (1961–2018). *Theoretical and Applied Climatology*, *144*(1-2), 691–710. doi:10.100700704-021-03560-w

Morecroft, M. D., & Keith, S. A. (2009). Plant ecology as an indicator of climate and global change. In *Climate Change* (pp. 297–305). Elsevier. doi:10.1016/B978-0-444-53301-2.00016-6

Mukul, S. A., Alamgir, M., Sohel, M. S. I., Pert, P. L., Herbohn, J., Turton, S. M., Khan, M. S. I., Munim, S. A., Reza, A. H. M. A., & Laurance, W. F. (2019). Combined effects of climate change and sea-level rise project dramatic habitat loss of the globally endangered Bengal tiger in the Bangladesh Sundarbans. *The Science of the Total Environment*, *663*, 830–840. doi:10.1016/j.scitotenv.2019.01.383 PMID:30738263

Nabi, G., Ullah, S., Khan, S., Ahmad, S., & Kumar, S. (2018). China-Pakistan Economic Corridor (CPEC): Melting glaciers—a potential threat to ecosystem and biodiversity. *Environmental Science and Pollution Research International*, *25*(4), 3209–3210. doi:10.100711356-017-1151-3 PMID:29290058

Nasiri, V., Sadeghi, S. M. M., Bagherabadi, R., Moradi, F., Deljouei, A., & Borz, S. A. (2022). Modeling wildfire risk in western Iran based on the integration of AHP and GIS. *Environmental Monitoring and Assessment*, *194*(9), 644. doi:10.100710661-022-10318-y PMID:35930117

Neumann, R. B., Moorberg, C. J., Lundquist, J. D., Turner, J. C., Waldrop, M. P., McFarland, J. W., Euskirchen, E. S., Edgar, C. W., & Turetsky, M. R. (2019). Warming effects of spring rainfall increase methane emissions from thawing permafrost. *Geophysical Research Letters*, *46*(3), 1393–1401. doi:10.1029/2018GL081274

Oram, N. J., De Deyn, G. B., Bodelier, P. L., Cornelissen, J. H., van Groenigen, J. W., & Abalos, D. (2020). Plant community flood resilience in intensively managed grasslands and the role of the plant economic spectrum. *Journal of Applied Ecology*, *57*(8), 1524–1534. doi:10.1111/1365-2664.13667

Petrişor, A.-I., Sirodoev, I., & Ianoş, I. (2020). Trends in the national and regional transitional dynamics of land cover and use changes in Romania. *Remote Sensing (Basel)*, *12*(2), 230. doi:10.3390/rs12020230

Piao, S., Ciais, P., Huang, Y., Shen, Z., Peng, S., Li, J., Zhou, L., Liu, H., Ma, Y., Ding, Y., Friedlingstein, P., Liu, C., Tan, K., Yu, Y., Zhang, T., & Fang, J. (2010). The impacts of climate change on water resources and agriculture in China. *Nature*, *467*(7311), 43–51. doi:10.1038/nature09364 PMID:20811450

Piao, S., Liu, Q., Chen, A., Janssens, I. A., Fu, Y., Dai, J., Liu, L., Lian, X., Shen, M., & Zhu, X. (2019). Plant phenology and global climate change: Current progresses and challenges. *Global Change Biology*, *25*(6), 1922–1940. doi:10.1111/gcb.14619 PMID:30884039

Pouyan, S., Rahmanian, S., Amindin, A., & Pourghasemi, H. R. (2022). Spatial and seasonal modeling of the land surface temperature using random forest. In *Computers in Earth and Environmental Sciences* (pp. 221–234). Elsevier. doi:10.1016/B978-0-323-89861-4.00035-X

Prăvălie, R., Patriche, C. V., Sirodoev, I., Bandoc, G., Dumitraşcu, M., & Peptenatu, D. (2016). Water deficit and corn productivity during the post-socialist period. Case study: Southern Oltenia drylands, Romania. *Arid Land Research and Management*, *30*(3), 239–257. doi:10.1080/15324982.2015.1091399

Prăvălie, R., Piticar, A., Roşca, B., Sfîcă, L., Bandoc, G., Tiscovschi, A., & Patriche, C. (2019). Spatio-temporal changes of the climatic water balance in Romania as a response to precipitation and reference evapotranspiration trends during 1961–2013. *Catena*, *172*, 295–312. doi:10.1016/j.catena.2018.08.028

Prăvălie, R., Sîrodoev, I., Nita, I.-A., Patriche, C., Dumitraşcu, M., Roşca, B., Tişcovschi, A., Bandoc, G., Săvulescu, I., Mănoiu, V., & Birsan, M.-V. (2022). NDVI-based ecological dynamics of forest vegetation and its relationship to climate change in Romania during 1987–2018. *Ecological Indicators*, *136*, 108629. doi:10.1016/j.ecolind.2022.108629

Praveen, B., & Sharma, P. (2019). A review of literature on climate change and its impacts on agriculture productivity. *Journal of Public Affairs*, *19*(4), e1960. doi:10.1002/pa.1960

Qian, C., Qiang, H., Wang, F., & Li, M. (2021). Optimization of Rocky Desertification Classification Model Based on Vegetation Type and Seasonal Characteristic. *Remote Sensing (Basel)*, *13*(15), 2935. doi:10.3390/rs13152935

Richardson, A. D., Keenan, T. F., Migliavacca, M., Ryu, Y., Sonnentag, O., & Toomey, M. (2013). Climate change, phenology, and phenological control of vegetation feedbacks to the climate system. *Agricultural and Forest Meteorology*, *169*, 156–173. doi:10.1016/j.agrformet.2012.09.012

Rivero, R. M., Mittler, R., Blumwald, E., & Zandalinas, S. I. (2022). Developing climate-resilient crops: Improving plant tolerance to stress combination. *The Plant Journal*, *109*(2), 373–389. doi:10.1111/tpj.15483 PMID:34482588

Şerban, C., Maftei, C., & Dobrică, G. (2022). Surface water change detection via water indices and predictive modeling using remote sensing imagery: A case study of Nuntasi-Tuzla Lake, Romania. *Water (Basel)*, *14*(4), 556. doi:10.3390/w14040556

Shen, F., Yang, L., Zhang, L., Guo, M., Huang, H., & Zhou, C. (2023). Quantifying the direct effects of long-term dynamic land use intensity on vegetation change and its interacted effects with economic development and climate change in jiangsu, China. *Journal of Environmental Management*, *325*, 116562. doi:10.1016/j.jenvman.2022.116562 PMID:36308967

Thuiller, W. (2007). Climate change and the ecologist. *Nature*, *448*(7153), 550–552. doi:10.1038/448550a PMID:17671497

Tucker, C. J., Pinzon, J. E., Brown, M. E., Slayback, D. A., Pak, E. W., Mahoney, R., Vermote, E. F., & El Saleous, N. (2005). An extended AVHRR 8-km NDVI dataset compatible with MODIS and SPOT vegetation NDVI data. *International Journal of Remote Sensing*, *26*(20), 4485–4498. doi:10.1080/01431160500168686

Vicca, S., Gilgen, A. K., Serrano, M. C., Dreesen, F., & Dukes, J. (2012). Urgent need for a common metric to make precipitation manipulation experiments comparable. *The New Phytologist*, *195*(3), 518–522. doi:10.1111/j.1469-8137.2012.04224.x PMID:22734795

Wang, M., He, G., Ishwaran, N., Hong, T., Bell, A., Zhang, Z., Wang, G., & Wang, M. (2020). Monitoring vegetation dynamics in East Rennell Island World Heritage Site using multi-sensor and multi-temporal remote sensing data. *International Journal of Digital Earth*, *13*(3), 393–409. doi:10.1080/17538947.2018.1523955

Wang, Y., Pineda-Munoz, S., & McGuire, J. L. (2023). Plants maintain climate fidelity in the face of dynamic climate change. *Proceedings of the National Academy of Sciences of the United States of America*, *120*(7), e2201946119. doi:10.1073/pnas.2201946119 PMID:36745797

Wei, X., Shao, M., Fu, X., Ågren, G. I., & Yin, X. (2011). The effects of land use on soil N mineralization during the growing season on the northern Loess Plateau of China. *Geoderma*, *160*(3-4), 590–598. doi:10.1016/j.geoderma.2010.11.007

Wei, Y., Sun, S., Liang, D., & Jia, Z. (2022). Spatial–temporal variations of NDVI and its response to climate in China from 2001 to 2020. *International Journal of Digital Earth*, *15*(1), 1463–1484. doi:10.1080/17538947.2022.2116118

Williams, A. P., Abatzoglou, J. T., Gershunov, A., Guzman-Morales, J., Bishop, D. A., Balch, J. K., & Lettenmaier, D. P. (2019). Observed impacts of anthropogenic climate change on wildfire in California. *Earth's Future*, *7*(8), 892–910. doi:10.1029/2019EF001210

Wu, G., Fang, Y., Jiang, Q., Cui, M., Li, N., Ou, Y., Diao, Z., & Zhang, B. (2023). Early identification of strawberry leaves disease utilizing hyperspectral imaging combing with spectral features, multiple vegetation indices and textural features. *Computers and Electronics in Agriculture*, *204*, 107553. doi:10.1016/j.compag.2022.107553

Zheng, X., Streimikiene, D., Balezentis, T., Mardani, A., Cavallaro, F., & Liao, H. (2019). A review of greenhouse gas emission profiles, dynamics, and climate change mitigation efforts across the key climate change players. *Journal of Cleaner Production*, *234*, 1113–1133. doi:10.1016/j.jclepro.2019.06.140

Zhou, J., Chen, J., Chen, X., Zhu, X., Qiu, Y., Song, H., Rao, Y., Zhang, C., Cao, X., & Cui, X. (2021). Sensitivity of six typical spatiotemporal fusion methods to different influential factors: A comparative study for a normalized difference vegetation index time series reconstruction. *Remote Sensing of Environment*, *252*, 112130. doi:10.1016/j.rse.2020.112130

Zhou, X., Geng, X., Yin, G., Hänninen, H., & Hao, F. (2020). Legacy effect of spring phenology on vegetation growth in temperate China. *Agricultural and Forest Meteorology*, *281*, 107845. doi:10.1016/j.agrformet.2019.107845

APPENDIX

Figure 8. A1 Multi temporal LULC maps; the zoomed in windows show the details of LULC changes

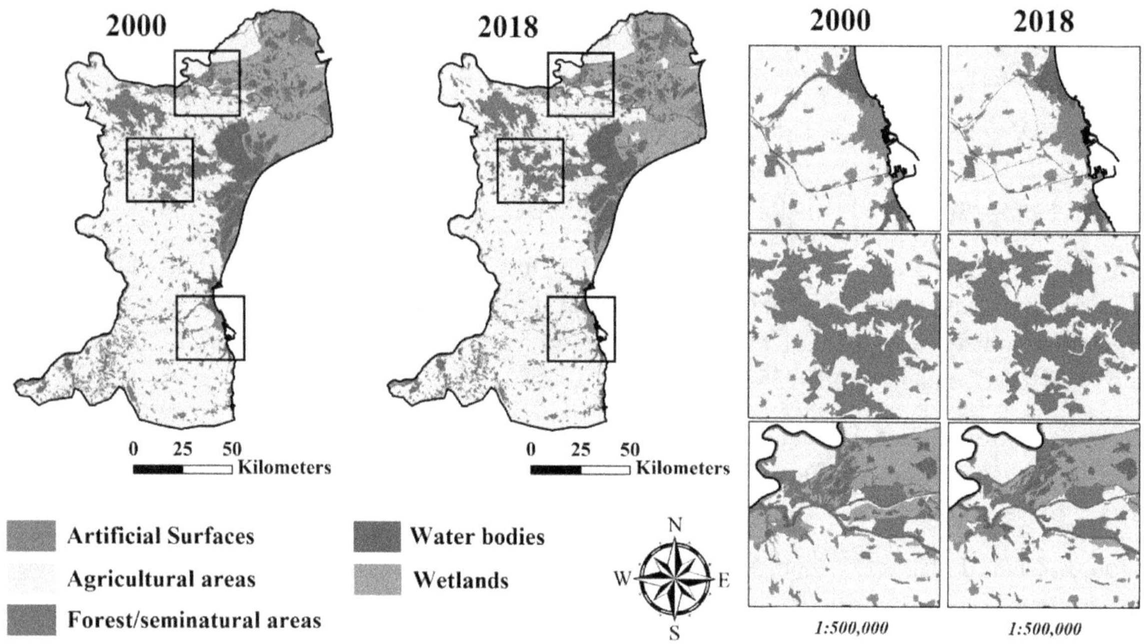

Table 2. LULC changes detection statistics. All values are rounded to full hectares.

LULC	Area (ha)		Change (ha)
	2000	2018	
Artificial surfaces	68,684	69,478	794
Agricultural areas	903,941	924,695	20754
Forest and semi natural areas	203,409	190,219	-13190
Wetlands	230,095	230,409	314
Water bodies	147,946	140,869	-7077

Chapter 9
An Assessment of Low Flow and Water Deficits on the Danube and Romanian Rivers During 1980 – 2020

Silvia Mihaela Chelcea
https://orcid.org/0009-0001-8253-9768
National Institute of Hydrology and Water Management, Romania

Adrian Alexandru Aldea
National Institute of Hydrology and Water Management, Romania

Maria Cristina Trifu
National Institute of Hydrology and Water Management, Romania

ABSTRACT

The daily streamflow data—at 93 gauges on Romanian rivers and other 17 on the Romanian part of the Danube River, covering the period 1980 – 2020—were used to assess the characteristics of low flows, deficits, and seasonality. It was necessary to consider several indicators, such as: meanflow (MQ), mean annual minimum n-day discharges (MAMn-day), low flow percentiles (Q80, Q95), base flow index (BFI), extreme values of annual minimum 7-day discharge (AM7) and the streamflow below a fixed Q80 threshold. For assessing the occurrence and variability of low flows, rate and index of seasonality was used. The statistical computation of indices was made using the LFSTAT application packages developed in the R program. Also, in order to have an overview, at the level of the Romania, of low flows magnitude, the variability of minimum discharges on different time durations (days), the relative contribution of base flow to total river flow and for highlighting the driest year in the period 1980 - 2020, there were produced several maps with the spatial distribution of low flow indices.

DOI: 10.4018/978-1-6684-8771-6.ch009

THE ASSESSMENT OF LOW FLOW AND WATER DEFICITS ON THE DANUBE AND ROMANIAN RIVERS FOR THE PERIOD 1980 – 2020

At a regional scale, in this case the Danube and Romanian rivers, an assessment of both low flow and water deficit, over the last 41 years (1980 – 2020), is crucial in providing useful information, both at national and European level, for proactive management of drought risks. Specific aspects of low flows and deficits are the subject of many current research initiatives in the context of climate change.

There are studies that focus on low water periods characterized by standard methods that include minimum annual flows, drought duration and streamflow deficit below a certain threshold (e.g., Gustard and Demuth, 2008; Koffler et al., 2016; Hosking and Wallis, 2005; Hisdal et al., 2004; Gauster and Laaha, 2016; Tallaksen and Van Lanen, 2004; Laaha et al., 2013; Smakhtin, 2001; Salinas et al, 2013). Based on a range of low flow indices calculated from observed streamflow records of approximately 800 gauges across Europe, Laaha et al. (2016) was analyzed European drought of 2015 from a hydrological perspective.

During the period from 1990 to 2017, many parts of Europe suffered from noticeable droughts, such as in 1991–1995, 1996–1997, 2000, 2003, 2004–2007, 2010, 2011, 2015 (Stagge et al., 2013; Spinoni et al., 2015; Van Lanen et al., 2016, Sutanto and van Lanen, 2020).

As regards the Romanian rivers basins, a study by Sutanto and Van Lanen (2020), based on an analysis of hydrological characteristics in Europe from 1990 to 2017, showed that in the eastern part of Europe, from Poland to Bulgaria, there are clear regions with severe hydrological and groundwater drought hazards. These regions, as well as north-western Russia, experienced the longest drought duration and the highest deficit volume compared to other regions. Based on summary statistics of drought characteristics, the most severe droughts occurred from 1992 to 1997. The European 2003 drought was considered severe, but it appeared not the most extreme, in term of duration and water deficits (Sutanto and van Lanen, 2020).

A comprehensive characterization of hydrological drought events, such as those of 2015 and 2003, requires several different indices (Tallaksen and Van Lanen, 2004; Laaha et al., 2013; Smakhtin, 2001; Salinas et al., 2013, Laaha et al., 2016). First, the magnitude of the low flow discharge is important; it may be characterized by annual minimum flows or flow quantiles with high exceedance probability. Second, the timing of low flow is important. It may be characterized by a monthly low flow index, such as the monthly 7-day minimum flow, or by a seasonality index such as the day of occurrence for the annual minimum. Third, a characterization of drought events when the flow is below a given threshold is important. These drought events may be characterized by their duration, deficit volume, or similar indices (Yevjevich, 1967; Hisdal et al., 2004, Laaha et al., 2016).

The seasonality of hydrological characteristics is also one of the key factors controlling the development and stability of natural ecosystems. In the context of climate change, the use of seasonality indices, such as a cyclic index and seasonality rate, is recommended to assess the occurrence of minimum flows and to evaluate the seasonal variability of the low flow regime. The value of different seasonality indices for low flow regionalization was examined by Laaha and Blöschl (2006).

The Danube River in the Romanian section, like all major rivers in Europe, has been affected by significant low-flow events in the last two decades in 2003, 2011, 2012, 2015, 2018 and 2019. According to the 2022 edition of the Copernicus Climate Change Service's (C3S*) European State of the Climate (ESOTC) report (https://climate.copernicus.eu/esotc/2022/river-discharge), for 10 months of the year 2022, river discharge was below average and, also, river discharge was the second lowest on record across

Europe, marking the sixth consecutive year of below-average flows. For Danube catchment, March marked the onset of an exceptionally dry year, with record low water levels recorded on multiple occasions.

In the present study, the daily streamflow data at 93 gauges on Romanian rivers and other 17 on the Romanian segment of the Danube River, covering the period 1980 – 2020, were used to assess characteristics of low flows, water deficits, and seasonality. To carry out this assessment, the several indicators, such as: meanflow, mean annual minimum n-day discharges, low flow percentiles, base flow index, extreme values of annual minimum 7-day discharge, streamflow deficit below a certain threshold, seasonality rate, and seasonality index, were estimated and analysed. A rank analysis of the annual series of mean flow MQ and low flow percentile Q95, as well as an analysis of the extreme values of the 7-day minimum estimated for different return periods, were also carried out in this study.

In order to provide an overview at the Romanian level of low flows and water deficit for the reference period 1980 - 2020, maps were made showing the distribution of the magnitude of low flows, the variability of minimum flows over different time periods (days), the relative contribution of base flow to total flow, the seasonality rate, and the degree of variability in the occurrence of low flows, as well as highlighting the driest years from a hydrological point of view.

The study also analysed periods of low flow below the fixed Q80 threshold and identified the largest streamflow deficits from 1980 to 2020.

STUDY AREA, DATA SETS, STATISTICAL ANALYSES, AND INDICES

Study Area

The Romanian hydrographic space is entirely integrated in the lower sector of the Danube River, which stretches from downstream of the Iron Gate to Black Sea and drains the area of the Balkan Mountains, the Southern Carpathians, the Curved Carpathians, the Eastern Carpathians, the Pre-Balkan, and the Moldavian Plateau, as well as the Lower Danube Plain. Thus, the surface of the basin increases by another 228580 km^2, to a total of 805300 km^2 at release. Draining the waters from the Carpathian and Balkan Mountains, the lower reaches of the Danube run between Bazias and the Black Sea for a length of 1072 km, which represents 38% of the total length of the river. The Romanian tributaries of the Danube in this sector are the rivers Jiu, Olt and Arges, which originate in the Southern Carpathians, as well as several smaller tributaries, including Desnatui, Calmatui and Mostistea, whose flows do not influence the hydrological regime of the Danube.

Data Sets

For study were selected 110 hydrometric stations on rivers from 14 first order river basins (according to the Water Cadastre Atlas of Romania, 1992) of the whole Romanian hydrographic network, as well as the lower Danube River sector, Romanian side (left bank of the Danube). The location of hydrometric stations is shown in Figure 1 (for inland rivers) and Figure 2 (for Danube River), respectively. Therefore, the daily streamflow data at 93 hydrometric stations (Table 1) on inland rivers and other 17 (Table 2) on the Danube River, covering the period 1980 - 2020, were used to assess characteristics of low flows, water deficits, and seasonality.

Figure 1. The spatial distribution of 93 hydrometric stations on Romanian rivers (inland rivers)

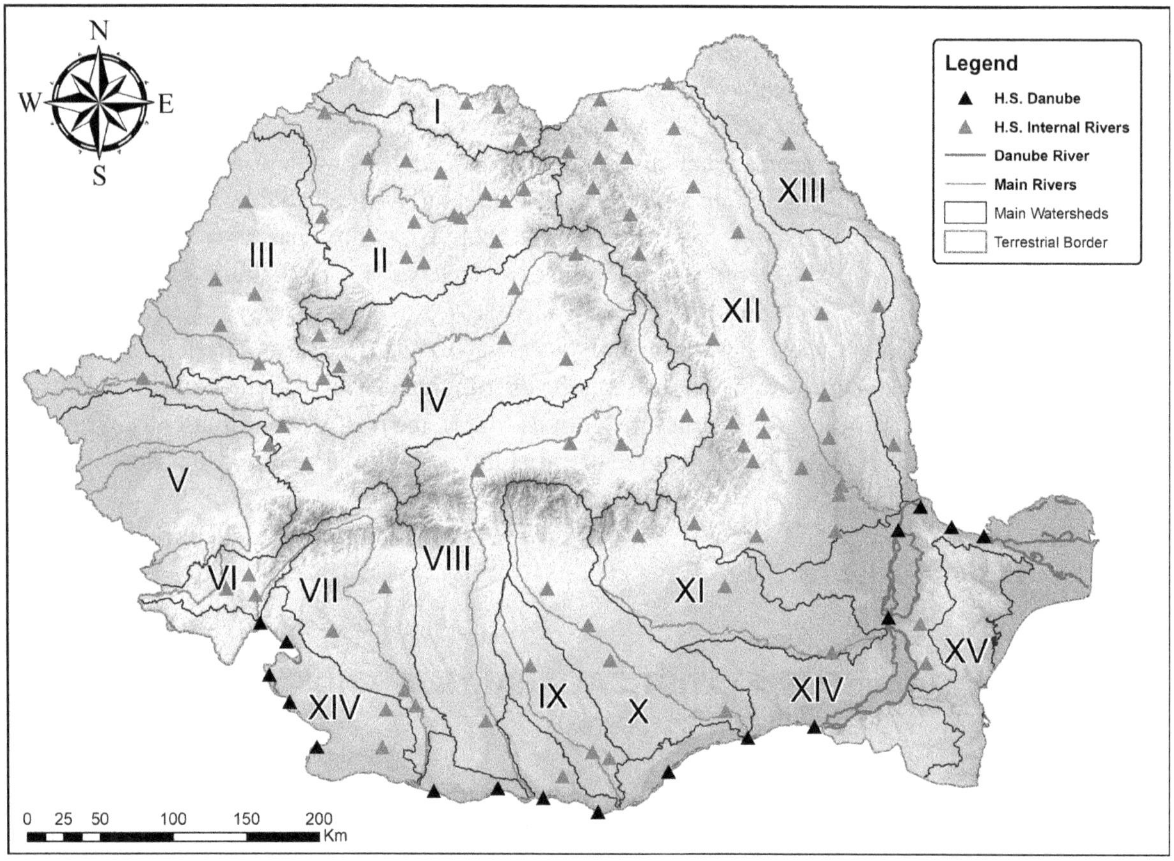

Figure 2. The spatial distribution of 17 hydrometric stations on Romanian side of Danube River

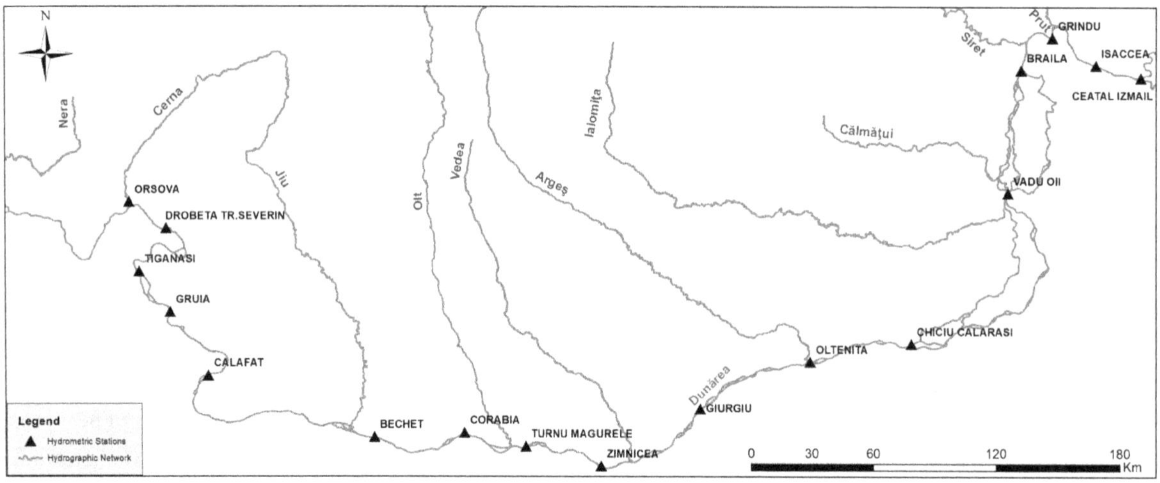

Table 1. The 93 hydrometric stations on Romanian rivers, used in the study

No.	River	Hydrometric stations on inland rivers	Catchment size (km²)	Altitude (m)
I – TISA watershed				
1	VISEU	BISTRA	1547	1020
2	RUSCOVA	LUHEI	185	1188
3	VISEU	POIANA BORSA	133	1268
II – SOMES-CRASNA watershed				
4	CRASNA	CRASNA	211	429
5	SALAJ	SALSIG	457	249
6	OLPRET	MAIA	101	394
7	BORSA	BORSA	179	452
8	SIEU	DOMNESTI	153	590
9	ILISUA	CRISTESTII CICEULUI	353	494
10	CAVNIC	COPALNIC	241	545
11	ALMAS	HIDA	556	410
12	SUCIU	SUCIU DE JOS	220	715
13	SALAUTA	SALVA	407	785
14	GADALIN	BONTIDA	283	390
15	ILVA	POIANA ILVEI	234	880
16	SOMESUL MARE	NEPOS	1138	959
17	SOMESUL MARE	BECLEAN	4323	711
18	SOMES	SATU MARE	15000	537
III – CRISURI watershed				
19	BELIU	BELIU	100	366
20	VL. ROSIE	POCOLA	255	453
21	FINETELOR	SARSIG	244	170
22	CRISUL ALB	BLAJENI	111	716
23	CRISUL ALB	GURAHONT	1581	515
24	CRISUL NEGRU	TINCA	2216	438
IV – MURES-ARANCA watershed				
25	LUT	BREAZA	268	479
26	GEOAGIU	TEIUS	185	851
27	FEERNIC	SIMONESTI	145	683
28	CERNA	TOPLITA	238	752
29	ABRUD	ABRUD	108	891
30	NIRAJ	CINTA	551	539
31	DOBRA	DOBRA	183	711
32	ARIES	SCARISOARA	203	1126

continued on following page

Table 1. Continued

No.	River	Hydrometric stations on inland rivers	Catchment size (km^2)	Altitude (m)
33	MURES	STANCENI	1532	967
34	MURES	ARAD	27036	681
		V – BEGA-TIMIS-CARAS watershed		
35	SASA	POIENI	93	763
		VI – NERA-CERNA watershed		
36	BELA RECA	MEHADIA	698	676
37	PRIGOR	PRIGOR	141	729
38	MEHADICA	CUPTOARE	120	727
		VII – JIU watershed		
39	RAZNIC	BREASTA	465	201
40	COSUSTEA	CORCOVA	420	482
41	BLAHNITA	TG. CARBUNESTI	220	467
42	JIU	PODARI	9253	446
		VIII – OLT watershed		
43	TESLUI	RESCA	509	162
44	SERCAIA	SERCAIA	346	728
45	CASIN	RUSENI	476	830
46	OLT	FELDIOARA	5614	820
47	OLT	SEBES	10850	748
		IX – VEDEA watershed		
48	TELEORMAN	TELEORMANU	1341	155
49	COTMEANA	CIOBANI	444	325
50	VEDEA	ALEXANDRIA	3246	195
		X – ARGES watershed		
51	POTOP	GURA FOII	194	348
52	NEAJLOV	MOARA DIN GROAPA	392	217
53	ARGESEL	MIOVENI	242	668
54	ARGES	BUDESTI	9229	389
		XI – IALOMITA watershed		
55	ISTAU	BABA ANA	114	282
56	DOFTANA	TESILA	288	1200
57	IALOMITA	SLOBOZIA	9154	365
		XII – SIRET watershed		
58	MOLDOVA	FUNDU MOLDOVEI	325	1083
59	MOLDOVITA	LUNGULET	144	977

continued on following page

Table 1. Continued

No.	River	Hydrometric stations on inland rivers	Catchment size (km^2)	Altitude (m)
60	NEAGRA	BROSTENI	338	1220
61	SUCEAVA	BRODINA	366	990
62	BISTRITA	DORNA GIUMALAU	742	1255
63	CARLIBABA	CARLIBABA	110	1253
64	TECUCEL	TECUCI	110	179
65	ZELETIN	GALBENI	402	268
66	RM. SARAT	TATARU	1060	295
67	TUTOVA	RADENI	172	281
68	CRASNA	VINETESTI	298	227
69	BARLAD	BACESTI	138	292
70	MILCOV	GOLESTI	395	410
71	SOLONET	PARHAUTI	214	467
72	SLANIC	CERNATESTI	415	595
73	SUSITA	CIURUC	179	588
74	SIRET	SIRET	1637	572
75	BASCA CHIOJDULUI	CHIOJDU	114	896
76	RASCA	BOGDANESTI	185	628
77	PUTNA	LEPSA	143	1022
78	ZABALA	NEREJU	244	1170
79	PUTNA	COLACU	1100	921
80	ASAU	ASAU	205	951
81	MOLDOVA	PRISACA DORNEI	666	1087
82	MOLDOVA	TUPILATI	4016	703
83	BISTRICIOARA	TULGHES	410	1073
84	NARUJA	HERASTRAU	127	1040
85	BUZAU	RACOVITA	5238	530
86	SIRET	LUNGOCI	36030	539
XIII – PRUT watershed				
87	CHINEJA	FARTANESTI	426	167
88	JIJIA	TODIRENI	1080	186
XIV – Small tributaries of the Danube				
89	TOPOLOG	SARAIU	264	181
90	CALMATUI	CRANGU	960	113
91	DESNATUI	DRAGOIA	216	170
92	VALEA DUNAREA	BALTAGESTI	110	124
93	BABOIA	AFUMATI	560	134

Table 2. The 17 hydrometric stations on Danube River in the Romanian section, used in the study

No.	Hydrometric stations on Danube River (From upstream to downstream)	River kilometre (km+m)
1	ORSOVA	953+300
2	DROBETA-TURNU SEVERIN	931
3	TIGANASI	877
4	GRUIA	851
5	CALAFAT	794+600
6	BECHET	679
7	CORABIA	629+500
8	TURNU MAGURELE	597
9	ZIMNICEA	553+500
10	GIURGIU	492+800
11	OLTENITA	429+700
12	CALARASI-CHICIU	375+250
13	VADU OII	238+800
14	BRAILA	169+400
15	GRINDU	139+600
16	ISACCEA	103
17	CEATAL IZMAIL	79+600

The name and the cadastral code (from I to XIV) of the main large basins to which the sub-basins of the analysed inland rivers belong, as well as the corresponding morphometric parameters (basin area and altitude), for the 93 hydrometric stations, are presented in Table 1. The morphometric parameter values are ranging in the intervals 113 ÷ 1268 m, respectively 93 ÷ 36030 km^2. The values of the basin area in the range 2216 - 36030 km^2 are found at 13 hydrometric stations located on large rivers, such as Somesul Mare, Somes, Crisul Negru, Mures, Jiu, Olt, Vedea, Arges, Ialomita, Buzau, Moldova and Siret. These have been considered for highlighting the direct/indirect contribution of these basins to the minimum flow on the Romanian sector of the Danube River basin. Thus, if the hydrometric stations Satu Mare (Somes River), Tinca (Crisul Negru River) and Arad (Mures River) show the indirect contribution of the flow on the Danube River, the hydrometric stations Podari (Jiu River), Alexandria (Vedea River), Budesti (Arges River), Slobozia (Ialomita River) and Lungoci (Siret River) show the direct contribution of the flow on the Danube River.

In the case of the Romanian section of the Danube, Table 2 shows the 17 hydrometric stations selected for the study, in order of length river kilometre, from upstream to downstream.

Statistical Analyses and Indices

Low streamflow is a seasonal phenomenon that occurs every year and defines the lowest (driest) parts of a continuous streamflow hydrograph (Smakhtin, 2001).

Low flow indices are often derived directly from the curve as low-flow exceedances. A flow exceedance is an index that expresses the proportion of time that a specified daily flow is exceeded during the period of record. The flow exceedance is often given in terms of percentiles (Tallaksen & van Lanen, 2004).

Low flow indices include statistics such as a particularly rare (hence extreme) quantile of the flow duration curve, e.g., the flow exceeded 95% of the time (Q95), the mean annual n-day minimum flow, MAM(n-day), or the value associated with a given return period (AM-n-dayT). Drought event characteristics are identified using the threshold level method or the sequent peak algorithm (Stahl, K. et al., 2020).

The indices studied in the paper serve as the basis for many decisions on the design or management of water resources during low-flow conditions. According to the WMO Manual on Low Flow Estimation and Prediction (Gustard & Demuth, 2008) and others (e.g., Koffler & Laaha, 2013; Tallaksen & van Lanen, 2004; Laaha & Blöschl, 2006b), these indices are described below, as follows:

- Mean flow (MQ) - is one of the most used statistics in hydrology and water resources planning. It can be estimated from a time series of gauged data by summing all daily discharges and dividing by the number of days in the record. It is normally calculated for complete calendar or hydrological years of data. It can also be calculated for specific months or seasons.
- Q95 is of the most common low-flow indices used operationally and is defined as the flow exceeded for 95 per cent of the time. It can be determined by ranking all daily discharges and finding the discharge exceeded by 95 percent of all values. This flow quantile and other (Q_{90}, Q_{80}, Q_{70}, etc.) also, can be derived from the flow-duration curve; these are frequently chosen as threshold level values in the definition of hydrological drought (streamflow deficit).

The median annual daily streamflow provides a conservative upper bound below which streamflows may be considered 'low', while flows exceeded 70 – 99 percent of the time are widely used as design flows (Smakhtin, 2001). Annual mean n-day low streamflows (the lowest average flows that occur for a consecutive n-day period) are often used to define low-streamflow indices for water resources management (Smakhtin, 2001).

- MAMn-day - Annual minima can be derived from a daily flow series by selecting the lowest flow every year and the mean of the minima calculated. Minima of different durations can be determined with 1, 7, 10, and 90 days being commonly used. The annual minima (AMn-day) can be used to determine a distribution function for estimating the frequency or return period of low flows. In temperate climates, the mean annual 7- day minimum discharge is numerically similar to Q95, for most flow records.

The annual minimum 7-day index (AM7) represents the magnitude of the low flow event of a year. It is the annual minimum of a smoothed hydrograph, obtained by using a central 7-day moving-average filter. The moving-average filter is applied to reduce short-term disturbances of the discharge record.

The extreme values are the prime interest, as is the case for minima values in low-flow analysis (Koffler and Laaha, 2013). Extreme value statistics are of particular importance to hydrology. The methodology was originally developed for floods but has recently received increasing attention for low flows. It allows the user to characterize the severity of an event by a statistical probability or return period i.e., the expected time interval until an event of the same magnitude occurs (Laaha, 2023).

Related to extreme values of low flow, another statistical analysis based on the lfstat package, it was applied in this study. The analysis consisted in calculating the occurrence of one event in T years, based on the *Extreme value -> T-year Event function*, for the annual minimum over a 7-day period (AM7). *Tyears* function fits an extreme value distribution using L-moments to the minima of a time series of discharges and subsequently estimates quantiles (then so-called T-years event) for given return periods. In the presence of zero flow observations a mixed distribution is fitted.

- The Base Flow Index (BFI) was developed during a low flow study in the UK (Institute of Hydrology, 1980). The index gives the ratio of base flow to total flow calculated from a hydrograph smoothing and separation procedure using daily discharges. The BFI is thus considered a measure of the river's runoff that derives from stored sources and as a general catchment descriptor it has found many areas of application, including low flow estimation and groundwater recharge assessment. Values of the index range from above 0.9 for a permeable catchment with a very stable flow regime to 0.15 – 0.20 for an impermeable catchment with a flashy flow regime (Tallaksen & Van Lanen, 2004).

The seasonality of low flows can be quantified based on two seasonality indices: Seasonality Rate (SR) and Seasonality Index (SI).

- Seasonality Rate (SR) is used to characterize the low flow regime of rivers, the occurrence of minimum discharges. It is the ratio of 95th quantile of duration curve of daily mean flows determined for two seasons within the hydrological year, such as in the present study: from 1 April to 30 November (IV – XI season) and from 1 December to 31 March (XII - III season). It is given in equation (1):

$$SR = Q95_{IV\text{-}XI} \,/\, Q95_{XII\text{-}III} \quad (1)$$

Where: SR<1, indicates the presence of low flow in the warm season (IV - XI); SR>1, indicates the presence of low flow in the cold season (XII - III).

- Seasonality Index (SI), it was developed to assess the mean date of occurrence and distribution of minimum flows, as well as hydrological drought, during the year. This index represents the seasonal distribution of low flow occurrence and is similar to the one used by Burn (1997), Young et al. (2000), Laaha & Blöschl (2006) and Cupak & Kaczor (2023). It is based on two parameters, θ and r, which are calculated from the Julian dates of all days of the observation period when discharges are equal or below Q95. The parameter θ represents a measure of the average seasonality of low flows by the average day of low flow occurrence in radians (the values range from 0 to 2π, where $\theta = 0$ relates to 1 January, $\theta = \pi/2$ relates to 1 April, $\theta = \pi$ relates to 1 July, $\theta = 3\pi/2$ relates to 1 October 1); it is calculated as follows:

$$\theta_j = \frac{D_j 2\pi}{365} \quad (2)$$

Where: D_j is the day of the occurrence, when flows are lower than Q95.

In turn, the parameter r is the mean days of occurrence and describes the seasonal variability of low flows and it is dimensionless indicator. Its values range between 0 (low seasonality and means that all low flow events are uniformly distributed over a year) and 1 (strong seasonality and means that all low flow events occurred exactly the same day of the year) (Burn, 1997; Laaha & Blöschl, 2006; Cupak & Kaczor, 2023).

The seasonal concentration index r is calculated as equation (3) (Laaha & Blöschl, 2006):

$$r = \sqrt{x_\theta^{\,2} + y_\theta^{\,2}} \tag{3}$$

Where: x_θ and y_θ are the arithmetic mean of Cartesian coordinate of a total of n single days j and are calculated as equation (4) (Laaha & Blöschl, 2006):

$$x_\theta = \frac{1}{n}\sum_j \cos\left(\theta_j\right) \text{and } y_\theta = \frac{1}{n}\sum_j \sin\left(\theta_j\right) \tag{4}$$

The angle of the mean directional vector is defined as equations (5) and (6) (Laaha & Blöschl, 2006):

$$\theta = arctan\left(\frac{y_\theta}{x_\theta}\right), \text{ 1}^{\text{st}} \text{ and 4}^{\text{th}} \text{ quadrant: } x>0 \tag{5}$$

$$\theta = arctan\left(\frac{y_\theta}{x_\theta}\right) + \pi, \text{ 2}^{\text{nd}} \text{ and 3}^{\text{rd}} \text{ quadrant: } x<0 \tag{6}$$

Lastly, the drought events were analyzed, wich can be defined by introducing a threshold below which the flow is regarded as being in a deficit.

A streamflow drought event is defined as a dry spell in the flow record when discharge is below some given threshold (Yevjevich, 1967). Each event has a beginning and an end. The deficits can be described by different characteristics like duration, severity, time of occurrence and spatial extent. A time series of deficit characteristics, for example the drought deficit duration, can be derived. Finally, another characteristic, a drought deficit index, e.g., the "mean drought deficit duration", can be estimated (Tallaksen & Van Lanen, 2004).

Streamflow deficits (Hisdal et al., 2004) are periods when the river is below a specific threshold defining a dry period or critical deficit. One of the methods for selecting and characterising deficits (lack of water in the riverbed) is the threshold-level method. This method is used to provide estimates of the frequency of low water periods and to design and bring to an operating level reservoir, where discharges from reservoirs are aimed to provide downstream abstractions (Gustard & Demuth, 2008).

The application of a threshold level allows both the definition of drought events as periods of below normal flow and the identification of periods of insufficient water supply for a specific demand. In the latter case, the threshold level is set to be equal to the flow demand, but in the first case it is considered to represent "normal" conditions, which means that it can be chosen freely, depending on the stream studied as well as the study demand (Gustard & Demuth, 2008). It is entirely usual to apply an objective and comparable value for the threshold level as an index of low flow.

The threshold is used when to define how the flow is in deficit. The deficit starts when the flow is below a certain threshold and ends as soon as the flow returns above that threshold. Thus, the beginning and end of a deficit period can be defined. The characteristics of deficits can be defined as follows:

- duration (d, days) is the period during which the discharge in the riverbed is below the threshold level and refers to the duration of the drought event;
- volume (v, m^3) or severity of drought event;
- intensity or magnitude of drought event (m, m^3/day) is the ratio of volume to the duration of the deficit;
- minimum discharge (Q_{min}, m^3/s) of each drought event;
- time of occurrence: year, month, and day of the beginning of the deficit period (drought event); date of the minimum discharge value.

In Fleig et al. (2006) it was shown that the moving average (MA) procedure is applicable to both fast and slow drainage regimes. The averaging interval can be easily optimized and a filter length of 7 days is suitable for several regime types. An advantage of the MA pooling method is that the problem of minor deficits is reduced.

Another objective of the present study is to identify the largest annual streamflow deficits. Therefore, a constant threshold was used, given by the Q80 low flow quantile [$P(Q \geq Q80) = 0.8$] computed for the entire reference period. The Q80 is used in many drought studies (e.g., Andreadis et al., 2005; Corzo Perez et al., 2011; Sheffield et al., 2009; Van Huijgevoort et al., 2014; Van Loon & Van Lanen, 2012, Laaha et al., 2016). Overall, low-flow thresholds usually adopted in drought studies range between the 70th and 95th percentiles (Hisdal et al. 2004, Corzo Perez et al. 2011).

In order to analyse and assess low flow and water deficits for inland rivers and Romanian part of Danube River, over the period 1980 - 2020, a standardised computation of low flow indicators was carried out, which consisted in determining the values of water quantities flowing through the streams, during the characteristic periods of river recharge exclusively from the aquifers, at weekly, monthly, seasonal, annual, multiannual or with a certain frequency of occurrence over the years. The meanflow and baseflow index were also computed.

Estimation and statistical analysis were performed using the lfstat package (Koffler & Laaha, 2013), that provides a collection of current functions to compute a range of low flow characteristics that are fully described in the WMO manual on low flow estimation and prediction (Gustard and Demuth, 2008). The package performs extreme-value statistics of both low flow discharges and drought characteristics, such as duration and deficit volume. The package uses a robust approach based on L-moments to fit extreme-value distributions (Hosking & Wallis, 2005).

Low flow indices and characteristics used in this study are described in the table below:

Although low flow indices are widely used internationally, many of those was not found in the management plans of Romanian basin administrations. Given the increasing monitoring and knowledge requirements, at regional and international levels, an assessment of low flow and water deficit based on these indicators, at the Romanian level, is necessary.

Table 3. Low flow indices and characteristics used in the study and estimated using the lfstat applications

Low flow indices	Function	Additional input parameter
BFI (Base Flow Index)	BFI	-
MQ (Mean flow)	meanflow	-
Q80, Q95 (quantiles from flow duration curve)	Qxx	xx
Mean Annual Minima (n-day)	MAM	n
Extreme value (n-day)	tyears	T-years event
SI (Seasonality Index)	seasindex (theta, D, r)	Q95
SR (Seasonality Ratio)	seasratio	Q95, breakday(s) = 01/04, 01/12
Low flow characteristics	**Function**	**Additional input parameter**
		pooling method = MA
		threslevel = 80
Streamflow deficit	streamdef	thresbreaks = fixed
		MAdays = 7
		durmax ≥ 30 days

The results of this study could improve strategies to manage hydrological extremes and contribute to the efficient management of water resources during periods of supply restrictions.

RESULTS

Spatial Distribution of Low Flow Indices

The aim of this study is not only to quantify some low flow indicators and deficit periods, but also to spatially represent them, in order to get an overview of low flow in Romania, over the reference period 1980 - 2020, as follows: the magnitude, variability for different time durations, the relative contribution of base flow to total flow, seasonality and degree of variability in occurrence of minimum discharges, as well as highlighting the driest years, from a hydrological point of view.

Firstly, the summary statistics of low flow indices estimated on based the lfstat package, over the 41 years period, at 93 stations on Romanian rivers and other 17 on the Romanian part of Danube River, are in Table 4 and Table 5, respectively.

Regarding the inland rivers, there were made the spatial distributions for the next indicators: mean flow (MQ), low flow percentiles (Q80 and Q95), mean annual minima of 7, 30, and 90-days, respectively MAM7, MAM30, and MAM90.

The specific discharge (discharge standardized by catchment area) is an important hydrological variable (Tallaksen & van Lanen, 2004). The spatial representations were made for identical classes of specific values of indicators, in $l/s*km^2$, in order to identify and compare at watershed or area level, the magnitude (Q80 and Q95 indices) and the variability over different durations (MAM7, MAM30, and MAM90) of low flows. In order to provide this, the estimated values of low flow indicators were standardised according to the area of each catchment.

Table 4. Summary statistics of low flow indices at 93 hydrometric stations on inland rivers

Variable	Observations	Minimum	Maximum	Mean	Std. deviation
MQ (m^3/s)	93	0.106	203	12.6	32.326
Q80 (m^3/s)	93	0.026	88.2	4.64	13.391
Q95 (m^3/s)	93	0.006	47.9	2.81	7.938
MAM7 (m^3/s)	93	0.017	62.0	3.34	9.538
MAM30 (m^3/s)	93	0.024	81.7	4.15	12.080
MAM90 (m^3/s)	93	0.038	97.7	5.28	14.910
BFI	93	0.26	0.84	0.53	0.126

Table 5. Summary statistics of low flow indices at 17 hydrometric stations on Danube River

Variable	Observations	Minimum	Maximum	Mean	Std. deviation
MQ (m^3/s)	17	5177	6403	5730	404.344
Q80 (m^3/s)	17	3224	4050	3597	280.986
Q95 (m^3/s)	17	2380	3070	2699	232.529
MAM7 (m^3/s)	17	2345	3163	2716	279.752
MAM30 (m^3/s)	17	2664	3443	3018	265.809
MAM90 (m^3/s)	17	3261	4076	3630	272.571
BFI	17	0.84	0.95	0.90	0.04

Also, a map with spatial distribution of the BFI values was made for all studied hydrometric stations (including those on the Danube River section).

Firstly, the mean specific discharge values over the reference period, for stations on inland rivers, were presented in Figure 3. For 36.6% of stations, the values below 5 l/s*km^2 it was highlighted. The lowest values, below 1 l/s*km^2, are found at just two stations on the tributaries of the Barlad River located in eastern part of the country, in the Siret (XII) watershed. Values between 1 l/s*km^2 and 5 l/s*km^2 can also be found in the lower basins of small tributary rivers of the Danube in the south, south-east and east of country, and in the Somes basin in the north-west.

The highest specific mean flow values are also highlighted at 12 stations located, most of them, on rivers from Tisa, Somes and Siret watersheds.

Concerning of the low specific flow (q80 and q95) (Figure 4 and Figure 5), its spatial distributions is similar to the one of mean specific flow. The specific values of these indices ranging from 0.114 to 14.60 l/s*km^2, in case of q80, and from 0.035 to 10.05 l/s*km^2 in case of q95. When comparing these two spatial representations, the number of hydrometric stations with lowest specific values (red colour symbol) increases from 31 stations (33.33% of stations on inland rivers), in case of q80, to 46 stations (49.46%), in case of q95.

Other spatial representations of low flow were made based on the specific values of mean annual minima over different time durations, such as mam7 (in Figure 6), mam30 (in Figure 7), and mam90 (in Figure 8). They emphasize the variability of specific low flow for time durations from 7 to 90 days,

Figure 3. The mean specific flow over the 1980 - 2020 period at 93 hydrometric stations on inland rivers

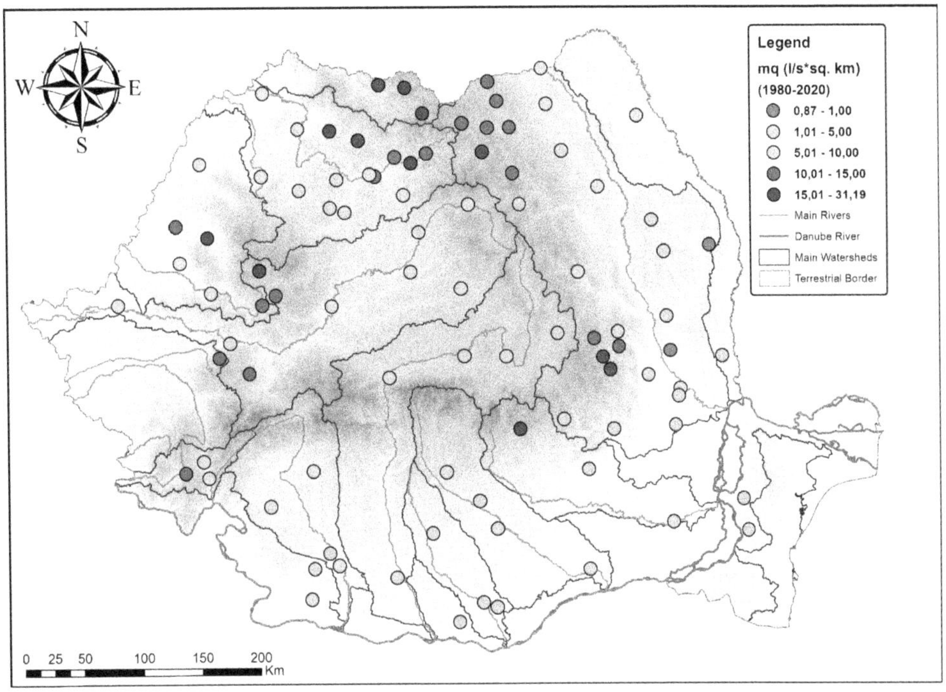

Figure 4. The low specific flow (q80) over the 1980 – 2020 period, at 93 hydrometric stations on Romanian rivers

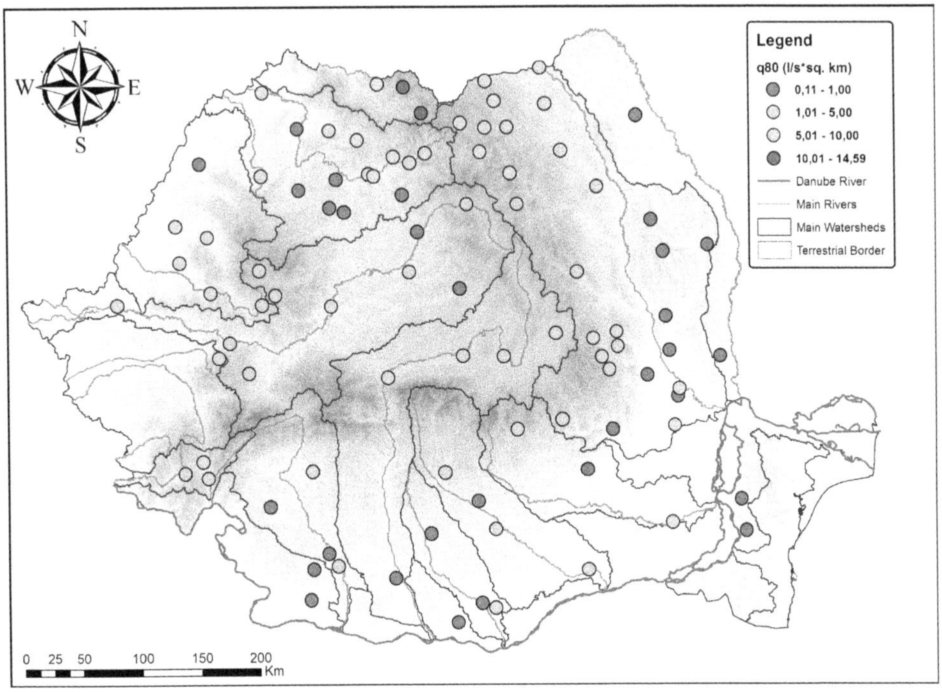

Figure 5. The low specific flow (q95) over the 1980 – 2020 period, at 93 hydrometric stations on Romanian rivers

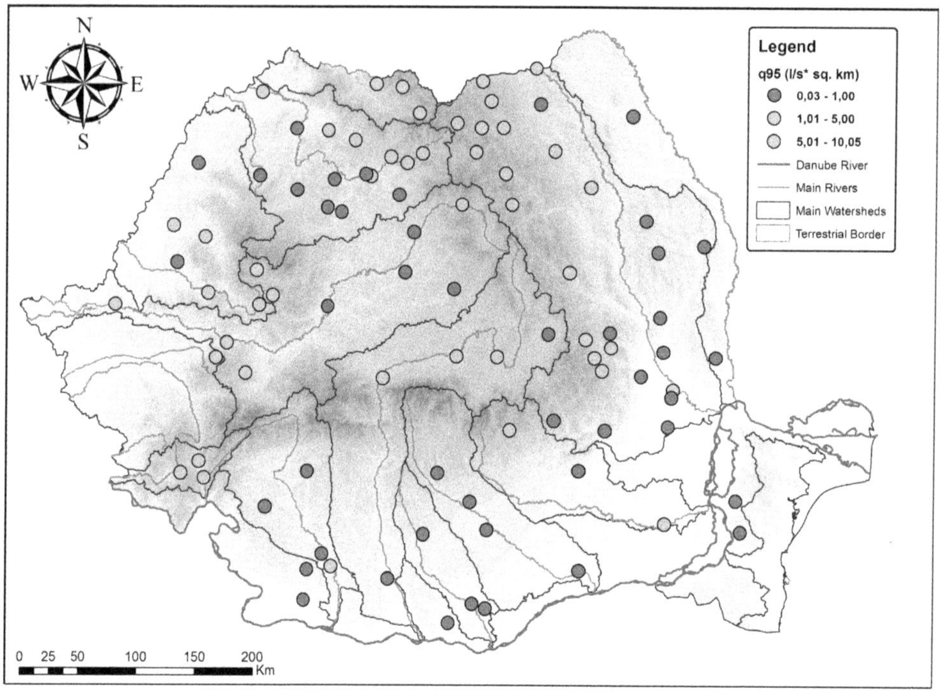

Figure 6. The low specific flow over 7 days time duration (mam7) at 93 hydrometric stations on inland rivers

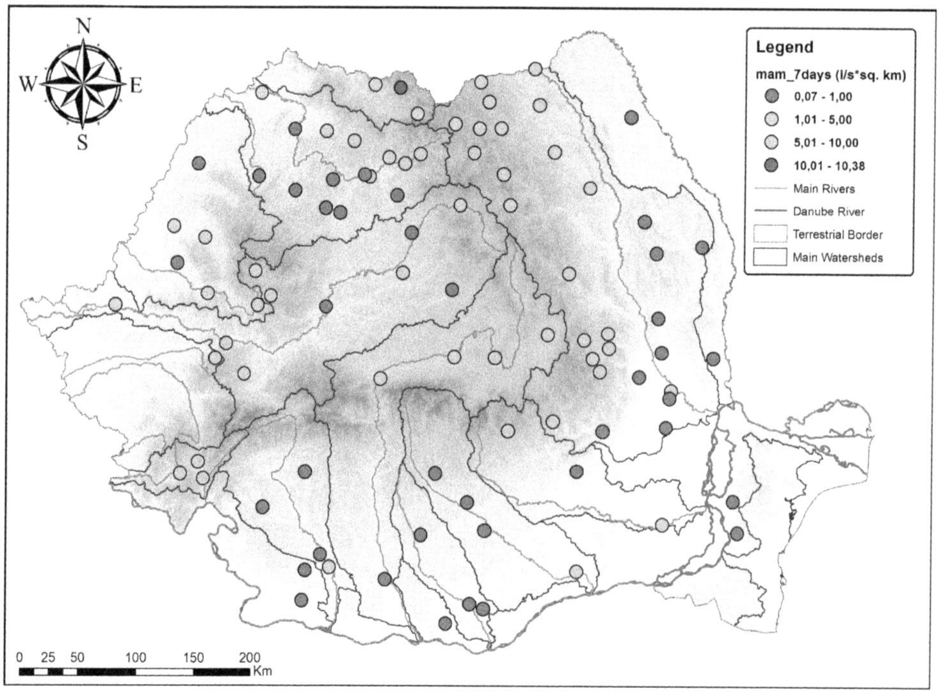

Figure 7. The low specific flow over 30 days time duration (mam30) at 93 hydrometric stations on inland rivers

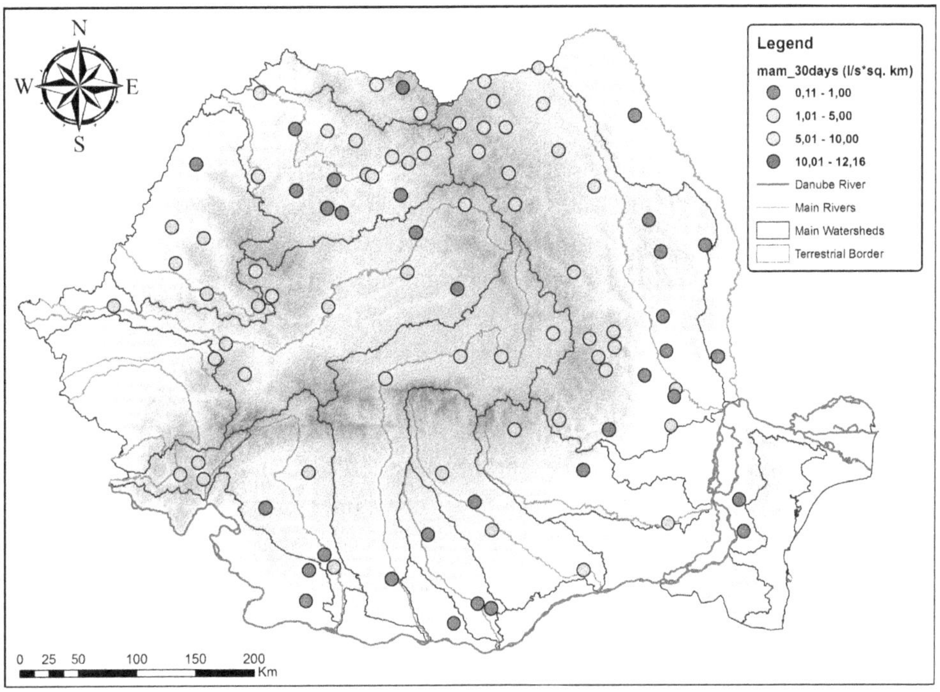

Figure 8. The low specific flow over 90 days time duration (mam90) at 93 hydrometric stations on inland rivers

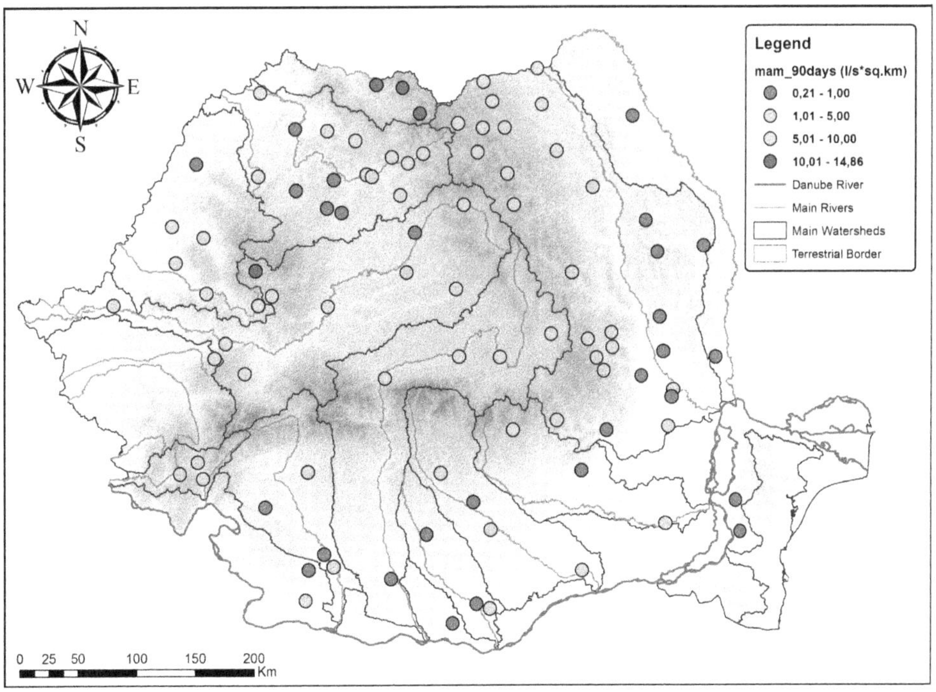

with values in the ranges: 0.07 – 10.38 l/s*km^2 (Figure 6), 0.11 – 12.16 l/s*km^2 (Figure 7), and 0.21 – 14.86 l/s*km^2 (Figure 8).

The percentage of values under 1 l/s*km^2 (red symbol) is declining from 43% in case of mam7 to 30.1% in case of mam90. This demonstrates a low variability of minimum discharges from one to three months period. This applies for about 30% of stations that are located both outside and inside the Carpathian Mountains.

The spatial representation of baseflow index is another result of the study. This index demonstrates the slow continuous contribution of groundwater to river flow. For streams and rivers located on permeable aquifers, the baseflow provides most of the flow in the river and, in summer months, the entire flow. Therefore, understanding and assessing baseflow is important for river water management.

For this purpose, an assessment of the base flow index of the reference period 1980 - 2020 was carried out at the 93 hydrometric stations on the inland rivers and 17 stations on the Danube River, using the BFI function of the LFSTAT application package. As a result, a map with the spatial distribution of the BFI values was produced (Figure 9). The considerable contribution of phreatic aquifer to the flow of Danube River is highlighted here, with percentages between 81% and 95% (dark green symbol). The same contribution is evident in small rivers, tributaries of Danube River.

Also, high percentage values of base flow ranging between 61% and 80% are highlighted at 25 hydrometric stations (23% of the total number) in the upper basin of Siret River, the lower basin of the Mures River, in the Olt River basin, on small tributaries of the Danube in the Dobrogea area, and at the terminus stations of the Somes, Mures, Jiu, Vedea and Ialomita basins (light green symbol in Figure 9). For about 57% of stations, the base flow represents between 31% and 60% of the total flow (orange symbol in Figure 9).

Figure 9. The base flow index (BFI) at 110 hydrometric stations on inland and Danube rivers in Romania

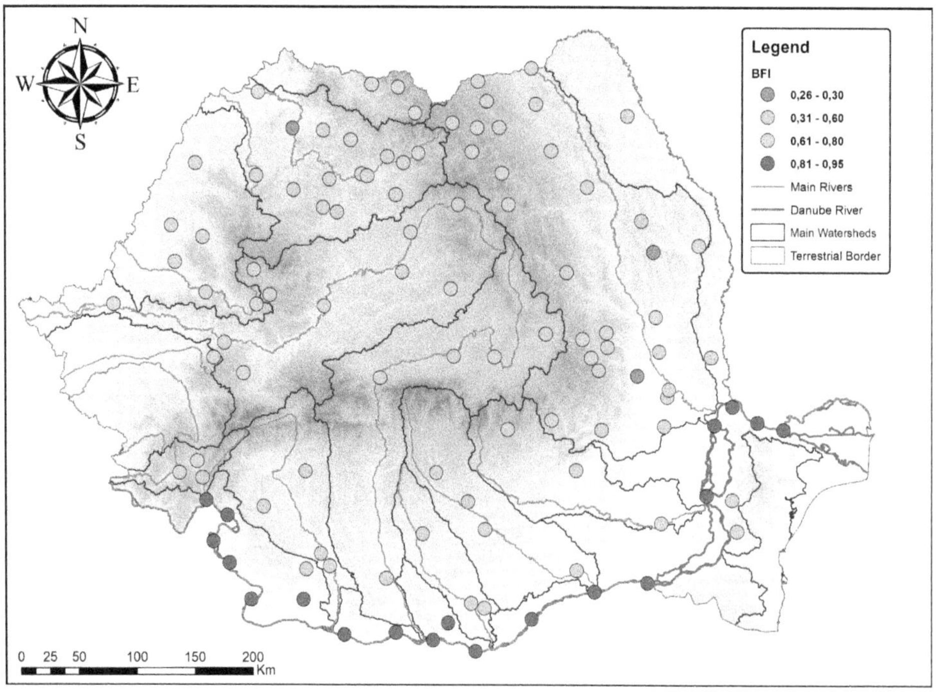

The lowest percentage values, up to 30% (red symbol), are only shown at three hydrometric stations: Salsig (from Somes basin), Golesti and Radeni (from Siret basin).

Analysis of Ranking Annual Values of MQ and Q95 Indices

In order to identify the driest years in the reference period, it was necessary to investigate the rank of the lowest value in the annual series of indicators concerning the magnitude of flow through the riverbed, primarily under average water conditions, the mean flow indicator (MQ), and of course, under absence of precipitation and high temperatures, the low flow percentile (Q95).

Therefore, the statistical function rank (value) was applied to the annual series of MQ and Q95 indicators, in order to highlight the years with the three lowest values in each annual series of indicators for the reference period 1980 – 2020.

The results of rank analysis are shown in Table 6 and Table 7, for inland rivers, and in Table 8 and Table 9, for Danube River. The years corresponding to 1st – 3rd rank values, representing the lowest values of MQ, as well as Q95, in the 41-year interval, were emphasized.

So, for inland rivers, the 1st rank values of indicators were identified in the years: 1990 (at 18 stations), 1987 (at 14 stations) and 2014 (at 13 stations), in case of MQ, as well as 1987 (at 26 stations) and 2012 (at 24 stations), in case of Q95.

The identification of the driest year of the reference period, derived on rank analysis of annual series of MQ and Q95 indices, at all hydrometric stations, was also carried out spatially (Figure 10 and Figure 11).

Thus, in the case of MQ index (Figure 10), the year 1990 is highlighted (at 35 hydrometric stations), that is found at most hydrometric stations on the Danube River, as well as at stations in the Siret, Crisuri and Jiu basins. In general, the 1990s (1990, 1992, 1993 and 1994) are the years with the lowest mean annual flow in the southern part of the country. Also, 2014 and 1987 stand out in the north-west (at stations from Tisa and Somes basins) and eastern part of the country (at stations from the Siret basin).

Regarding the low flow percentile Q95, the spatial distribution of the years with the lowest index values at all stations, was produced (Figure 11). Once again, 1987 stands out with the lowest value of Q95, for the eastern part of the country, at stations from Siret basin. On the other hand, for the north-west, west and south-east of country, 2012 is highlighted, which is also found at the bordering stations of large river basins, such as: Satu Mare (Somes basin), Gurahont (Crisul Alb basin), Arad (Mures basin), Slobozia (Ialomita basin), as well as at the hydrometric stations Feldioara on the Olt River and Tg. Carbunesti on the Jiu River.

Concerning of Danube River, for the river segment Orsova - Ceatal Izmail, the rank analysis revealed the year 1990 with the lowest value of the MQ (Table 8 and Figure 10) and the year 2003, in case of Q95 (Table 9 and Figure 11). Regarding to the 2nd and 3rd rank values, the years 2003, 2020 and 2011, in the case of MQ, as well as 1985, 1992 and 2011, in the case of the Q95 index, also stand out.

Clearly, throughout the Romanian part of Danube River, 1990 (1st rank value), 2003 and 2020 (2nd rank values) were emphasized, related to the lowest mean annual flow in the reference period. Also, 2003 was emphasized with the lowest value of Q95, on the river segment Orsova - Oltenita, while further downstream, up to Ceatal Izmail, 1992 and 1990 are highlighted (Table 9).

Table 6. The years with the three lowest MQ values over the 1980 – 2020 period, at hydrometric stations on inland rivers

No.	Hydrometric stations on inland rivers	Rank (MQ) / Year		
		1	2	3
1	BISTRA	2014	2012	1987
2	LUHEI	2014	2015	2009
3	POIANA BORSA	2017	2014	2015
4	CRASNA	2014	1990	1994
5	SALSIG	2014	2012	1991
6	MAIA	2014	2003	1995
7	BORSA	2020	1995	2014
8	DOMNESTI	2015	2014	2012
9	CRISTESTII CICEULUI	2014	2020	2012
10	COPALNIC	2014	2003	2012
11	HIDA	2014	2020	1990
12	SUCIU DE JOS	2014	2003	2015
13	SALVA	2014	2003	2020
14	BONTIDA	1995	1990	1991
15	POIANA ILVEI	1990	2014	1983
16	NEPOS	2003	2014	2012
17	BECLEAN	2003	2014	2012
18	SATU MARE	2012	2014	2003
19	GURAHONT	2015	2020	1990
20	TINCA	1990	2014	2011
21	BELIU	2014	1990	2020
22	BLAJENI	2015	2017	2020
23	POCOLA	1990	2014	2020
24	SARSIG	2012	2014	1990
25	ARAD	2012	1990	2017
26	ABRUD	2017	1992	2015
27	BREAZA	2014	2015	2017
28	CINTA	1990	2015	2012
29	DOBRA	1983	1984	1994
30	SCARISOARA	1993	1983	2014
31	SIMONESTI	2017	2020	2013
32	STANCENI	2017	1990	1992
33	TEIUS	2012	1992	1990
34	TOPLITA	1993	1984	1983

continued on following page

Table 6. Continued

No.	Hydrometric stations on inland rivers	Rank (MQ) / Year		
		1	2	3
35	POIENI	1992	1991	1986
36	CUPTOARE	2002	2012	2001
37	MEHADIA	2002	1992	1994
38	PRIGOR	1992	1995	2017
39	PODARI	1992	1993	1990
40	BREASTA	1994	1993	1992
41	CORCOVA	2002	2001	2017
42	TG. CARBUNESTI	1990	2001	1993
43	RESCA	2020	1993	1994
44	SERCAIA	2013	1993	2020
45	RUSENI	1987	1990	1986
46	FELDIOARA	1990	2013	1987
47	SEBES-OLT	2013	1986	1987
48	TELEORMANU	1994	1990	2002
49	CIOBANI	1994	1992	1993
50	ALEXANDRIA	1993	1990	1994
51	GURA FOII	1994	1992	1993
52	MOARA DIN GROAPA	2020	2002	1993
53	MIOVENI	1992	1993	1994
54	BUDESTI	1994	1993	1989
55	BABA ANA	2020	2019	1987
56	TESILA	1990	1987	1992
57	SLOBOZIA	1989	1990	1994
58	FUNDU MOLDOVEI	1987	1986	1990
59	LUNGULET	1987	1990	1994
60	BROSTENI	1990	2009	2012
61	BRODINA	1990	1987	1994
62	DORNA GIUMALAU	1987	2003	1990
63	CARLIBABA	1987	2013	2012
64	TECUCI	2012	2020	2001
65	GALBENI	2009	2012	1995
66	TATARU	1987	1990	1989
67	RADENI	2009	2020	2007
68	VINETESTI	2020	2012	2018

continued on following page

Table 6. Continued

No.	Hydrometric stations on inland rivers	Rank (MQ) / Year		
		1	2	3
69	BACESTI	1995	2012	1990
70	GOLESTI	1987	1990	1994
71	PARHAUTI	1990	1987	1986
72	CERNATESTI	1987	1989	2020
73	CIURUC	1987	1986	2020
74	SIRET	1990	2012	2016
75	CHIOJDU	1990	1989	1987
76	BOGDANESTI	1987	1986	1990
77	LEPSA	1990	1987	1986
78	NEREJU	1987	1990	2020
79	COLACU	1987	1990	1986
80	ASAU	1987	1986	2013
81	PRISACA DORNEI	1987	1990	1994
82	TULGHES	1990	1987	1986
83	HERASTRAU	1990	1987	1986
84	LUNGOCI	1990	1987	1986
85	RACOVITA	1990	1987	1986
86	TUPILATI	1990	1987	1994
87	FARTANESTI	2012	2009	2008
88	TODIRENI	2020	2016	2012
89	SARAIU	2014	2011	2007
90	CRANGU	2012	2013	2009
91	DRAGOIA	1993	1992	1994
92	BALTAGESTI	1984	1981	1980
93	AFUMATI	2002	2012	2013

Extreme Value Statistics of 7-Day Annual Minima (AM7)

Extreme value analysis applied to the series of 7-day annual minimum values of hydrometric stations located on inland rivers, led to frequency distribution models of AM7 series based on best-fit testing of the distribution function. The results of the analysis showed that two of the five distribution functions applied to the series of extremes (Weibull - wei, GEV - gev, Lognormal - ln3, Gumbel – gum, and Pearson type III - pe3), were the best fit, namely the Pearson type III and Weibull distributions. Thus, return levels of the AM7 index were estimated for return periods, on average, once every 5, 10, 20, 30, and 100 years, with probabilities of occurrence of 0.2, 0.1, 0.05, 0.0333, and 0.01, respectively.

The graphical representations of the frequency distribution patterns of the 7-day annual minimum series (AM7) determined by testing the five distribution functions, following which the return levels of

Table 7. The years with the three lowest Q95 values over the 1980 – 2020 period, at hydrometric stations on inland rivers

No.	Hydrometric stations on inland rivers	Rank (Q95) / Year		
		1	2	3
1	BISTRA	1984	1987	2012
2	LUHEI	1992	1983	1984
3	POIANA BORSA	1984	2005	1987
4	CRASNA	2012	1994	2019
5	SALSIG	2012	2020	2017
6	MAIA	2003	2012	1990
7	BORSA	2012	2014	2013
8	DOMNESTI	2012	1992	1987
9	CRISTESTII CICEULUI	2012	2003	2011
10	COPALNIC	2012	2003	2015
11	HIDA	1990	1994	2003
12	SUCIU DE JOS	2012	2017	2003
13	SALVA	1987	2012	2011
14	BONTIDA	2012	1990	1995
15	POIANA ILVEI	1992	1983	1994
16	NEPOS	1987	1984	2012
17	BECLEAN	1983	1986	2012
18	SATU MARE	2012	1983	2003
19	GURAHONT	2012	2009	2003
20	TINCA	2009	2012	2003
21	BELIU	2003	2009	2020
22	BLAJENI	1993	2015	1986
23	POCOLA	2009	2007	1992
24	SARSIG	2012	2007	2009
25	ARAD	2012	1986	1987
26	ABRUD	2000	2013	2011
27	BREAZA	2017	2015	2012
28	CINTA	1987	2017	1986
29	DOBRA	1993	2003	2011
30	SCARISOARA	2012	1987	1984
31	SIMONESTI	2012	2018	2016
32	STANCENI	1987	2012	1984
33	TEIUS	1987	2012	2011
34	TOPLITA	2012	2017	2015

continued on following page

Table 7. Continued

No.	Hydrometric stations on inland rivers	Rank (Q95) / Year		
		1	2	3
35	POIENI	1992	1993	1991
36	CUPTOARE	1987	2000	2001
37	MEHADIA	1987	2012	2003
38	PRIGOR	2017	2018	2015
39	PODARI	2000	1993	1990
40	BREASTA	1999	2013	1985
41	CORCOVA	2003	2000	2002
42	TG. CARBUNESTI	2012	2011	1994
43	RESCA	1985	2007	2012
44	SERCAIA	2000	2012	1992
45	RUSENI	1987	2015	1991
46	FELDIOARA	2012	1990	2013
47	SEBES-OLT	2015	1994	1993
48	TELEORMANU	1993	1990	2007
49	CIOBANI	1992	1990	1982
50	ALEXANDRIA	1990	1993	1985
51	GURA FOII	1994	1992	2003
52	MOARA DIN GROAPA	1998	2020	2002
53	MIOVENI	1987	1986	2012
54	BUDESTI	1993	1994	1989
55	BABA ANA	2020	2019	1987
56	TESILA	1987	2001	2000
57	SLOBOZIA	2012	1990	1989
58	FUNDU MOLDOVEI	2012	1987	1985
59	LUNGULET	1987	2012	1985
60	BROSTENI	2010	2009	2014
61	BRODINA	1987	2012	1980
62	DORNA GIUMALAU	1984	1989	2012
63	CARLIBABA	2020	1989	1985
64	TECUCI	2012	2016	2020
65	GALBENI	1995	2012	1987
66	TATARU	1987	2012	2008
67	RADENI	2009	2015	2020
68	VINETESTI	2020	2012	2016

continued on following page

Table 7. Continued

No.	Hydrometric stations on inland rivers	Rank (Q95) / Year		
		1	2	3
69	BACESTI	2015	2012	2000
70	GOLESTI	1987	1989	2009
71	PARHAUTI	2015	1990	2016
72	CERNATESTI	2012	2002	1987
73	CIURUC	1987	2012	1990
74	SIRET	1987	2015	2016
75	CHIOJDU	1990	1987	2008
76	BOGDANESTI	1987	2012	1986
77	LEPSA	1987	1985	1984
78	NEREJU	1987	1993	2001
79	COLACU	1987	2012	1989
80	ASAU	1987	2001	2013
81	PRISACA DORNEI	1987	1985	2006
82	TULGHES	1987	2012	2001
83	HERASTRAU	1985	1987	1986
84	LUNGOCI	1987	1986	1994
85	RACOVITA	1987	1989	2012
86	TUPILATI	1987	1986	2012
87	FARTANESTI	2012	1986	2009
88	TODIRENI	2012	2017	2020
89	SARAIU	2020	1980	2001
90	CRANGU	2012	2013	2008
91	DRAGOIA	2013	2012	2003
92	BALTAGESTI	1988	1992	1993
93	AFUMATI	2002	2003	2008

the AM7 index were estimated for different return periods, are shown in Figure 12, for six hydrometric stations on inland rivers.

For Danube River, the estimation of the return levels of the AM7 index, corresponding to certain return periods, was also illustrated through the graphical representation of five types of distribution models, in case of Isaccea hydrometric station (Figure 13). Thus, based on the Gumbel distribution function, the return levels of AM7 were estimated for return periods, in average, once every 5, 10, 20, 30 and 100 years.

In order to provide an overview of the future situation of low flow in Romania, in the context of current global climate change, the estimated extreme values of AM7 series for different return periods (5, 10, 20, 30, and 100 years, respectively) were compared with the MAM7 index of the reference period 1980 - 2020, at all hydrometric stations, thus estimating the decrease rate of AM7 extreme values (Table 10 and Table 11).

Table 8. The years with the three lowest MQ values over the 1980 – 2020 period, at hydrometric stations on Romanian side of Danube River

No.	Hydrometric stations on DUNAREA River	Rank (MQ) / Year		
		1	2	3
1	ORSOVA	1990	2003	2011
2	DROBETA	1990	2003	2011
3	TIGANASI	1990	2003	2011
4	GRUIA	1990	2003	2020
5	CALAFAT	1990	2003	2020
6	BECHET	1990	2003	1993
7	CORABIA	1990	2020	2003
8	TURNU MAGURELE	1990	2020	2003
9	ZIMNICEA	1990	2003	2020
10	GIURGIU	1990	2003	2020
11	OLTENITA	1990	2003	2020
12	CALARASI-CHICIU	1990	2003	2020
13	VADU OII	1990	2020	2017
14	BRAILA	1990	2020	2012
15	GRINDU	1990	2020	2003
16	ISACCEA	1990	2020	2003
17	CEATAL IZMAIL	1990	2020	1993

The results show a downward trend in AM7 extreme values at all hydrometric stations, but more pronounced for some stations located on inland rivers (bolded values in Table 10). The decrease rate of AM7 extreme values were ranging as follows: -22% ÷ -91% for 5-year return period; -30% ÷ -96% for 10-year return period; -34% ÷ -99% for 20-year return period; -36% ÷ -100% for 30-year return period, and -39% ÷ -100% in case of 100-year return period. The lowest values are found for several hydrometric stations on rivers in the Somes, Crisuri, Mures, Jiu, Arges, Siret and small tributaries in the lower Danube basin.

The same analysis comparing the return levels of the AM7 index with the MAM7 index (for the reference period) was also applied to the Danube River sections. The results showed lower decrease rates of AM7 return levels compared to those obtained for inland rivers, with values averaging between -19% for the 5-year return period and -39% for the 100-year return period (Table 11).

The return period of the MAM7 value for the reference period was also determined based on the best frequency distribution model of the AM7 index series from hydrometric stations on the inland rivers (Table 10). Following the analysis, return periods between 1.45 years (Afumati gauge - Baboia River) and 2 years (Bistra gauge - Viseu River, Scarisoara gauge - Aries River, Carlibaba gauge - Carlibaba River, Fundu Moldovei gauge - Moldova River) were estimated for the inland rivers.

In case of Danube River, the return period of MAM value was estimated, based on the Gumbel distribution model, for once every 1.75 years at all 17 hydrometric stations.

Table 9. The years with the three lowest Q95 values over the 1980 – 2020 period, at hydrometric stations on Romanian side of Danube River

No.	Hydrometric stations on Danube River	rank (Q95) / Year		
		1	2	3
1	ORSOVA	2003	1992	1985
2	DROBETA	2003	1985	1992
3	TIGANASI	2003	1985	1992
4	GRUIA	2003	1985	1992
5	CALAFAT	2003	1985	2018
6	BECHET	2003	1985	1992
7	CORABIA	2018	2003	2019
8	TURNU MAGURELE	2003	1992	2011
9	ZIMNICEA	2003	1992	2011
10	GIURGIU	2003	1992	2011
11	OLTENITA	2003	2018	1992
12	CALARASI-CHICIU	1992	2003	1986
13	VADU OII	1992	2003	2018
14	BRAILA	1992	2003	1990
15	GRINDU	1992	1990	2003
16	ISACCEA	1990	2003	1992
17	CEATAL IZMAIL	1990	2003	1992

Following the analysis of the estimated return periods of the MAM7 index in the reference period, it was found that its lowest return periods (1.45 – 1.65 years) correspond to those stations, only on inland rivers, for which there were sharp decreases in the return levels of AM7 (from 50 to 100%).

A statistical analysis of the frequency of AM7 extreme value rate at stations on inland rivers, in case of return periods once every 5, 10, 20, 30 and 100 years, respectively, exemplified by a boxplot (Figure 14), also revealed a sharp decrease in the extreme AM7 values for 50% of the stations, where the rate values ranging between -33.6% to -62.6% for the 5-year return period, and decreasing to percentages between -62.1% and -84.2% for the 100-year return period.

Low Flow Seasonality

In order to assess the low flow seasonality, seasratio (Seasonality Rate) and seasindex (Seasonality Index) functions available in lfstat package, were used. The seasonality indices were determined for the period 1980 – 2020 at 110 hydrometric stations on Romanian rivers and Danube and then it was spatially represented (Figures 15 – 17).

The spatial representation of seasonality rate (SR) for all 110 hydrometric stations (Figure 15), based on five classes of values, shows a low variability in terms of minimum flow recording, indicating predominantly a minimum flow regime in the warm for 67% of stations season (values in the range 0.13 - 0.90). This mainly concerns hydrometric stations on inland rivers located both outside and inside the

Table 10. The decrease rate of AM7 extreme value for different return periods at 93 hydrometric stations on inland rivers

No.	Hydrometric stations on inland rivers	MAM7	AM7 extreme value decrease rate				
		Return period over 1980 - 2020	Return periods (years)				
			5	10	20	30	100
1	BISTRA	2.00	-23%	-35%	-43%	-48%	-57%
2	LUHEI	1.90	-28%	-40%	-47%	-51%	-58%
3	POIANA BORSA	1.70	-30%	-37%	-41%	-43%	-45%
4	CRASNA	1.60	**-67%**	**-78%**	**-83%**	**-84%**	**-87%**
5	SALSIG	1.50	**-83%**	**-91%**	**-94%**	**-96%**	**-97%**
6	MAIA	1.50	**-84%**	**-90%**	**-90%**	**-90%**	**-91%**
7	BORSA	1.50	**-86%**	**-94%**	**-98%**	**-99%**	**-100%**
8	DOMNESTI	1.60	**-65%**	**-76%**	**-81%**	**-83%**	**-86%**
9	CRISTESTII CICEULUI	1.60	**-69%**	**-81%**	**-86%**	**-88%**	**-91%**
10	COPALNIC	1.55	**-57%**	**-66%**	**-71%**	**-72%**	**-74%**
11	HIDA	1.60	**-64%**	**-75%**	**-81%**	**-82%**	**-85%**
12	SUCIU DE JOS	1.80	-41%	-55%	-64%	-68%	-75%
13	SALVA	1.85	-43%	-58%	-68%	-72%	-80%
14	BONTIDA	1.50	**-91%**	**-96%**	**-99%**	**-99%**	**-100%**
15	POIANA ILVEI	1.70	-46%	-58%	-64%	-67%	-70%
16	NEPOS	1.90	-28%	-39%	-47%	-51%	-58%
17	BECLEAN	1.65	-35%	-42%	-46%	-47%	-49%
18	SATU MARE	1.80	-37%	-48%	-55%	-58%	-64%
19	GURAHONT	1.70	-48%	-58%	-64%	-66%	-69%
20	TINCA	1.70	-39%	-50%	-56%	-59%	-64%
21	BELIU	1.60	**-72%**	**-82%**	**-87%**	**-89%**	**-91%**
22	BLAJENI	1.60	-46%	-54%	-59%	-61%	-63%
23	POCOLA	1.55	-54%	-60%	-63%	-63%	-64%
24	SARSIG	1.55	**-71%**	**-80%**	**-84%**	**-85%**	**-87%**
25	ARAD	1.80	-31%	-40%	-46%	-49%	-53%
26	ABRUD	1.70	-48%	-61%	-69%	-72%	-77%
27	BREAZA	1.60	**-78%**	**-91%**	**-96%**	**-97%**	**-99%**
28	CINTA	1.55	-34%	-39%	-41%	-41%	-42%
29	DOBRA	1.50	-32%	-41%	-45%	-47%	-51%
30	SCARISOARA	2.00	-29%	-43%	-54%	-59%	-72%
31	SIMONESTI	1.50	**-78%**	**-84%**	**-88%**	**-88%**	**-90%**
32	STANCENI	1.80	-22%	-30%	-34%	-37%	-40%
33	TEIUS	1.75	**-56%**	**-71%**	**-81%**	**-85%**	**-92%**

continued on following page

Table 10. Continued

No.	Hydrometric stations on inland rivers	MAM7	AM7 extreme value decrease rate				
		Return period over 1980 - 2020	Return periods (years)				
			5	10	20	30	100
34	TOPLITA	1.70	-36%	-44%	-49%	-51%	-53%
35	POIENI	1.70	-30%	-38%	-43%	-44%	-47%
36	CUPTOARE	1.70	-48%	-59%	-66%	-69%	-73%
37	MEHADIA	1.70	-46%	-58%	-65%	-68%	-73%
38	PRIGOR	1.65	-38%	-46%	-50%	-51%	-54%
39	PODARI	1.65	-43%	-53%	-58%	-60%	-64%
40	BREASTA	1.60	-59%	-67%	-71%	-72%	-74%
41	CORCOVA	1.60	**-65%**	**-75%**	**-79%**	**-81%**	**-83%**
42	TG. CARBUNESTI	1.55	**-65%**	**-72%**	**-75%**	**-75%**	**-76%**
43	RESCA	1.90	-29%	-42%	-52%	-57%	-67%
44	SERCAIA	1.80	-38%	-51%	-59%	-63%	-70%
45	RUSENI	1.80	-44%	-59%	-68%	-72%	-80%
46	FELDIOARA	1.70	-25%	-31%	-35%	-36%	-39%
47	SEBES	1.70	-27%	-33%	-37%	-38%	-41%
48	TELEORMANU	1.90	-26%	-37%	-45%	-48%	-56%
49	CIOBANI	1.90	-36%	-51%	-63%	-68%	-80%
50	ALEXANDRIA	1.95	-27%	-39%	-48%	-53%	-62%
51	GURA FOII	1.80	**-54%**	**-75%**	**-97%**	**-100%**	**-100%**
52	MOARA DIN GROAPA	1.90	-34%	-47%	-56%	-61%	-69%
53	MIOVENI	1.70	**-63%**	**-79%**	**-88%**	**-91%**	**-96%**
54	BUDESTI	1.60	-50%	-59%	-63%	-65%	-67%
55	BABA ANA	1.70	-26%	-32%	-36%	-38%	-39%
56	TESILA	1.80	-37%	-51%	-61%	-65%	-74%
57	SLOBOZIA	1.60	-36%	-43%	-46%	-47%	-49%
58	FUNDU MOLDOVEI	2.00	-30%	-45%	-56%	-61%	-73%
59	LUNGULET	1.80	-42%	-57%	-66%	-70%	-78%
60	BROSTENI	2.00	-22%	-31%	-38%	-42%	-49%
61	BRODINA	1.90	-31%	-43%	-51%	-55%	-62%
62	DORNA GIUMALAU	2.00	-23%	-36%	-46%	-52%	-64%
63	CARLIBABA	2.00	-29%	-43%	-53%	-57%	-68%
64	TECUCI	1.65	**-69%**	**-82%**	**-91%**	**-93%**	**-97%**
65	GALBENI	1.70	**-64%**	**-79%**	**-86%**	**-89%**	**-94%**
66	TATARU	1.60	**-63%**	**-74%**	**-79%**	**-81%**	**-84%**
67	RADENI	1.50	**-82%**	**-94%**	**-96%**	**-98%**	**-99%**

continued on following page

Table 10. Continued

No.	Hydrometric stations on inland rivers	MAM7	AM7 extreme value decrease rate				
		Return period over 1980 - 2020	Return periods (years)				
			5	10	20	30	100
68	VINETESTI	1.60	**-72%**	**-86%**	**-93%**	**-95%**	**-98%**
69	BACESTI	1.65	**-61%**	**-72%**	**-78%**	**-81%**	**-83%**
70	GOLESTI	1.50	**-69%**	**-76%**	**-76%**	**-78%**	**-78%**
71	PARHAUTI	1.65	**-53%**	**-63%**	**-69%**	**-71%**	**-73%**
72	CERNATESTI	1.60	**-67%**	**-78%**	**-83%**	**-85%**	**-87%**
73	CIURUC	1.60	**-54%**	**-62%**	**-67%**	**-68%**	**-71%**
74	SIRET	1.70	-41%	-51%	-56%	-59%	-62%
75	CHIOJDU	1.65	**-67%**	**-82%**	**-90%**	**-93%**	**-98%**
76	BOGDANESTI	1.70	-41%	-50%	-55%	-57%	-60%
77	LEPSA	1.90	-33%	-46%	-57%	-61%	-71%
78	NEREJU	1.70	-34%	-43%	49%	-51%	-54%
79	COLACU	1.80	-41%	-55%	-64%	-68%	-75%
80	ASAU	1.90	-34%	-48%	-58%	-63%	-73%
81	PRISACA DORNEI	1.70	-31%	-38%	-42%	-43%	-45%
82	TULGHES	1.70	-37%	-47%	-53%	-55%	-58%
83	HERASTRAU	1.90	-33%	-47%	-57%	-62%	-72%
84	LUNGOCI	1.75	-40%	-52%	-60%	-63%	-69%
85	RACOVITA	1.65	**-54%**	**-65%**	**-71%**	**-73%**	**-76%**
86	TUPILATI	1.75	-45%	-58%	-66%	-69%	-75%
87	FARTANESTI	1.75	-42%	-54%	-62%	-65%	-70%
88	TODIRENI	1.60	**-58%**	**-69%**	**-74%**	**-76%**	**-78%**
89	SARAIU	1.65	**-60%**	**-71%**	**-78%**	**-80%**	**-83%**
90	CRANGU	1.75	-49%	-62%	-70%	-73%	-79%
91	DRAGOIA	1.65	**-50%**	**-61%**	**-66%**	**-67%**	**-71%**
92	BALTAGESTI	1.85	-41%	-57%	-68%	-74%	-84%
93	AFUMATI	1.45	**-90%**	**-96%**	**-99%**	**-99%**	**-100%**

Carpathian Mountains, as well as stations on the Danube River. The lowest values of seasonality rate, such as the values in the first class (0.13 - 0.50), represented by a red symbol, are found at stations on tributaries of the Barlad, Jiu, Mures, Crisuri and Somes rivers. For these rivers, the minimum flows recorded in the warm season (8-month interval) are between 13% and 50% of the minimum flows recorded in the cold season (4-month interval).

In terms of seasonality of the low flows in the cold season, the seasonality rate values ranging in the interval 1.11 - 2.51 are found in a much smaller percentage, at about 18% of all hydrometric stations. These correspond to river basins generally located in the Carpathian and sub-Carpathian areas, where the lowest discharges are recorded in winter season.

Table 11. Decrease rate of AM7 extreme values for different return periods at 17 hydrometric stations on Danube River

Hydrometric stations on Danube River	Decrease rate (%) of AM7 extreme value vs. MAM7 (1980 – 2020)				
	Return period (years)				
	5	10	20	30	100
ORSOVA	-18%	-24%	-29%	-31%	-36%
DROBETA	-19%	-25%	-30%	-32%	-37%
TIGANASI	-19%	-25%	-30%	-32%	-37%
GRUIA	-18%	-24%	-29%	-31%	-36%
CALAFAT	-18%	-24%	-29%	-31%	-36%
BECHET	-18%	-25%	-29%	-31%	-37%
CORABIA	-19%	-25%	-30%	-32%	-38%
TURNU MAGURELE	-18%	-24%	-29%	-31%	-36%
ZIMNICEA	-17%	-23%	-27%	-29%	-34%
GIURGIU	-18%	-24%	-28%	-30%	-35%
OLTENITA	-19%	-26%	-31%	-33%	-39%
CHICIU-CALARASI	-19%	-25%	-30%	-33%	-38%
VADU OII	-18%	-24%	-28%	-30%	-35%
BRAILA	-18%	-24%	-28%	-30%	-35%
GRINDU	-18%	-24%	-28%	-30%	-35%
ISACCEA	-18%	-24%	-29%	-31%	-36%
CEATAL IZMAIL	-19%	-25%	-30%	-32%	-38%

Regarding the seasonality index (SI), its representation was made by applying a circular statistic based on two parameters, θ and r, calculated from the Julian calendar data of all days during the observation period when flows were less than or equal to Q95. The values of this index are highlighted, on the graph, by points determined by a vector through which, with an angle θ, it is estimated, on average, the day of the year (D) when a low flow occurs, and the degree of variability of its occurrence is determined by the length of the vector r. On the graph are also represented the calendar months and the delimitation of the two seasons (IV - XI and XII - III). The seasonal distribution of the minimum discharges occurrence, in case of all hydrometric stations, based on the SI index values computed for the reference period, is shown in Figure 16.

A high degree of seasonality in the occurrence of low flows is therefore revealed. This high seasonality was associated with r parameter values greater than 0.6 (> 60%) for 68% of stations. Thus, the occurrence of minimum flows in August, September and October is shown for 63 hydrometric stations, and in December and January for only 12 stations.

Also, the SI index plot shows points related to r-values ranging from 0.5 to 0.6 for 17% of stations, revealed a lower degree of seasonality in the occurrence of minimum flows from August to December. For the other 15% percent, a low seasonality of the occurrence of minimum flows is evident, which is justified by the approximately unit values of the seasonality rate determined at these stations.

Figure 10. The year with the lowest value of MQ over the 1980 – 2020 period, at 110 hydrometric stations on inland and Danube rivers in Romania

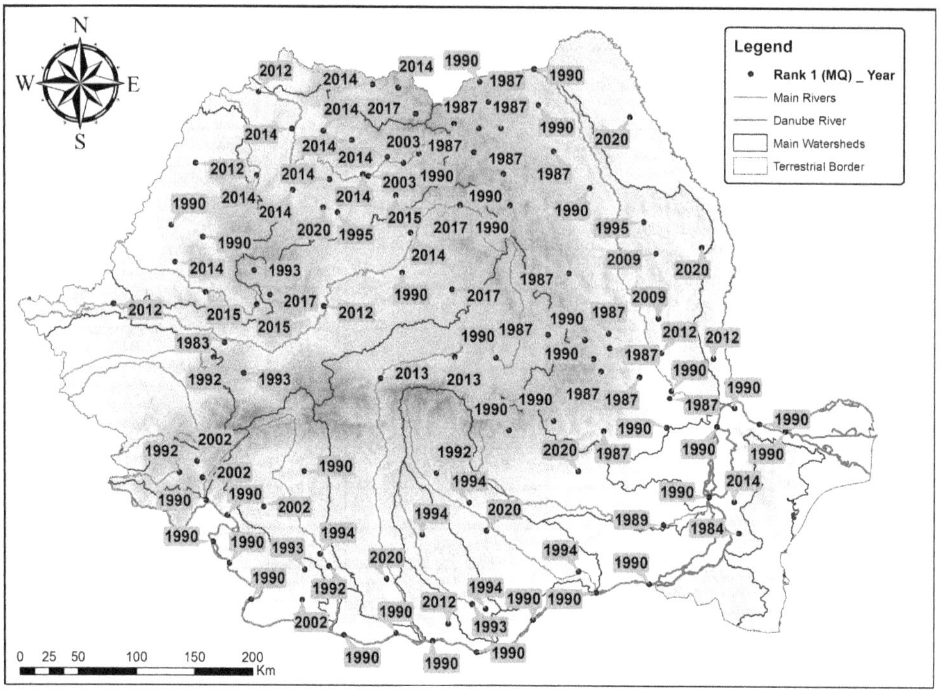

Figure 11. The year with the lowest value of Q95 over the 1980 – 2020 period, at 110 hydrometric stations on inland and Danube rivers in Romania

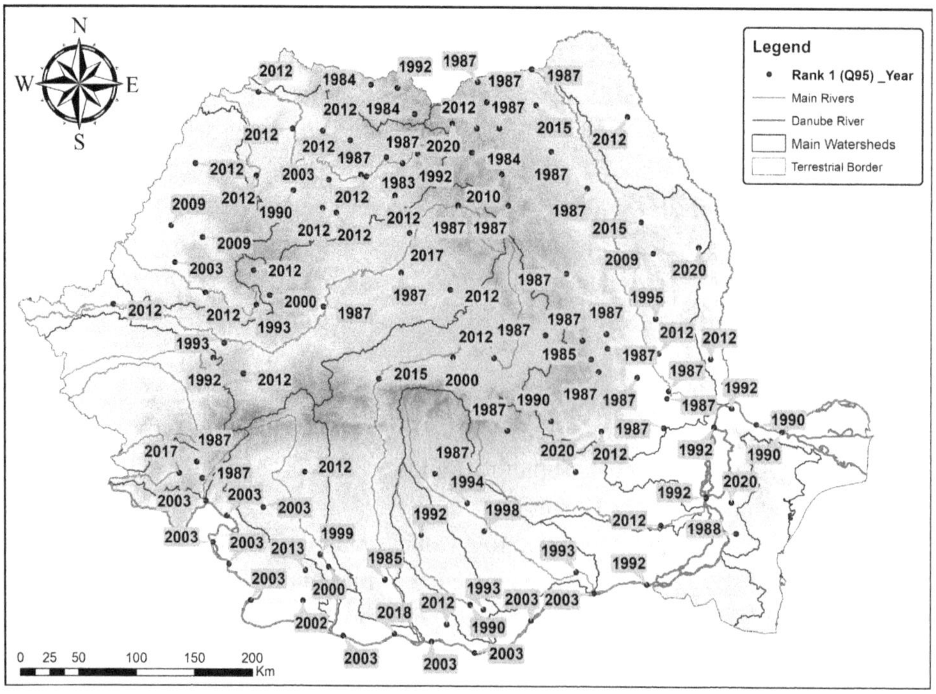

Figure 12. Distribution models of AM7 series at hydrometric stations on inland rivers

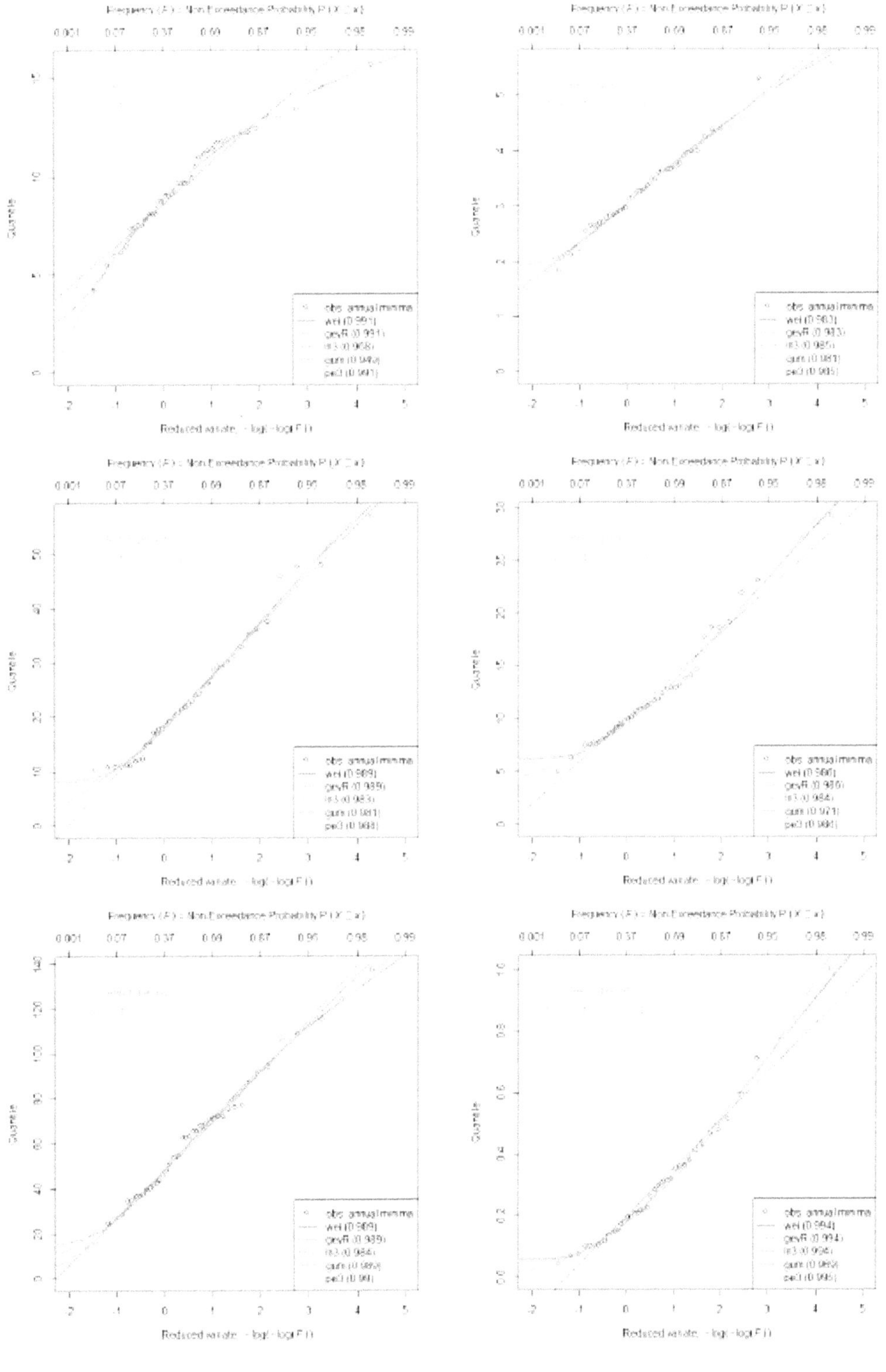

Figure 13. Distribution models of AM7 seria at Isaccea hydrometric station on the Danube River

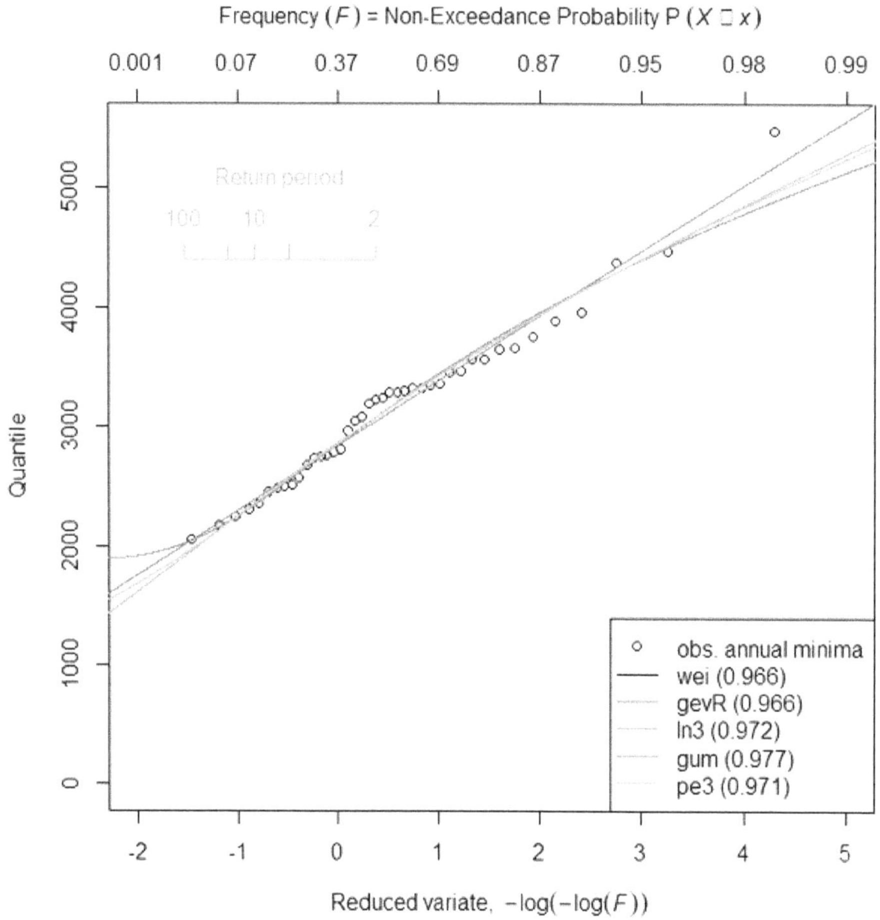

Furthermore, the mean day of occurrence of low flows was highlighted in the August-November period for 82 hydrometric stations, and for the other stations, in December and January.

Degree of variability in the seasonality of low flows, represented by the r parameter, was also highlighted by a spatial representation at all hydrometric stations, in Figure 17.

In terms of spatial distribution over the country, a high degree of seasonality greater than 50%, was found for the most hydrometric stations (96 stations) (including Danube River section), while for the others (14 stations), were determined low values of seasonality degree ($r < 50\%$). Concerning the Romanian part of the Danube River, the lowest flows occurred in October (purple symbol, in Figure 16), with a high degree of seasonality, more than 70%, which is highlighted in Figure 17.

Low Flow Periods

Regarding the water deficit identified during the low flow periods, an evaluation of its characteristics was performed based on the threshold level method, and a moving average filter of 7 days. Low flow periods are often defined using a threshold-based methodology, 10th-30th percentiles being most com-

Figure 14. Frequency of decrease rate of the AM7 extreme values estimated for return periods once every 5, 10, 20, 30, and 100 years respectively, at stations on inland rivers

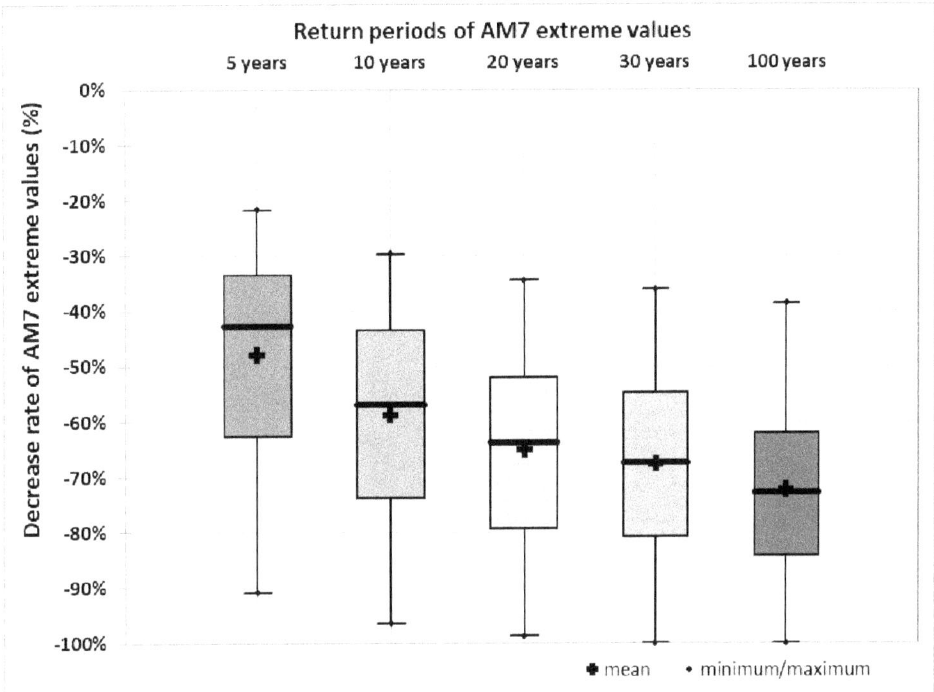

Figure 15. The seasonality rate (SR) at 110 hydrometric stations on inland and Danube rivers in Romania

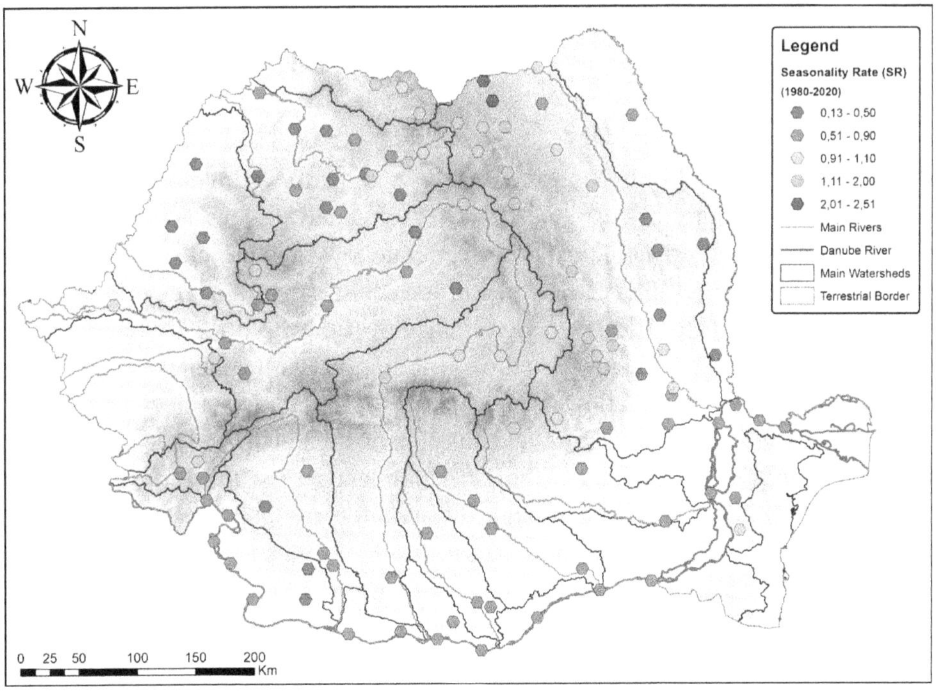

Figure 16. The seasonality index (SI) at 110 hydrometric stations on inland and Danube rivers in Romania

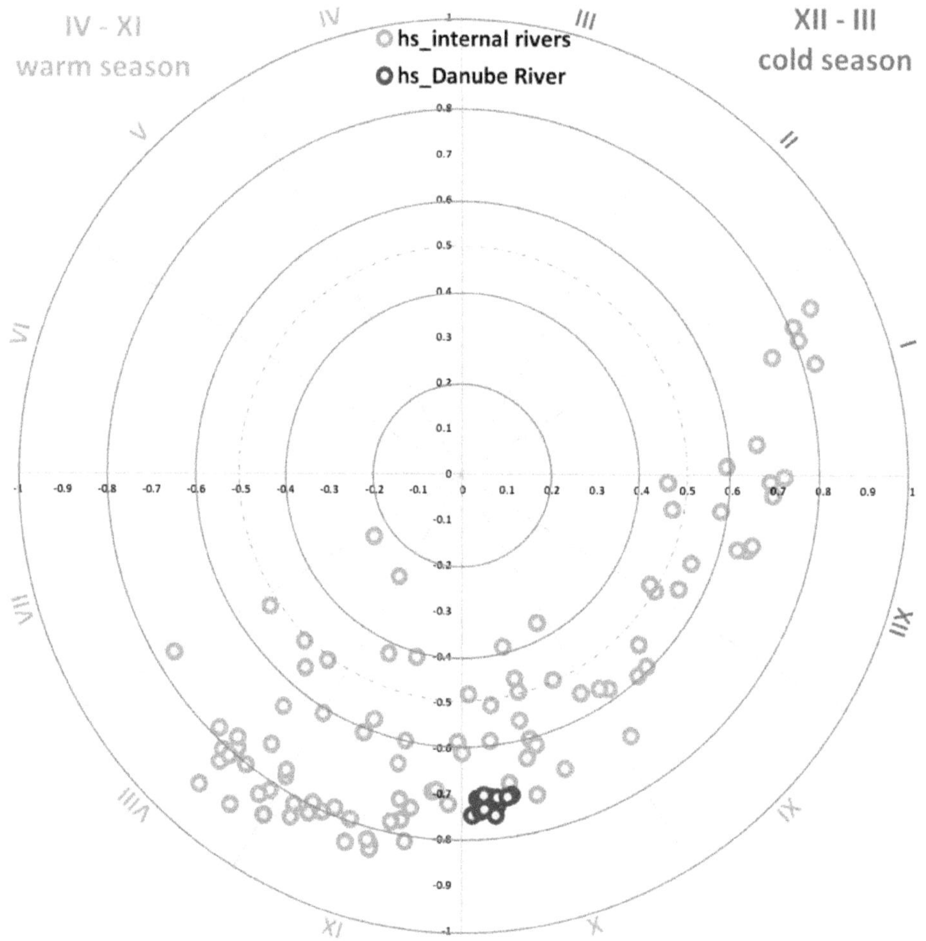

monly used as reference thresholds. In case of the present study, for estimation of dry events, the 20th (Q80) percentile of the flow duration curve, i.e., the flow exceeded 80% of the time, was used as a fixed threshold level. This index is used an indicator for moderate low flow periods.

According to these hypotheses, the series of maximum annual streamflow deficits with more than 30 days durations, and their characteristics, were determined over the 1980 – 2020 reference period, at 93 hydrometric stations on inland rivers and 17 hydrometric stations on the Danube River section, Orsova - Ceatal Izmail. Also, considering the three main characteristics of a low flow event, it was identified the years with the highest magnitude, volume, and duration of streamflow deficits. Deficits lasting more than 30 consecutive days were highlighted in Table 12, for inland rivers, and Table 13, for Danube River. These tables show the following: number of drought events below the Q80 threshold and the year with the highest streamflow deficit, according to its magnitude (max M), volume (max V), and duration (max D).

It is found that the number of annual maximum streamflow deficits with more than than 30 days durations, for the fixed threshold Q80, is 27 deficits in the 41 years interval, in average, at hydrometric stations on inland rivers, which means that in approximately two out of three years, there was a stream-

Figure 17. The degree of seasonality (r) at 110 hydrometric stations on inland and Danube rivers in Romania

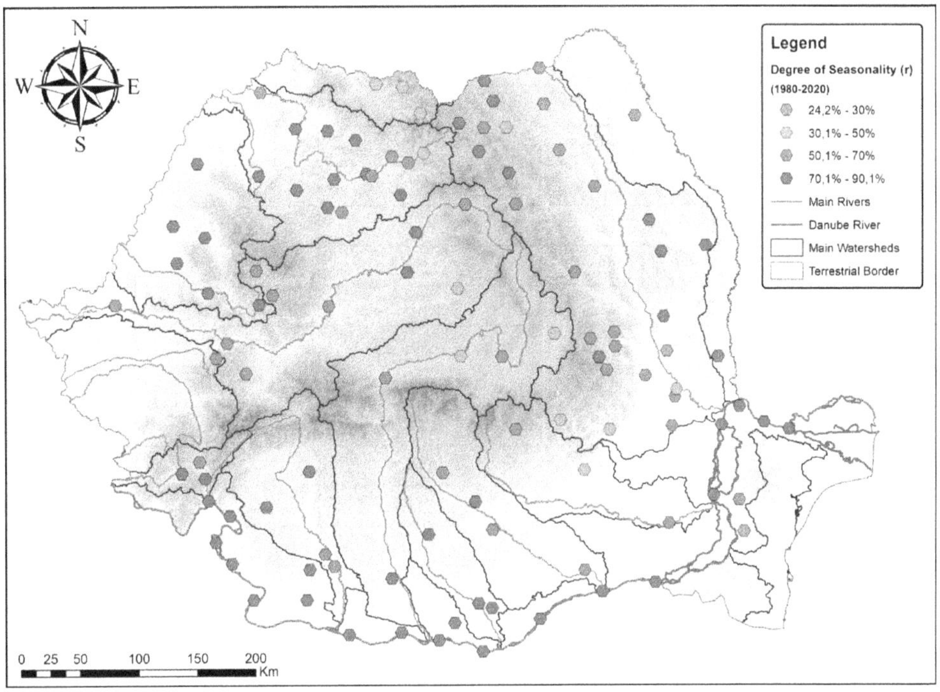

flow deficit with a duration greater than 30 days. The duration of these dry events ranges from 89 days (Tataru station - Ramnicu Sarat River) to 513 days (Afumati station – Baboia River).

Assessing these, it was found that for 56% of the total number of hydrometric stations (on inland rivers), the largest deficits occurred during the last two decades, in the following years: 2000 - 2003, 2011 - 2013, 2015, 2017, 2019 and 2020 (Table 12).

The histogram of the durations corresponding to the maximum annual deficits, for inland rivers, shows that the duration of deficits ranged from 100 to 150 days for 35% of stations, from 150 to 200 days for 33% stations, and from 200 to 250 days for 24% of stations, respectively (Figure 18).

Moreover, on the inland rivers, the highest number of maximum annual streamflow deficits was identified in the periods 1983 - 1984, 1986 – 1987, at Tisa, Somes, Crisuri, Mures, Olt, and Siret basins, as well as in the years 1990 – 1997, on rivers in the Bega, Jiu, Vedea, Arges, and Ialomita basins. In case of the smaller tributaries of the Danube, the highest number of low flow events occurred in the intervals 2001 - 2003, 2011 - 2013, as well as in 2019 – 2020 (Table 12).

Following the evaluation of drought events with a duration of more than 30 consecutive days, the year 2003, on the Orsova - Bechet sector, and the year 1986, on the Isaccea - Ceatal Izmail sector, are highlighted with the longest durations. At the same time, the years 2019 and 1983 stand out in terms of the maximum duration of streamflow deficit on the Corabia - Giurgiu and Chiciu - Braila sectors (Table 13).

A representation, by duration and volume, of streamflow deficits below the Q80 threshold, with a duration greater than 30 days, at the hydrometric stations Orsova, Calarasi - Chiciu and Isaccea, can be

Table 12. Years with the highest streamflow deficits over the period 1980 - 2020, at hydrometric stations on inland rivers

No.	Hydrometric stations on inland rivers	Low flow events (D ≥ 30 days)	max M	max V	max D	
			Year	Year	Year	days
1	BISTRA	29	1987	1983 - 1984	1983 - 1984	193
2	LUHEI	29	1992	2015	2005 - 2006	126
3	POIANA BORSA	29	1983 - 1984	1983 - 1984	1988 - 1989	133
4	CRASNA	25	1994	2012	2012	120
5	SALSIG	32	2012	2012	2003	163
6	MAIA	25	2014	2003	2003	128
7	BORSA	30	2017	2019 - 2020	2019 - 2020	119
8	DOMNESTI	33	1992	2012	2012	205
9	CRISTESTII CICEULUI	26	2012	2011	2011	168
10	COPALNIC	23	2012	2012	2012	130
11	HIDA	28	2003	1983 - 1984	1983 - 1984	118
12	SUCIU DE JOS	25	2003	1983	1983	106
13	SALVA	27	1986 - 1987	2012	2012	223
14	BONTIDA	27	1990	2012 - 2013	2012 - 2013	155
15	POIANA ILVEI	30	1983	1983	1988 - 1989	142
16	NEPOS	24	1986 - 1987	1983 - 1984	1983 - 1984	146
17	BECLEAN	23	2012	1983 -1984	1983 - 1984	201
18	SATU MARE	29	1986 - 1987	2012 - 2013	2012 - 2013	213
19	GURAHONT	26	2003	2000	2000	141
20	TINCA	26	2009	2012	2000	139
21	BELIU	30	2003	2009	2000	120
22	BLAJENI	21	2017	2012	1983 - 1984	111
23	POCOLA	21	1992	2012	2000	136
24	SARSIG	30	2007	2012	2019	131
25	ARAD	34	2012 - 2013	2012 - 2013	2012 - 2013	91
26	ABRUD	29	2011	2000	2000	152
27	BREAZA	32	2017	2012 - 2013	2012 - 2013	144
28	CINTA	29	2017	2000 - 2001	2000 - 2001	122
29	DOBRA	25	2003	1984 - 1985	1983 - 1984	191
30	SCARISOARA	26	2012	1983 - 1984	2000	129
31	SIMONESTI	31	2017 - 2018	2017 - 2018	2016 - 2017	187
32	STANCENI	29	2019 - 2020	1986 - 1987	1983 - 1984	200
33	TEIUS	26	1987	2000	2000	189
34	TOPLITA	28	2012	2012	1993	236

continued on following page

Table 12. Continued

No.	Hydrometric stations on inland rivers	Low flow events (D ≥ 30 days)	max M	max V	max D	
			Year	Year	Year	days
35	POIENI	28	1992	1991 - 1992	1991 - 1992	178
36	CUPTOARE	28	2000 - 2001	2000 - 2001	2000 - 2001	145
37	MEHADIA	26	1987	2012	2012	228
38	PRIGOR	26	1994	2012 - 2013	2012 - 2013	174
39	PODARI	29	1985	2000 - 2001	2011 - 2012	180
40	BREASTA	27	1999	1994	1994	222
41	CORCOVA	31	2000	2000	1993	249
42	TG. CARBUNESTI	28	2012	2000 - 2001	2000 - 2001	188
43	RESCA	30	2012	2012	1984	139
44	SERCAIA	27	2000	2012 - 2013	2012 - 2013	231
45	RUSENI	27	2015	1986 - 1987	1990 - 1991	173
46	FELDIOARA	24	2013	2019 - 2020	2000 - 2001	235
47	SEBES	28	1987	2012 - 2013	2012 - 2013	203
48	TELEORMANU	34	1993	2000	1992	212
49	CIOBANI	32	1982	1992 - 1993	1992 - 1993	148
50	ALEXANDRIA	33	1990	1990	2002	179
51	GURA FOII	27	1994	1992	1995	138
52	MOARA DIN GROAPA	19	1998	1996 - 1997	1996 - 1997	335
53	MIOVENI	26	1986	2000 - 2001	2000 - 2001	231
54	BUDESTI	26	1994 1995	1994 - 1995	1994 - 1995	273
55	BABA ANA	28	2020	2019 - 2020	2019 - 2020	168
56	TESILA	31	1986 - 1987	2000 - 2001	1992 - 1993	203
57	SLOBOZIA	19	1989	1990	1990	352
58	FUNDU MOLDOVEI	16	2012	1983 - 1984	1983 - 1984	103
59	LUNGULET	26	2011 - 2012	1986 - 1987	1986 - 1987	200
60	BROSTENI	24	2009	2011 - 2012	2011 - 2012	190
61	BRODINA	24	1986 - 1987	2011 - 2012	2011 - 2012	174
62	DORNA GIUMALAU	25	1983 - 1984	1983 - 1984	1983 - 1984	89
63	CARLIBABA	25	1988 - 1989	1983 - 1984	1983 - 1984	210
64	TECUCI	28	2012	2012	2012	129
65	GALBENI	25	1987	2012	2015	155
66	TATARU	30	2012	2012	2012	158
67	RADENI	31	2019	2015	2012 - 2013	155
68	VINETESTI	32	2020	2020	2020	144
69	BACESTI	27	2012 - 2013	2012 - 2013	2015	230

continued on following page

Table 12. Continued

No.	Hydrometric stations on inland rivers	Low flow events (D ≥ 30 days)	max M	max V	max D	
			Year	Year	Year	days
70	GOLESTI	29	2009	2012	2012	196
71	PARHAUTI	24	2015	1990	1986 - 1987	194
72	CERNATESTI	24	2012	1987	1990	184
73	CIURUC	28	1986 - 1987	1986 - 1987	1986 - 1987	181
74	SIRET	26	2016	2015 - 2016	1990 - 1991	172
75	CHIOJDU	30	1987	1990	2000 - 2001	150
76	BOGDANESTI	30	1986 - 1987	1986 - 1987	1986 - 1987	148
77	LEPSA	28	1984 -1985	1986 - 1987	1990 - 1991	131
78	NEREJU	25	1986 - 1987	1986 - 1987	2011 - 2012	230
79	COLACU	28	2012	1986 - 1987	1986 - 1987	219
80	ASAU	26	1986 - 1987	1986 - 1987	1986 - 1987	213
81	PRISACA DORNEI	28	1986 - 1987	1986 - 1987	1983 - 1984	160
82	TULGHES	24	1986 - 1987	1986 - 1987	1988 - 1989	206
83	HERASTRAU	28	1989 - 1990	1988 - 1989	1988 - 1989	143
84	LUNGOCI	27	1994	2000 - 2001	2000 - 2001	183
85	RACOVITA	23	1987	1987	1986 - 1987	143
86	TUPILATI	28	1986 - 1987	1986 - 1987	1990 - 1991	139
87	FARTANESTI	29	2009	2012	2012	138
88	TODIRENI	24	2012 - 2013	2015 - 2016	2015 - 2016	366
89	SARAIU	31	2020	2020	2011	144
90	CRANGU	27	2012 - 2013	2012 - 2013	2012 - 2013	311
91	DRAGOIA	29	2013	2012	2012	185
92	BALTAGESTI	24	1995	1979 - 1980	1983 - 1984	211
93	AFUMATI	19	2013	2001 - 2003	2001 - 2003	513

found in Figure 19. Thus, the highest annual maximum streamflow deficits over the reference period, their onset and duration are highlighted here. It is observed that low flow events below the Q80 threshold with the longest durations are generally concur with the largest volumes.

Therefore, on the Romanian Danube sector, downstream from Orsova station, the analysis of highest annual maximum streamflow deficits revealed that 1986, 1983, 2003, 2018, and 2019, are among the years with the longest duration of streamflow deficit (from 108 days at Gruia station, in 2003, to 149 days at Ceatal Izmail, in 1986).

CONCLUSION

This study evaluated the low flow and water deficit characteristics on inland rivers and the Danube in Romania, for 110 hydrometric stations, providing an overview of this specific phase of hydrological regime for the period 1980 -2020.

Figure 18. Frequency of the duration of streamflow deficits at hydrometric station on inland rivers

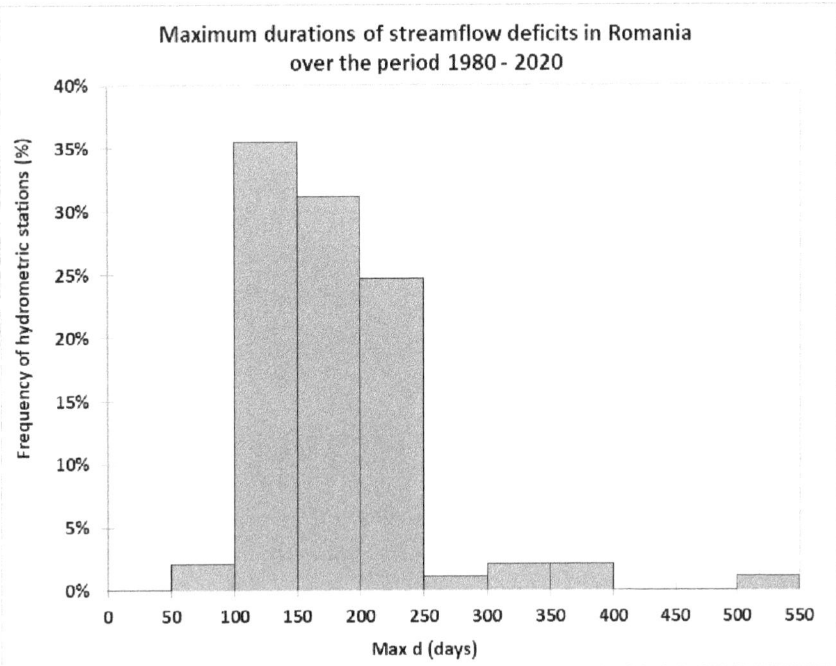

Table 13. Years with the highest streamflow deficits over the period 1980 - 2020, at hydrometric stations on Danube River

No.	Hydrometric stations on Danube River	Low flow events (D ≥ 30 days)	max M	max V	max D	
			Year	Year	Year	days
1	ORSOVA	26	2003	2003	2003	118
2	DROBETA	26	2003	2003	2003	116
3	TIGANASI	26	1985	2003	2003	114
4	GRUIA	28	1985	2003	2003	108
5	CALAFAT	28	2003	2003	2003	112
6	BECHET	25	2003	2003	2003	111
7	CORABIA	26	2018	2003	2019	120
8	TURNU MAGURELE	27	2003	2003	2019	117
9	ZIMNICEA	27	2003	2003	2019	117
10	GIURGIU	29	2003	2003	2019	115
11	OLTENITA	28	1992	2003	2018	139
12	CHICIU-CALARASI	29	1992	2003	1983	127
13	VADU OII	29	1992	1992	1983	127
14	BRAILA	28	1992	2003	1983	129
15	GRINDU	27	1992	2003	1983	132
16	ISACCEA	28	1990	2003	1986	148
17	CEATAL IZMAIL	28	1990	2003	1986	149

Figure 19. The highest annual maximum streamflow deficits over the period 1980 – 2020, at Orsova, Chiciu-Calarasi, and Isaccea stations on Danube River

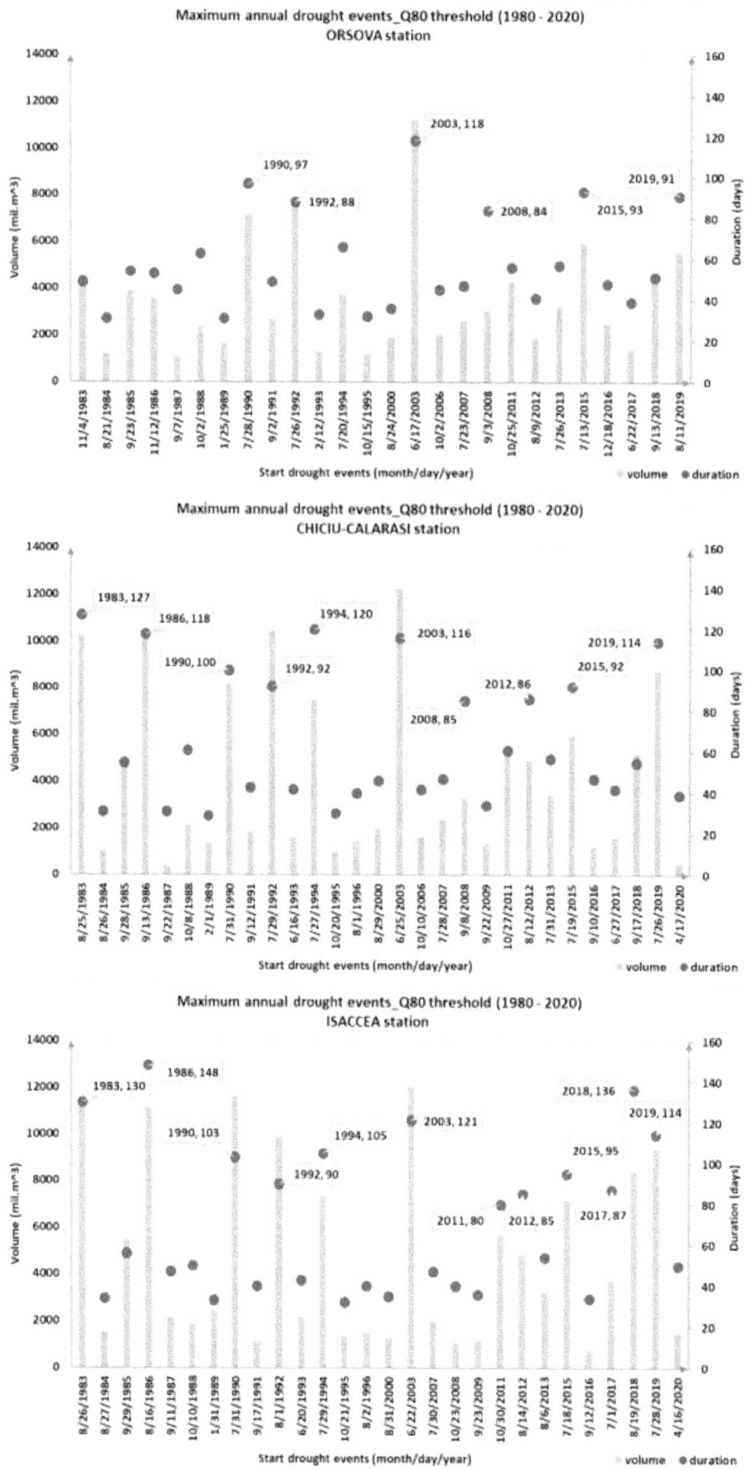

The spatial distribution of the indicators measuring the magnitude of low specific flow, q80 and q95, showed the same spatial pattern particularly evident for values below 1 l/s*km2. These specific low flow values were highlighted at hydrometric stations on inland rivers, increasing from approximately 33% in case of Q80 to 50% in case of Q95. The results also showed a low variability of minimum flows over a period of one to three months, which demonstrates the presence of dry periods without precipitation, for about 30% of the stations located both inside (in the north-west and centre of the country, in Somes and upper Mures watersheds) and outside the Carpathian Mountains.

In terms of relative contribution of base flow to the total flow in rivers, the BFI index revealed values ranging between 31% and 60%, for about 57% of stations. A considerable contribution of the phreatic aquifer to the Danube River flow was represented by values between 81% and 95%, as well as in the case of some small tributaries with BFI values for about 83%.

The rank analysis of annual values of mean flow and Q95 low flow indicators emphasized, as follows: the year 1990 with the lowest value of MQ index (for 45% of stations), for the most hydrometric stations on the Danube River, as well for stations in the Siret, Crisuri, and Jiu watersheds; also, 1990, 1992, 1993 and 1994 years with the lowest value of mean flow for the southern half of the country; the lowest value of Q95 low flow indicator in the year 1987, for the eastern part of the Romania (in the Siret basin), and 2012, for the north-west, west and south-east of the country, as well as at the terminus stations of large rivers, such Somes, Crisul Alb, Mures and Ialomita, and at stations on the Olt and Jiu rivers.

Concerning the Danube River segment, from Orsova station to Ceatal Izmail station, the rank analysis revealed that the years 1990 and 2003 had the lowest (1st rank) value of mean annual discharge, respectively the lowest value of Q95. Regarding the 2nd and 3rd rank values, the years highlighted were: 2003, 2020 and 2011, in case of mean flow indicator, as well as 1985, 1992 and 2011, in case of Q95 index.

The distribution of the rate of fall in AM7 return levels (compared to MAM7 in the reference period) showed a pronounced decrease for 50% of the stations, where the rate values ranged from -33.6% to -62.6% for the 5-year return period, and fell to percentages ranging from -62.1% to -84.2% for the 100-year return period. In case of Danube River, the results showed lower decrease rates of AM7 return levels, with values averaging between -19% for the 5-year return period and -39% for the 100-year return period.

As a result of the assessment of low flow seasonality, it can be concluded that there is generally a high degree of seasonality in the occurrence of low flows in the warm season, especially in August, September, and October, for 67% of stations, which is also evidenced by the seasonality rate (values ranging between 0.13 to 0.90) that indicates, predominantly, a low flow regime in the warm season, for the same percent. The lowest values (values ranging between 0.13 to 0.50) of seasonality rate, were found at stations on the tributaries of Barlad, Jiu, Mures, Crisuri and Somes rivers. In the case of Danube River, the occurrence of minimum flows occurs in October, with a high degree of seasonality (70 - 75%).

The study also highlighted the highest water deficits below the Q80 threshold, which for 56% of the total number of hydrometric stations (on inland rivers) have been recorded in the last two decades: 2000 - 2003, 2011 - 2013, 2015, 2017, 2019 and 2020. The durations of the highest maximum annual deficits had values ranging between 100 to 150 days for 35% percent of stations, between 150 to 200 days for 33% stations, and between 200 to 250 days for 24% of stations. In the case of the Danube River, downstream of the Orşova station, the years 1986, 1983, 2003, 2018, and 2019 are among the years with the longest duration of streamflow deficit (e.g., 3 months in 2003 and 5 months in 1986).

REFERENCES

Andreadis, K., Clark, E., Wood, A., Hamlet, A., & Lettenmaier, D. (2005). Twentieth-century drought in the conterminous United States. *Journal of Hydrometeorology*, *6*(6), 985–1001. doi:10.1175/JHM450.1

Burn, D. H. (1997). Regionalisation of catchments for regional flood frequency analysis. *Journal of Hydrologic Engineering*, *1997*(2), 6–82.

Corzo Perez, G. A., van Huijgevoort, M. H. J., Voß, F., & van Lanen, H. A. J. (2011). On the spatio-temporal analysis of hydrological droughts from global hydrological models. *Hydrology and Earth System Sciences*, *15*(9), 2963–2978. doi:10.5194/hess-15-2963-2011

Cupak, A., & Kaczor, G. (2023). Determination of Seasonal Indices for the Regionalization of Low Flows in the Upper Vistula River Basin. *Water (Basel)*, *15*(2), 246. doi:10.3390/w15020246

Fleig, A. K., Tallaksen, L. M., Hisdal, H., & Demuth, S. (2006). A global evaluation of streamflow drought characteristics. *Hydrology and Earth System Sciences*, *10*(4), 535–552. doi:10.5194/hess-10-535-2006

Gustard, A., & Demuth, S. (2008). *Manual on low-flow estimation and prediction. Open Hydrology Report, Vol. No. 50*. World Meteorological Organization.

Hisdal, H., Tallaksen, L., Clausen, B., Peters, E., & Gustard, A. (2004). Hydrological drought characteristics, Hydrological drought, Processes and estimation methods for streamflow and groundwater. In *Developments in Water Science, 48* (pp. 139–198). Elsevier Science B. V.

Hosking, J. R. M., & Wallis, J. R. (2005). *Regional Frequency Analysis: An Approach Based on L-Moments*. Cambridge University Press.

Koffler, D., & Laaha, G. (2013). *LFSTAT - Low-Flow Analysis in R, Geophysical Research Abstracts* (Vol. 15). EGU General Assembly.

Laaha, G. (2023). A mixed distribution approach for low-flow frequency analysis – Part 1: Concept, performance, and effect of seasonality. *Hydrology and Earth System Sciences*, *27*(3), 689–701. doi:10.5194/hess-27-689-2023

Laaha, G., & Blöschl, G. (2006). Seasonality indices for regionalizing low flows. *Hydrological Processes*, *20*(18), 3851–3878. doi:10.1002/hyp.6161

Laaha, G., & Blöschl, G. (2006b). Seasonality indices for regionalizing low flows. *Hydrological Processes*, *20*(18), 3851–3878. doi:10.1002/hyp.6161

Laaha, G., Demuth, S., Hisdal, H., Kroll, C. N., van Lanen, H. A. J., Nester, T., Rogger, M., Sauquet, E., Tallaksen, L. M., Woods, R., & Young, A. (2013). Prediction of low flows in ungauged basins. In G. Bloeschl, M. Sivapalan, T. Wagener, A. Viglione, & H. Savenije (Eds.), *Runoff Prediction in Ungauged Basins: Synthesis across Processes, Places and Scales* (pp. 163–188). Cambridge University Press. doi:10.1017/CBO9781139235761.011

Laaha, G., Gauster, T., Delus, C., Vidal, J.-P., & the UNESCO FRIEND-Water low flow and drought group Team (2016). *The European 2015 drought from a hydrological perspective, EGU (European Geosciences Union)*, General Assembly, Vienna, Austria.

Laaha, G., Parajka, J., Viglione, A., Koffler, D., Haslinger, K., Schöner, W., Zehetgruber, J., & Blöschl, G. (2016). A three-pillar approach to assessing climate impacts on low flows. *Hydrology and Earth System Sciences*, *20*(9), 3967–3985. doi:10.5194/hess-20-3967-2016

Salinas, J. L., Laaha, G., Rogger, M., Parajka, J., Viglione, A., Sivapalan, M., & Blöschl, G. (2013). Comparative assessment of predictions in ungauged basins – Part 2: Flood and low flow studies. *Hydrology and Earth System Sciences*, *17*(7), 2637–2652. doi:10.5194/hess-17-2637-2013

Sheffield, J., Andreadis, K., Wood, E., & Lettenmaier, D. (2009). Global and continental drought in the second half of the twentieth century: Severity-area-duration analysis and temporal variability of large scale events. *Journal of Climate*, *22*(8), 1962–1981. doi:10.1175/2008JCLI2722.1

Smakhtin, V. U. (2001). Low flow hydrology: A review. [Elsevier.]. *Journal of Hydrology (Amsterdam)*, *240*(3-4), 147–186. doi:10.1016/S0022-1694(00)00340-1

Spinoni, J., Naumann, G., Vogt, J. V., & Barbosa, P. (2015). (2015). The biggest drought events in Europe from 1950 to 2012. *Journal of Hydrology. Regional Studies*, *3*, 509–524. doi:10.1016/j.ejrh.2015.01.001

Stagge, J. H., Tallaksen, L. M., Kohn, I., Stahl, K., & van Loon, A. (2013). *A European Drought Reference Database: Design and Online Implementation*. (DROUGHT-R&SPI, technical Report; No. 12). EU. https://edepot.wur.nl/291635

Stahl, K., Vidal, J.-P., Hannaford, J., Tijdeman, E., Laaha, G., Gauster, T., & Tallaksen, L. M. (2020). The challenges of hydrological drought definition, quantification, and communication: an interdisciplinary perspective. *Proc. IAHS*. IEEE. 10.5194/piahs-383-291-2020

Sutanto, S. J., & Van Lanen, H. A. J. (2020). Hydrological Drought Characteristics Based on Groundwater and Runoff Across Europe, *Proc. IAHS*. IEEE. 10.5194/piahs-383-281-2020

Tallaksen, L. M., & Van Lanen, H. A. J. (2004). *Hydrological drought: processes and estimation methods for streamflow and groundwater. Developments in water science*. Elsevier.

Van Huijgevoort, M. H. J., Van Lanen, H. A. J., Teuling, A. J., & Uijlenhoet, R. (2014). Identification of changes in hydrological drought characteristics from a multi-GCM driven ensemble constrained by observed discharge. *Journal of Hydrology (Amsterdam)*, *512*, 421–434. doi:10.1016/j.jhydrol.2014.02.060

Van Lanen, H. A. J., Laaha, G., Kingston, D. G., Gauster, T., Ionita, M., Vidal, J.-P., Vlnas, R., Tallaksen, L. M., Stahl, K., Hannaford, J., Delus, C., Fendekova, M., Mediero, L., Prudhomme, C., Rets, E., Romanowicz, R. J., Gailliez, S., Wong, W. K., Adler, M.-J., & Van Loon, A. F. (2016). Hydrology needed to manage droughts: The 2015 European case. *Hydrological Processes*, *30*(17), 3097–3104. doi:10.1002/hyp.10838

Van Loon, A. F., & Van Lanen, H. A. J. (2012). A process-based typology of hydrological drought. *Hydrology and Earth System Sciences*, *16*(7), 1915–1946. doi:10.5194/hess-16-1915-2012

Water Cadastre Atlas of Romania. (1992).

Yevjevich, V. (1967). *An Objective Approach to Definitions and Investigations to Continental Hydrologic Droughts*. Colorado State University.

Young, A. R., Round, C. E., & Gustard, A. (2000). Spatial and temporal variations in the occurrence of low flow events in the UK. *Hydrology and Earth System Sciences*, *2000*(4), 35–45. doi:10.5194/hess-4-35-2000

Chapter 10
Tools and Techniques Used in the Modeling of Hydrometeorological Events:
Understanding the Scientific Process of Predicting Such Events

Zahid Nabi
https://orcid.org/0000-0001-6724-6345
Central University of Jammu, India

Shivali Kundan
Central University of Jammu, India

Neha Verma
Central University of Jammu, India

Dinesh Kumar
Central University of Jammu, India

ABSTRACT

Hydrometeorological events that result from the interactions between the Earth's atmosphere and the water cycle encompass a wide range of phenomena, including precipitation, flooding, droughts and other weather-related incidents and pose a significant challenge to human beings. Through the integration of numerical weather prediction models, hydrological models, and remote sensing technologies, researchers gain a comprehensive understanding of atmospheric dynamics, water movement, and their interactions with the Earth's surface thus unfolding the complexities of hydrometeorological events. The chapter highlights the significance of modeling such events making the use of Numerical modeling, data assimilation, ensemble forecasting and other tools and techniques.

DOI: 10.4018/978-1-6684-8771-6.ch010

INTRODUCTION TO HYDROMETEOROLOGICAL EVENTS

Hydrometeorological events are natural occurrences that result from the interactions between the Earth's atmosphere and the water cycle. These events encompass a wide range of phenomena, including precipitation, flooding, droughts, cyclones, thunderstorms, and other weather-related incidents. A list of various hydrometeorological events has been given in Figure 1.1. These events usually bring misery to people and result in damage to life and infrastructure. With the scars of climate change beginning to appear worldwide, there has been a substantial increase in the frequency of such events worrying both the scientific researchers and the local populace. This has led to increased scientific efforts to understand, predict, and minimize the damage caused by such events (Jentsch et al. 2007; Pachauri et al. 2014; Matthies and Quevauviller, 2022).

Figure 1. A list of various hydrometeorological events

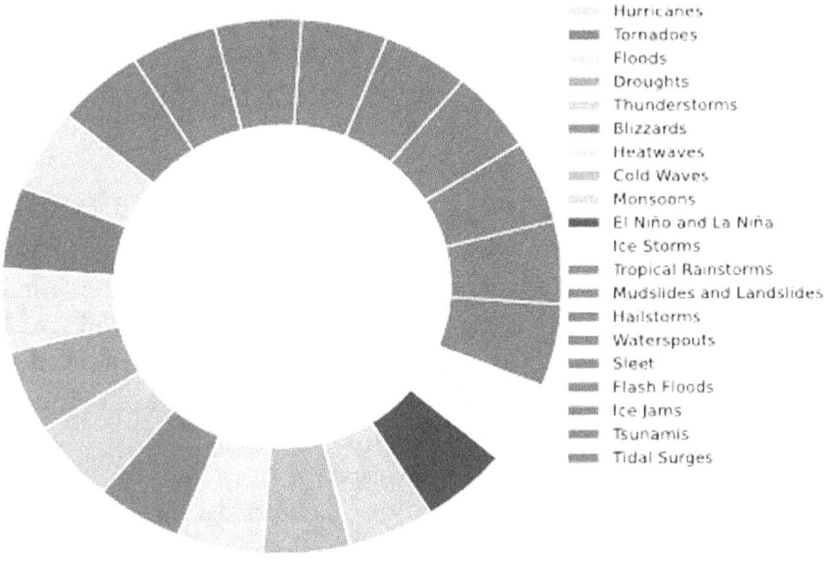

MOST FATAL HYDROMETEOROLOGICAL EVENTS AND THE DANGERS THEY POSE

Hydrometeorological events, which encompass a broad range of atmospheric and water-related phenomena, have been responsible for significant loss of life globally. Understanding the patterns and impacts of these events is crucial for enhancing preparedness, response, and mitigation strategies. This section aims to shed light on hydrometeorological events that are most fatal and damaging by drawing insights from historical data and authoritative sources. Given below in Figure 2.1 is a chart that shows the average number of annual deaths caused by the most profound hydrometeorological events and afterward, each type of event has been elaborated on in detail.

Figure 2. Average global deaths caused by different hydrometeorological events annually

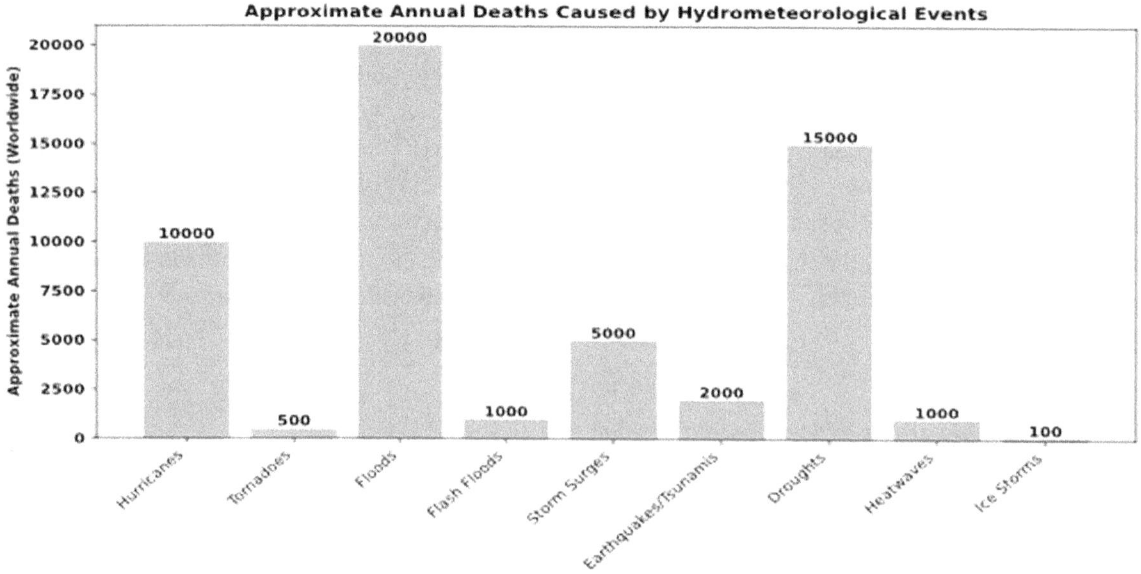

Tornadoes

Tornadoes, with their violent and rotating columns of air, pose a significant threat to life. While they typically impact localized areas, tornadoes can cause substantial casualties by demolishing structures and uprooting trees, leading to fatalities. According to historical data compiled by meteorological agencies and disaster response organizations, tornadoes have been responsible for a significant number of deaths globally. Tornadoes, which are the most intense wind phenomenon on Earth (AMS, 2020), occur mainly between 20 ° and 60 ° of latitude (Goliger and Milford, 1998).

In the United States, which experiences a high frequency of tornadoes, an average of approximately 70 fatalities occurs each year due to these intense storms, based on data from the National Oceanic and Atmospheric Administration (NOAA). It is well known that central and eastern USA present favorable conditions for tornadic storms (Brooks et al., 2003; Taszarek et al., 2020)

However, the toll can vary significantly from year to year, with certain outbreaks resulting in a considerably higher number of casualties. For instance, the infamous Tri-State Tornado of 1925 claimed the lives of over 695 people, making it one of the deadliest tornadoes in recorded history.

Globally, tornado-related deaths are also documented in regions beyond the United States. While tornadoes are more commonly associated with the central United States, other countries, such as Bangladesh and Argentina, have also experienced deadly tornado outbreaks. The variability in tornado-related fatalities underscores the importance of understanding regional vulnerabilities, implementing effective early warning systems, and fostering community resilience.

Analyzing the historical data further reveals that the majority of tornado-related deaths result from the destruction of buildings. Tornadoes can level homes, schools, and businesses, and individuals caught in the path of these storms face a heightened risk of injury and death. Improved construction standards, enhanced tornado forecasting capabilities, and community education campaigns are essential components of comprehensive strategies aimed at reducing tornado-related fatalities.

Floods and Flash Floods

Flooding, whether caused by prolonged rainfall or sudden flash floods, has historically led to a high number of fatalities. The rapid onset of flash floods, in particular, can catch communities off guard, limiting evacuation options and increasing the risk of casualties. A flash flood is a rapid response to a severe thunderstorm that occurs in a short period of time (usually only a few minutes) (Elkhrachy, 2015).

Flash floods as per the United States' (US) National Weather Service as "A rapid and extreme flow of high water into a normally dry area, or a rapid water level rise in a stream or creek above a predetermined flood level, beginning within six hours of the causative event (e.g., intense rainfall, dam failure, ice jam) (NWS Glossary). According to global disaster databases and meteorological agencies, floods account for a substantial portion of weather-related fatalities.

The World Health Organization (WHO) estimates that floods cause approximately 7,000 deaths globally each year, with millions more affected by the associated health and social consequences. The distribution of these fatalities is uneven, with developing regions facing a disproportionate burden due to inadequate infrastructure, limited resources for disaster preparedness, and dense populations residing in vulnerable areas.

Flash floods, characterized by their sudden and intense nature, contribute significantly to the toll of flood-related deaths. In the United States alone, flash floods cause an average of 84 fatalities annually, according to the National Weather Service (NWS). The unpredictability and swiftness of flash floods make them particularly challenging for communities to respond to effectively.

Internationally, flash floods have claimed thousands of lives in various incidents. For instance, the flash floods in Uttarakhand, India, in 2013 resulted in over 5,000 deaths, highlighting the catastrophic potential of these rapid-onset events. Similarly, flash floods in developing countries, such as those in sub-Saharan Africa and Southeast Asia, often result in significant casualties, exacerbating the challenges faced by vulnerable populations. Understanding the data on flood-related deaths is crucial for developing targeted strategies to reduce the impact of these disasters. Improved early warning systems, community education programs, and investments in resilient infrastructure can play pivotal roles in minimizing casualties. Analyzing historical events allows for the identification of patterns and vulnerable areas, informing policies and initiatives that enhance preparedness and response capabilities.

Thus floods and flash floods continue to pose a substantial threat to human life globally. By examining the data on fatalities associated with these events, societies can work towards implementing evidence-based measures that save lives and build resilience in the face of the escalating challenges posed by water-related disasters.

Storm Surges

Storm surge is the abnormal rise in seawater level during a storm, measured as the height of the water above the normal predicted astronomical tide. The surge is caused primarily by a storm's winds pushing water onshore Storm surges occur when strong winds generated by tropical cyclones or hurricanes push seawater toward the shore, causing a temporary rise in sea level as shown in Figure 2.2. The combination of high winds and low atmospheric pressure associated with these storms intensifies the surge effect, leading to the inundation of coastal areas. The height and extent of storm surges depend on various factors, including the storm's intensity, speed, size, and the local topography of the coastal area.

Figure 3. Water level differences for storm surge, storm tide, and a normal (predicted) high tide as compared to sea level
(source: oceanservice.noaa.gov)

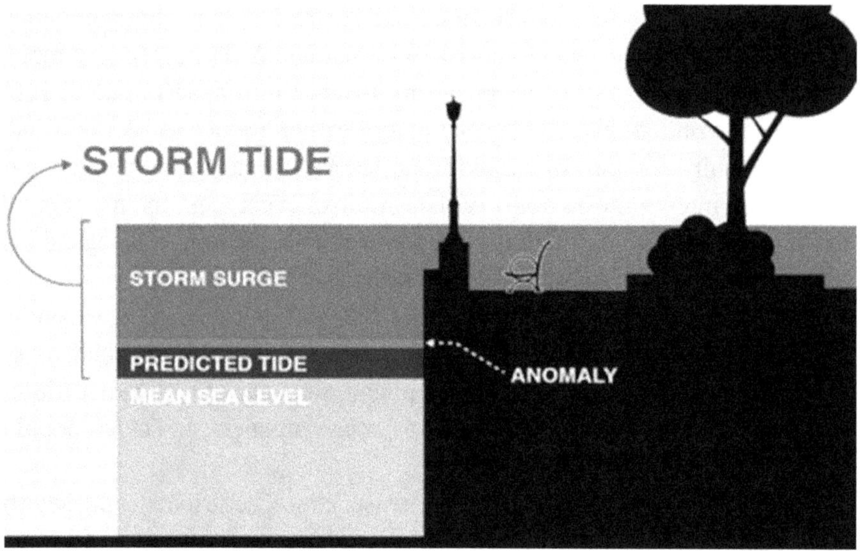

Storm surges, a formidable consequence of tropical cyclones or hurricanes, present a significant hazard to coastal regions, often leading to inundation of low-lying areas. The amalgamation of intense winds and rising seas pose a severe threat to coastal populations, resulting in casualties, especially in densely inhabited areas. This section will further explore the dynamics of storm surges, emphasizing the risks they pose and the historical data on casualties associated with these coastal phenomena.

Historical data reveals the devastating impact of storm surges on coastal communities. The Atlantic hurricane season, in particular, has witnessed numerous instances of storm surge-related casualties. For example, Hurricane Katrina in 2005 led to a catastrophic storm surge in New Orleans, causing extensive flooding and resulting in over 1,200 deaths. Similarly, the storm surge associated with Hurricane Sandy in 2012 led to widespread damage and casualties along the East Coast of the United States.

Globally, densely populated coastal regions in Asia, such as those in Bangladesh and the Philippines, have experienced significant storm surge-related fatalities during powerful typhoons. The vulnerability of these areas is exacerbated by factors such as high population density, inadequate infrastructure, and limited resources for effective disaster preparedness and response.

Earthquakes/Tsunamis

Earthquakes, as geological events, have the potential to cause significant devastation. However, their association with tsunamis, large ocean waves triggered by seismic activity, introduces an additional layer of threat, particularly for coastal regions. The 2004 Indian Ocean tsunami stands as a stark and tragic example of the deadly synergy between seismic events and the hydrometeorological impacts of tsunamis.

Earthquakes result from the sudden release of energy in the Earth's crust, leading to the propagation of seismic waves. When an earthquake occurs beneath the ocean, it can displace large volumes of water, setting off a tsunami. These tsunamis travel across the open ocean at high speeds and may only become noticeable and reach their full destructive potential when they approach coastal areas.

The 2004 Indian Ocean tsunami was one of the deadliest natural disasters in recorded history. Triggered by a massive undersea earthquake off the coast of Sumatra, Indonesia, with a magnitude of 9.1–9.3, the resulting tsunami radiated across the Indian Ocean, affecting coastal regions in 14 countries. The impact was catastrophic, with waves reaching heights of up to 100 feet (30 meters). The death toll exceeded 230,000 people, making it one of the deadliest tsunamis in history.

In the aftermath of such devastating events, the understanding of seismic and oceanic interactions has grown, leading to improved early warning systems and enhanced preparedness measures. The establishment of the Indian Ocean Tsunami Warning System (IOTWS) is a testament to global efforts to mitigate the impact of future tsunamis. The system involves a network of seismographic and sea-level monitoring stations, enabling the timely detection of earthquakes and the issuance of alerts to at-risk coastal communities.

Despite advancements in detection and warning systems, the 2004 Indian Ocean tsunami highlights the need for ongoing international collaboration and investment in resilient infrastructure for coastal regions. Moreover, public education and community preparedness remain crucial elements in minimizing the loss of life during tsunami events.

In summary, the association between earthquakes and tsunamis poses a significant threat to coastal regions, often resulting in substantial loss of life. The 2004 Indian Ocean tsunami serves as a poignant reminder of the destructive potential of this synergy, emphasizing the importance of ongoing efforts to enhance early warning systems, invest in resilient infrastructure, and educate communities on tsunami.

Heatwaves

Heatwaves, characterized by prolonged periods of excessively high temperatures, present a significant threat to human health, particularly as climate change exacerbates their frequency and intensity. The impacts of heat waves extend beyond discomfort, with heat-related illnesses and fatalities being major concerns. Vulnerable populations, including the elderly and individuals with pre-existing health conditions, face heightened risks during such extreme heat events.

One of the immediate dangers during heatwaves is the increased risk of heat-related illnesses, ranging from heat stress to more severe conditions such as heat exhaustion and heatstroke. The human body may struggle to dissipate excess heat, leading to dehydration, electrolyte imbalances, and potentially fatal organ failure if not addressed promptly.

Elderly individuals, often more susceptible to heat-related stress due to reduced thermoregulatory capacity, are at a heightened risk during heatwaves. Likewise, people with pre-existing health conditions such as cardiovascular diseases and respiratory disorders face exacerbated symptoms and an elevated risk of complications. The socio-economic factors, including inadequate access to cooling facilities or air conditioning, can further magnify the risks for vulnerable populations.

Heat-related fatalities can occur directly or indirectly. Direct fatalities result from severe heat-related illnesses, while indirect fatalities may stem from exacerbations of pre-existing conditions, cardiovascular events, or complications from other health issues that are triggered or worsened by extreme heat. The risk of mortality is particularly pronounced in urban areas where the urban heat island effect can lead to even higher temperatures.

Public health interventions during heatwaves are crucial to mitigating the risks. These may include the establishment of cooling centers, public awareness campaigns, and targeted outreach to vulnerable populations. Urban planning that incorporates green spaces, improved building design, and heat-resilient infrastructure can contribute to reducing the overall impact of heat waves.

Moreover, early warning systems and effective communication strategies are vital to alerting communities and individuals about impending heatwaves, enabling proactive measures to prevent heat-related illnesses and fatalities. The integration of climate-resilient health systems, along with community-based programs that address social determinants of vulnerability, is essential for building resilience in the face of increasing heat extremes.

Heatwaves therefore represent a serious threat to public health, with the potential for heat-related illnesses and fatalities, especially among vulnerable populations. As climate change intensifies the frequency and severity of heat waves, comprehensive strategies that prioritize public health, community resilience, and adaptive urban planning become increasingly crucial to mitigate the impact of extreme heat events.

MODELING OF HYDROMETEOROLOGICAL EVENTS

Modeling of hydrometeorological events is a critical tool used by meteorologists and hydrologists to predict and understand the behavior of various weather and water-related phenomena. Modeling makes use of mathematical equations to simulate the physical processes that govern the atmosphere, the land surface, and the oceans, and to predict the occurrence of events such as floods, droughts, and storms. The modeling of hydrometeorological events involves using mathematical and computational methods to simulate and predict the behavior of atmospheric and hydrological processes that influence weather-related phenomena. These models are crucial tools in understanding and forecasting events such as precipitation, floods, droughts, hurricanes, and other weather-related incidents. Modeling hydrometeorological events is a complex task due to the intricate interactions between the atmosphere, land, and water bodies. It requires substantial computational resources, high-quality data, and a good understanding of the underlying physical processes. Moreover, uncertainties in initial conditions, parameterizations, and external factors like volcanic eruptions can affect the accuracy of predictions. The following methods and techniques are employed to carry out the modeling of hydrometeorological events (Sherwood, 2018; Raymond et al., 2020b; Xu et al., 2021; Yuan et al., 2020)

Droughts

Droughts, characterized by prolonged periods of water scarcity, can have far-reaching and cascading impacts on various aspects of human life. Although droughts do not typically result in immediate fatalities, the cumulative effects over time can lead to water shortages, crop failures, and related health issues, contributing to malnutrition and other indirect causes of death. This section explores the multifaceted consequences of drought and their implications for human well-being.

One of the primary consequences of drought is the scarcity of water resources. Prolonged dry spells can lead to the depletion of rivers, lakes, and groundwater reservoirs, affecting both urban and rural communities. The shortage of water for drinking, sanitation, and agriculture becomes a critical issue, forcing communities to adopt stringent water conservation measures and often relying on inadequate or contaminated water sources.

Agriculture, a sector highly dependent on water, is particularly vulnerable during droughts. Insufficient rainfall can result in crop failures, diminishing yields, and food production. This, in turn, can lead to food shortages, increased food prices, and malnutrition. The impact is not only local but can extend to global food markets, affecting vulnerable populations in various parts of the world.

Drought-related crop failures also have implications for livelihoods, especially in agrarian societies. Farmers may face economic hardships, debt, and displacement as they grapple with the loss of income and the inability to sustain their agricultural activities. This can contribute to a cycle of poverty, exacerbating social and economic inequalities.

While direct fatalities may not be the immediate consequence of droughts, the indirect effects can be severe. Malnutrition, waterborne diseases, and the socio-economic repercussions of crop failures contribute to an increased risk of morbidity and mortality, especially among vulnerable populations such as children and the elderly.

In addressing the impacts of drought, comprehensive strategies are required, encompassing water management, agricultural resilience, and public health interventions. Early warning systems, drought-resistant crop varieties, and sustainable water resource management play crucial roles in mitigating the long-term consequences of droughts and reducing their impact on human health and well-being.

DIFFERENT METHODS USED IN THE MODELING OF HYDROMETEOROLOGICAL EVENTS

This section briefly sums up the various methods, tools, and techniques that are operationally used for the modeling of hydrometeorological events.

Numerical Methods

Numerical methods are mathematical techniques used to approximate the continuous equations governing atmospheric and hydrological processes. They involve discretizing the equations into algebraic equations that can be solved on a computer. Finite difference, finite element, and spectral methods are commonly employed in hydrometeorological modeling. These methods divide the model domain into a grid or mesh and approximate the equations within each grid cell or element. By solving these approximated equations iteratively, numerical methods provide solutions that represent the behavior of the atmosphere and hydrological systems over time and space (Miralles et al., 2014)

Atmospheric Models

Atmospheric models, also known as Numerical Weather Prediction (NWP) models, focus on simulating the behavior of the Earth's atmosphere. They use a set of mathematical equations, known as the fundamental equations of fluid dynamics and thermodynamics, to describe the atmospheric processes. These models are typically based on a three-dimensional grid, dividing the atmosphere into cells, and simulate the movement of air masses and the exchange of heat, moisture, and momentum between grid points (Mesinger and Arakawa, 1976).

Atmospheric models are used for short to medium-range weather forecasting, typically ranging from a few hours to a few days into the future. They require initial conditions, such as current atmospheric variables (temperature, humidity, wind, pressure) derived from weather observations, to initiate the simulation. As time progresses, the model predicts how atmospheric variables will evolve, allowing meteorologists to forecast weather patterns and phenomena (Norton and Suppe, 2001).

Hydrological Models

Hydrological models focus on simulating the movement and distribution of water within the hydrological cycle. These models consider processes such as precipitation, evaporation, infiltration, runoff, and groundwater flow. They are used to study the behavior of rivers, lakes, and other water bodies, as well as the availability of water resources in a given area (Devia et al., 2015).

Hydrological models can range from simple lumped models that represent an entire basin as a single unit to more complex distributed models that divide the basin into smaller sub-catchments and simulate hydrological processes at a finer spatial resolution. These models often use historical precipitation and temperature data, land cover information, and topographic data to estimate water flow and storage within a watershed (Sood and Smakhtin, 2015).

Coupled Models

Coupled models integrate both atmospheric and hydrological components to study the interactions between the atmosphere and the Earth's surface. These models simulate feedback mechanisms between atmospheric processes (e.g., rainfall) and hydrological processes (e.g., runoff) that occur on different time scales. For example, precipitation in the atmosphere can affect river discharge and groundwater levels in the hydrological system, while the availability of water on the Earth's surface can influence evaporation and atmospheric moisture content. Coupled models are crucial for understanding and predicting complex interactions between the atmosphere and the hydrological cycle, especially in regions where these interactions have significant impacts on weather patterns and water resources (Walko et al., 2020).

Climate Models (General Circulation Models - GCMs)

Climate models, also known as General Circulation Models (GCMs), are designed to simulate the Earth's climate system over much longer periods, ranging from decades to centuries. These models consider various components of the Earth system, including the atmosphere, oceans, land surface, ice, and biosphere.

Climate models are used to project future climate scenarios and investigate the potential impacts of factors such as greenhouse gas concentrations, solar radiation variations, and volcanic eruptions on the Earth's climate. They are essential tools for climate scientists studying long-term trends, climate change, and the potential consequences for ecosystems and human societies (Grotch and MacCracken, 1991).

Hybrid Models

Hybrid models combine different types of models to take advantage of their individual strengths. For example, some models may couple atmospheric models with hydrological models to study the interactions between the two systems. These models allow for the examination of feedback mechanisms between atmospheric processes (e.g., rainfall) and hydrological processes (e.g., runoff), capturing their influences on each other. Hybrid models provide a more comprehensive understanding of how atmospheric and hydrological processes interact and impact weather-related phenomena.

Data Assimilation

Data assimilation is the process of incorporating real-world observations into the model to improve its accuracy and reliability. Weather stations, satellites, radar, and other observation sources provide valuable data on atmospheric conditions, precipitation, soil moisture, river flow, and other variables. Data assimilation techniques use statistical methods, such as Kalman filters or variational methods, to combine these observations with the model's initial conditions and ongoing simulations. By assimilating the observations, the model updates its state variables, corrects biases, and better represents the current state of the atmosphere and hydrological systems. This improves the accuracy of short-term forecasts and enhances the model's predictive capabilities (Rabier, 2005; Lorenc, 1995).

Parameterization

Parameterization is a technique used to represent sub-grid-scale processes in the atmosphere and hydrological systems. Many physical processes, such as turbulence, cloud formation, and surface roughness, occur at scales smaller than what can be resolved by the models. Parameterization involves developing simplified equations or empirical relationships that approximate these processes and their effects on larger-scale phenomena. For example, in atmospheric modeling, parameterization schemes represent cloud microphysics, convection, radiation, and boundary layer processes. In hydrological modeling, parameterization is used for processes such as evapotranspiration, infiltration, and runoff generation. Parameterization enables the models to simulate the impacts of these processes on larger scales, even though they cannot explicitly resolve them (Alizadeh, 2022; Randall, 1989).

Ensemble Modeling

Ensemble modeling involves running multiple simulations with slightly different initial conditions or model configurations. These simulations create an ensemble of possible outcomes, providing a range of predictions that account for uncertainties in the data and model formulation. Ensemble modeling allows for the quantification of uncertainty in weather forecasts and projections of hydrometeorological events. By analyzing the spread and distribution of the ensemble members, forecasters and researchers can assess the likelihood of extreme weather events, such as heavy rainfall or drought, and make more informed decisions (Murray, 2018).

Downscaling

Downscaling is the process of obtaining high-resolution local weather information from coarser-resolution global climate or atmospheric models. Global models, such as General Circulation Models (GCMs), provide information at larger scales, typically hundreds of kilometers, and may not capture fine-scale regional or local variations. Downscaling techniques bridge this gap by using Regional Climate Models (RCMs) or Statistical Downscaling Models (SDMs). RCMs simulate the regional climate by taking the output from the global models as boundary conditions and incorporating regional-scale features. SDMs, on the other hand, use statistical relationships between large-scale climate variables and local observations to estimate the fine-scale climate parameters. Downscaling techniques provide more detailed information for local decision-making and impact assessments (Benestad, 2016).

Remote Sensing and GIS

Remote sensing techniques use data from satellites, aircraft, and ground-based sensors to observe the Earth's surface and atmosphere. Satellites provide valuable information on land surface conditions, precipitation, temperature, humidity, and other variables at a global scale. Geographic Information Systems (GIS) are used to store, analyze, and visualize spatial data, such as topography, land cover, and river networks. Remote sensing and GIS data play a crucial role in hydrometeorological modeling by providing input data for initial conditions, model calibration/validation, and assessing the spatial distribution of hydrometeorological variables (Navalgund et al., 2007).

Machine Learning and Artificial Intelligence

Machine learning and artificial intelligence techniques are increasingly integrated into hydrometeorological modeling. These techniques involve using algorithms to automatically learn patterns and relationships from large datasets and real-time observations. Machine learning algorithms, such as neural networks, decision trees, and support vector machines, can improve model performance, data assimilation, and prediction accuracy. They can handle complex nonlinear relationships and identify patterns that may not be captured by traditional physical equations. Machine learning is applied in tasks such as precipitation nowcasting, flood prediction, and climate pattern recognition.

Supercomputing and High-Performance Computing

Hydrometeorological models require substantial computational power due to their complexity and the large amount of data they handle. Supercomputers and high-performance computing facilities provide the necessary computational resources to run these models efficiently. These systems allow for parallel processing, which enables the models to solve complex equations and handle the vast amounts of data required for accurate predictions. Supercomputing and high-performance computing technologies are essential for producing timely and accurate forecasts and supporting advanced research in hydrometeorology.

In summary, hydrometeorological modeling involves a range of techniques, including numerical methods for solving equations, data assimilation for incorporating observations, parameterization to represent sub-grid-scale processes, ensemble modeling to account for uncertainties and hybrid approaches to combine different models effectively. Additionally, machine learning, remote sensing, GIS, and advanced computing technologies contribute to improving model performance, enhancing data analysis, and advancing our understanding of weather-related phenomena.

CONCLUSION

Hydrometeorological events, stemming from the intricate interactions between the Earth's atmosphere and the water cycle, hold profound implications for ecosystems, societies, and economies worldwide.

As the consequences of climate change increasingly manifest themselves, the frequency and intensity of these events have escalated, underscoring the urgency for robust and accurate hydrometeorological modeling. This chapter delved into the multifaceted world of hydrometeorological modeling, shedding light on the diverse methodologies employed to comprehend, predict, and mitigate the impacts of these natural phenomena. The book chapter commenced with an introduction, highlighting the wide array of hydrometeorological events, including floods, droughts, cyclones, and thunderstorms, and their devastating consequences for both the scientific community and local populations. With climate change at the forefront of global concerns, researchers and practitioners alike face the daunting task of developing strategies to minimize the damage caused by these events. This chapter reveals that the path toward sustainable solutions is paved by a comprehensive understanding of the complex interactions between atmospheric, hydrological, and climatic processes.

Throughout the chapter, various modeling techniques were unveiled, each serving as a powerful tool in the arsenal of meteorologists and hydrologists. Numerical methods, such as finite difference, finite element, and spectral methods, constitute the foundation for approximating the continuous equations governing atmospheric and hydrological processes. These methods, enabled by high-performance computing, offer vital insights into the behavior of weather-related phenomena over time and space. From the atmospheric models (Numerical Weather Prediction - NWP) that concentrate on simulating the Earth's atmosphere to hydrological models that focus on water movement and distribution, the chapter underscored the significance of understanding the intricacies of these interconnected systems.

The integration of coupled models, combining atmospheric and hydrological components, was emphasized as pivotal for studying feedback mechanisms and interactions between the atmosphere and Earth's surface. Such models facilitate a more comprehensive comprehension of weather patterns and water resources, particularly in regions where these interactions exert profound influences.

Climate models (General Circulation Models - GCMs), operating over longer periods, play a vital role in projecting future climate scenarios and examining the consequences of factors like greenhouse gas concentrations and volcanic eruptions. By encompassing various Earth system components, these models offer invaluable insights into long-term trends and climate change impacts, serving as indispensable tools for climate scientists.

Hydrometeorological modeling, though a potent and invaluable asset, presents its share of challenges. The ever-evolving climate necessitates an ongoing refinement of models to remain relevant and reliable. Uncertainties surrounding initial conditions, parameterizations, and external factors like volcanic eruptions emphasize the need for robust data assimilation techniques to enhance model accuracy. Furthermore, the incorporation of machine learning and artificial intelligence in modeling demands a delicate balance between traditional physical equations and innovative data-driven approaches.

As the book chapter concludes, it is evident that hydrometeorological modeling is a multidisciplinary endeavor that amalgamates numerical methods, data assimilation, parameterization, and innovative technologies like remote sensing and GIS. Embracing ensemble modeling and downscaling techniques empowers researchers to comprehend uncertainties and cater to localized decision-making. The pursuit of hybrid models allows for a holistic understanding of the intricate interactions between atmospheric and hydrological processes. Leveraging supercomputing and high-performance computing bolsters the capabilities of these models and augments their accuracy.

In the face of escalating hydrometeorological events, it is crucial to continually advance and refine modeling techniques to mitigate their adverse impacts effectively. The insights and methodologies outlined in this chapter lay the foundation for future research, paving the way for resilient communities that can adapt to and combat the challenges posed by climate change. As we continue to deepen our understanding of the Earth's complex systems, hydrometeorological modeling stands as an indispensable pillar in safeguarding our planet's future and well-being.

REFERENCES

Alizadeh, O. (2022). Advances and challenges in climate modeling. *Climatic Change*, *170*(1-2), 18. doi:10.100710584-021-03298-4

Benestad, R. (2016). Downscaling climate information. In Oxford research encyclopedia of climate science. doi:10.1093/acrefore/9780190228620.013.27

Brooks, H. E., Lee, J. W., & Craven, J. P. (2003). The spatial distribution of severe thunderstorm and tornado environments from global reanalysis data. *Atmospheric Research*, *67–68*, 73–94. doi:10.1016/S0169-8095(03)00045-0

Devia, G. K., Ganasri, B. P., & Dwarakish, G. S. (2015). A review of hydrological models. *Aquatic Procedia*, *4*, 1001–1007. doi:10.1016/j.aqpro.2015.02.126

Elkhrachy, I. (2015). Flash flood hazard mapping using satellite images and GIS tools: A case study of Najran City, Kingdom of Saudi Arabia (KSA). *The Egyptian Journal of Remote Sensing and Space Sciences*, *18*(2), 261–278. doi:10.1016/j.ejrs.2015.06.007

Goliger, A. M., & Milford, R. V. (1998). A review of worldwide occurrence of tornadoes. *Journal of Wind Engineering and Industrial Aerodynamics*, *74–76*, 111–121. doi:10.1016/S0167-6105(98)00009-9

Grotch, S. L., & MacCracken, M. C. (1991). The use of general circulation models to predict regional climatic change. *Journal of Climate*, *4*(3), 286–303. doi:10.1175/1520-0442(1991)004<0286:TUOGCM>2.0.CO;2

Jentsch, A., Kreyling, J., & Beierkuhnlein, C. (2007). A new generation of climate-change experiments: Events, not trends. *Frontiers in Ecology and the Environment*, *5*(7), 365–374. doi:10.1890/1540-9295(2007)5[365:ANGOCE]2.0.CO;2

Lorenc, A. C. (1995). *Atmospheric data assimilation*. Meteorological Office.

Matthies-Wiesler, F., & Quevauviller, P. (Eds.). (2022). *Hydrometeorological Extreme Events and Public Health*. John Wiley & Sons. doi:10.1002/9781119259350

Mesinger, F., & Arakawa, A. (1976). *Numerical methods used in atmospheric models.*

Miralles, D. G., Holmes, T. R. H., De Jeu, R. A. M., Gash, J. H., Meesters, A. G. C. A., & Dolman, A. J. A. J. (2014). Mega-heatwave temperatures due to combined soil desiccation and atmospheric heat accumulation. *Nature Geoscience*, *7*(5), 345–349. doi:10.1038/ngeo2141

Murray, S. A. (2018). The importance of ensemble techniques for operational space weather forecasting. *Space Weather*, *16*(7), 777–783. doi:10.1029/2018SW001861

Navalgund, R. R., Jayaraman, V., & Roy, P. S. (2007). Remote sensing applications: An overview. *Current Science*, 1747-1766.

Norton, S., & Suppe, F. (2001). Why atmospheric modeling is good science. *Changing the atmosphere: Expert knowledge and environmental governance*, 67-105.

Pachauri, R. K., Allen, M. R., Barros, V. R., Broome, J., Cramer, W., Christ, R., & van Ypserle, J. P. (2014). *Climate change 2014: synthesis report. Contribution of Working Groups I, II, and III to the Fifth Assessment Report of the Intergovernmental Panel on Climate Change*. IPCC.

Rabier, F. (2005). Overview of global data assimilation developments in numerical weather-prediction centers. *Quarterly Journal of the Royal Meteorological Society: A journal of the atmospheric sciences, applied meteorology and physical oceanography, 131*(613), 3215-3233.

Randall, D. A. (1989). Cloud parameterization for climate modeling: Status and prospects. *Atmospheric Research*, *23*(3-4), 345–361. doi:10.1016/0169-8095(89)90025-2

Raymond, C., Matthews, T., & Horton, R. M. (2020). The emergence of heat and humidity is too severe for human tolerance. *Science Advances, 6*(19).

Sherwood, S. C. (2018). How important is humidity in heat stress? *Journal of Geophysical Research. Atmospheres*, *123*(21), 11–808. doi:10.1029/2018JD028969

Sood, A., & Smakhtin, V. (2015). Global hydrological models: A review. *Hydrological Sciences Journal*, *60*(4), 549–565. doi:10.1080/02626667.2014.950580

Taszarek, M., Allen, J. T., Púčik, T., Hoogewind, K. A., & Brooks, H. E. (2020). Severe Convective Storms across Europe and the United States. Part II: ERA5 Environments Associated with Lightning, Large Hail, Severe Wind, and Tornadoes. *Journal of Climate*, *33*(23), 10263–10286. doi:10.1175/JCLI-D-20-0346.1

Walko, R. L., Band, L. E., Baron, J., Kittel, T. G., Lammers, R., Lee, T. J., Ojima, D., Pielke, R. A. Sr, Taylor, C., Tague, C., Tremback, C. J., & Vidale, P. L. (2000). Coupled atmosphere–biophysics–hydrology models for environmental modeling. *Journal of Applied Meteorology*, *39*(6), 931–944. doi:10.1175/1520-0450(2000)039<0931:CABHMF>2.0.CO;2

Xu, F., Chan, T. O., & Luo, M. (2021). Different changes in dry and humid heat waves over China. *International Journal of Climatology*, *41*(2), 1369–1382. doi:10.1002/joc.6815

Yuan, J., Stein, M. L., & Kopp, R. E. (2020). The evolving distribution of relative humidity is conditional upon daily maximum temperature in a warming climate. *Journal of Geophysical Research: Atmospheres, 125*(19), e2019JD032100.

Section 2

Case Studies

Chapter 11
Artificial Recharge to the Groundwater Resources Using Rainfall Water Collection:
A Case Study of Kabul, Afghanistan

Hafizullah Rasouli
Kabul University, Afghanistan

Ashok Vaseashta
https://orcid.org/0000-0002-5649-0067
International Clean Water Institute, USA

ABSTRACT

Afghanistan has a very dry climate with most of the land area suffering from the adverse effects of hydroclimatic extremes. The vast amounts of snow and ice in mountain regions might have greater economic, environmental, and community implications due to such extreme hydrometeorological events. Overflow due to snowmelt from mountain regions is also of great significance for continued water supply. The presence of snow in any basin strongly affects the moisture that is deposited at the surface and is accessible for future runoff. The groundwater research presented here is used to quantify artificial recharge due to rainfall water that is falling from the rooftop of the Faculty of Geoscience building, Kabul University, Af+ghanistan. The main objective of this research is to study the effects of rainfall on groundwater fluctuations, as per collected data from the study site, especially in the spring and rainy seasons (January to April). In addition, we have studied harvesting methods, borehole lithography, and the tectonic setting of the Kabul basin to supplement our research objectives.

INTRODUCTION

Historically, Afghanistan is a country that is known to have a very dry climate. Although people have adapted to the timing and quantity of water arising from mountain snow and glacier melting for prediction of water resources with 3nagement. One of the diverse impacts of climate change is on the spatio-

DOI: 10.4018/978-1-6684-8771-6.ch011

temporal distribution of hydrological systems throughout the country, Due to rugged terrain, most of the country depends on water from mountain snow and ice in the mountain lakes and hence adapts to the timing and quantity of water from such freshwater resources. Thus, the estimation, simulation, and prediction of the spatiotemporal variability of freshwater resources are of great importance to water resources management and forecasting (Rasouli, H., et al., 2015).

Due to a large regional dependence on water from lakes of snow and ice from the mountain regions on the hydrological systems, the impact of climate change has far-reaching impacts on the local economy, environment, and community (Hamdard M. H., 2022). Snow is a critical vital water resource for the region, but due to varied precipitations, it is released gradually in the form of meltwater over a longer time period which provides steady streams throughout dry periods to achieve the necessary water supply. In addition to a water shortage, extreme hydroclimatic and associated hydrogeological events can also cause overflow due to excess ice and snow melting from mountain regions causing flash floods in the region. The presence of snow in water basins strongly affects the moisture that is deposited at the surface and is accessible for future runoff (Rasouli, H., 2022a, Belhassan, K., 2020b). The hydrology of mountain regions is largely dependent on the climate conditions of the regions, as mountains usually receive large amounts of precipitation, which may be stored in the form of snow (Rasouli, H., 2022b). Mountains also represent unique areas for the detection of climate change and the assessment of climate-related impacts. Climate change contributes to the increased variability of river runoff due to changes in the timing and intensity of precipitation and snow melting (Rasouli H., et al., 2021c, Rasouli, H.et al., 2021a).

The key waterways drain during the snow melting times (from January to May), rainy periods (March to April), and occasionally through quick overflowing times (May to August), the main elevations of snow cover are Parwan Maintains series, Wardak, Loger, Baba, Spingher, Salang, Kohkurugh, Koha Safi, Hindu Kush Mountains ranges in Afghanistan (, Belhassan, K., 2011), as well as here is certain cold provinces for example; Bamyan, Wardak, Loger, Badakhshan, Pangesher, Parwan, some parts of Kabul, snow covers these areas from September to November and its storing is used for water in Afghanistan (Rasouli, H. and Safi, A. G., 2021a, Belhassan, K., 2020a). Similarly, on the north side of Afghanistan, there are specific glaciers; Pamir Badakhshan, Mymai Badakhshan, and Panjsher Mountains range which are part of the Hindu Kush mountains series in Afghanistan. These are the main sources for Panjsher, Helmand, and Koner Rivers (Shamal S., Rasouli H., 2018, Rasouli H., 2020c). In particular provinces, we use rivers as a mean for irrigation and water supply (Drinking water), such as; Bamyan, Panjsher, Wardak, Parwan, Helmand and Kandahar, Kapesa, but in some of these provinces for instance; Wardak, Parwan, Panjsher spending from spring, and in several provinces for drinking and irrigation benefit from Kariz water, and some other provinces benefit from wells (Rasouli H., 2021b., Belhassan, K., 2020b). Currently 4.14 million people are living in Kabul City, the increase in the population in Kabul leads to unplanned houses, water shortage, and environmental and other social-economic problems. Recently, the rapid population growth and climate change impacts have placed new stresses on environmental pollution, especially air pollution and water resource depletion. Over the period of 1960 to 2008, the mean annual temperature high of Afghanistan increased by 0.6°C with an average increase of 0.13°C per decade; and the mean annual precipitation decreased at an average rate of 0.5mm per month with an average decrease of 2% per decade (Arian, H. et al., 2015., Rasouli H., 2019).

From a geological standpoint, the number of sediments in Kabul Basins arises because of climate change, as different types of sediments are transported from Paghman, Aliabad, Asmayie, Qorugh, and Logger mountains, and such sediments get stacked year after year. These sediments are carried from surrounding mountains by water at diverse periods of time and are made of different thicknesses. In the upper and middle parts of Kabul, several terraces have accumulated some heavy and light minerals, and these minerals originate from the mother rocks located in the surrounding mountain areas of the Kabul Basin (Arian, H et al., 2015). Furthermore, most of the rivers conjoining in Kabul Basin originate at different locations and flow from west to east, thus the sediments in Kabul Basin consist of Tertiary formations (Eocene and Oligocene) – i.e., the age of sediments between the upper and middle parts of Kabul is about (20 - 45) millions of years (WHO, 2009, Belhassan, K., 2011b). All surrounding mountains of Kabul Province consist of metamorphic rocks (Rasouli, H.et al., 2023.), but the Paghman Mountains have originated from adapted metamorphic rocks. From the Radiometry method, it was established that the lifetime of these mountains is about (928±8) million years. The older one is located in the Khair Khana Mountains and the younger one is in the Shawaky and Qorugh mountains ranges (Rasouli, H. and Safi, A. G., 2021).

2. OVERVIEW OF TECTONIC SETTING AND PRECIPITATION IN AFGHANISTAN

The Kabul Province is divided into 22 areas and each area is further divided into subparts. Generally, Kabul is an old province whose area is about 344 km^2 the New Kabul Province area is 740 km^2, with the combined Kabul Province area being ~1084 km^2 (Rasouli, H., 2022a). The areas of Kabul Province consist of Kharkhana, Hazara Baghal, Khajrawash, south side Charasiab, east side Polycharkhay, Tankgaru, and Phaghman mountains. The Kabul block encompasses from the west side by the Paghman fault, and from the East side by the Altimor fault, and this stable mass is made up by the middle Alpine geo-synclinal folds. The bedrocks of this block consist of metamorphic rocks, and the upper part of this block is covered by vend- Jurassic and chalk (Rasouli, H., et al., 2023). From the tectonic viewpoint, the east side is the Indian plate, the northeast side is Pamir, Karakoram, and Kohistan, from northeast side of Eurasian plate, and the southeast side by Central Afghan blocks (Figure 1).

Figure 1. Tectonic setting of Kabul basins

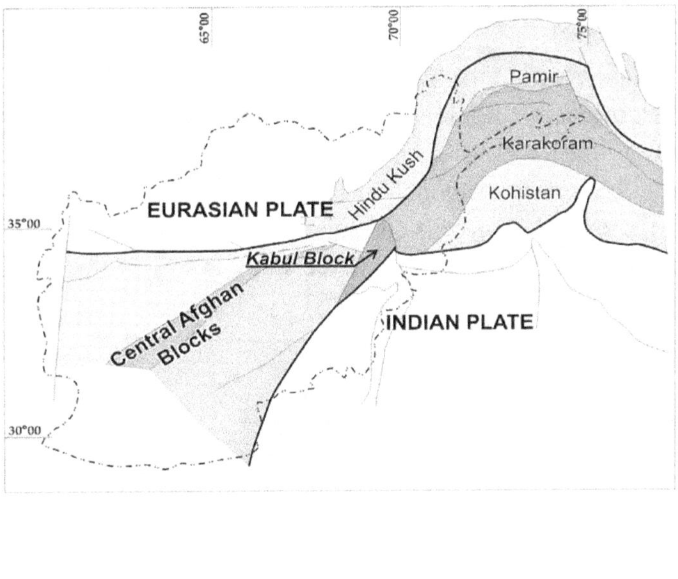

Precipitation in Afghanistan

In Afghanistan, the amount of precipitation depends on the location. Generally, in the north side of Afghanistan, there is more precipitation, such as in Panjsher, and Badakhshan, and the precipitation is lower in the south side such as Kandahar, Helmand, Qalat, Aruzgan. Also, some central provinces such as Bamyan, Wardak, Ghazni, have more precipitation (figure 2). Hence the research focuses on rainwater monitoring and water harvesting, an idea that can be applied across the country, to mitigate the effects of extreme hydroclimatic conditions.

Figure 2. Precipitation amount and distribution in Afghanistan. (a): from worldofmaps.net) and (b): Grieser et al. (2006)

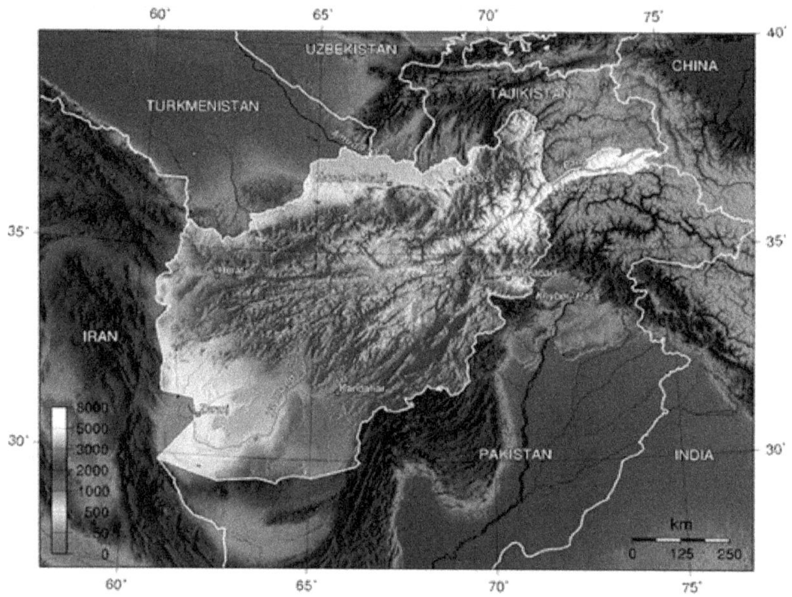

2.1 The Aim of Research

The aim of this research is to analyze different data on rainwater that falls on a designated test site, which for this study is the rooftop of the Faculty of Geoscience, Kabul University. The water is infiltrated into the groundwater for water harvesting. Although this is an experimental test site, the concept is intended for broader use within the country. Furthermore, this might also shed some light on the groundwater infiltration taking place throughout the country. Since water is scarce, this investigation may provide additional water resources.

3. METHODS AND MATERIALS

For this investigation, the rainfall data is analyzed using water collected from the rooftop of the Faculty of Geoscience building. We calculated the amount of water that fell and infiltrated the groundwater and into the wells. We used different types of well designs, which involve the parts of the area of the rooftop, hod, filter, rings, and well, as shown schematically in later sections.

4. RESULTS AND DISCUSSIONS

In Afghanistan, the amount of precipitation depends on the location. We determined from this investigation that the amount of artificial discharge from the rooftop of our test site in Kabul city, and according to the lithology of the earth materials the amount of infiltration is different than what is normal for this area. Generally, using homogeny and rounded materials, the amount of infiltration is more but in the fine and heterogenic materials, the amount of infiltration is lower. This may also provide some additional information concerning other regions in Afghanistan about the infiltration that is occurring from riverbeds, and lateral, rooftop, and irrigation sites.

4.1 Borehole Lithology or Stratigraphy

The borehole lithology in Kabul City has different kinds of materials according to the depth of stratigraphy. This material is directly related to the types of rocks that are located in the surrounding mountains of Kabul City. Generally, the surrounding mountains of Kabul City are metamorphic and igneous rocks. These mountains weather at different periods of geology and by different processes such as gravity, winds, and different types of runoff and streams transported to the basins and plains of Kabul, and mask different types of layers and terraces after one another's. The age of these sediments generally consists of Tertiary (Eocene and Oligocene), with the age of sediments between the upper and middle parts of Kabul about (20 - 45) million years and is called Tertiary formations (Table 1). From the Radiometry method, they found the lifetime of these mountains is about (928±8) million years ago. The oldest of such formations is located in the Khair Khana Mountains and the younger one is in the Shawaky and Qorugh Mountains ranges (Rasouli, H. and Safi, A. G., 2021, WHO, 2011).

Table 1. Borehole lithology

DEPTH		LITHOLOGICAL DESCRIPTION	COMMENTS
From (m)	To (m)	(grain size, texture, sorting, percentage, color, rock type, etc.)	(Microfossils etc.)
0 m	11m	Gravel Clay -Gravel 40%, Clay 60% - (Quartzite, Gneiss) Silicates Rock – Mostly Gray color	
11m	12m	Clayey Gravel - Clay 20%, Gravel 80% - (Quartzite, Gneiss) Silicates Rock – Mostly Black color	
12m	41m	Gravel (100%) (Quartzite, Gneiss) Silicates Rock – Mostly Black color	
41m	45m	Clayey Gravel - Clay 20%, Gravel 80% - (Quartzite, Gneiss) - Silicates Rock – Black	
45m	49m	Gravel (100%) - (Quartzite, Gneiss) - Silicates Rock- mostly black	
49m	50m	Clayey Gravel - Clay 30%, Gravel 70% - (Quartzite, Gneiss) - Silicate Rock - Black	
50m	81m	Gravel Clay -Gravel 10%, Clay 90% - (Quartzite, Gneiss) Silicates Rock – Mostly Gray color	

4.2 Precipitation

The precipitation seasons in Afghanistan are winter (snowfall), and spring (rainfall), the mean monthly precipitation in Badam Bagh, Darulaman, Gulkhana, Kabul, Kariz Mire, Paghman, Qargha for average during (1960-84) - (2006-13) is shown in (fig. 3). This data provides some estimates of the monthly average precipitation changes according to the seasons of the year for the regions listed in the table. Generally, more precipitation occurs in two seasons of the year winter and spring.

Figure 3. Average precipitation in Kabul around areas (2006 – 2013)

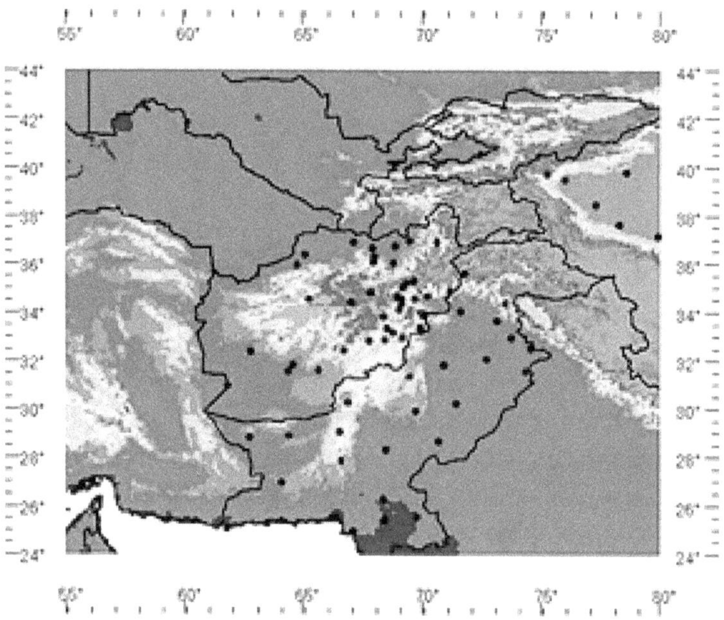

4.3 Rain Frequency Analysis

In this study, we used rainfall frequency at the Badam Bagh, Darulaman, Gul Khana, Kabul, Kariz Mir, Paghman, and Qargha regions to assess that the amount of precipitation is different from what it should be. Figure 4 below presents the data for these locations, which clearly shows that the distribution of precipitation is different in these regions. The data shown is for the data taken shows the amount of precipitation for specific days for the duration of 8 years (figure 4).

4.4 Harvesting Method

In this investigation, we used the collection of precipitations around the Kabul Basins as the harvesting method, because in Afghanistan this is one of the better ways for collecting artificial recharge for water harvesting. More water is collected during the rainy spring season and in the snowfall season of winter. Because most of the precipitations in Afghanistan are in these two seasons (fig. 5). The infiltration of

Figure 4. Rainfall frequency analysis in Kabul Basins

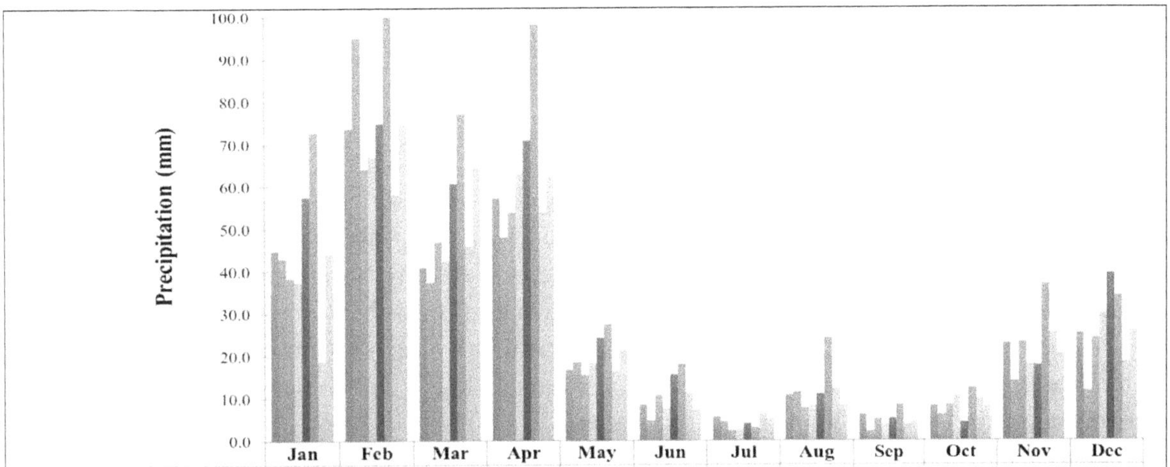

	Jan	Feb	Mar	Apr	May	Jun	Jul	Aug	Sep	Oct	Nov	Dec
Badam Bagh	44.7	73.7	40.8	57.3	16.7	8.4	5.5	10.7	6.0	8.1	22.8	25.1
Darulaman	43.1	95.1	37.4	48.1	18.4	4.6	4.5	11.3	2.1	6.0	13.8	11.7
Gul Khana	38.4	64.1	46.9	53.9	15.4	10.7	2.5	7.5	5.1	8.4	23.0	24.1
Kabul	37.4	66.9	42.6	63.1	18.2	7.5	2.5	8.3	3.8	10.7	17.8	29.8
Kariz Mir	57.5	74.8	60.7	70.8	24.2	15.5	4.0	10.8	5.2	4.2	17.6	39.4
Paghman	72.7	105.8	77.1	98.1	27.4	17.9	3.1	24.2	8.4	12.3	36.9	34.0
Qargha	18.7	58.2	45.9	53.8	16.3	11.2	6.3	12.2	3.8	9.7	25.2	18.4
Average (1960-1984)-(2006-2013)	44.0	74.5	64.7	62.2	21.5	7.2	5.1	8.3	4.5	8.1	20.4	25.9

Figure 5. Show the harvesting method for collection precipitation and infiltration to the groundwater at the different locations

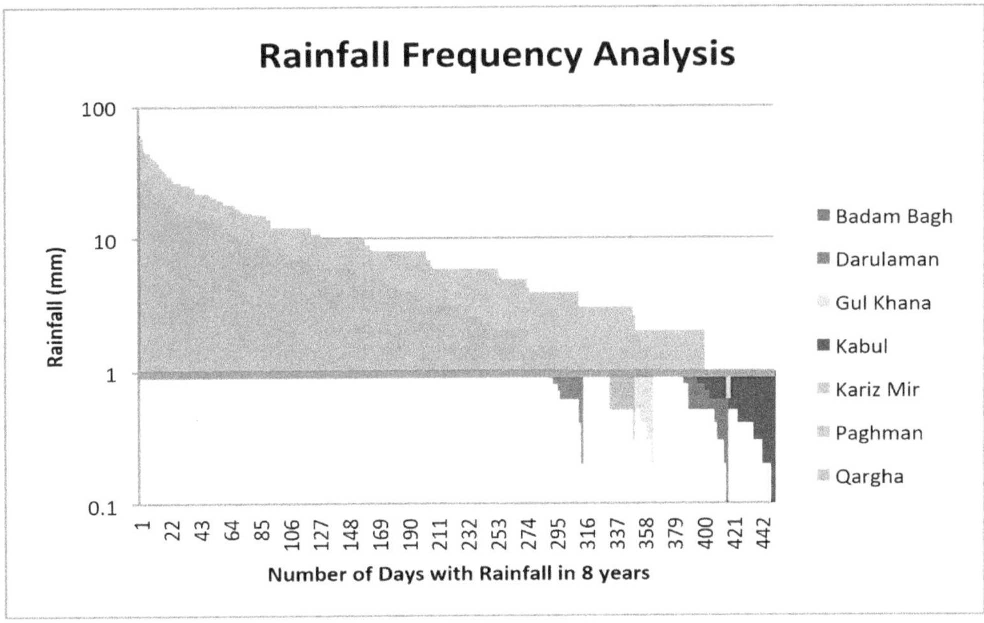

water from the harvesting method depends on the lithology of stratigraphy of these areas, since there is more gravel, and the infiltration rate is higher in those regions. However, there is more sand, and the infiltration rate is medium, but in those areas, the stratigraphy involves clay since the infiltrations are lower.

4.5 Kabul Main Sub-basins

The areas of Kabul province consist of Kharkhana, Hazara Baghal, Khajrawash, south side Charasiab, east side Polycharkhay, Tankgaru and Phaghman mountains. The Kabul province is divided into 22 wards and every ward is further divided into different parts. Generally, Kabul's old province has an area of about 344 km^2 and Kabul's new province has an area of 740 km^2, with a total province area of ~1084 km^2. Using this investigation, we estimated that the Kabul province is 13 km from north to south and 22 km from east to west larger in size (Figure 6), (Rasouli, H., 2022a). Tables 2 and 3 show recharge components of the North Kabul and Logar sub-basins.

Figure 6. Location map of Kabul Sub-basins, Afghanistan

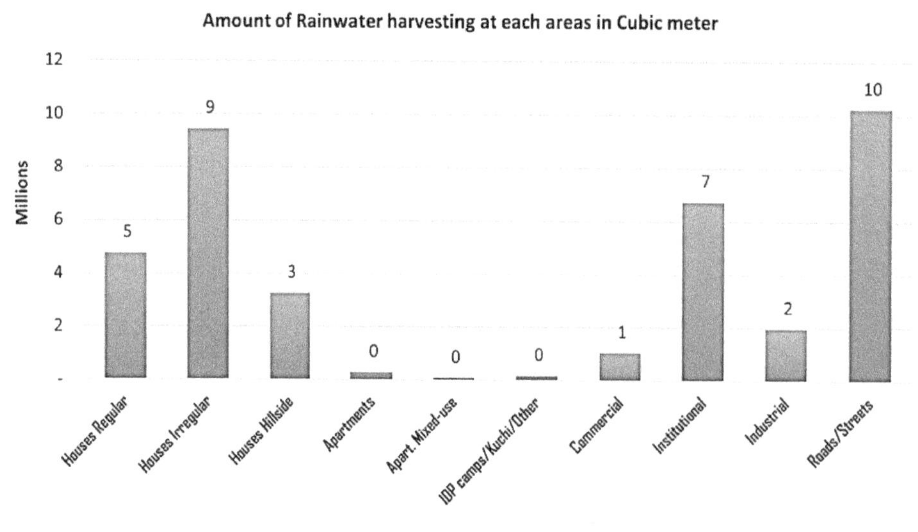

4.6. Data From Main Users

To get meaningful data, we obtained the annual discharge data from the main users, for example, Afghanistan Urban Water Supply and Sewerage Corporation (AUWSSC), Commercial well total discharge, 30% of missing commercial well survey, Bagrami irrigation wells total discharge, and Greenery well, as shown in Table 4.

4.7 Groundwater

In Afghanistan, the groundwater sources are vital, because the drinking and irrigation water is supplied from the underground water. In the last several years, climate change has impacted groundwater-surface,

Table 2. North Kabul Sub-basins (2019)

<table>
<tr><td colspan="4">1. Recharge Components</td></tr>
<tr><td colspan="3">Recharge from River</td><td>15.5 MCM/year</td></tr>
<tr><td colspan="3">Recharge from other basin</td><td>340*10^3 CM/year</td></tr>
<tr><td colspan="3">Total Aquifer Recharge</td><td>15.84 MCM/year</td></tr>
<tr><td colspan="4">2. Discharge Components</td></tr>
<tr><td colspan="3">Evaporation Losses</td><td>MC/year</td></tr>
<tr><td rowspan="3">Groundwater Pumping</td><td>Residential Pumping</td><td>31.88 MCM/year</td><td rowspan="3">41.97MCM/year</td></tr>
<tr><td>Water supply</td><td>10.09 MCM/year</td></tr>
<tr><td></td><td></td></tr>
<tr><td colspan="3">Total Aquifer Discharge</td><td>41.97MCM/year</td></tr>
<tr><td colspan="3">Change in Storage</td><td>26.13 MCM</td></tr>
</table>

and air pollution directly has affected the ground and surface water quality and quantity. In Afghanistan, there are large fluctuations in groundwater levels during the year and, sometimes, from year to year. During winter and spring, the level of groundwater recovers while it drops during the summer and fall seasons (i.e., from June to October), and hence the groundwater is due to the infiltration into the ground from normal flows in rivers and its banks. The mean groundwater surface in Kabul Province is about 30 m but, in some places, it may be 60 -70 m. The main reasons for the drawdown are due to climate change in Afghanistan (Figure 7).

In Afghanistan we have three types of rivers, first type flows in all four seasons of the year from glaciers area (Hindu Kush mountains range) of Afghanistan, such as Helmand, Kunar, and Panjsher, and the second one flows only in snow-melting seasons and winter snowfall and rains such as Kabul, Loger, Paghman rivers, but the end and third one belongs to sudden floods such as mountains valleys, that occur in summer seasons and recharging at the flow time to the groundwater (figure 8). In the Kabul sub-basins, the groundwater source primarily comes from the surface water (rivers, streams ways), and these rivers and streams continue to recharge the groundwater. Furthermore, the aquifers that are located in the riverbed and banks charge with surface water during the snow melting season (spring season) and snow season fluctuations. All groundwater in Afghanistan is related to valleys because we can find

Table 3. Loger sub-basins (2019)

1. Recharge Components			
Recharge from Precipitation			1.88 MCM/year
Recharge from River			21.17 MCM/year
Recharge from channel			1.16 MCM/year
Recharge from Irrigation			3.85 MCM/year
Total Aquifer Recharge			**32.85 MCM/year**
2. Discharge Components			
Evaporation Losses			MC/year
Groundwater Pumping	Water Supply	8 MCM/year	27.557 MCM/year
	Agriculture	11 MCM/year	
	Residential	7.88 MCM/year	
	Domestic Consumption	677*10^3MCM/year	
Total Aquifer Discharge			**27.557MCM/year**
Change in Storage			**5.293 MCM**

Table 4. Total annual discharge of Groundwater main users

No	Main users	Annual Discharge (m^3)
1	AUWSSC	16000000
2	Commercial wells total discharge	9006336.72
3	30% of missing of commercial well survey	2701901.016
4	Bagrami irrigation wells total discharge	11000000
5	Greenery well	2800000
6	Total Discharge	41508237.74

more thickness of sediments that make different types of aquifers. Due to steady population growth, it is necessary to have steady and constant water resources.

4.8. 3D- Modelling of Kabul Sub-Basins

In addition, in this investigation, we used 3D modeling and ground lithology of the ground from the profile, and we found, using different types of layers such as loams, sands and gravels, conglomerates and sandstone, and Neogen, different types of sediments (figure 9). From this model, we can find more

Figure 7. Groundwater table at the different locations of Kabul basins

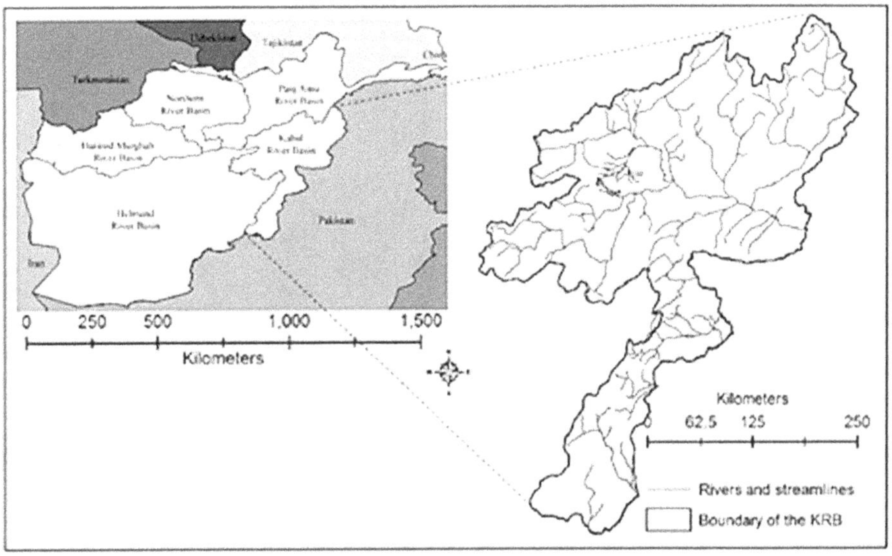

Figure 8. The location map of the different aquifers in the Kabul basins

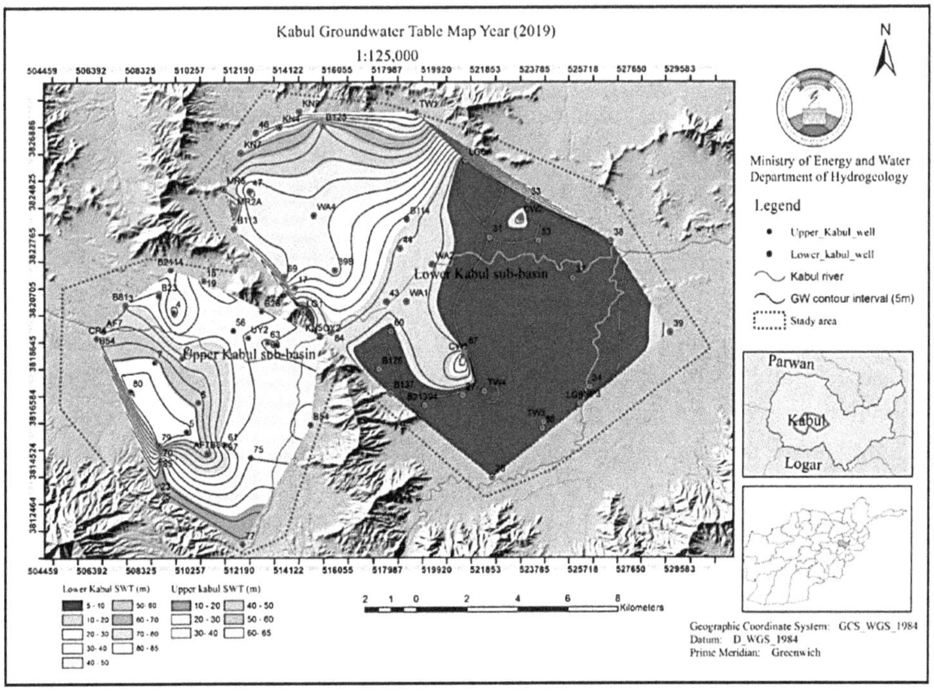

groundwater peculated in the sedimentary sediments and sedimentary rocks that cover the earth's surface to the average depth of 1000m, and the thickness of sediments corresponds to the morphology and slopes of the ground at these locations. Generally, in the plain areas, the thickness of sediments is higher, but in the sloppy areas, the thickness of sediments is less. In the last 30 years of drought, the groundwater Static

Figure 9. The 3D model of Kabul Basins

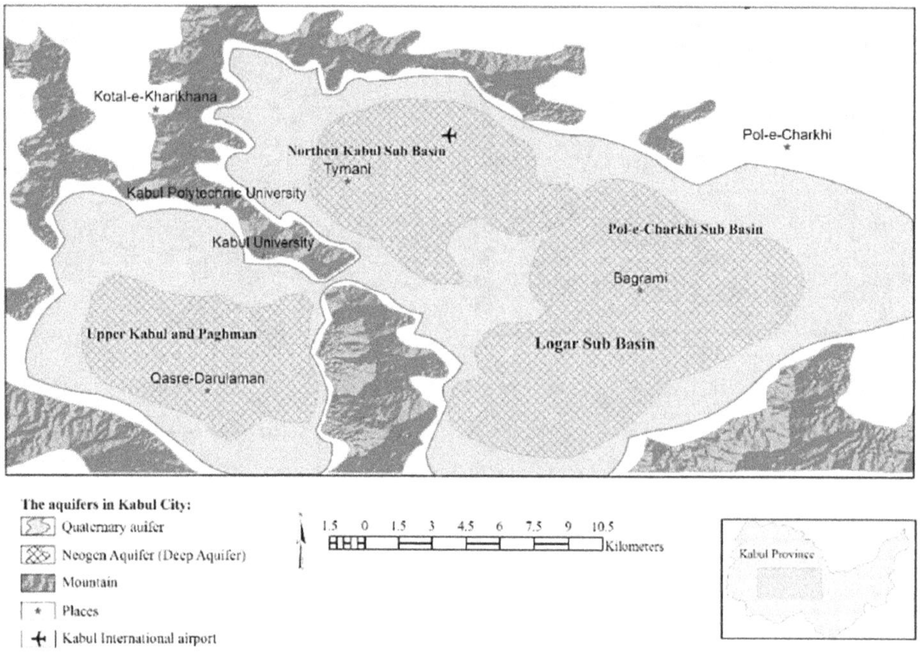

Water Level (SWL) has been drawn down to about 55m. Due to the lack of precipitation in Afghanistan for many years, the recovery of the groundwater will not be complete. However, in Afghanistan, the groundwater fluctuations are much smaller. Thus, the surface streams are much smaller and arise only due to the snow-melting at higher elevations during the appropriate season.

4.9 Recharge Components in Kabul Basins

For the last three decades, climate change has had close effects on air temperature, precipitation, surface water flows (Table 5), and groundwater resources in Afghanistan. It especially affects human lives, because more populations are using water for drinking and irrigation systems. Some rivers such as Kabul, Loger, and Paghman are dry, and we cannot use them for irrigation. For this reason, the groundwater level varies in depth from the earth's surface and its fluctuations according to the seasons of the years. In Kabul City for the last three decades, the groundwater has been drawing down year by year, and now the average groundwater level is 60 to 90m. For this reason, the artificial recharge system has been started in Kabul, including recharges from big and small dams, harvesting methods, and Chack dams for water collection during the snow melting seasons and collection of rooftop waters from wells in the spring and winter seasons.

4.10 Artificial Recharge at Kabul University

In one of these projects, we used a rooftop of the Faculty of Geoscience, Kabul University, Afghanistan. This design involves the parts of the area of rooftop of Geoscience Faculty, hod, filter, rings, and well,

Table 5. The recharge components in Kabul basins

<table>
<tr><td colspan="3">Recharge Component</td><td></td></tr>
<tr><td colspan="3">Recharge from Precipitation</td><td>7.8 MCM/year</td></tr>
<tr><td colspan="3">Recharge from River</td><td>3.264 MCM/year</td></tr>
<tr><td colspan="3">Recharge from Qargha- artificial Dam</td><td>106717.7CM/year</td></tr>
<tr><td colspan="3">Total Aquifer Recharge</td><td>11.17 MCM/year</td></tr>
<tr><td colspan="4">2. Discharge Components</td></tr>
<tr><td colspan="3">Evaporation Losses</td><td>MCM/year</td></tr>
<tr><td colspan="3">Groundwater Lateral Outflow</td><td>340588.8 CM/year</td></tr>
<tr><td rowspan="2">Groundwat Pumping</td><td>Residential Pumping</td><td>22.6 MCM/year</td><td rowspan="2">25.4 MCM/year</td></tr>
<tr><td>Irrigation Pumping</td><td>2.8 MCM/year</td></tr>
<tr><td colspan="3">Total Aquifer Discharge</td><td>25.7 MCM/year</td></tr>
<tr><td colspan="3">Change in Storage</td><td>-14.57 MCM/year</td></tr>
</table>

which we was designed, as shown in fig.11-12. The main reason for this activity is to assess the lithology and groundwater recharge of Kabul University. The area of Kabul University consists of different material clay, silt, sand, loam, clayey loam, but generally the Kabul University area consists of fine materials such as clay and silt and the infiltration of waters are much lower because due to the permeability between pore space is very smaller (figure 10).

Figure 10. Location map of Kabul University, Afghanistan

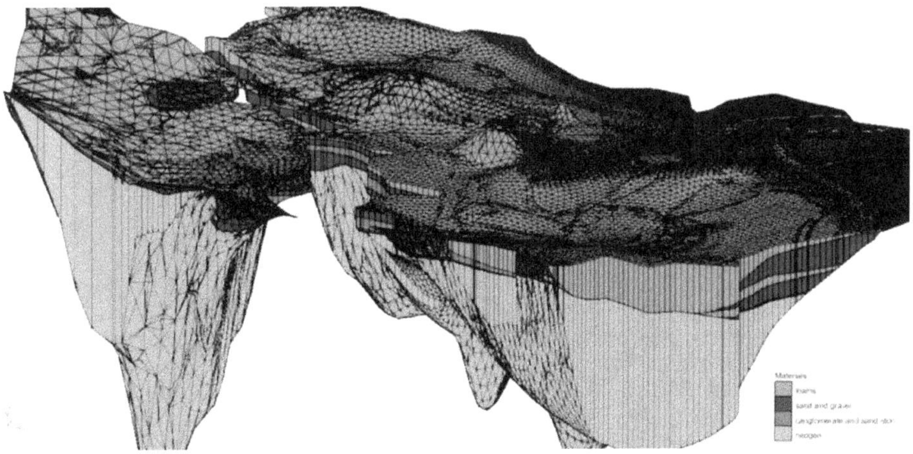

4.11 Area of Rooftop and Well Depth

We used different areas of the rooftop and its rainwater infiltration to well depth from the earth's surface, and well depth pertains to the area of the rooftop. In those areas, for the larger rooftop, the well-depth infiltration is more than for smaller areas. In all areas of the Geoscience Faculty, there are metal plates and all rainwater falling from rooftop is completely percolated into the ground from shallow well (table 6).

Table 6. Roof area and wells depth from the earth's surface for this research

Roof Area (m^2)	Well depth from ground level (m)	No of Rings
200-400	4.4	12
400-600	6	17
600-800	9	23
800-1000>	18	46

4.12 Artificial Recharge to the Groundwater

4.12.1 Geoscience Faculty Building

The Geoscience Faculty building rooftops is ~1000 m^2 (figures 11 and 12). Also, it is known that in Afghanistan we have four seasons such as spring, summer, fall, and winter, and most of the precipitation occurs in winter and spring seasons. For this investigation, we used the precipitation arising in the winter and spring seasons.

Figure 11. Building of Geoscience faculty, Kabul University, Afghanistan

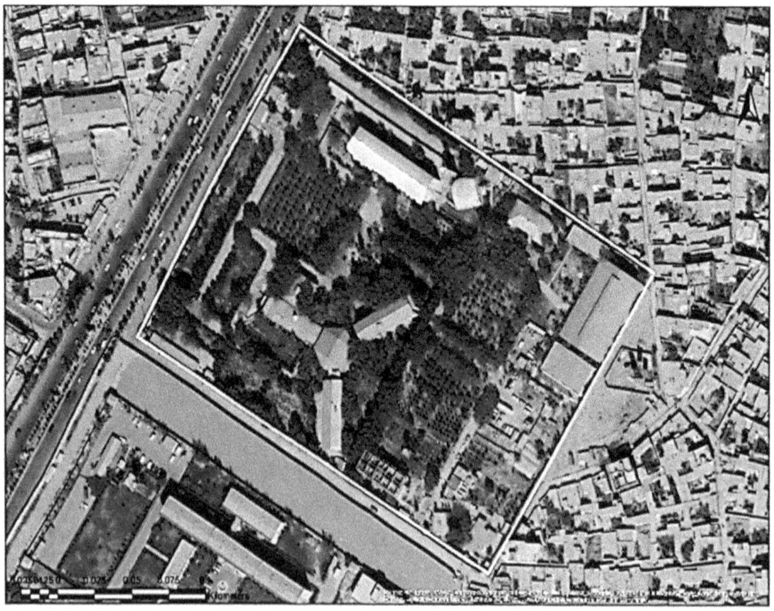

The upper rooftop consists of a metal plate and there is no water infiltrating the under roof to any room, or into the building and all precipitations waters come down by hod to the ground surface and is collected from coarse sand, and gravels such as granules and pebbles to the shallow well. The cross-section of hod we can find in Figures 13 (a,b)

Figure 12. The schema of rooftop falling water to the groundwater influence

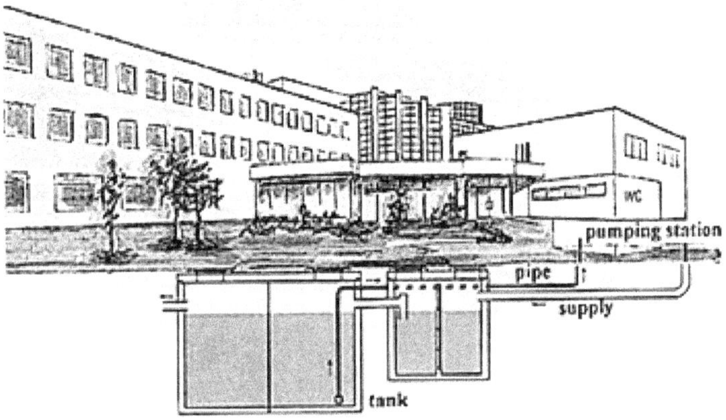

Figure 13. (a): The hod from the rooftop for transfer of water to the groundwater. (b): Scheme of a cross-section of hod that was used in the Geoscience Faculty

4.12.2 Well and Filter Design

To avoid some solid toxins and some surface debris, we use different types of filters to minimize the life cycle of materials (Vaseashta, 2015). Also, for surface collecting water underground, we must drill shallow wells. As a prototype, for different types of filters, we drilled shallow wells at the depths of 20 m and 1.2 m in diameter. The outflow capacity in clay/ clayey loam is about 0.55 m^3/hr. Also, the outflow capacity in gravel/sand is 0.36 m^3/hr. The total outflow capacity is 0.91 m^3/hr and the total outflow capacity in 24 hr is 21.85 m^3/hr (figure 14).

Furthermore, for this investigation, we used brick with plastering on the outer earth surface as filter design, with 1:6 cement mortar, and 10% of these walls are open, and from the way of pipe falling water to the Splash pad. This way the water percolates to the different layers of coarse Sand to approximately ~15%, and under these layers of gravels of granules, and at the bottom of these layers we can use from 40 – 50% pebbles for infiltrating rain waters (figure 15).

Figure 14. Different types of filters that we used in this research

Figure 15. The cross section of the filter according to the fractions of stratigraphy

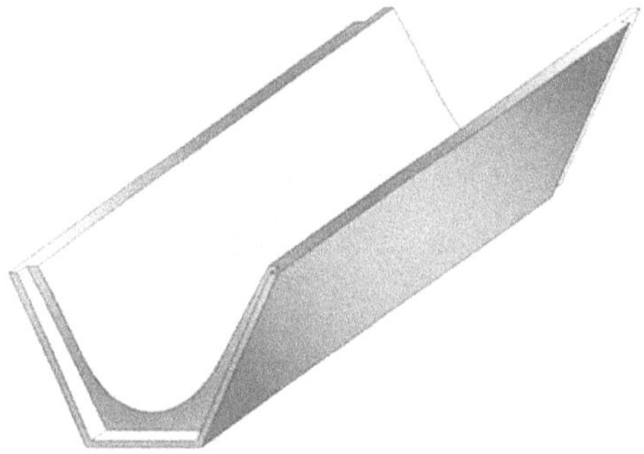

4.12.3 Geoscience Faculty

The Geoscience faculty of Kabul University is located in Kabul, Afghanistan. The main area of Geoscience faculty is 3599.5 m^2, and its top roof is covered by an iron tent. The building has 24 hods with typically 11 hods at the north side of the building, 11 hods at the south side, 1 hod at the west side, and 1 hod at the east side of the Geoscience faculty building (figure 16).

4.13. Estimation of Rainwater

The borehole drillings are done in the vicinity of the Geoscience Faculty. The total area of the Geoscience Faculty roof is 3600m^2 which facilitates rainwater infiltration to the ground. As the precipitation of Kabul occurs as per season of the year, and the maximum rainfall starts from winter to the spring seasons, and the minimum precipitation is about 10mm/day, but the maximum precipitation is 20mm/day, and if we look at the maximum precipitation in this case about 36m^3, i.e., the precipitation that comes down from the roof to the earth surface. The design presented above can be reproduced at other sites and can serve as a source of water, although from a historical standpoint, the cisterns have been used for underground water storage for thousands of years. However, in the case of Afghanistan, due to years of conflict, these alternate sources of rainwater harvesting provide a glimpse of water sustainability in view of the rising population in start contracts with the extreme weather conditions.

Figure 16. The schema of the Geoscience Faculty building

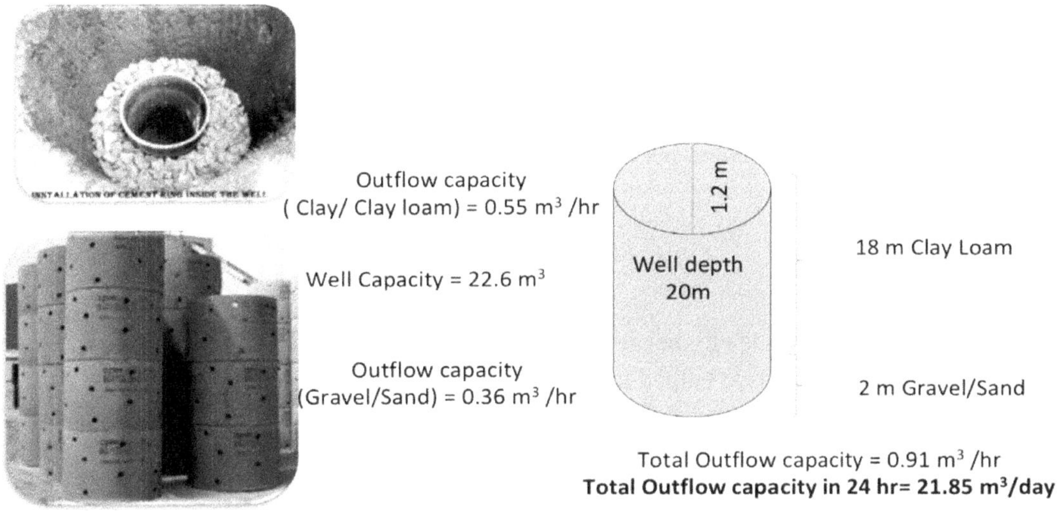

CONCLUSION

The study described here is used to determine artificial recharge in view of extreme hydroclimatic conditions, as is experienced by Afghanistan. The study used the Faculty of Geoscience rooftop, Kabul University, Afghanistan. The most precipitation in Afghanistan occurs in the winter and spring seasons, and in the winter season we have snowfall, but in spring seasons we have only rainfall. The annual rainfall has been ~66mm, and the minimum is ~12.82 mm. The rooftop area is 1000 m^2, upper rooftop consists of a metal plate and water does not infiltrate inside of the building, and all the precipitations waters proceeds from the hod to the ground surface, where it is collected from coarse sand, gravel such as granules and pebbles to the shallow well. To avoid some solid toxins and surface debris, we must use different types of filters. For the collection of surface water underground we have drilled shallow wells, and different types of filters have been used. First of all, we drilled shallow at depths of 20m and 1.2m in diameter. In filter design, we used brick with plastering on the outer earth surface, with 1:6 cement mortar, and 10% of these walls are open, and from way of pipe falling water to the Splash pad. This water percolates to the different layers of coarse Sand ~15%, under this layer gravels of granuls, and at the bottom of these layers, we can use from 40 – 50% pebbles for infiltrating rain waters. The results obtained here suggest that conducting this study on a test site provides the scientific basis for carrying out water harvesting devices, elsewhere in the country. Due to extreme hydroclimatic conditions, it is possible that after some years Afghanistan may face water shortage problems and even ground water drawdowns. Since, during the last three decades the groundwater has drawn down year by year in Kabul City, and now the average groundwater level is from 60 to 90m. The prototype studied here provides a viable alternative to water harvesting. For these reasons, the artificial recharge for example recharges from big and small dams, harvesting methods, Chack dams at that water collected at the snow melting seasons, and collection of rooftop waters from wells at the spring and winter seasons started in Kabul. As well as will be more agricultural lands and deserts will be faced with drought problems and food. The main reasons are rapidly increasing populations and climate change because currently, 4.14 million people are living in Kabul City. The increase in population in Kabul leads to unplanned houses, water shortages, and environmental and other social-economic problems. Recently, the rapid population growth and climate change impacts have placed new stresses on environmental pollution, especially air pollution and water resource depletion. Over period the of 1960 to 2008, the mean annual temperature high of Afghanistan increased by 0.6°C with an average increase of 0.13°C per decade; and the mean annual precipitation decreased at an average rate of 0.5mm per month with an average decrease of 2% per decade.

ACKNOWLEDGMENT

The authors extend warm gratitude to the anonymous reviewers for their valuable comments, as well as to other faculty members who have helped to write and publish this paper.

Conflicts of Interest: The authors declare no conflicts of interest regarding the publication of this paper.

REFERENCES

Arian, H. Kayastha, R. B., Bikas C. B., Ahuti S., Rasouli, H., Armstrong, R., (2015). Application of the Snowmelt Runoff Model in the Saang River Basin, Afghanistan Using MODIS *Satelite Data. Journal of Hydrology and Meteorology, 9.* / doi:10.3126/jhm.v9i1.15586

Belhassan, K. (2011a). Hydro-geological Context of Groundwater Mikkes and different variations of Its springs flow (Morocco). Research Journal of Earth Science 3(1), 15 – 26. IDOSI Publications.

Belhassan, K., (2011b). *Relationship flow- Rainfall in the stream Mikkes (Morocco).* IDOSI Publications.

Belhassan, K. (2020b). Relationship between River and Groundwater: Water table Piezometry of the Mikkes Basin (Morocco. *International Journal of Water Resources and Environmental Science, 9* (1).

Grieser, J., Gommes, R., & Cofield, S. (2006), On the Estimation of Monthly Precipitation Fields in Afghanistan, A communication uploaded on Researchgate by the authors. https://www.researchgate.net/publication/267723218_On_the_estimation_of_monthly_precipitation_fields_in_Afghanistan#fullTextFileContent

Hamdard, M. H., Soliev, I., Rasouli, H., Kløve, B., & Belhassan, K. (2022). Groundwater Quality Assessment in Chak Karstic Sedimentary Basin, Wardak Province, Afghanistan. *Central Asian Journal of Water Research, 8*(2), 110–127. Advance online publication. doi:10.29258/CAJWR/2022-R1.v8-2/110-127.eng

Rasouli, H. (2019). A Study on Some River Sediments, Hydrology and Geological Characteristics In Chak Sedimentary Basin, Wardak, Afghanistan. *International Journal of Geology, Earth & Environmental Sciences.*

Rasouli, H. (2020a). *Application of soil physical and chemical parameters and it's Comparing in Kabul Sedimentary basins.*, doi:10.24327/ijrsr.2020.1102.5095

Rasouli, H. (2020c). Well Design and Stratigraphy of Sheerkhana Deep Well in Chak District, Wardak, Afghanistan. *International Journal of Geology, Earth & Environmental Sciences.* https://www.cibtech.org/jgee.htm2020.

Rasouli H., (2021b). Analysis of Groundwater Quality in Jabal Sarage and Charikar Districts, Parwan, Afghanistan. *Journal of Geological Research, 3*. doi:10.30564/jgr.v3i4.3717

Rasouli, H. (2022a). Climate Change Impacts on Water Resource and Air Pollution in Kabul Sub-basins, Afghanistan. *Advances in Geological and Geotechnical Engineering Research, 04*. https://ojs.bilpublishing.com/index.php/agger

Rasouli, H. (2022b). Methods of Well Construction Complication, Design, and Developing for Sixteen Observation and Test Wells at the Eight Locations of Zarang district, Nimroz, Afghanistan. *International Journal of Earth Science Knowledge and Applications*, *4*(3), 426–448.

Rasouli, H. Sarwari, M. H., Khairuddin R.· Said A. H., (2020a). Geological Study of Tangi Mahipar Mountain Range along Kabul Jalalabad Road. *Afghanistan*. doi:10.4236/ojg.2020.1010044

Rasouli, H., Kayastha, R. B., Bikas C. B., Ahuti S., Arian, H., Armstrong, R., (2015). Estimation of Discharge from Upper Kabul River Basin, Afghanistan Using the Snowmelt Runoff Model. *Journal of Hydrology and Meteorology, 9*. . doi:10.3126/jhm.v9i1.15584

Rasouli H., Qureshi R., Belhassan K., (2021c). Investigations on River Sediments in Chak Sedimentary Basin, Wardak Province, Afghanistan. *Journal of Geological Research, 3*.. doi:10.30564/jgr.v3i4.3574

Rasouli, H., & Safi, A. G. (2021a). Geological, Soil and Sediment Studies in Chelsaton Sedimentary Basin, Kabul, Afghanistan. *International Journal of Geosciences*, *12*(2), 170–193. doi:10.4236/ijg.2021.122011

Rasouli, H., Vaseashta, A., & Belhassan, K. (2023). Mechanical Analysis of Khair Abad Village, Surskhrud District, Nangarhar Province, Afghanistan. *International Journal of Earth Sciences Knowledge and Applications.*, *5*(1), 103–118. http://www.ijeska.com/index.php/ijeska

Rasouli, H., Vaseashta, A., & Hamdard, M. H. (2023). Sedimentological Study of Chack Hydropower Reservoir, Wardak, Afghanistan. *International Journal of Earth Sciences Knowledge and Applications*, *5*(1), 21–32. http://www.ijeska.com/index.php/ijeska

Shamal, S., & Rasouli, H. (2018). Comparison between pH, EC, $CaCO_3$, and mechanical analysis of Qala Wahid and Company Areas soil, Kabul, Afghanistan. *International Journal of Scientific Research*, *8*(5). Advance online publication. doi:10.21275/ART20197318

Vaseashta, A. (2015). Life Cycle Analysis of Nanoparticles Risk, Assessment, and Sustainability. DEStech Publications, Inc.

WHO. (2009). *WHO Handbook on Indoor Radon: A Public. Health Perspective*. World Health Organization.

WHO. (2011). *Uranium in drinking water. Background document, for development of WHO Guidelines for drinking-water quality*. WHO.

Chapter 12
On the Modeling of Groundwater Features Due to Extreme Arid Conditions in Nimroz Province, Afghanistan

Hafizullah Rasouli
Kabul University, Afghanistan

Ashok Vaseashta
https://orcid.org/0000-0002-5649-0067
International Clean Water Institute, USA

ABSTRACT

Due to extreme arid conditions in Afghanistan, it is critical to conduct research by modeling groundwater features. The studies conducted in Nimroz province on the groundwater flow directions, aquifer types, inflow, outflow, evaporation, and geology of the region are presented. Samples were collected from each profile and transported to our laboratory for characterization. The objective was to identify parameters in different locations of Nimroz and Zarang, for potential use as a water resource, since no such data exists due to years of conflict in the region. From this study, we assess the features of lithology, such as silty sand, clay with gravel, sandy silt, gravel, boulder, and clay with gravel, in addition to the groundwater quality, which is saline, due to evaporation in this region. Generally, the thickness of aquifers is about 5-25m, and the groundwater has evaporated to 5-10m depth with an annual precipitation of about 50-100 mm. It is further observed that the groundwater direction is from east to west. The results suggest that the modeling can be used for other regions.

INTRODUCTION

Afghanistan is known for its high mountains, which are mainly located in the northeast part of Afghanistan, as part of the Mustage series, Wakhan, central and east Hindukush, and Hinduraj, Wakhan, located east to west and is 6505m in height. The east Hindukush Mountain is in the north-northeast with a height

DOI: 10.4018/978-1-6684-8771-6.ch012

between 5500-6000m, and in some parts, the height is ~7000 m, with a total length of approximately 350 km (Arian, et al. 2015, Hamdard, et al. 2022), while the Hinduraj little extends to only about 130 km from northeast to southwest (Abdullah, et al. 1979, Avouac and Burov, 1996). The middle mountain region consists of ParaPamizus, Turband Turkistan, with its height increasing to the southeast. The ParaPamizus height is about 3000 – 3500m and its total length is 550 km, while Turban and Turkistan are more than 4000m in height (Banks, et at. 2002, Bohannon, R.G., 2005). The low mountain region consists of the west of Badakhshan to Polykhomri River, and the main mountain region consists of Rustaq, Simi, Turband, and Turkistan with a mean height is 700 – 900m (Bohannon and Turner, 2007, Böckh, 1971). Generally, in Nimroz province the plateau and plain areas are located on both sides i.e., the southeast and northeast, of the Amu River, which has a height of 300 – 350 m. above sea level (a.s.l), while the mountain areas in Nimroz province are located near Iran's borders with plateau and desert areas. In Afghanistan, about 75% of the mountains are located from northeast to southwest, generally, in the northeast the height of the mountains continues to widen, which makes different forms of alluvial fans (Broshears, et al. 2005, Lave and Avouac, 2001), but the height of mountains is located near the Parwan, Qatghan, Wakhan, and Nuristan and it's the skeleton of Hindukush mountains in Afghanistan, which has watershed between Amu Indus rivers. This series of mountains continues all the way to Tajikistan, China, and India. The total length of the Hindukush Mountain is ~800 km and has different types of valley system, which is separated East, Central, and West of Hindukush. Generally, the middle mountains of Afghanistan are in the west and southwest parts of Hindukush, and the stretch of these mountains is east to west. On the west side of this mountain range is ParaPamizus, which has an average height of ~3000-3500 m.a.s.l, and a total length of 550 km (Koons, 1989, Malgary, 1987). On the south of this mountain range is Turbanda Turkistan, which is the highest, with over 4000 m.a.s.l. In addition, the Hazara Jat mountains consist of thin valleys with an average height of ~1500-4000 m.a.s.l. In the Southeast of Afghanistan, the Suleiman Mount range extends from east to west (Lave and Avouac, 2001, Molnar, 1990). In addition, in the south, north, and west of Afghanistan, there are plain areas and deserts, as shown in Figure. 1.

The climate of Afghanistan is semi-arid to extremely arid, as well as it is a mountainous country. In the central and northern parts of the country are high mountains, but in the southern part, viz. Farah, and Herat provinces of Afghanistan are plain areas and deserts (Malgary, 1987). Afghanistan is a country dominated by a dry climate, with most of the area impacted by the effect of global climate change, especially on hydrological systems, and more so due to mountain snow and glacier melting, which can modify the timing and amount of in mountain watersheds (Molnar, 1990, Montgomery, 1994). Therefore, accurate stream flow simulation and forecast are of great importance to water resources management and Planning (Montgomery, 1994, Munsell, 1999). The key watercourses drain during the snow melting times (from January to May), rainy periods (March to April), and occasionally through quick overflowing terms (May to August). In fact, the main elevations of snow cover are Parwan maintains series, Wardak, Loger, Baba, Spingher, Salang, Kohkurugh, Koha Safi, Hindu Kush mountains ranges in Afghanistan (Paetzold *et al.* 2005, Nakata, 1972, Ruleman, et al. 2007), as well as in certain cold province, for example, Bamyan, Wardak, Loger, Badakhshan, Pangesher, Parwan, some parts of Kabul, in these snow cover from September to November and it is used for storing water in Afghanistan (Souriau, et al. 1988, Wheeler, et al. 2005). Likewise, in the north sides of Afghanistan, there are specific glaciers Pamir Badakhshan, Mymai Badakhshan, and Panjsher mountains range belonging to the Hindu Kush mountains series in Afghanistan. Those are the main sources for Panjsher, Helmand, and Koner Rivers. In this province, we use rivers primarily for irrigation and drinking (potable) water supply, such as for Bamyan, Panjsher, Wardak, Parwan, Helmand, and Kandahar, Kapesa, but in some sections of the specific province, for example, Wardak, Parwan, Panjsher spending from spring, and in several provinces for drinking and irrigation waters benefit from Kariz, and more other province using from a well (Rasouli and Safi, 2021a).

In these basins, all floods and sediments are transported from different points of the Loger, and Asmayey Mountains by sudden floods, and are accumulated in the Kabul River at different thicknesses in different locations of this basin. The Chelsaton Basin consists of the Quaternary (Pleistocene) and Neogen geological periods, different sediments are deposited after one another and forms of morphology, which we can see at the different reliefs. Types of sediments in this basin directly belong to the kinds of rocks located in the surrounding mountains. In these sediments, we can see Garnete, Biotite, and Muscovite mineral particles. The surrounding mountains of this basin are formed from metamorphic rocks like Schist, Gneiss, and Slate, which are primarily crystalline (Rasouli et al., 2020a).

Hydrogeological and geological studies are very important for these sedimentary basins, because all villages, health centers, and industries use groundwater from wells, Kariz, and springs, to have a good quality of water for quality of life (Vaseashta et al, 2021a, Vaseashta et al 2021b, Vaseashta, 2022). The Panjsher, Sanlang, Gurband, and Shetal Rivers are the main rivers that flow between these basins, and more groundwater recharges from the riverbank and bed, especially in the snow-melting season. The aquifers of this sedimentary basin are located prolonging these rivers and their tributaries. The aquifers are between different sizes of sediments (sands and gravel). The hydrogeology of Jalalabad Sedimentary Basins belongs to the different aquifers that are located in prolonged mountain ranges in the longitudinal valleys. The thickness and depth of aquifers are related to the slope and distance from the mountain range, generally, near mountain and slope areas there are gravels and angular materials but far from mountains are rounded and fine materials like bolder, cobbles, pebbles, granules, sands, and silts. Generally, for drinking, the water of shallow and deep wells benefited. More to the point, in some areas spring water originates from the fracture zones where pipe scam for gravity pumping systems and distributed install and benefited from the water in villages (Rasouli, 2020b, Rasouli, 2021b).

The Grade acidic belongs to the pH of the soil, and the formation of acids belongs to the chemical characteristics of soils. Also, related to the elements and components that are in the air and after rain washing the air and infiltrating the soil (Rasouli, 2019). Activities of some animals and different plants also will be acidification of soils. In addition, the mutual effect of biochemical, solution particles of rocks and minerals that are located between soils and its absorption by colloids done some reactions of Cation exchange capacity and basic exchange capacity in some parts of soils mass. The groundwater moves from aquifer layers at different speeds and washes soluble materials from different layers. Some soluble loads from parent materials produced different elements and components (Shamal S., Rasouli, 2018). Industrial activities, for example, burning coal and oil in the predictable materials factory, vehicles, and permanent melting of plastics and metals result in a greater number of different gases going to the atmosphere with different types of precipitations again come down to the earth's surface and product different toxic elements between soils (Rasouli, 2020c, Rasouli, et al., 2015). For many years wars we could not do much research work on soil, air, and water pollution (Rasouli, et al., 2021c).

The basins can be described as valley-fill basins, which are filled with Quaternary and Tertiary sediments, gravels, and sedimentary rocks. Quaternary sediments are typically less than 80 m thick in the valleys. The underlying Tertiary sediments have been estimated to be as much as 1000 m thick in the valley Center. The gravel and sand were deposited mainly in the river channels. Described as the Lataband Formation, the Quaternary terrace sediments are from younger Pleistocene-age overlying conglomerates. The surrounding mountains are primarily composed of Paleoproterozoic gneiss and Late Permian through Late Triassic sedimentary rocks. The inter-basin ridges, composed of metamorphic complex rocks, are Paleoproterozoic gneiss. The Khengal and basement rocks are overthrusted by schist mélange, which has been called the Cottagay Series, in the northern Kohe Safi range. The sediments of the Khengal series started from the Jurassic and it belongs to the Thythes Ocean in Afghanistan (Rasouli, et al., 2021c). Triassic clays and Paleozoic schist form the impermeable substratum of this aquifer. These carbonate formations burrow under the Mio-Plio-Quaternary cover in the basins, which forms a deep confined aquifer. The depth of the Miocene Marls forming the impermeable roof of this aquifer is about 1500 m in contact with Prerif Ridges at the drilling point. The fracture rocks constitute the groundwater reservoirs. The main parameters for the migration of fluids in fractured rocks are the main geological characteristics of the fracturing, drainage, topography, and rainfall. The hydrogeological context of different regional structures implies the existence of groundwater tables. E1 Hajeb–Ifrane Tabular is a free water table circulating in the Limestones and Dolomites. It is supplied directly by the precipitation (Rasouli, 2022a).

From the viewpoint of geology, the Nimroz province is one of the plain areas, where all sediments belong to the alluvial and conglomerates belong to Carboniferous, and wind sands from Neogen and Quaternary, and the boundary of Afghanistan and Iran we can find metamorphic rocks belong to old Paleogen and Cretaceous periods. In this province, we can find some outcrops from red color marble called Maksia mountains, and around Helmand River we can find some lakes at the end of year's drought these lakes are dry, and we can find the remnants of salts around these lakes. In Nimroz province wind-soft sediments are at a thickness of 12-25m and in these sediments, we can find some aquifers whose yields are about 200 – 400 liters/Min. Some of these aquifers consist of fresh water, but more of these aquifers are salty waters. As well as in Nimroz province we can find Limestone and Carbonate Rocks, its thickness of about 50-500m, in these rocks we can find Karst aquifers and their yield is about 50-200 liters/min, in some places the groundwater in these rocks is free, but in some places, it is having saline water. Some wells are drilled at a nearby saline land, with an average well depth of ~150 m (Rasouli, 2022b).

STUDY AREA

The Nimroz province is one of the boundary provinces bordering Iran and has a substantial economic role for Afghanistan. The climate of Afghanistan is arid to semi-arid, and this type of climate extends all the way to the south of Nimroz, Helmand, and Farah and applies to all of Afghanistan. In these regions, the amount of precipitation is very low and annually between 70-80 mm, in the spring season there is rain and in the winter season, there is a lot of snow precipitation. When the temperature is high, there is a lot of evaporation, especially in the southwest part of Afghanistan, such as Registan, Margo, Bakwa, Khasherud, Systan, and Jahanom deserts. The Nimroz province from the viewpoint of geology is a desert and plain area, where all sediments come from the conglomerates due to carboniferous and wind sand from the Quaternary at the border areas. Additionally, there are, Nimroz province metamorphic rocks that are attributed to the Paleogen and Tertiary (Tünnemeier, et al., 2005), as shown in Figure 1.

METHOD AND MATERIALS

In this study, we conducted studies of different groundwater sources and their features to develop models, such as groundwater flow directions, aquifer types, inflow, outflow, groundwater balance,

Figure 1. Geological map of Nimroz province

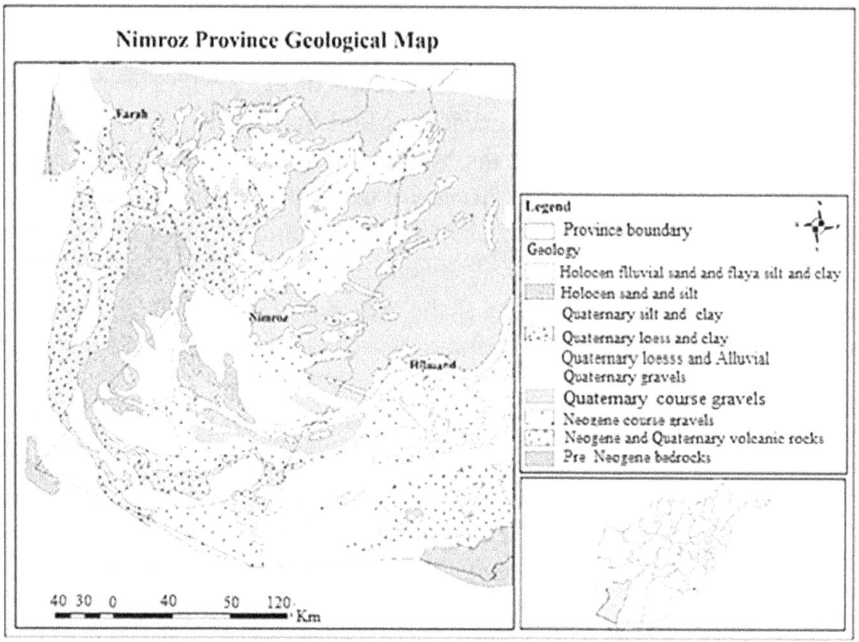

and groundwater evaporation, to relate them to the geology of the Nimroz province in Afghanistan. For this study, several dissimilar samples were collected representing each location profile and were transported to the research laboratory for analysis. As a routine, for most of the groundwater samples collected from Nimroz and other locations, a series of characterizations are conducted that include pH, turbidity, elemental analysis, and electroconductivity. However, to develop 3D models of groundwater flows in Nimroz province, we used borehole drilling in different types of aquifers to collect samples from different depths of stratigraphy. In Nimroz province groundwater in aquifers flows through the alluvial, sands, gravels, as well as limestone, however, the aquifers are not good for groundwater storage. From the viewpoint of climate, since the Nimroz province is generally warm, there is more evaporation. During the course of this investigation, the aquifers that were considered in Nimroz province were all located along the Helmand River and the thickness of sediments along this river is, generally, large and as a result, most of the groundwater is stored in these aquifers. Since these aquifers represent a rather small area, the model does not represent the entire Nimroz province, but a similar model can be used for other locations as well. In other areas, the sediments are fine materials and are not permeable for groundwater storage. It is instructive to note here that in Nimroz province, most lithology consists of gravel, sand, and desert areas. Most of the work reported here was conducted in the laboratories of the Geoscience faculty at Kabul University.

RESULT AND DISCUSSION

In this research, we used groundwater models and analyzed different samples of groundwater and soil in a laboratory, from different parts of Nimroz province, Afghanistan.

Temperature in Nimroz

The Nimroz province is located in the southern part of Afghanistan and the amount of precipitation is much lower due to extreme hydrometeorological conditions, it's about 50-100 mm, in this reason, the temperature of this province is very higher and the balance between precipitation and evapotranspiration is very different, meaning the amount of evaporation is higher than precipitation. In this province, the drawdown is much higher, and all water is year to year come down. Even in some places because of higher temperatures and topography, it is equal to zero recharge (Figure 2).

Figure 2. The amount of maximum and minimum temperatures in Nimroz province

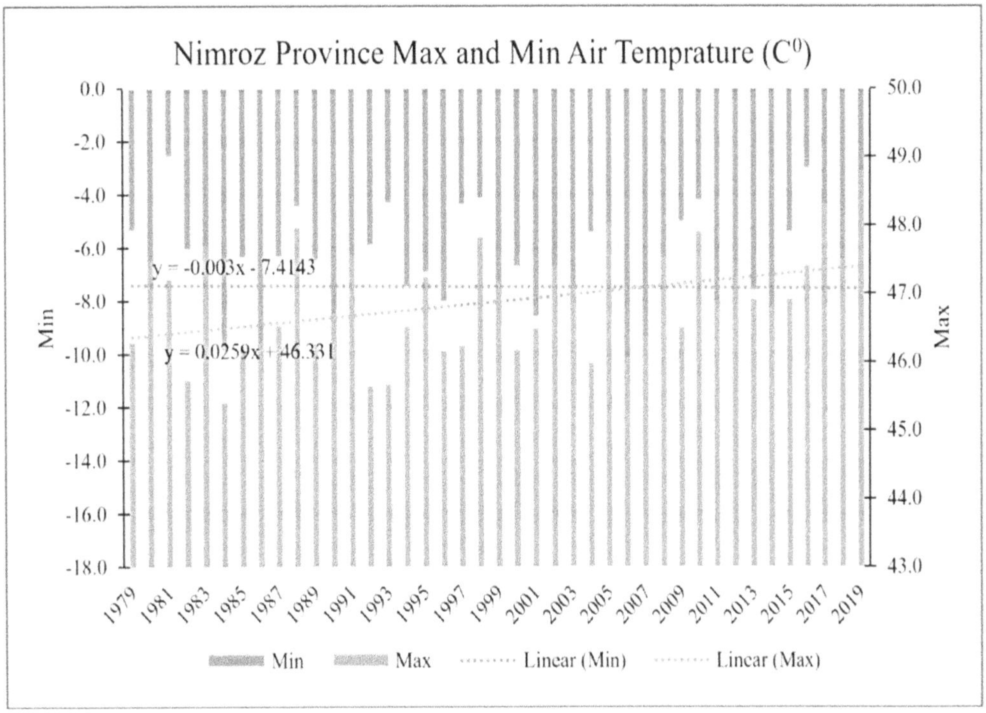

Groundwater Evaporation

The evaporation from groundwater is one of the very important parts of the balance equation. The evaporation from the groundwater surface belongs to the temperature of regions and the depth of the groundwater surface, and for these, we must first calculate the temperature of regions and the depth of the groundwater surface. According to the research about groundwater, it shows that groundwater evaporation occurs at the 5m depth of groundwater-surface, but more than 5m depth will occur the evaporation but it's very low. According to the drilling and geophysical methods in Nimroz province, the thickness of the aquifer is 5-25 m, and the surface of groundwater is lower than 5m in depth. In this study, the groundwater quality in Nimroz province is saline and we cannot use it for drinking and irrigation. From that, we can find the amount of evaporation from the groundwater surface in Nimroz 5-10 m of groundwater in Nimroz province (Figure 3). The air temperature in Nimroz province is very high and the water quality is saline.

Figure 3. The proportion of groundwater level and groundwater temperature

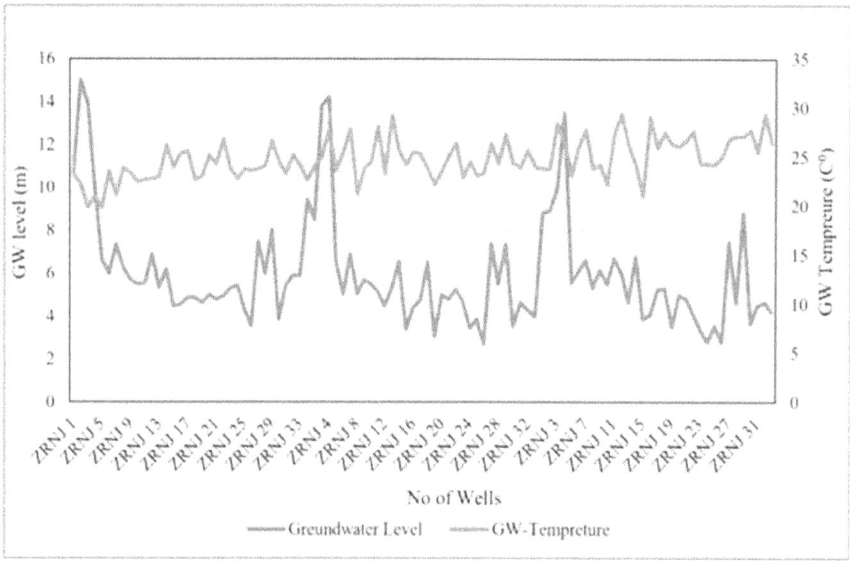

Precipitation in Nimroz

The high temperature in Nimroz province and the very low annual precipitation, but the higher evaporation, very higher effect on the groundwater saline. Only by the course of rivers and canals, we can find fresh water. In this research, because of the non-existence of previous data, we cannot calculate the recharge to the aquifers in the exact way. When we calculated the annual discharge at the two stations of Helmand River the amount of annual discharge was 5379 million cubic meters. From the viewpoint of geology in Nimroz province the annual recharge from Helmand River two to the aquifers was 4%, and this is about 215 million cubic meters recharge from Helmand River to the next aquifers.

The annual precipitation in Nimroz province is different according to the years, and the most precipitation is from 1991-1997 which is more than 150mm, but in other years is lower than 150mm. this is shown in Afghanistan because of climate change the amount of precipitation is lower (figure 4). From that, we can find because of lower precipitation the depth of groundwater is much lower and its fluctuations belong to the seasons of the year also.

Excavation from Groundwater

In Nimroz province one of the very important factors of groundwater is excavation from deep wells for irrigation, and its use is much more for irrigation systems (Table 1).

Recharge From Rivers

The Helmand River is one of the very important rivers flowing in Nimroz province and it is one of the very important rivers that recharge the groundwater, also those agricultural lands near this use this river.

Figure 4. Show the annual Precipitation (1979-2019) in Nimroz Province

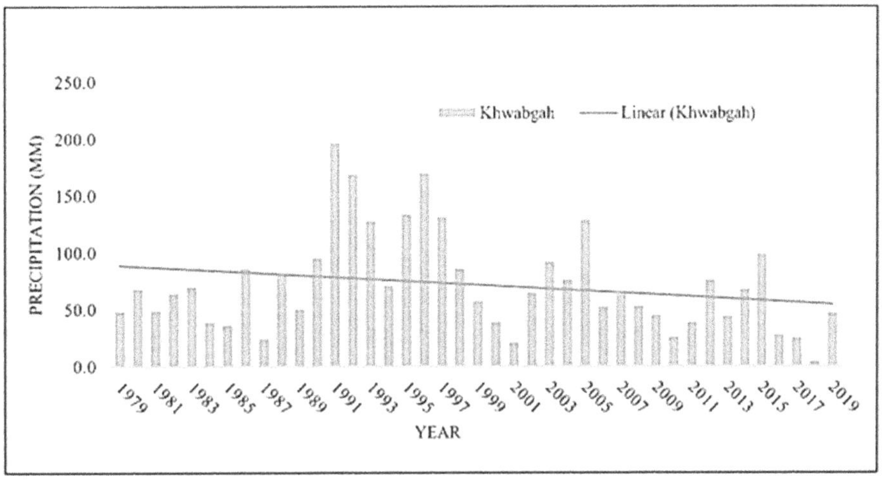

Table 1. Land cover in the Nimroz province

		Aggregated Provincial Land Cover Statistics (sq km)											
Province	District	Irrigated Ag. Land	Rainfed Ag. Land	Fruit Trees	Vineyards	Barren Land	Sand Cover	Forests & Shrubs	Rangeland	Permanent Snow	Built-Up	Water Bodies & Marshland	TOTAL
		3A,3A1,3B,3C	4A, 4B	2A	2B	8A	8B, 8C	6A,6B, 6B1,6C	7	13	1A, 1B	9A,9B,10A, 10B,11,12	
Nimroz	Chakhansur	208.98	0.00	0.00	0.34	8,808.51	257.40	1.49	11.00	0.00	4.11	586.00	9,877.82
	Charburjak	60.14	0.00	0.00	0.21	15,212.07	4,013.21	3.26	39.53	0.00	2.59	1,628.55	20,959.57
	Kang	33.95	0.02	0.00	1.26	167.55	0.94	1.03	12.06	0.00	4.26	941.60	1,162.67
	Khasherud	606.23	0.00	0.08	0.17	6,746.51	97.66	0.00	14.81	0.00	16.71	364.36	7,846.53
	Zarang	41.06	0.00	0.05	1.87	645.16	295.67	2.06	20.52	0.00	15.71	170.19	1,192.30
	TOTAL (sqkm)	950.37	0.02	0.13	3.85	31,579.80	4,664.87	7.84	97.92	0.00	43.39	3,690.69	41,038.89
	TOTAL %	2.3%	0.0%	0.0%	0.0%	77.0%	11.4%	0.0%	0.2%	0.0%	0.1%	9.0%	100.0%

The river flowing from those places has having very good aquifer for groundwater recharge. Figure 5 shows the annual discharge from 1971-1978 of Helmand River at the Malakhan and Chaharburjak stations.

The Daily Groundwater Discharge Expenditure People

One of the other factors in the discharge and expenditure of groundwater is the daily expenditure of groundwater by people in urban areas for drinking water. According to some reports in Nimroz province about 200,000 population live, and in urban areas per person uses 80 liters/day, but in suburban areas is about 60 liters/day.

Figure 5. The annual discharge of the Helmand River at the Malakhan and Chaharburjak stations

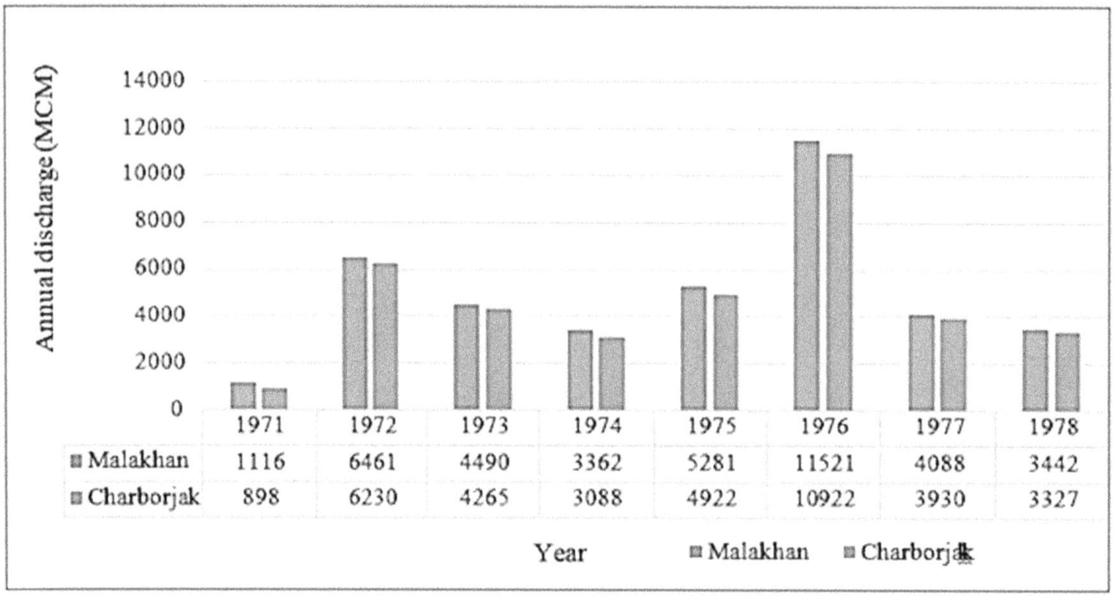

Groundwater Aquifers

Generally, the groundwater accumulated in aquifers belongs to the alluvial, sands, gravels, as well as in limestone but it's not good aquifers for storage groundwater. From the viewpoint of climate, the Nimroz province is warm, and it has more evaporation. When we did this research in Nimroz province all aquifers located along of Helmand River and the thickness of sediments along Helmand River is more and all groundwater is storage in these aquifers, but these aquifers are in a very small area and it's not complete for all Nimroz province. But in other areas, the sediments are fine materials and in it's not permeable for groundwater storage. As well as because of more evaporation the groundwater is saline, and we cannot use it for drinking and irrigation.

From the calculation of the balance and potential of groundwater in Nimroz province we find and estimate the recharge to the groundwater is 285 million meters cubic and the discharge from groundwater is 4.290 million meters cubic. These we can find in table 1.

Groundwater Resource in Nimroz

Generally, from viewpoint of aquifers are divided into four types:

- The alluvial sand and gravel aquifers,
- The sand and gravel aquifers,
- Limestone aquifers,
- Aquifage and consolidated rocks.

Alluvial Sand and Gravel Aquifers

These aquifers consist of both sides of Helmand River and the lithology of these rocks consists of gravel and sands. The aquifers of this interval are about 200-400Li/min and its depth is estimated at about 25m.

Sandy Gravel Aquifers

The regions of Nimroz province are covered by sands and gravels, and the types of materials consist of unconsolidated sands, clayey sands, and gravels and it's about horizontal aquifers. These aquifers are one the high aquifers and their yields are about 200-400 liter/min from a viewpoint of quality these aquifers are saline their depth is 0-100m, and the groundwater fluctuation according to the seasons of the year (figure 6).

Figure 6. The Neogen sediments and the directions of winds in Nimroz province

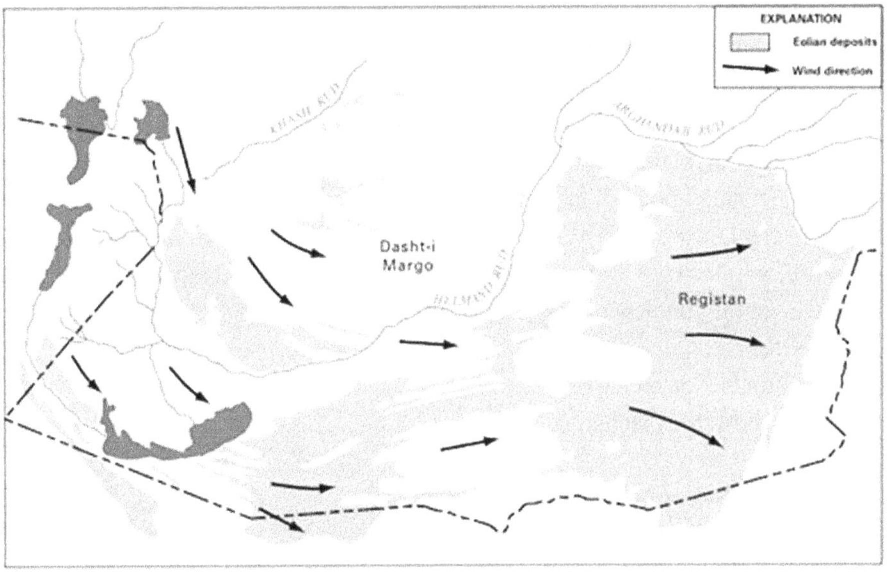

Limestone Aquifer

This kind of aquifer is very small and is located on the north side of Nimroz, the lithology of these kind of rocks are changeable at the very next space, and between them, we can find unconsolidated lime, conglomerates, shale, sandstone, the maximum groundwater yield is 100-300 liters/min., and it concludes of saline groundwater. The groundwater quality in these aquifers is toxic and we cannot use it for drinking water. The mean depth of these aquifers is about 25m, the yield is about 400 liters/min and the Total Dissolved Solid (TDS) is 15000 mg/L.

Groundwater Balance

The assessment of balance and potential in Nimroz province belongs to geological conditions, topography, climate, and hydrology, and from these parameters, we can calculate the potential and balance of groundwater in Nimroz province. One of the very important managements of groundwater we can find from groundwater balance, and without the groundwater balance we cannot do groundwater management. In this case, we must know the annual groundwater recharge and discharge, and without this, we cannot find groundwater management. The assessment and exchange of water at one special limit, for staying one water mass at the one cycle of water it's called groundwater balance. From this definition, we can find that all groundwater at the one special limit of time in this region is used for different usage and or reservoirs. It is more belongs to usage by industries such as factories and urban areas. In this case, will change the groundwater quality for different uses such as irrigation and water supply. From these toxic aquifers maybe the neighboring aquifers percolate and as a result, we cannot use the water for drinking purposes and other usages. In this case, it is very important to balance groundwater and avoid such problems. For groundwater balance in every region, we can use the following chart (figure 7) which addresses the groundwater cycles, as well as the output and input of groundwater from the aquifer layers.

The groundwater balance is very important for research and planning some ministries and organizations in Afghanistan such as the Ministry of Rural Rehabilitation and Development (MRRD), Ministry of Mining and Petroleum (MMP), Municipality, Ministry of Water and Power (MWP). As well as by Groundwater balance we can do the rule of groundwater for the future. For the calculation of groundwater balance, we must collect data about all features such as recharge, discharge of aquifers, hydrogeological characteristics, coefficient of transmissivity, thickness of aquifers, amount of precipitation, evapotranspiration, temperature, and surface water streams. For groundwater balance, we need the equation of groundwater balance, and it consists of two sections:

- Parameter of groundwater recharge,
- Parameter of groundwater discharge,

The equation of groundwater balance is the following:

$$R_r + R_c + R_t + R_i + I_g = O_g + E_t + T_p + \Delta S$$

Where the R_r, R_c, R_i, R_t are recharge from the reservoir, agricultural irrigation respectively, canal irrigation and stream flow by rain, Ig, and Og are stream flow output and stream flow output, E_t is Evapotranspiration from groundwater, Tp is excavated from wells, Δs is the changes in groundwater reservoir.

Method of Recharge From Irrigation

One of the very important methods to recharge groundwater we can recharge from irrigation. This method is one of the very important for balance equations. This belongs to visitation cover, intensity of temperature, evaporation, temperature of soil, as well as the depth of groundwater level also in this we studied. According to the investigation annually 2.12 million meters of cubic groundwater are excavated from groundwater. According to (Table 3), about 83.1 million meters of cubic groundwater were excavated from groundwater in Nimroz province.

Figure 7. Parameters of input and output balance groundwater

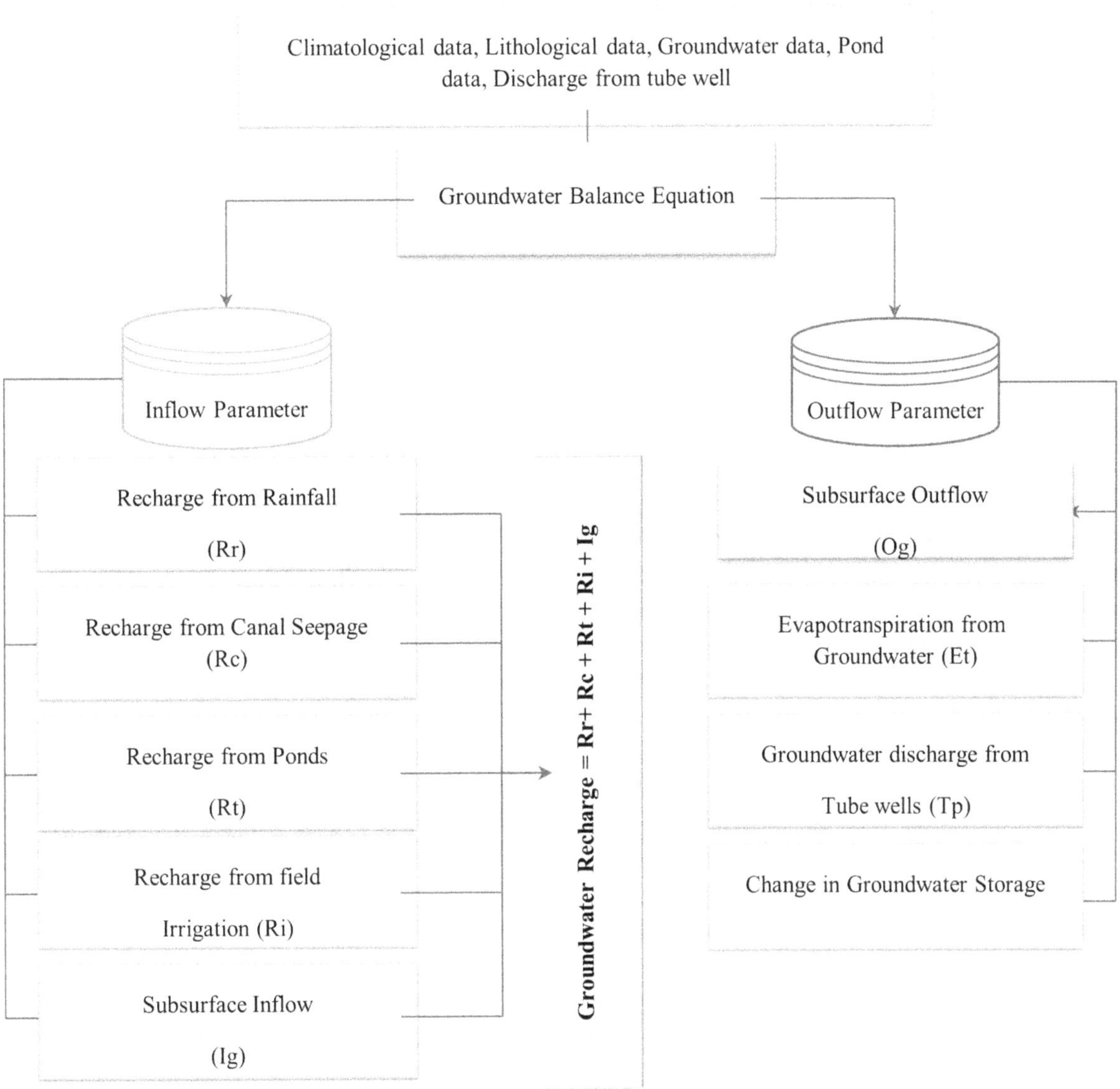

Table 2. Show the balance of groundwater in the Nimroz province

Groundwater Balance Mm³/y	Outflow				Inflow			Nimroz Province
	Recharge Effective Precipitation Mm³/y	Recharge from Rivers Mm³/y	Recharge from Irrigations Mm³/y	Evaporation from GW Mm3/y	Discharge by People Mm³/y	GW Outflow Mm³/y	Discharge for Irrigation Mm³/y	
Nimroz	69	215	1.83	220	4.8	53.4	12.2	-4.57

Table 3. Show the annual recharge in 2015 for the irrigation of the groundwater

Source of irrigation	Type of Corp	Water table below ground surface		
		<10m	10-25m	>25m
Groundwater	Non-paddy	25	15	5
Surface water	Non-paddy	30	20	10
Groundwater	Paddy	45	35	20
Surface water	Paddy	50	40	25

Groundwater Flow Direction

The groundwater flow and direction belong to topography and relief of the earth's surface, as well as belonging to the relief of bedrocks and gradient of bedrocks. Also, the flow directions of groundwater occurring to the lower amount of groundwater means that at those places the amount of groundwater flows to the smaller amount of groundwater. The groundwater flow direction and velocity belong to many parameters, but the main parameter consists of sediment sorting, homogeneity, heterogeneity, angularity, grade of rounding, and ground morphology. The groundwater flow direction not only belongs to the relief and morphology of the earth but there are other factors also such as water pressure and the confined aquifers, but the main factors are relief of the earth. As in (figure 8) we can see the flow direction of groundwater in Nimroz province is from east to west, but in the south side of Nimroz province, as we can see on the other sides of Nimroz province.

First, we must all data from Nimroz province analysis, and after the head of observation wells are subtracted from one another's for projection of these maps we use GIS, as we can see in this all the water outlets are in the Kang district, belong to the Nimroz province. The cross-section of hydraulic is

Figure 8. Flow directions of groundwater at Nimroz province

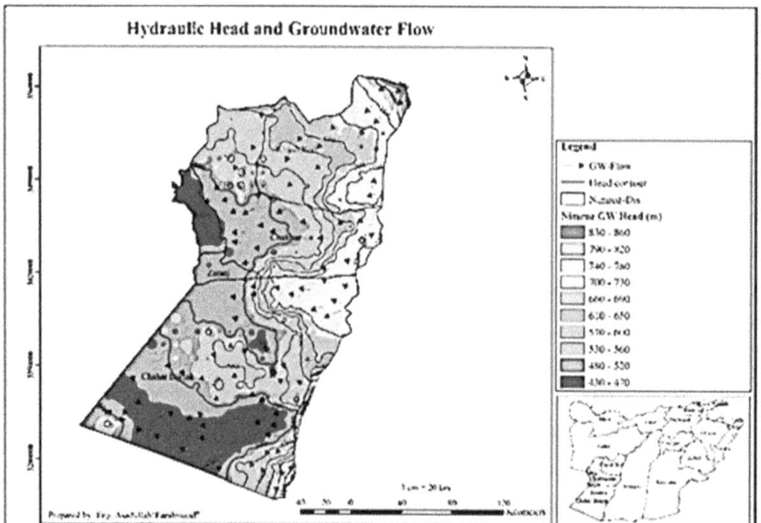

0023.0 which we found by GIS, and the Electroconductivity of aquifers is 600 m^2/day, and the amount of annual flow direction outlet is 4.53 million meters cubic.

Groundwater Modeling in Nimroz

The increasing population in the country, the limitation of surface water, and the more usage of aquifers are the main preseasons of natural resources. In addition, the drawdown of Static Water Level (STW), agricultural activities, industrial and urbanized, the different types of contaminations of aquifers, to avoid continuous quality and quantity of contaminations, and protection groundwater are the main basements of Afghanistan. In this case for better understanding we used a mathematical model, and at the pave the ground we can use it as a tool for the management of groundwater.

The mathematical model is the same as the hydrogeological one, in this, we can use the physical and mathematical models. Two ways in these models we can use Conceptual and mathematical models, Conceptual model is the exact figure of this system. The mathematical model is the total of formulas we can use for groundwater, and different active physical processes in the period of aquifers. The groundwater model is one of the very important parts of groundwater aquifers. The groundwater model is one of the tools to change the data in the system of groundwater. The groundwater is one of the very important agents for hydrological stresses (e.g., Irrigation and pump tests we can show in the form of numbers.

Groundwater Modeling in Zarang

Generally, after determining the aims of the model, the second step is in the model process and the second important part is the conceptual model. The conceptual model is one of the very important and simplified forms of actual data. This model is one of the special systems of hydrogeology and belongs to the exact graphical at the one cross-sectional of geology and hydrogeology of main systems. The conceptual model is based on mathematics, and it is very good information for field primary data and the interpretation of hydrogeological basement. In this research, we used the stratigraphical model at the Zarange city, center of Nimroz. In this study, we used data from the drilling logging, electrical geophysics, and groundwater from 33 rings of observation wells projection (Figure 9).

As well as in this research we found the three-dimensional cross section in the Nimroz province. In this research we show the stratigraphy at the Nimroz province and in this stratigraphy, we found the different types of materials such as clay, clayey with sand, sand with gravel, and sand (Figure 10).

In this study, we found the three-dimensional cross-section of the profile at the Nimroz province. Furthermore, the research shows the stratigraphy in the Nimroz province and also the stratigraphy. Furthermore, we found different types of materials such as clay, clayey with sand, sand with gravel, and sand (Figure 11).

Lithological Model in Nimroz Province

This model shows different layers of sedimentary materials such as sandy silt, clay with gravel, sandy gravel, gravel, and boulder, clay with gravel, and silty sand (Figure 12). In this research, the thickness of sediments is different because it belongs to the paleo-geographical flows at different periods of geology. The main reason is climate change because in the previous times, the thickness of sediments was higher

Figure 9. Logging drilling wells and three-dimensional model of stratigraphy in Zarange, with a profile of (cross-sectional A-A)

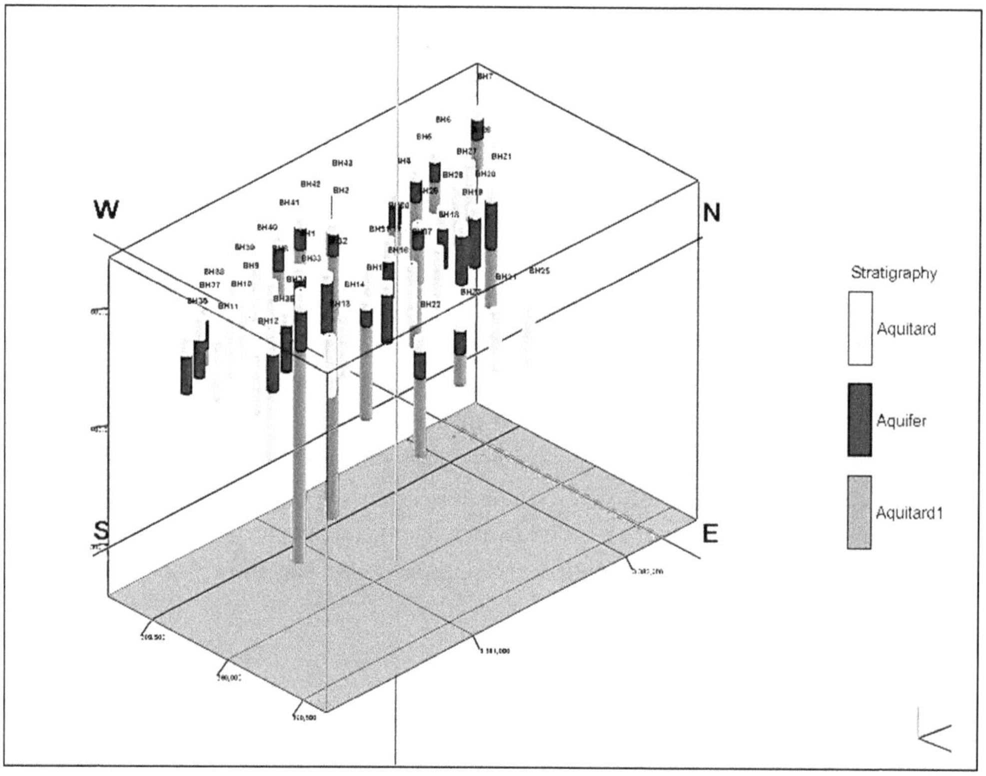

Figure 10. Modeling of the dimensional cross section at the Nimroz province

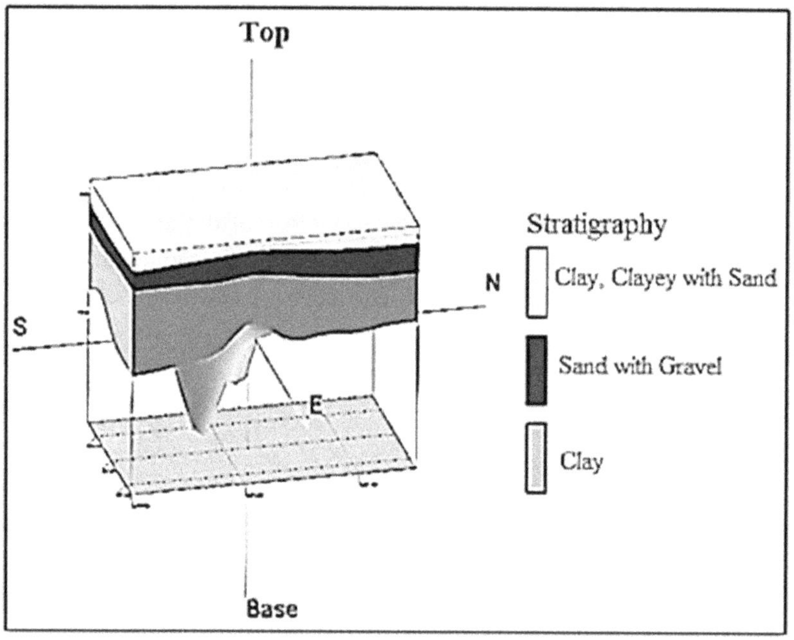

Figure 11. Stratigraphical model in Nimroz province

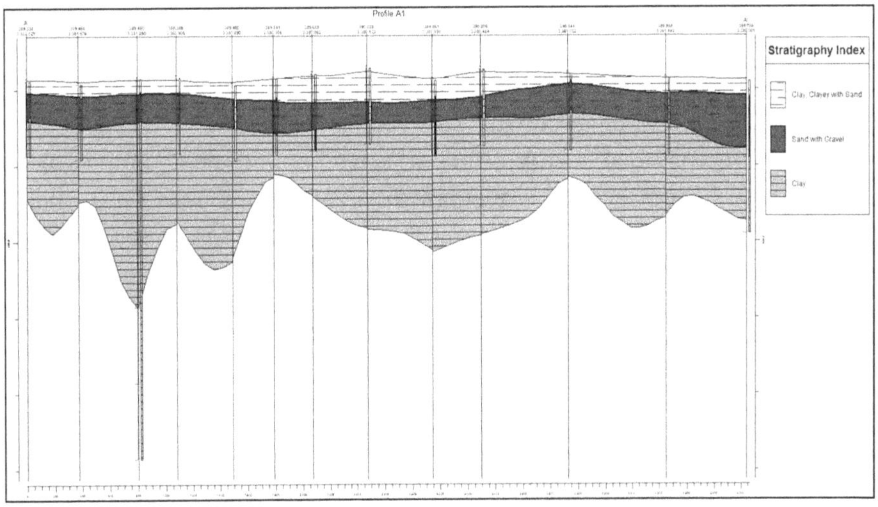

Figure 12. Three-dimensional cross section of profile at the Nimroz province

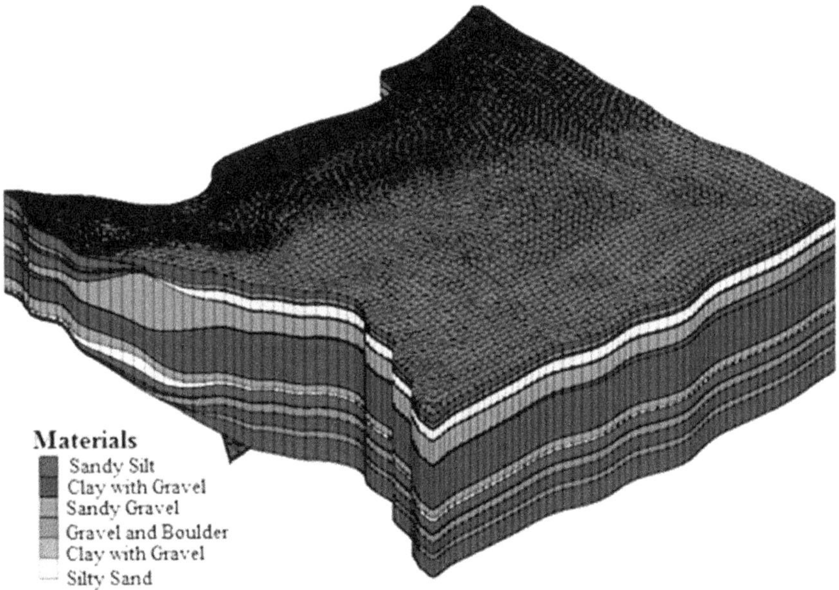

but in the three decades we do not have in Afghanistan, and in the last three decades the thickness of sedimentary layers is small.

CONCLUSION AND DISCUSSION

In Nimroz province, in the north side of Afghanistan, generally, this area near the recharge areas such as the north side plain areas has a positive balance, and fare areas from recharge areas such as the south district have a negative balance. Generally, in south-side provinces, it has a potential of 1.775 BCM/year. In Nimroz province a total of 161872 hectares of agricultural land are by groundwater irrigation and it's near the 9.435 million meter cubic annually excavation for agriculture, generally, from border provinces annually 4.146 million mater cubic groundwater outlets to the Iran country. The aquifer layers of this province consist of Quaternary sediments such as sand and gravel. The aquifer layers generally consist of cementation sediments and quality from the viewpoint of porosity and permeability is lower about 80% than other layers. From data in Nimroz province, more lacks are at the drilling condition and the balance of groundwater in these provinces is negative. In Nimroz province those areas fare from recharged areas, in addition to the drawdown of groundwater but also the groundwater quality is also saline. The main reasons for these are more evaporation, lower precipitation, and climate change at the end of three decades.

Acknowledgments

The authors extend warm gratitude to the anonymous reviewers for their valuable comments, as well as to other faculty members who have helped to write and publish this paper.

Conflicts of Interest

The authors declare no conflicts of interest regarding the publication of this paper.

REFERENCES

Abdullah, S. H., & Chmyriov, V. M. (Eds.). (1997). *Map of mineral resources of Afghanistan Kabul,* Ministry of Mines and Industries of the Democratic Republic of Afghanistan, Department of Geological and Mineral Survey. V/O Technoexport USSR, scale 1:500,000.

Arian, H., Kayastha, R. B., Bikas, C. B., Ahuti, S., Rasouli, H., & Armstrong, R. (2015). Application of the Snowmelt Runoff Model in the Salang River Basin, Afghanistan Using MODIS Satellite Data. *Journal of Hydrology and Meteorology, 9*(1), SOHAM, Nepal. . doi:10.3126/jhm.v9i1.15586

Avouac, J.P. & Burov, E. (1996). Erosion as a driving mechanism of intracontinental mountain growth. *J. Geophys. Res., 101*(17).

Banks, D., & Soldal, O. (2002). Towards a policy for sustainable use of groundwater by non-governmental organizations in Afghanistan. *Hydrogeology Journal, 10*(3), 377–392. doi:10.100710040-002-0203-y

Böckh, E.G., (1971). *Report on the groundwater resources of the city of Kabul.* Report for Bundesanstalt für Geowissenschaften und Rohstoffe BGR file number 0021016.

Bohannon, R. G. (2005). Geologic map of quadrangle 3468, Chak-e-Wardak (509) and Kabul (510) quadrangles. *Afghan Open-File Report*, (509/510), 2005–1001.

Bohannon, R. G., & Turner, K. J. (2007). *Geologic map of quadrangle 3468, Chak Wardak- Syahgerd (509) and Kabul (510) quadrangles*. Open- File Report 2005–1107–A. 1 sheet. Afghanistan: U.S. Geological Survey.

Broshears, R. E., Akbari, M. A., Chornack, M. P., Mueller, D. K., & Ruddy, B. C. (2005). Inventory of ground-water resources in the Kabul Basin, Afghanistan. *U.S. Geological Survey Scientific Investigations Report*, 2005–5090. doi:10.3133ir20055090

Hamdard, M. H., Soliev, I., Rasouli, H., Kløve, B., & Belhassan, K. (2022). Groundwater Quality Assessment in Chak Karstic Sedimentary Basin, Wardak Province, Afghanistan. *Central Asian Journal of Water Research*, *8*(2), 110–127. doi:10.29258/CAJWR/2022-R1.v8-2/110-127.eng

Koons, P. (1989). The topographic evolution of collisional mountain belts: A numerical look at the Southern Alps, New Zealand. *American Journal of Science*, *289*(9), 1041–1069. doi:10.2475/ajs.289.9.1041

Lave, J. & Avouac, J.P. (2001). Fluvial incision and tectonic uplift across the Himalayas of central Nepal. *Journal Of Geophysical Research, 106*(26).

Malgary, F. (1987). Mineralogy book. Geosciences faculty Kabul University, Kabul Afghanistan.

Molnar, P. A. (1990). Review of the seismicity and rates of active under thrusting deformation at the Himalayas. *J. Himalayan Geol.*, *I*, 131–154.

Montgomery, D. R. (1994). Valley incision and the uplift of mountain peaks. *J. Geophys. Res., 99*(13), 913-13, 921.

Munsell Color Co. (2023). *Soil color charts*. Munsell Color Company, Inc., USA.

Nakata, T., (1972). Geomorphic history and crustal movements of the foothills of the Himalayas, Sci. Rep. Tohoku Univ. Ser.

Paetzold, S. (2005). *Arbeitung zur Durchfuerung von Boden Analyses* (10th ed.). Institut fuer Bodenkunde, University Bonn.

Rasouli, H. (2019). A Study on Some River Sediments, Hydrology and Geological Characteristics in Chak Sedimentary Basin, Wardak, Afghanistan. *International Journal of Geology, Earth & Environmental Sciences*, *9*(2).

Rasouli, H. (2020b). *Application of soil physical and chemical parameters and its Comparison in Kabul Sedimentary basins*. doi:10.24327/ijrsr.2020.1102.5095

Rasouli, H. (2020c). Well, Design and Stratigraphy of Sheer Khana Deep Well in Chak District, Wardak, Afghanistan. *International Journal of Geology, Earth & Environmental Sciences, 10*(2) May-August, pp.54-68/Rasouli

Rasouli H., (2021b). Analysis of Groundwater Quality in Jabal Sarage and Charikar Districts, Parwan, Afghanistan. *Journal of Geological Research, 3*(04). doi:10.30564/jgr.v3i4.3717

Rasouli, H. (2022a). Climate Change Impacts on Water Resource and Air Pollution in Kabul Sub-basins, Afghanistan. *Advances in Geological and Geotechnical Engineering Research, 04*(1). https://ojs.bilpublishing.com/index.php/agger

Rasouli, H. (2022b). Methods of Well Construction Complication, Design, and Development for Sixteen Observation and Test Wells at the Eight Locations of Zarang district, Nimroz, Afghanistan. *International Journal of Earth Science Knowledge and Applications*, *4*(3), 426–448.

Rasouli, H., Kayastha, R. B., Bikas C. B., Ahuti S., Arian, H., Armstrong, R., (2015). Estimation of Discharge from Upper Kabul River Basin, Afghanistan Using the Snowmelt Runoff Model. *Journal of Hydrology and Meteorology, 9*(1). . doi:10.3126/jhm.v9i1.15584

Rasouli, H., Qureshi, R., & Belhassan, K. (2021c). Investigations on River Sediments in Chak Sedimentary Basin, Wardak Province. *Afghanistan. Journal of Geological Research, 3*(4). . doi:10.30564/jgr.v3i4.3574

Rasouli, H., & Safi, A. G. (2021a). Geological, Soil and Sediment Studies in Chelsaton Sedimentary Basin, Kabul, Afghanistan. *International Journal of Geosciences*, *12*(2), 170–193. doi:10.4236/ijg.2021.122011

Rasouli, H., Sarwari, M. H., Khairuddin, R., & Said, A. H. (2020a). Geological Study of Tangi Mahipar Mountain Range along Kabul-Jalalabad Road. *Afghanistan*, *10*(10), 971–980. doi:10.4236/ojg.2020.1010044

Ruleman, C. A., Crone, A. J., Machette, M. N., Haller, K. M., & Rukstales, K. S. (2007). *Geological Map and database of probable and possible Quaternary faults in Afghanistan.* U.S Survey Open-File Report 2007–1103. US Survey.

Shamal, S., & Rasouli, H. (2018, May). Comparison between pH, EC, $CaCO_{3,}$ and mechanical analysis of Qala Wahid and Company Areas soil, Kabul, Afghanistan. *International Journal of Scientific Research*, *8*(5). doi:10.21275/ART20197318

Souriau, M. E., Ruddiman, W. F., & Froelich, P. N. (1988). Influence of late Cenozoic Mountain building on ocean geochemical cycles. *Geology*, *16*, 649–653.

Summerfield, M. A., & Hulton, N. J. (1994). Hulton, Natural controls of fluvial denudation rates in Major world drainage basins. *Journal of Geophysical Research*, *99*(B7), 13,871–13,883. doi:10.1029/94JB00715

Torge, T., Georg, H., & Thomas, H., (2003). *Hydrogeology of the Kabul Basin Part I: Geology, aquifer characteristics, climate, and hydrography.* Foreign office of the Federal Republic of Germany.

Tünnemeier. T. (2005). Hydrogeology of Kabul Basin Part 1. Geology, Aquifer characteristics, climate, and hydrography (BGR). Afghanistan.

Vaseashta, A. (2021a). Introduction to Water Safety, Security, and Sustainability. In A. Vaseashta & C. Maftei (Eds.), *Water Safety, Security and Sustainability. Advanced Sciences and Technologies for Security Applications*. Springer. doi:10.1007/978-3-030-76008-3_1

Vaseashta, A., Duca, G., & Travin, S. (Eds.). (2022). *Handbook of Research on Water Sciences and Society* (Vols. 1–2). IGI Global. doi:10.4018/978-1-7998-7356-3

Vaseashta, A., Gevorgyan, G., Kavaz, D., Ivanov, O., Jawaid, M., & Vasović, D. (2021b). Exposome, Biomonitoring, Assessment, and Data Analytics to Quantify Universal Water Quality. In A. Vaseashta & C. Maftei (Eds.), *Water Safety, Security and Sustainability. Advanced Sciences and Technologies for Security Applications*. Springer. doi:10.1007/978-3-030-76008-3_4

Wheeler, R. L., Bufe, C. G., Johnson, M. L., & Dart, R. L. (2005). Seismotectonic map of Afghanistan, with annotated bibliography. *U.S. Geological Survey Open-File Report*, 2005–1264. doi:10.29258/CAJWR/2022-R1.v8-2/110-127.eng

Chapter 13
The 2013 North Indian Floods:
A Case Study

Zahid Nabi
https://orcid.org/0000-0001-6724-6345
Central University of Jammu, India

Shivali Kundan
Central University of Jammu, India

Neha Verma
Central University of Jammu, India

Dinesh Kumar
Central University of Jammu, India

ABSTRACT

The 2013 North Indian floods were a devastating hydrometeorological event that affected several states in northern India, primarily Uttarakhand and Himachal Pradesh. The floods were triggered by heavy rainfall combined with cloud bursts, cloudburst-induced landslides, and the melting of glaciers in the region. The event resulted in extensive flooding, landslides, and loss of life and infrastructure. This book chapter aims to explore the causes, impacts, and aftermath of the 2013 North Indian floods, highlighting both the devastation brought upon by the disaster and the remarkable resilience displayed by the affected communities. It emphasizes the need for improved disaster management strategies, sustainable development practices, and community resilience to minimize the impact of future floods and protect vulnerable regions.

INTRODUCTION

Flood hazard is a serious threat to at least 20 million people worldwide, which claims around 20,000 lives every year (Kellens et al., 2013; Yin and Li, 2001). Flash floods stand apart from many other forms of flooding due to their swift onset, transpiring within a timeframe of 2 to 6 hours following intense

DOI: 10.4018/978-1-6684-8771-6.ch013

rainfall, the breach of a dam, failure of a levee, rapid snowmelt, or ice-related events. Flash floods are one of the most destructive types of natural hazards. The main threat arises from the speed with which they develop and the depth and power of the resulting flows (Bandyopadhyay and Gyawali, 1994; Sene and Sene, 2016). The year 2013 marked a tragic chapter in the history of North India as the region was ravaged by one of the deadliest floods it had ever witnessed. Starting in June 2013, heavy monsoon rains caused widespread flooding and landslides in various states, including Uttarakhand, Himachal Pradesh, and parts of Uttar Pradesh, Rajasthan, and Haryana (Joseph et al., 2015; Sati and Gahalaut, 2013). The floods in the parts of the north-west Himalayan region caused severe damage in the Uttarakhand state of India and some parts of western Nepal (Sati and Gahalaut, 2013).

Between the 13th and 17th of June 2013, a series of torrential downpours wreaked havoc on several states in northern India, causing catastrophic flooding in Uttarakhand as shown in Figures 1.1 and 1.2, and giving rise to one of the most devastating floods in recorded history (Dobhal et al. 2013). The deluge triggered massive landslides, resulting in an alarming death toll exceeding 5,000 lives lost and inflicting significant damage to properties. Apart from the severe impact on Uttarakhand, this calamitous event also reverberated across other regions of India, including Himachal Pradesh, Haryana, Delhi, and Uttar Pradesh, as well as affecting western Nepal and parts of Tibet (Cho et al., 2016; Dubey et al., 2014). Heavy rainfall in the early summer monsoon causes rapid snow melting, sediment transport, and debris flow, thus impacting soil profile (Rasouli et al, 2023) and water quality (Vaseashta et al, 2021). This process triggers lake outbursts, landslides, and river erosion in downstream areas. The sudden discharge of water quickly spreads over the river valleys and surrounding regions, therefore causing devastating floods (Al Fatah,2019)

Figure 1. Before and after flood picture of Kedernath Uttrakhand
(Numerical identifiers delineate distinct regions in the following manner: "1" marks the presence of a solitary watercourse originating from a glacier, "2" indicates a noticeable accumulation of debris obstructing the water body's path, "3" represents the bifurcation of water into two separate streams, and "4" illustrates the water's exclusive progression through narrow channels.) (**Source**: *http://www.downtoearth.org.in/news/floods-in-uttarakhand-explained-41451)*

ORIGIN OF FLOOD

Summer is a monsoon season for most of India. By the 16th of June in 2013, the Indian summer monsoon had enveloped the entire country, arriving a month earlier than its usual schedule. During this time, the northern Indian state of Uttarakhand, located on the southern slopes of the Himalayan Ranges, encountered intense rainfall. This torrential downpour, coupled with frequent landslides, inflicted severe destruction resulting in a significant loss of lives.

This calamity of floods was set in motion by a combination of factors, including a well-marked cyclonic circulation that developed over the Bay of Bengal. This system moved westwards and rapidly gained strength due to the moisture it received from both the Bay of Bengal and the Arabian Sea. Adding to the intensity were powerful western disturbances from the north (Joseph et al., 2015).

As a result of this convergence of weather phenomena, Uttarakhand and its neighboring areas experienced relentless heavy rainfall, surpassing 375% of the benchmark rainfall during a typical monsoon season. The continuous downpour, combined with the melting of the Chorabari Glacier at an altitude of 3,800 meters, caused the Mandakini River to overflow. The ensuing floods wreaked havoc near Gobindghat, Kedar Dome, Rudraprayag district, and extended to Himachal Pradesh and Western Nepal. The impact was also felt in other nearby regions, including Delhi, Haryana, Uttar Pradesh, and parts of Tibet (Singh et al., 2014).

As shown in satellite imagery in Figure 2.1 the affected areas, characterized by their rugged terrain, dense forests, and snow-capped mountains, presented significant challenges in terms of accessibility.

Figure 2. A comparison of satellite images taken before (on May 28, 2012) and after (on June 21, 2013) (Source: (P. K. Champati Ray et.al 2015)

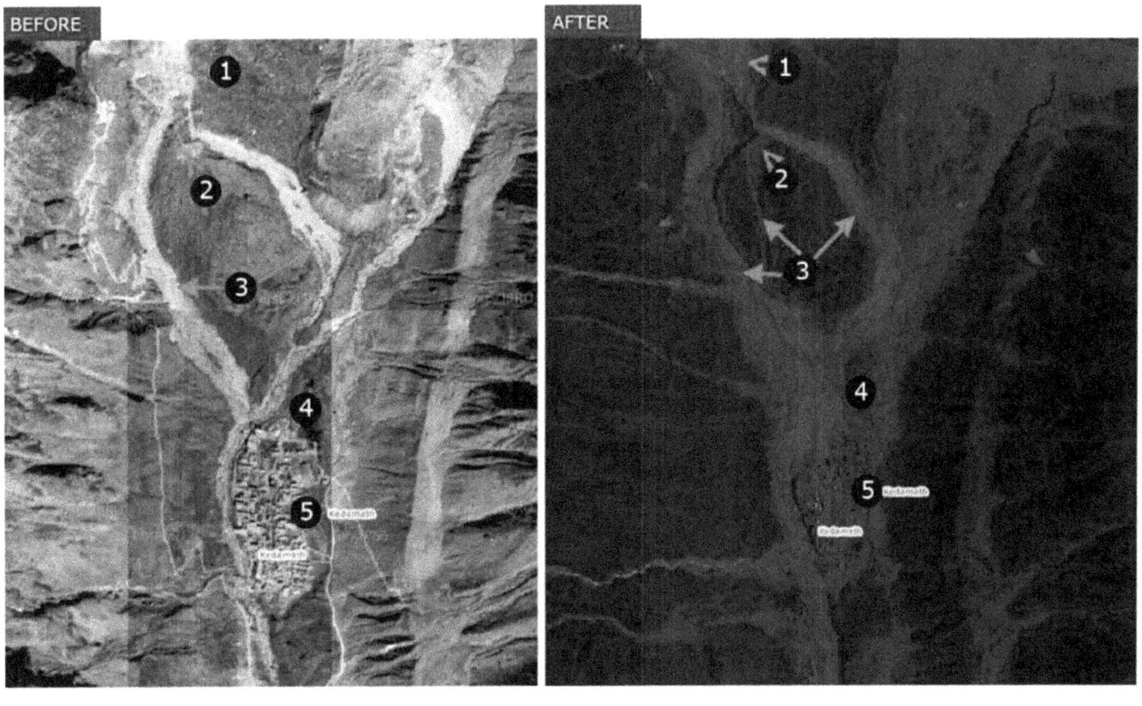

These regions are not only renowned for their natural beauty but also host numerous important Hindu and Sikh pilgrimage sites, as well as popular tourist spots and trekking trails.

Unfortunately, prior warnings issued by the India Meteorological Department regarding the heavy rainfall were not adequately disseminated, leading to many people being caught off guard. This lack of awareness resulted in substantial loss of life and widespread property damage.

POSSIBLE REASONS THAT TRIGGERED THE FLOODS

This section lists several reasons that resulted in the aforesaid floods.

Intense Monsoon Rains

Few aspects of climate variability carry as much potential to significantly affect society as alterations in monsoon precipitation, given that monsoons serve as the lifeline for approximately two-thirds of the global population (Wang et al., 2012). The 2013 North Indian floods were primarily driven by exceptionally heavy and prolonged monsoon rainfall. The monsoon season typically brings substantial precipitation to the region, but in 2013, the intensity and duration of the rainfall were unprecedented. The heavy downpours resulted in excessive water runoff and overwhelmed the river systems, leading to widespread flooding. The continuous and intense rainfall over several days saturated the soil, reducing its capacity to absorb water, and further exacerbating the flooding (Dube et al., 2014). Widespread heavy rainfall occurred over Uttarakhand during 16–18 June 2013 and some stations in Uttarakhand reported extremely heavy rainfall (exceeding 24.5 cm) on 17 and 18 June.

Topography and Drainage System

The topographical characteristics of the affected regions, particularly Uttarakhand, played a significant role in amplifying the impact of the floods. The region is characterized by steep slopes, narrow valleys, and a complex network of rivers and streams. These geographical features accelerate the flow of water, increasing the runoff and potential for flash floods. The limited space for water to spread out horizontally intensifies the force and depth of the floodwaters. Additionally, the intricate river systems, with numerous tributaries and confluences, contribute to the rapid accumulation and propagation of floodwater downstream. Water-induced disasters like flash floods and debris flows rank among the primary natural calamities in the Himalayas, contributing to over 70% of total economic losses and accounting for more than half of all casualties (Sati, 2013.

Glacial Lake Outburst Floods (GLOFs)

The Himalayan region, including Uttarakhand, is home to numerous glaciers and glacial lakes. As global temperatures rise, these glaciers are melting at an accelerated rate. Consequently, glacial lakes are forming as a result of the accumulation of meltwater. If the natural dams that contain these glacial lakes become weak or breached, a glacial lake outburst flood (GLOF) can occur. The sudden release of massive volumes of water from these glacial lakes significantly contributes to

Figure 3. The Indian Meteorological Department image (17th June 2013) showed the heavy rainfall on the higher 53 Uttarakhand, Himachal, and Nepal Himalaya

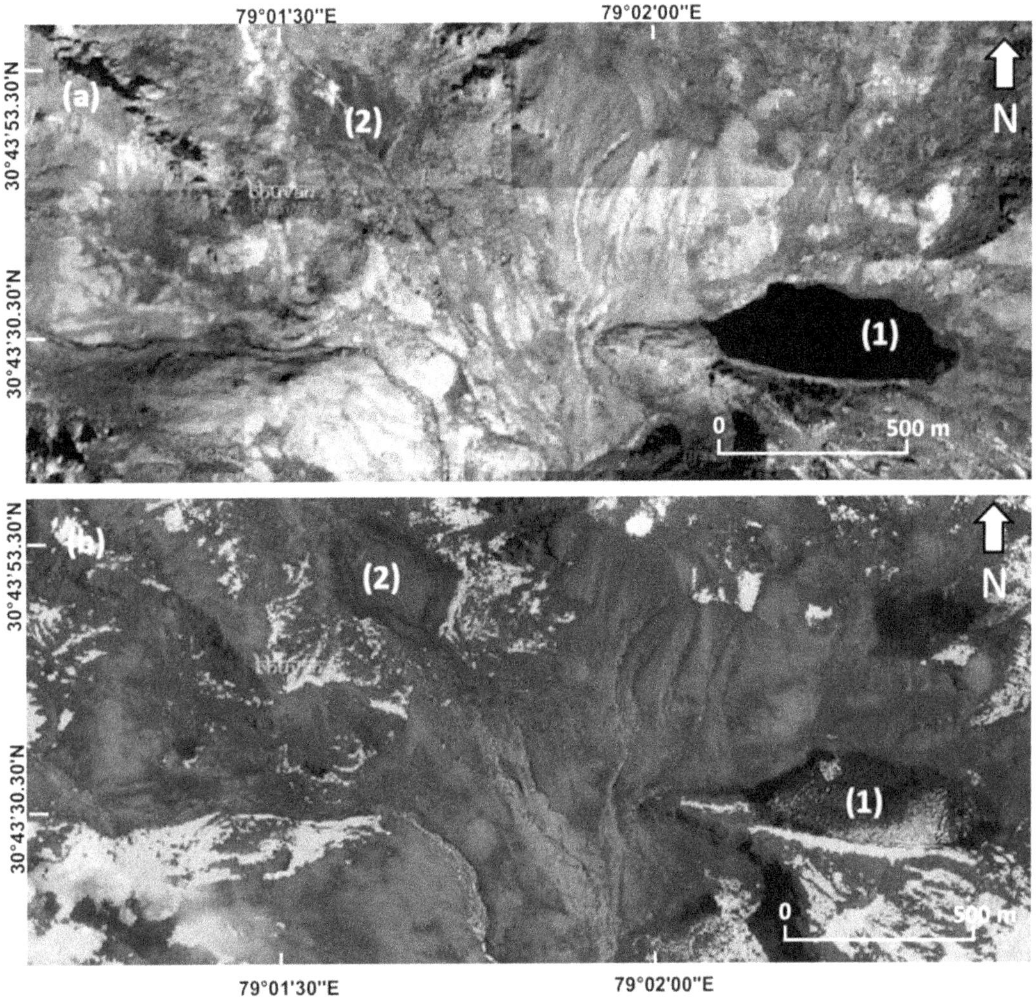

the floodwaters downstream. The 2013 North Indian floods experienced some GLOFs that further intensified the floods (Schwanghart et al., 2016).

Deforestation and Environmental Degradation

The loss of forest cover and environmental degradation in the affected regions have played a crucial role in exacerbating the floods. Deforestation, often driven by human activities such as logging, agriculture expansion, and infrastructure development, reduces the natural absorption capacity of the soil. Trees and vegetation help regulate the water cycle by intercepting rainfall, reducing surface runoff, and enhancing infiltration. With deforestation, the land becomes more prone to erosion, and water is unable to permeate the soil adequately, leading to increased runoff and flash flooding (Hasnat et al., 2018).

Land Use Changes and Urbanization

Unplanned land use changes and rapid urbanization have contributed to the severity of the floods. The conversion of natural areas, including forests, wetlands, and floodplains, into urban and industrial zones reduces the available space for water absorption and increases surface runoff. The replacement of permeable surfaces with impermeable ones, such as concrete and asphalt, further limits water infiltration and enhances the speed at which water reaches the river systems. As a result, floodwaters accumulate more rapidly and intensify downstream (Avashia et al., 2020).

Inadequate Water Management and Dam Operations

Improper water management practices, including mismanagement of reservoirs, dams, and barrages, can contribute to flooding. During periods of heavy rainfall, dam operators face the challenge of balancing water storage to prevent dam failure due to excessive inflow. Poor water release strategies, including sudden and excessive discharge downstream, can worsen flood situations. In some cases, inadequate maintenance and outdated infrastructure of dams and reservoirs can lead to breaches or failures, amplifying the floodwaters downstream.

Climate Change and Variability

Climate change and variability play a role in altering rainfall patterns and increasing the frequency and intensity of extreme weather events. Rising global temperatures lead to increased atmospheric moisture, which can result in more intense rainfall during monsoon seasons. Multiple research studies indicate that the warming experienced in the Himalayas has significantly exceeded the global average temperature increase of 0.74°C over the past century (Du et al., 2004; Solomon et al., 2007). The changing climate conditions might have contributed to the heavy and prolonged rainfall experienced during the 2013 North Indian floods (Cho et al., 2016).

Lack of Early Warning Systems and Preparedness

Inadequate early warning systems and limited preparedness measures further exacerbated the impacts of the floods. Timely and accurate forecasting of extreme rainfall events is essential for issuing warnings and initiating evacuation procedures. However, the existing early warning systems in the region had limitations in terms of accuracy and coverage. Insufficient dissemination of timely information to the vulnerable population hindered their ability to take preventive measures. Moreover, inadequate infrastructure, including evacuation routes and shelters, posed challenges in effectively responding to the floods.

It's important to note that the causes and triggers of floods are often interconnected and can vary across different regions. Multiple factors can act in combination, amplifying the impact of the floods. The 2013 North Indian floods were likely influenced by a complex interplay of these factors, and their contributions may vary depending on the specific geographical locations affected.

In summary, the 2013 North Indian floods were primarily triggered by intense and prolonged monsoon rainfall, topographical vulnerability, glacial lake outbursts, environmental degradation, and a lack of effective early warning systems. These factors combined to create a catastrophic flood situation, causing immense loss of life, displacement of people, and widespread destruction of infrastructure. The event served as a wake-up call to improve disaster preparedness, sustainable land management, and climate change adaptation measures in the affected regions.

RESPONSE OF LOCAL COMMUNITIES AND ITS IMPLICATIONS POST-DISASTER

(a) Local Communities and Volunteers: Local communities played a crucial role in responding to the floods and supporting those affected. In the immediate aftermath of the disaster, local residents displayed extraordinary acts of bravery and compassion. Many individuals volunteered their time and resources to assist in search and rescue operations, providing much-needed help to those stranded or in distress. They formed local committees and set up community kitchens to provide cooked meals to the affected population. The spirit of solidarity and unity was evident as neighbors helped each other evacuate, shared available resources, and offered emotional support to those in need. Their resilience and willingness to lend a helping hand demonstrated the strength and determination of the local communities.

(b) Humanitarian Aid and NGOs: Numerous national and international humanitarian organizations and NGOs stepped in to provide critical assistance to the flood-affected regions. These organizations mobilized resources, relief supplies, and personnel to deliver emergency aid and support the affected communities. They focused on providing immediate relief, including food, clean drinking water, medical aid, and hygiene kits. NGOs set up temporary shelters and distributed essential items such as blankets, clothing, and basic household supplies. These organizations worked in coordination with local authorities and community leaders to identify the most vulnerable populations and ensure that aid reached those who needed it the most.

(c) Government Response and Relief Efforts: The government responded swiftly to the disaster, deploying emergency response teams, including the armed forces, national disaster response forces, and local administration. These teams were involved in search and rescue operations, medical assistance, and providing immediate relief to the affected areas. The government set up relief camps and evacuation centers to provide temporary shelters and basic amenities to displaced individuals. Medical teams were dispatched to treat the injured and prevent the outbreak of waterborne diseases. The government also coordinated efforts to restore essential services, including electricity, water supply, and communication networks. Local authorities worked with government agencies to assess the damage, distribute relief materials, and facilitate the rehabilitation process.

(d) Rehabilitation and Reconstruction: Following the initial response phase, the focus shifted to the rehabilitation and reconstruction of the flood-affected areas. The government, along with NGOs and community-based organizations, embarked on long-term recovery plans. Rehabilitation efforts involved rebuilding damaged infrastructure, including roads, bridges, schools, hospitals, and public facilities. Special attention was given to rebuilding homes for the displaced population. The government provided financial assistance, grants, and loans to affected individuals and families to support them in rebuilding their lives. Livelihood support programs, such as skill training, employment generation, and small business initiatives, were implemented to help communities recover their sources of income. Reconstructing damaged public services, such as healthcare centers and educational institutions, was prioritized to ensure the resumption of essential services and restore a sense of normalcy in the affected areas.

(e) Community Resilience and Solidarity: The floods brought communities together, showcasing their resilience and solidarity in the face of adversity. Affected communities organized themselves to initiate local initiatives for recovery and rebuilding. They formed self-help groups, community-based organizations, and committees to coordinate relief efforts, share resources, and address the specific needs of vulnerable groups. Community-driven cleaning campaigns were conducted to clear debris, restore water sources, and reduce the risk of waterborne diseases. Efforts were made to develop local early warning systems, enhancing community preparedness for future disasters. These community-led initiatives demonstrated the strength, resilience, and determination of the people to rebuild their lives and protect their communities from future risks.

(f) Learning and Preparedness: The 2013 North Indian floods served as a significant learning experience for all stakeholders involved. Authorities, policymakers, and communities recognized the importance of enhanced disaster preparedness and response mechanisms. Lessons learned from the floods led to discussions and actions to strengthen early warning systems, improve forecasting capabilities, and ensure timely dissemination of information to vulnerable communities. The government implemented measures to enhance the resilience of critical infrastructure, including dams, embankments, and bridges. Emphasis was placed on incorporating climate change adaptation strategies into disaster management plans. Awareness campaigns and training programs were conducted to educate communities about disaster preparedness, evacuation procedures, and the importance of maintaining an emergency kit. The experience of the 2013 floods prompted individuals, communities, and institutions to be better prepared and more resilient in the face of future disasters.

The response of people following the 2013 North Indian floods demonstrated the power of collective action, resilience, and the human spirit in overcoming adversity. The collaboration between local communities, NGOs, and government agencies was crucial in providing immediate relief, facilitating recovery, and building back stronger. The experiences and lessons from this disaster continue to inform and shape disaster management strategies in the region.

POLICY RECOMMENDATIONS

In the aftermath of the Uttarakhand floods in 2013, various policy recommendations and measures were proposed to mitigate the negative effects of such disasters in the future. Some of these recommendations focused on improving disaster preparedness, response capabilities, and sustainable development practices. Here are a few key areas where policy changes and recommendations were suggested:

Early Warning Systems

Strengthening and expanding early warning systems was identified as a crucial step. This involves the use of technology to monitor weather patterns, river levels, and other indicators to provide timely warnings to communities at risk. Strengthen communication networks in remote and vulnerable areas to ensure that timely warnings reach communities at risk. Developing public awareness campaigns to educate local residents about the importance of heeding early warnings and understanding evacuation procedures.

Land Use Planning and Zoning

There were calls for improved land use planning and zoning regulations, particularly in ecologically sensitive areas. Restricting or regulating construction in vulnerable zones, such as riverbanks and steep slopes, was suggested to minimize the impact of future disasters. Also, it was proposed to conduct comprehensive land use assessments to identify areas prone to landslides, floods, or other natural disasters, and incorporate this information into urban and regional planning.

Infrastructure Resilience

Enhancing the resilience of critical infrastructure, including roads, bridges, and buildings, was emphasized. This includes using engineering practices that take into account the region's susceptibility to landslides and floods.

Community Awareness and Training

Initiatives to raise awareness among local communities about the risks of living in vulnerable areas and providing training on evacuation procedures and disaster response were recommended. Community engagement is crucial for building resilience at the grassroots level.

Environmental Conservation

Recognizing the importance of environmental conservation and sustainable practices in the fragile Himalayan ecosystem, there were suggestions to incorporate environmental considerations into development projects. This includes measures to protect forests, regulate tourism, and address climate change.

Coordination and Governance

Improving coordination among various government agencies, disaster response teams, and NGOs was highlighted. Effective governance and coordination are essential for a swift and efficient response during and after a disaster.

Capacity Building

Building the capacity of local governments, emergency responders, and community organizations to handle disasters effectively was a key recommendation. This involves providing training, resources, and equipment to enhance the overall response capabilities.

CONCLUSION

In conclusion, the 2013 North Indian floods stand as a tragic chapter in the region's history, marked by devastating loss of life, displacement, and extensive damage to infrastructure. The floods were triggered by a combination of factors, including intense monsoon rains, topographical vulnerability, glacial lake outburst floods, environmental degradation, and inadequate early warning systems. However, amidst the devastation, the response of local communities, humanitarian organizations, and the government showcased the power of collective action, resilience, and solidarity. Local communities and volunteers displayed remarkable bravery and compassion, providing immediate assistance and support to those in need. Humanitarian aid organizations and NGOs mobilized resources and relief supplies, working hand-in-hand with local authorities to deliver emergency aid. The government responded swiftly, deploying emergency teams, setting up relief camps, and initiating rehabilitation efforts. The focus shifted towards rehabilitation and reconstruction, with efforts aimed at rebuilding infrastructure, restoring essential services, and supporting the affected population in rebuilding their lives.

The 2013 North Indian floods served as a wake-up call, leading to valuable lessons learned in disaster preparedness, early warning systems, and sustainable land management. The experience prompted individuals, communities, and institutions to prioritize resilience and preparedness for future disasters. The response to the floods demonstrated the strength of community resilience, solidarity, and the spirit of overcoming adversity.

Moving forward, it is crucial to continue building upon the lessons learned, strengthening early warning systems, enhancing infrastructure resilience, and promoting sustainable land management practices. By fostering collaboration between communities, NGOs, and government agencies, North India can better prepare for future disasters, mitigate their impact, and protect the lives and livelihoods of its people.

REFERENCES

Al Fatah, J. (2019). *Flash Floods of 2013 in Northern India: main reasons and potential prevention of damage*.

Avashia, V., & Garg, A. (2020). Implications of land use transitions and climate change on local flooding in urban areas: An assessment of 42 Indian cities. *Land Use Policy*, *95*, 104571. doi:10.1016/j.landusepol.2020.104571

Bandyopadhyay, J., & Gyawali, D. (1994). Himalayan water resources: Ecological and political aspects of management. *Mountain Research and Development*, *14*(1), 1–24. doi:10.2307/3673735

Cho, C., Li, R., Wang, S. Y., Yoon, J. H., & Gillies, R. R. (2016). The anthropogenic footprint of climate change in the June 2013 northern India flood. *Climate Dynamics*, *46*(3-4), 797–805. doi:10.100700382-015-2613-2

Dobhal, D. P., Gupta, A. K., Manish, M., & Khandelwal, D. D. (2013). Kedarnath disaster: Facts and plausible causes. *Current Science*, *105*(2), 171–174.

Du, M., Kawashima, S., Yonemura, S., Zhang, X., & Chen, S. (2004). Mutual influence between human activities and climate change in the Tibetan Plateau during recent years. *Global and Planetary Change*, *41*(3-4), 241–249. doi:10.1016/j.gloplacha.2004.01.010

Dube, A., Ashrit, R., Ashish, A., Sharma, K., Iyengar, G. R., Rajagopal, E. N., & Basu, S. (2014). Forecasting the heavy rainfall during Himalayan flooding—June 2013. *Weather and Climate Extremes*, *4*, 22–34. doi:10.1016/j.wace.2014.03.004

Guri, P. K., Champati Ray, P. K., & Patel, R. C. (2015). Spatial prediction of landslide susceptibility in parts of Garhwal Himalaya, India, using the weight of evidence modeling. *Environmental Monitoring and Assessment*, *187*(6), 1–25. doi:10.100710661-015-4535-1 PMID:25944750

Hasnat, G. T., Kabir, M. A., & Hossain, M. A. (2018). Major environmental issues and problems in South Asia, particularly Bangladesh. Handbook of environmental materials management, 1-40.

Joseph, S., Sahai, A. K., Sharmila, S., Abhilash, S., Borah, N., Chattopadhyay, R., Pillai, P. A., Rajeevan, M., & Kumar, A. (2015). North Indian heavy rainfall event during June 2013: Diagnostics and extended range prediction. *Climate Dynamics*, *44*(7-8), 2049–2065. doi:10.100700382-014-2291-5

Joseph, S., Sahai, A. K., Sharmila, S., Abhilash, S., Borah, N., Chattopadhyay, R., Pillai, P. A., Rajeevan, M., & Kumar, A. (2015). North Indian heavy rainfall event during June 2013: Diagnostics and extended range prediction. *Climate Dynamics*, *44*(7-8), 2049–2065. doi:10.100700382-014-2291-5

Kellens, S., Terpstra, T., & De Maeyer, P. (2013). Perception and communication of flood risks: A systematic review of empirical research. *Risk Analysis*, *33*(1), 24–49. doi:10.1111/j.1539-6924.2012.01844.x PMID:22651128

Kotal, S. D., Roy, S. S., & Bhowmik, S. R. (2014). Catastrophic heavy rainfall episode over Uttarakhand during 16–18 June 2013–observational aspects. *Current Science*, 234–245.

Rasouli, H., & Vaseashta, A. (2023). Investigation of Physicochemical Properties of Qalay Abdul Ali Soil, Kabul, Afghanistan. *Advances in Geological and Geotechnical Engineering Research*, *5*(3), 55–68. doi:10.30564/agger.v5i3.5773

Sati, S. P., & Gahalaut, V. K. (2013). The fury of the floods in the north-west Himalayan region: The Kedarnath tragedy. *Geomatics, Natural Hazards & Risk*, *4*(3), 193–201. doi:10.1080/19475705.2013.827135

Sati, V. P. (2013). Extreme weather-related disasters: A case study of two flash floods hit areas of Badrinath and Kedarnath Valleys, Uttarakhand Himalaya, India. *Journal of Earth Science and Engineering*, *3*(8), 562.

Schwanghart, W., Worni, R., Huggel, C., Stoffel, M., & Korup, O. (2016). Uncertainty in the Himalayan energy–water nexus: Estimating regional exposure to glacial lake outburst floods. *Environmental Research Letters*, *11*(7), 074005. doi:10.1088/1748-9326/11/7/074005

Sene, K., & Sene, K. (2016). Flash floods. *Hydrometeorology: forecasting and applications*, 273-312.

Singh, D., Horton, D. E., Tsiang, M., Haugen, M., Ashfaq, M., Mei, R., & Diffenbaugh, N. S. (2014). Severe precipitation in Northern India in June 2013: Causes, historical context, and changes in probability. *Bulletin of the American Meteorological Society*, *95*(9), S58–S61.

Solomon, S., Qin, D., Manning, M., Chen, Z., Marquis, M., & Averyt, K. B. (Eds.). (2007). *Contribution of Working Group I to the Fourth Assessment Report of the Intergovernmental Panel on Climate Change*. Cambridge University Press.

Vaseashta, A., Gevorgyan, G., Kavaz, D., Ivanov, O., Jawaid, M., & Vasović, D. (2021). Exposome, Biomonitoring, Assessment and Data Analytics to Quantify Universal Water Quality. In A. Vaseashta & C. Maftei (Eds.), *Water Safety, Security and Sustainability. Advanced Sciences and Technologies for Security Applications*. Springer. doi:10.1007/978-3-030-76008-3_4

Wang, B., Liu, J., Kim, H. J., Webster, P. J., & Yim, S. Y. (2012). Recent change of the global monsoon precipitation (1979–2008). *Climate Dynamics*, *39*(5), 1123–1135. doi:10.100700382-011-1266-z

Yin, H., & Li, C. (2001). Human impact on floods and flood disasters on the Yangtze River. *Geomorphology*, *41*(2-3), 105–109. doi:10.1016/S0169-555X(01)00108-8

Compilation of References

Mesinger, F., & Arakawa, A. (1976). *Numerical methods used in atmospheric models.*

Miralles, D. G., Holmes, T. R. H., De Jeu, R. A. M., Gash, J. H., Meesters, A. G. C. A., & Dolman, A. J. A. J. (2014). Mega-heatwave temperatures due to combined soil desiccation and atmospheric heat accumulation. *Nature Geoscience, 7*(5), 345–349. doi:10.1038/ngeo2141

Murray, S. A. (2018). The importance of ensemble techniques for operational space weather forecasting. *Space Weather, 16*(7), 777–783. doi:10.1029/2018SW001861

Navalgund, R. R., Jayaraman, V., & Roy, P. S. (2007). Remote sensing applications: An overview. *Current Science*, 1747-1766.

Norton, S., & Suppe, F. (2001). Why atmospheric modeling is good science. *Changing the atmosphere: Expert knowledge and environmental governance*, 67-105.

Pachauri, R. K., Allen, M. R., Barros, V. R., Broome, J., Cramer, W., Christ, R., & van Ypserle, J. P. (2014). *Climate change 2014: synthesis report. Contribution of Working Groups I, II, and III to the Fifth Assessment Report of the Intergovernmental Panel on Climate Change*. IPCC.

Rabier, F. (2005). Overview of global data assimilation developments in numerical weather-prediction centers. *Quarterly Journal of the Royal Meteorological Society: A journal of the atmospheric sciences, applied meteorology and physical oceanography, 131*(613), 3215-3233.

Randall, D. A. (1989). Cloud parameterization for climate modeling: Status and prospects. *Atmospheric Research, 23*(3-4), 345–361. doi:10.1016/0169-8095(89)90025-2

Raymond, C., Matthews, T., & Horton, R. M. (2020). The emergence of heat and humidity is too severe for human tolerance. *Science Advances, 6*(19).

Sherwood, S. C. (2018). How important is humidity in heat stress? *Journal of Geophysical Research. Atmospheres, 123*(21), 11–808. doi:10.1029/2018JD028969

Sood, A., & Smakhtin, V. (2015). Global hydrological models: A review. *Hydrological Sciences Journal, 60*(4), 549–565. doi:10.1080/02626667.2014.950580

Taszarek, M., Allen, J. T., Púčik, T., Hoogewind, K. A., & Brooks, H. E. (2020). Severe Convective Storms across Europe and the United States. Part II: ERA5 Environments Associated with Lightning, Large Hail, Severe Wind, and Tornadoes. *Journal of Climate, 33*(23), 10263–10286. doi:10.1175/JCLI-D-20-0346.1

Walko, R. L., Band, L. E., Baron, J., Kittel, T. G., Lammers, R., Lee, T. J., Ojima, D., Pielke, R. A. Sr, Taylor, C., Tague, C., Tremback, C. J., & Vidale, P. L. (2000). Coupled atmosphere–biophysics–hydrology models for environmental modeling. *Journal of Applied Meteorology*, *39*(6), 931–944. doi:10.1175/1520-0450(2000)039<0931:CABHMF>2.0.CO;2

Xu, F., Chan, T. O., & Luo, M. (2021). Different changes in dry and humid heat waves over China. *International Journal of Climatology*, *41*(2), 1369–1382. doi:10.1002/joc.6815

Yuan, J., Stein, M. L., & Kopp, R. E. (2020). The evolving distribution of relative humidity is conditional upon daily maximum temperature in a warming climate. *Journal of Geophysical Research: Atmospheres, 125*(19), e2019JD032100.

Abbas, A., Ekowati, D., Suhariadi, F., & Fenitra, R.M. (2023). Health implications, leaders societies, and climate change: A global review. *Ecological footprints of climate change: Adaptive approaches and sustainability*, 653-675.

Abbot, J., & Marohasy, J. (2014). Input selection and optimization for monthly rainfall forecasting in Queensland, Australia, using artificial neural networks. *Atmospheric Research*, *2014*(138), 166–178. doi:10.1016/j.atmosres.2013.11.002

Abdullah, S. H., & Chmyriov, V. M. (Eds.). (1997). *Map of mineral resources of Afghanistan Kabul,* Ministry of Mines and Industries of the Democratic Republic of Afghanistan, Department of Geological and Mineral Survey. V/O Technoexport USSR, scale 1:500,000.

Acharki, S., Singh, S. K., do Couto, E. V., Arjdal, Y., & Elbeltagi, A. (2023). Spatio-temporal distribution and prediction of agricultural and meteorological drought in a Mediterranean coastal watershed via GIS and machine learning. *Physics and Chemistry of the Earth Parts A/B/C*, *131*, 103425. doi:10.1016/j.pce.2023.103425

Acosta-Coll, M., Ballester-Merelo, F., Martinez-Peiró, M., & De la Hoz-Franco, E. (2018). Real-time early warning system design for pluvial flash floods—A review. *Sensors (Basel)*, *18*(7), 2255. doi:10.339018072255 PMID:30002358

Afuye, G. A., Kalumba, A. M., & Orimoloye, I. R. (2021). Characterisation of vegetation response to climate change: A review. *Sustainability (Basel)*, *13*(13), 7265. doi:10.3390u13137265

Ahmed, A. A. M., Jui, S. J. J., Chowdhury, M. A. I., Ahmed, O., & Sutradha, A. (2023). The development of a dissolved oxygen forecast model using a hybrid machine learning algorithm with hydro-meteorological variables. *Environmental Science and Pollution Research International*, *30*(3), 7851–7873. doi:10.100711356-022-22601-z PMID:36045185

Al Fatah, J. (2019). *Flash Floods of 2013 in Northern India: main reasons and potential prevention of damage.*

Al Kafy, A., Bakshi, A., Saha, M., Al Faisal, A., Almulhim, A. I., Rahaman, Z. A., & Mohammad, P. (2023). Assessment and prediction of index-based agricultural drought vulnerability using machine learning algorithms. *The Science of the Total Environment*, *867*, 161394. doi:10.1016/j.scitotenv.2023.161394 PMID:36634773

Alam, S., & Masum, F. (2021). Rapid mapping of flood inundation using Google Earth Engine: A case study of Bangladesh. *Journal of Water and Climate Change*, *12*(1), 238–252. doi:10.2166/wcc.2020.190

Alfieri, L., Velasco, D., & Thielen, J. (2011). Flash flood detection through a multi-stage probabilistic warning system for heavy precipitation events. *Advances in Geosciences*, *29*, 69–75. doi:10.5194/adgeo-29-69-2011

Alizadeh, O. (2022). Advances and challenges in climate modeling. *Climatic Change*, *170*(1-2), 18. doi:10.100710584-021-03298-4

Alkaraki, K. F., & Hazaymeh, K. (2023). A comprehensive remote sensing-based Agriculture Drought Condition Indicator (CADCI) using machine learning. *Environmental Challenges*, *11*, 100699. doi:10.1016/j.envc.2023.100699

Alley, W. M. (1984). The Palmer Drought Severity Index: Limitations and Assumptions. *Journal of Climate and Applied Meteorology*, *23*(7), 1100–1109. doi:10.1175/1520-0450(1984)023<1100:TPDSIL>2.0.CO;2

Almamalachy, Y. S., Al-Quraishi, A. M. F., & Moradkhani, H. (2019). Agricultural drought monitoring over Iraq utilizing MODIS products. In Springer water (pp. 253–278). Springer. doi:10.1007/978-3-030-21344-2_11

Alshammari, R. K., Alrwais, O., & Aksoy, M. S. (2022). Machine Learning Applications to Dust Storms: A Meta-Analysis. *Aerosol and Air Quality Research*, *22*(12), 220183. doi:10.4209/aaqr.220183

Alsumaiei, A. A., & Alrashidi, M. S. (2020). Hydrometeorological drought forecasting in hyper-arid climates using nonlinear autoregressive neural networks. *Water (Basel)*, *12*(9), 2611. doi:10.3390/w12092611

Al-Waili, S. M., Abd Latif, Z., & Salleh, S. A. (2023). GIS-Based Decision Support System and Analytical Hierrachical Process for Integrated Flood Management. *International Journal of Intelligent Systems and Applications in Engineering*, *11*(4s), 392–399.

Amani, M., Ghorbanian, A., Ahmadi, S. A., Kakooei, M., Moghimi, A., Mirmazloumi, S. M., Moghaddam, S. H. A., Mahdavi, S., Ghahremanloo, M., Parsian, S., Wu, Q., & Brisco, B. (2020). Google earth engine cloud computing platform for remote sensing big data applications: A comprehensive review. *IEEE Journal of Selected Topics in Applied Earth Observations and Remote Sensing*, *13*, 5326–5350. doi:10.1109/JSTARS.2020.3021052

ANCPI. (n.d.). *Geoportal ANCPI*. ANCPI. https://geoportal.ancpi.ro/portal/home/

Anderegg, L. D., Anderegg, W. R., Abatzoglou, J., Hausladen, A. M., & Berry, J. A. (2013). Drought characteristics' role in widespread aspen forest mortality across Colorado, USA. *Global Change Biology*, *19*(5), 1526–1537. doi:10.1111/gcb.12146 PMID:23504823

Andreadis, K., Clark, E., Wood, A., Hamlet, A., & Lettenmaier, D. (2005). Twentieth-century drought in the conterminous United States. *Journal of Hydrometeorology*, *6*(6), 985–1001. doi:10.1175/JHM450.1

Andreu, A., Kustas, W., Polo, M., Carrara, A., & González-Dugo, M. (2018). Modeling Surface Energy Fluxes over a Dehesa (Oak Savanna) Ecosystem Using a Thermal Based Two-Source Energy Balance Model (TSEB) I. *Remote Sensing (Basel)*, *10*(4), 567. doi:10.3390/rs10040567

Arian, H. Kayastha, R. B., Bikas C. B., Ahuti S., Rasouli, H., Armstrong, R., (2015). Application of the Snowmelt Runoff Model in the Saang River Basin, Afghanistan Using MODIS *Satelite Data. Journal of Hydrology and Meteorology, 9.* / doi:10.3126/jhm.v9i1.15586

Aristizábal, E., Arango Carmona, M. I., Gómez, F. J., López Castro, S. M., De Villeros Severiche, A., & Riaño Quintanilla, A. F. (2020). Hazard analysis of hydrometeorological concatenated processes in the Colombian Andes. In *Advances in Natural Hazards and Hydrological Risks: Meeting the Challenge: Proceedings of the 2nd International Workshop on Natural Hazards (NATHAZ'19),* (pp. 7-10). Springer International Publishing. 10.1007/978-3-030-34397-2_2

Aroca, R., Porcel, R., & Ruiz-Lozano, J. M. (2007). How does arbuscular mycorrhizal symbiosis regulate root hydraulic properties and plasma membrane aquaporins in Phaseolus vulgaris under drought, cold, and salinity stresses? *The New Phytologist*, *173*(4), 808–816. doi:10.1111/j.1469-8137.2006.01961.x PMID:17286829

Associated Programme on Flood Management (Ed.). (2009). Integrated flood management: Concept paper. World Meteorological Organization: Associated Programme on Flood Management. APFM.

Avashia, V., & Garg, A. (2020). Implications of land use transitions and climate change on local flooding in urban areas: An assessment of 42 Indian cities. *Land Use Policy*, *95*, 104571. doi:10.1016/j.landusepol.2020.104571

Avouac, J.P. & Burov, E. (1996). Erosion as a driving mechanism of intracontinental mountain growth. *J. Geophys. Res., 101*(17).

Ayehu, G., Tadesse, T., Gessesse, B., Yigrem, Y. M., & Melesse, A. (2020). Combined Use of Sentinel-1 SAR and Landsat Sensors Products for Residual Soil Moisture Retrieval over Agricultural Fields in the Upper Blue Nile Basin, Ethiopia. *Sensors (Basel)*, *20*(11), 3282. doi:10.339020113282 PMID:32526894

Bacov, A. (1996). *Hydrotechnical arrangements of hillslopes for soil erosion control*. Mirton Publishing.

Bajracharya, S. R., Khanal, N. R., Nepal, P., Rai, S. K., Ghimire, P. K., & Pradhan, N. S. (2021). Community assessment of flood risks and early warning system in Ratu Watershed, Koshi Basin, Nepal. *Sustainability (Basel)*, *13*(6), 3577. doi:10.3390u13063577

Baldwin, J., Noorali, S., & Vaseashta, A. (2019, July). *Wide Spectrum Bio-threats Identification and Classification*. Conference: Chemical, Biological, Radiological, and Nuclear (CBRN) Defense - Modernizing the Future Fight: Accelerate & Adapt, Delaware.

Bandyopadhyay, J., & Gyawali, D. (1994). Himalayan water resources: Ecological and political aspects of management. *Mountain Research and Development*, *14*(1), 1–24. doi:10.2307/3673735

Baniya, B., Tang, Q., Pokhrel, Y., & Xu, X. (2019). Vegetation dynamics and ecosystem service values changes at national and provincial scales in Nepal from 2000 to 2017. *Environmental Development*, *32*, 100464. doi:10.1016/j.envdev.2019.100464

Banks, D., & Soldal, O. (2002). Towards a policy for sustainable use of groundwater by non-governmental organizations in Afghanistan. *Hydrogeology Journal*, *10*(3), 377–392. doi:10.100710040-002-0203-y

Bao, Y., Xiong, T., & Hu, Z. (2014). Multi-step-ahead time series prediction using multiple-output support vector regression. *Neurocomputing*, *2014*(129), 482–493. doi:10.1016/j.neucom.2013.09.010

Bărbulescu, A. (2016). *Studies on Time Series Applications in Environmental Sciences* (*Vol. 103*). Springer International Publishing. doi:10.1007/978-3-319-30436-6

Barbulescu, A., & Maftei, C. (2021). *STATISTICAL APPROACH OF THE BEHAVIOR OF HAMCEARCA RIVER (ROMANIA). 703*.

Bărbulescu, A., Gherghina, C., & Maftei, C. (2010). Evaluation of Hurst exponent for precipitation time series. *LATEST TRENDS on COMPUTERS*, *II*, 590–595.

Bayanudin, A. A., & Jatmiko, R. H. (2016). Orthorectification of Sentinel-1 SAR (Synthetic Aperture Radar) Data in Some Parts of South-eastern Sulawesi Using Sentinel-1 Toolbox. *IOP Conference Series. Earth and Environmental Science*, *47*, 012007. doi:10.1088/1755-1315/47/1/012007

Bayat, M., & Tavakkoli, O. (2020). Application of machine learning in flood forecasting. *Future Technology*, *1*(1), 1–6. doi:10.55670/fpll.futech.1.1.1

Beamish, A., Raynolds, M. K., Epstein, H., Frost, G. V., Macander, M. J., Bergstedt, H., Bartsch, A., Kruse, S., Miles, V., Tanis, C. M., Heim, B., Fuchs, M., Chabrillat, S., Shevtsova, I., Verdonen, M., & Wagner, J. (2020). Recent trends and remaining challenges for optical remote sensing of Arctic tundra vegetation: A review and outlook. *Remote Sensing of Environment*, *246*, 111872. doi:10.1016/j.rse.2020.111872

Behrouz, M. S., Zhu, Z., Matott, L. S., & Rabideau, A. J. (2020). A new tool for automatic calibration of the Storm Water Management Model (SWMM). *Journal of Hydrology (Amsterdam)*, *581*, 124436. doi:10.1016/j.jhydrol.2019.124436

Beilicci, E.& Beilicci, R. (2015a). Proceedings from: SGEM2015. *Flash floods - causes, effects, possibilities of modeling with hydroinformatic tools.*

Beilicci, E.& Beilicci, R. (2015b). A comparative analysis of two advanced hydroinformatic tools for rainfall–runoff phenomena modeling. *Scientific Bulletin of the Politehnica University of Timişoara, Transactions on Hydrotechnics, 60*(74), 55-58.

Beilicci, E., & Beilicci, R. (2007). The water flow modellation through unpermanent hydrographic network. *Proceedings from: 32nd Internationally Attended Scientific Conference of the Military Technical Academy.* MTA.

Beilicci, E., Beilicci, R., Girbaciu, A., & Girbaciu, C. (2013). Proceedings from: 6th International Conference on Environmental and Geological Science and Engineering. *Rainfall – Runoff modeling using advanced hydroinformatic tools.*

Belhassan, K. (2011a). Hydro-geological Context of Groundwater Mikkes and different variations of Its springs flow (Morocco). Research Journal of Earth Science 3(1), 15 – 26. IDOSI Publications.

Belhassan, K., (2011b). *Relationship flow- Rainfall in the stream Mikkes (Morocco).* IDOSI Publications.

Belhassan, K. (2020b). Relationship between River and Groundwater: Water table Piezometry of the Mikkes Basin (Morocco. *International Journal of Water Resources and Environmental Science, 9* (1).

Benestad, R. (2016). Downscaling climate information. In Oxford research encyclopedia of climate science. doi:10.1093/acrefore/9780190228620.013.27

Benzougagh, B., Meshram, S. G., Fellah, B. E., Masters, M., Dridri, A., Sadkaoui, D., Mimich, K., & Khedher, K. M. (2022). Combined use of Sentinel-2 and Landsat-8 to monitor water surface area and evaluated drought risk severity using Google Earth Engine. *Earth Science Informatics, 15*(2), 929–940. doi:10.100712145-021-00761-9

Benz, S. A., & Blum, P. (2019). Global detection of rainfall-triggered landslide clusters. *Natural Hazards and Earth System Sciences, 19*(7), 1433–1444. doi:10.5194/nhess-19-1433-2019

Bergström, S. (2019). *HBV (Hydrologiska Byråns Vattenbalansavdelning), Model Item, OpenGMS.* GeoModeling. https://geomodeling.njnu.edu.cn/modelItem/5ad110bc-5655-438d-908d-ca8c6452c95f

Bhaga, T. D., Dube, T., Shekede, M. D., & Shoko, C. (2023). Investigating the effectiveness of Landsat-8 OLI and Sentinel-2 MSI satellite data in monitoring the effects of drought on surface water resources in the Western Cape Province, South Africa. *Remote Sensing Applications: Society and Environment, 32*, 101037. doi:10.1016/j.rsase.2023.101037

Bhattacharya, R. K., Chatterjee, N. D., & Das, K. (2020). An integrated GIS approach to analyze the impact of land use change and land cover alteration on groundwater potential level: A study in Kangsabati Basin, India. *Groundwater for Sustainable Development, 11*, 100399. doi:10.1016/j.gsd.2020.100399

Birsan, M. V., Zaharia, L., Chendes, V., & Branescu, E. (2014). Seasonal trends in Romanian streamflow. *Hydrological Processes, 28*(15), 4496–4505. doi:10.1002/hyp.9961

Birsan, M.-V., Micu, D.-M., Nita, A., Mateescu, E., & Szep, R. (2019). Spatio-temporal changes in annual temperature extremes over Romania (1961–2013). *Romanian Journal of Physics, 64*(7-8), 816.

Blumenstock, G. (1942). *Drought in the United States Analyzed using the Theory of Probability* (Technical Bulletin No. 819). USDA.

Böckh, E.G., (1971). *Report on the groundwater resources of the city of Kabul.* Report for Bundesanstalt für Geowissenschaften und Rohstoffe BGR file number 0021016.

Bohannon, R. G., & Turner, K. J. (2007). *Geologic map of quadrangle 3468, Chak Wardak- Syahgerd (509) and Kabul (510) quadrangles.* Open- File Report 2005–1107–A. 1 sheet. Afghanistan: U.S. Geological Survey.

Bohannon, R. G. (2005). Geologic map of quadrangle 3468, Chak-e-Wardak (509) and Kabul (510) quadrangles. *Afghan Open-File Report*, (509/510), 2005–1001.

Borga, M., Gaume, E., Creutin, J. D., & Marchi, L. (2008). Surveying flash flood response: Bournas, A., & Baltas, E. (2022). Investigation of the gridded flash flood Guidance in a Peri-Urban basin in the greater Athens area, Greece. *Journal of Hydrology (Amsterdam)*, *610*, 127820.

Bowman, D. M., Perry, G. L., & Marston, J. (2015). Feedbacks and landscape-level vegetation dynamics. *Trends in Ecology & Evolution*, *30*(5), 255–260. doi:10.1016/j.tree.2015.03.005 PMID:25837918

Boyer. (2002). *Khronostat – HydroSciences Montpellier*. Hydro Sciences. http://www.hydrosciences.org/index.php/2020/09/04/khronostat/

Bradley, B. A., Jacob, R. W., Hermance, J. F., & Mustard, J. F. (2007). A curve fitting procedure to derive inter-annual phenologies from time series of noisy satellite NDVI data. *Remote Sensing of Environment*, *106*(2), 137–145. doi:10.1016/j.rse.2006.08.002

Brooks, H. E., Lee, J. W., & Craven, J. P. (2003). The spatial distribution of severe thunderstorm and tornado environments from global reanalysis data. *Atmospheric Research*, *67–68*, 73–94. doi:10.1016/S0169-8095(03)00045-0

Broshears, R. E., Akbari, M. A., Chornack, M. P., Mueller, D. K., & Ruddy, B. C. (2005). Inventory of groundwater resources in the Kabul Basin, Afghanistan. *U.S. Geological Survey Scientific Investigations Report*, 2005–5090. doi:10.3133ir20055090

Brown, M. E., Escobar, V., Moran, S., Entekhabi, D., O'Neill, P. E., Njoku, E. G., Doorn, B., & Entin, J. K. (2013). NASA's Soil Moisture Active Passive (SMAP) Mission and Opportunities for Applications Users. *Bulletin of the American Meteorological Society*, *94*(8), 1125–1128. doi:10.1175/BAMS-D-11-00049.1

Brunner, G.W. (2016). *HEC-RAS River Analysis System: Hydraulic Reference Manual, Version 5.0.* US Army Corps of Engineers–Hydrologic Engineering Center.

Brunner, M. I., Slater, L., Tallaksen, L. M., & Clark, M. (2021). Challenges in modeling and predicting floods and droughts: A review. *WIREs. Water*, *8*(3), e1520. doi:10.1002/wat2.1520

Buckreuss, S., Schättler, B., Fritz, T., Mittermayer, J., Kahle, R., Maurer, E., Böer, J., Bachmann, M., Mrowka, F., Schwarz, E., Breit, H., & Steinbrecher, U. (2018). Ten Years of TerraSAR-X Operations. *Remote Sensing (Basel)*, *10*(6), 873. doi:10.3390/rs10060873

Buishand, T. A. (1984). Tests for detecting a shift in the mean of hydrological time series. *Journal of Hydrology (Amsterdam)*, *73*(1), 51–69. doi:10.1016/0022-1694(84)90032-5

Burn, D. H. (1997). Regionalisation of catchments for regional flood frequency analysis. *Journal of Hydrologic Engineering*, *1997*(2), 6–82.

Butler, E., Reid, J. K., & Berwick, V. K. (1964). Magnitude and frequency of floods in the United States: Part 10, the Great Basin. In *Water Supply Paper*. U.S. Government Printing Office. (Original work published 1684), doi:10.3133/wsp1684

Calvin, K., Dasgupta, D., Krinner, G., Mukherji, A., Thorne, P. W., Trisos, C., Romero, J., Aldunce, P., Barrett, K., Blanco, G., Cheung, W. W. L., Connors, S., Denton, F., Diongue-Niang, A., Dodman, D., Garschagen, M., Geden, O., Hayward, B., Jones, C., & Péan, C. (2023). IPCC, 2023: Climate Change 2023: Synthesis Report. Intergovernmental Panel on Climate Change (IPCC). doi:10.59327/IPCC/AR6-9789291691647

Candiago, S., Remondino, F., De Giglio, M., Dubbini, M., & Gattelli, M. (2015). Evaluating Multispectral Images and Vegetation Indices for Precision Farming Applications from UAV Images. *Remote Sensing (Basel)*, *7*(4), 4026–4047. doi:10.3390/rs70404026

Cao, Z., Wang, S., Luo, P., Xie, D., Zhu, W. (2022). Watershed Ecohydrological Processes in a Changing Environment: Opportunities and Challenges. *Water, 14*(9), 1502. Re. doi:10.3390/w14091502

Carter, R. W., & Godfrey, R. G. (1960). *Storage and flood routing: Manual of hydrology Part 3.* Research Gate.

Chaudhry, S., & Sidhu, G. P. S. (2022). Climate change regulated abiotic stress mechanisms in plants: A comprehensive review. *Plant Cell Reports*, *41*(1), 1–31. doi:10.100700299-021-02759-5 PMID:34351488

Cheema, M. J. M., & Bastiaanssen, W. G. (2012). Local calibration of remotely sensed rainfall from the TRMM satellite for different periods and spatial scales in the Indus Basin. *International Journal of Remote Sensing*, *33*(8), 2603–2627. doi:10.1080/01431161.2011.617397

Chen, A., Huang, L., Liu, Q., & Piao, S. (2021). Optimal temperature of vegetation productivity and its linkage with climate and elevation on the Tibetan Plateau. *Global Change Biology*, *27*(9), 1942–1951. doi:10.1111/gcb.15542 PMID:33528057

Chen, J., Shao, C., Jiang, S., Qu, L., & Zhao, F. (2019). Effects of changes in precipitation on energy and water balance in a Eurasian meadow steppe. *Ecological Processes*, *8*, 1–15. doi:10.3390/pr8010001

Chen, X., Chen, W., & Xu, M. (2022). Remote-Sensing Monitoring of Postfire Vegetation Dynamics in the Greater Hinggan Mountain Range Based on Long Time-Series Data: Analysis of the Effects of Six Topographic and Climatic Factors. *Remote Sensing (Basel)*, *14*(13), 2958. doi:10.3390/rs14132958

Chen, X., Li, X., Jiang, B., Su, J., Zheng, X., & Wang, G. (2023). Prediction of spring agricultural drought using machine learning algorithms in the southern Songnen Plain, China. *Land Degradation & Development*, *34*(13), 3836–3849. doi:10.1002/ldr.4720

Chen, X., Zhao, P., Ouyang, L., Zhu, L., Ni, G., & Schäfer, K. V. R. (2020). Whole-plant water hydraulic integrity to predict drought-induced Eucalyptus urophylla mortality under drought stress. *Forest Ecology and Management*, *468*, 118179. doi:10.1016/j.foreco.2020.118179

Chen, Z., Liu, J., Qian, Z., Li, L., Zhang, Z., Feng, G., Ruan, S., & Sun, G. (2023). Monitoring Land Degradation through Vegetation Dynamics Mathematical Modeling: Case of Jornada Basin (in the US). *Remote Sensing (Basel)*, *15*(4), 978. doi:10.3390/rs15040978

Chicco, D., Warrens, M. J., & Jurman, G. (2021). The coefficient of determination R-squared is more informative than SMAPE, MAE, MAPE, MSE and RMSE in regression analysis evaluation. *PeerJ. Computer Science*, *7*, e623. doi:10.7717/peerj-cs.623 PMID:34307865

Chien, S., McLaren, D., Doubleday, J., Tran, D., Tanpipat, V., & Chitradon, R. (2019). Using taskable remote sensing in a sensor web for Thailand flood monitoring. *Journal of Aerospace Information Systems*, *16*(3), 107–119. doi:10.2514/1.I010672

Cho, C., Li, R., Wang, S. Y., Yoon, J. H., & Gillies, R. R. (2016). The anthropogenic footprint of climate change in the June 2013 northern India flood. *Climate Dynamics*, *46*(3-4), 797–805. doi:10.100700382-015-2613-2

Chow, V. T. (1959). *Open Channel Hydraulics.* McGraw-Hill.

Cismaru, C., Gabor, V., Bartha, I., & Scripcariu, D. (2004). *Gestiunea Secetelor | PDF.* Performantica. https://ro.scribd.com/document/450217576/Gestiunea-secetelor

Clark, R. A., Gourley, J. J., Flamig, Z. L., Hong, Y., & Clark, E. (2014). CONUS-wide evaluation of National Weather Service flash flood guidance products. *Weather and Forecasting*, *29*(2), 377–392. doi:10.1175/WAF-D-12-00124.1

Climate Centre. (2015). *Floods and mudslides trigger renewed concern over glacier-melt hazard across Central Asia.* Climate Centre. https://www.climatecentre.org/1801/floods-and-mudslides-trigger-renewed-concern-over-glacier-melt-hazard-across-central-asia/

CORINE Land Cover—Copernicus Land Monitoring Service. (n.d.). [Land Section]. Copernicus Land Monitoring. https://land.copernicus.eu/pan-european/corine-land-cover

Corral, C., Berenguer, M., Sempere-Torres, D., Poletti, L., Silvestro, F., & Rebora, N. (2019). Comparison of two early warning systems for regional flash flood hazard forecasting. *Journal of Hydrology (Amsterdam)*, *572*, 603–619. doi:10.1016/j.jhydrol.2019.03.026

Corzo Perez, G. A., van Huijgevoort, M. H. J., Voß, F., & van Lanen, H. A. J. (2011). On the spatio-temporal analysis of hydrological droughts from global hydrological models. *Hydrology and Earth System Sciences*, *15*(9), 2963–2978. doi:10.5194/hess-15-2963-2011

Crawford, N. H., & Linsley, R. K. (1966). *Digital simulation in hydrology: Stanford watershed model IV*. Dept. of Civil Engineering, Stanford University.

Cui, L., Pang, B., Zhao, G., Ban, C., Ren, M., Peng, D., Zuo, D., & Zhu, Z. (2022). Assessing the Sensitivity of Vegetation Cover to Climate Change in the Yarlung Zangbo River Basin Using Machine Learning Algorithms. *Remote Sensing (Basel)*, *14*(7), 1556. doi:10.3390/rs14071556

Cunge, J. (2010). On The Subject of A Flood Propagation Computation Method (Muskingum Method). *Journal of Hydraulic Research, April 1969*, 205–230. doi:10.1080/00221686909500264

Cupak, A., & Kaczor, G. (2023). Determination of Seasonal Indices for the Regionalization of Low Flows in the Upper Vistula River Basin. *Water (Basel)*, *15*(2), 246. doi:10.3390/w15020246

Cydzik, K., & Hogue, T. S. (2009). Modeling postfire response and recovery using the hydrologic engineering center hydrologic modeling system (HEC-HMS) 1. *Journal of the American Water Resources Association*, *45*(3), 702–714. doi:10.1111/j.1752-1688.2009.00317.x

da Silva, V. S., Salami, G., da Silva, M. I. O., Silva, E. A., & Monteiro, J. J. (2020). Methodological evaluation of vegetation indexes in land use and land cover (LULC) classification. *Geology, Ecology, and Landscapes*, *4*(2), 159–169. doi:10.1080/24749508.2019.1608409

Dalezios, N. R., Dunkel, Z., & Eslamian, S. (2017). Meteorological Drought Indices: Definitions. In Handbook of Drought and Water Scarcity. Principles of Drought and Water Scarcity. Routledge Handbooks Online. doi:10.1201/9781315404219-3

Dalrymple, T. (1960). Flood-frequency analyses, Manual of Hydrology: Part 3. In Water Supply Paper (1543-A). U.S. G.P.O., doi:10.3133/wsp1543A

Danish Hydraulic Institute (DHI). (2011). MIKE 11 - A modeling system for rivers and channels, Short introduction and tutorial. Horsholm, Denmark.

Darvishi, Y., & Arekhi, S. (2023). Analysis of Spatial Changes in the Landscape of Cites (Case Study: Kermanshah City). *Landscape Ecology*, *1*(1), 79–95. doi:10.22034/el.2023.170942 de Lima Moraes, A. G., Bocardo-Delgado, E., Bowling, L. C., Daneshvar, F., Pinto, J., Watkins, A. H., & Cherkauer, K. A. (2020). Assessment of Arequipa's hydrometeorological monitoring infrastructure to support water management decisions. *Journal of Contemporary Water Research & Education*, *171*(1), 27–48. doi:10.1111/j.1936-704X.2020.3343.x PMID:34790284

Deo, R. C., & Şahin, M. (2015). ¸Sahin, M. (2015). Application of the artificial neural network model for prediction of monthly standardized precipitation and evapotranspiration index using hydrometeorological parameters and climate indices in Eastern Australia. *Atmospheric Research*, *161*, 65–81. doi:10.1016/j.atmosres.2015.03.018

Devia, G. K., Ganasri, B. P., & Dwarakish, G. S. (2015). A review of hydrological models. *Aquatic Procedia*, *4*, 1001–1007. doi:10.1016/j.aqpro.2015.02.126

DeVries, B., Huang, C., Armston, J., Huang, W., Jones, J. W., & Lang, M. W. (2020). Rapid and robust monitoring of flood events using Sentinel-1 and Landsat data on the Google Earth Engine. *Remote Sensing of Environment*, *240*, 111664. doi:10.1016/j.rse.2020.111664

Di Sacco, A., Hardwick, K. A., Blakesley, D., Brancalion, P. H., Breman, E., Cecilio Rebola, L., Chomba, S., Dixon, K., Elliott, S., Ruyonga, G., Shaw, K., Smith, P., Smith, R. J., & Antonelli, A. (2021). Ten golden rules for reforestation to optimize carbon sequestration, biodiversity recovery and livelihood benefits. *Global Change Biology*, *27*(7), 1328–1348. doi:10.1111/gcb.15498 PMID:33494123

Di Silvio, G. (1998). *Soil erosion and conservation - lecture notes*. International Institute for Hydraulic and Environmental Engineering.

Di, Y., Ding, W., Mu, Y., Small, D. L., Islam, S., & Chang, N. B. (2015, April). Developing machine learning tools for long-lead heavy precipitation prediction with multi-sensor data. In *2015 IEEE 12th International Conference on Networking, Sensing and Control* (pp. 63-68). IEEE. 10.1109/ICNSC.2015.7116011

Dibike, Y. B., Velickov, S., Solomatine, D., & Abbott, M. B. (2001). Model induction with support vector machines: Introduction and applications. *Journal of Computing in Civil Engineering*, *2001*(15), 208–216. doi:10.1061/(ASCE)0887-3801(2001)15:3(208)

Dobhal, D. P., Gupta, A. K., Manish, M., & Khandelwal, D. D. (2013). Kedarnath disaster: Facts and plausible causes. *Current Science*, *105*(2), 171–174.

Dobrica, G. (2023). *Evaluarea secetei hidrologice in bazinul hidrografic al lacului Nuntasi-Tuzla* [PhD thesis, Ovidius University of Constanta, Romania].

Dracup, J. A., Lee, K. S., & Paulson, E. G. Jr. (1980). On the definition of droughts. *Water Resources Research*, *16*(2), 297–302. doi:10.1029/WR016i002p00297

Drobot, R. (2007). *Methodology to determine torrential hydrographical basins where exist localities exposed to the danger of flash floods*. Technical University of Civil Engineering.

Duan, W., Maskey, S., Chaffe, P. L., Luo, P., He, B., Wu, Y., & Hou, J. (2021). Recent advancement in remote sensing technology for hydrology analysis and water resources management. *Remote Sensing (Basel)*, *13*(6), 1097. doi:10.3390/rs13061097

Dube, A., Ashrit, R., Ashish, A., Sharma, K., Iyengar, G. R., Rajagopal, E. N., & Basu, S. (2014). Forecasting the heavy rainfall during Himalayan flooding—June 2013. *Weather and Climate Extremes*, *4*, 22–34. doi:10.1016/j.wace.2014.03.004

Dube, T., Shekede, M. D., & Massari, C. (2022). Remote Sensing for Water Resources and Environmental Management. *Remote Sensing (Basel)*, *15*(1), 18. doi:10.3390/rs15010018

Du, M., Kawashima, S., Yonemura, S., Zhang, X., & Chen, S. (2004). Mutual influence between human activities and climate change in the Tibetan Plateau during recent years. *Global and Planetary Change*, *41*(3-4), 241–249. doi:10.1016/j.gloplacha.2004.01.010

Dumitrescu, A., & Birsan, M.-V. (2015). ROCADA: A gridded daily climatic dataset over Romania (1961–2013) for nine meteorological variables. *Natural Hazards*, *78*(2), 1045–1063. doi:10.100711069-015-1757-z

Duveiller, G., Hooker, J., & Cescatti, A. (2018). the mark of vegetation change on Earth's surface energy balance. *Nature Communications*, *9*(1), 679. doi:10.103841467-017-02810-8 PMID:29463795

EEA. E. E. A., & JRC, J. R. C. (2011). *CLC 2018—Copernicus Land Monitoring Service* [Land item]. Copernicus Land Monitoring. https://land.copernicus.eu/pan-european/corine-land-cover/clc2018

El Hajj, M. M., Baghdadi, N., Zribi, M., & Bazzi, H. (2017). Synergic Use of Sentinel-1 and Sentinel-2 Images for Operational Soil Moisture Mapping at High Spatial Resolution over Agricultural Areas. *Remote Sensing (Basel)*, *9*(12), 1292. doi:10.3390/rs9121292

Elias, E. H., Rango, A., Steele, C. M., Mejia, J. F., & Smith, R. (2015). Assessing climate change impacts on water availability of snowmelt-dominated basins of the Upper Rio Grande basin. *Journal of Hydrology. Regional Studies*, *3*, 525–546. doi:10.1016/j.ejrh.2015.04.004

Elkhrachy, I. (2015). Flash flood hazard mapping using satellite images and GIS tools: A case study of Najran City, Kingdom of Saudi Arabia (KSA). *The Egyptian Journal of Remote Sensing and Space Sciences*, *18*(2), 261–278. doi:10.1016/j.ejrs.2015.06.007

Emily Cassidy. (2023). *Heavy Rains Hit Drought-Stricken Horn of Africa*. NASA, Earth Observatory. https://earthobservatory.nasa.gov/images/151208/heavy-rains-hit-drought-stricken-horn-of-africa

Erena, S. H., Worku, H., & De Paola, F. (2018). Flood hazard mapping using FLO-2D and local management strategies of Dire Dawa city, Ethiopia. *Journal of Hydrology. Regional Studies*, *19*, 224–239. doi:10.1016/j.ejrh.2018.09.005

Erickson, T. B., Brooks, J., Nilles, E. J., Pham, P. N., & Vinck, P. (2019). Environmental health effects attributed to toxic and infectious agents following hurricanes, cyclones, flash floods and major hydrometeorological events. *Journal of Toxicology and Environmental Health. Part B, Critical Reviews*, *22*(5-6), 157–171. doi:10.1080/10937404.2019.1654422 PMID:31437111

Eskander, S. M. S. U., & Fankhauser, S. (2020). Reduction in greenhouse gas emissions from national climate legislation. *Nature Climate Change*, *10*(8), 750–756. doi:10.103841558-020-0831-z

ESRI. (2017). *World Topographic Map—Overview*. ESRI. https://www.arcgis.com/home/item.html?id=7dc6cea0b1764a1f9af2e679f642f0f5

Etemad-Shahidi, A., & Mahjoobi, J. (2009). Comparison between m50 model tree and neural networks for prediction of significant wave height in lake superior. *Ocean Engineering*, *2009*(36), 1175–1181. doi:10.1016/j.oceaneng.2009.08.008

EU. (2012). *COM (2012)46—Implementation of the Soil Thematic Strategy and ongoing activities—EU monitor*. Eu-monitor. https://www.eumonitor.eu/9353000/1/j9vvik7m1c3gyxp/viwytay5yfzt

European Commission. (2012). *COM/2012/0673, COM/2012/0673*. Europea. https://eur-lex.europa.eu/legal-content/EN/TXT/?uri=CELEX%3A52012DC0673&qid=1692729471689 Directive 2000/60/EC,

European Environment Agency. (2021). *Meteorological and hydrological droughts in Europe* [Indicator Assessment]. EEA. https://www.eea.europa.eu/data-and-maps/indicators/river-flow-drought-3/assessment

Falkenmark, M., & Wang-Erlandsson, L. (2021). A water-function-based framework for understanding and governing water resilience in the Anthropocene. *One Earth*, *4*(2), 213–225. doi:10.1016/j.oneear.2021.01.009

Fang, Q., Wang, G., Zhang, S., Peng, Y., Xue, B., Cao, Y., & Shrestha, S. (2022). A novel ecohydrological model by capturing variations in climate change and vegetation coverage in a semi-arid region of China. *Environmental Research*, *211*, 113085. doi:10.1016/j.envres.2022.113085 PMID:35307372

Fan, M., Zhang, L., Liu, S., Yang, T., & Lu, D. (2023). Investigation of hydrometeorological influences on reservoir releases using explainable machine learning methods. *Frontiers in Water*, *5*, 1112970. doi:10.3389/frwa.2023.1112970

Fanta, S. S., & Tadesse, S. T. (2022). Application of HEC–HMS for runoff simulation of Gojeb Watershed, Southwest Ethiopia. *Modeling Earth Systems and Environment*, *8*(4), 4687–4705. doi:10.100740808-022-01397-4

Farhadi, H., & Najafzadeh, M. (2021). Flood risk mapping by remote sensing data and random forest technique. *Water (Basel)*, *13*(21), 3115. doi:10.3390/w13213115

Fatulová, E., Majerčáková, O., Houšková, B., Bardarska, G., Alexandrov, V., Kulířová, P., Gayer, J., Molnár, P., Fiala, K., Tamás, J., Zupan, M., Sušnik, A., Gregorič, G., & Bokal, S. (2015). *Guidelines for the preparation of the Drought Management Plans: Development and implementation of risk-based Drought Management Plans in the context of the EU Water Framework Directive - as part of the River Basin Management Plans*. Global Water Partnership Central and Eastern Europe.

FEMA. (2022). *Guidance for Flood Risk Analysis and Mapping*. General Hydrologic Considerations.

Fengjin, X., & Qiufeng, L. (2023). An evaluation of vegetation loss due to the super typhoon Sarika in Hainan Island of China. *Natural Hazards*, *115*(2), 1677–1695. doi:10.100711069-022-05613-3

Fernández-Nóvoa, D., García-Feal, O., González-Cao, J., de Gonzalo, C., Rodríguez-Suárez, J. A., Ruiz del Portal, C., & Gómez-Gesteira, M. (2020). MIDAS: A new integrated flood early warning system for the Miño River. *Water (Basel)*, *12*(9), 2319. doi:10.3390/w12092319

Fleig, A. K., Tallaksen, L. M., Hisdal, H., & Demuth, S. (2006). A global evaluation of streamflow drought characteristics. *Hydrology and Earth System Sciences*, *10*(4), 535–552. doi:10.5194/hess-10-535-2006

Flohn, H. (1970). Climatology—Descriptive or Physical Science? *World Meteorological Organization Bulletin*, *19*(4), 223–229.

Fu, B., Lan, F., Yao, H., Qin, J., He, H., Liu, L., Huang, L., Fan, D., & Gao, E. (2022). Spatio-temporal monitoring of marsh vegetation phenology and its response to hydro-meteorological factors using CCDC algorithm with optical and SAR images: In case of Honghe National Nature Reserve, China. *The Science of the Total Environment*, *843*, 156990. doi:10.1016/j.scitotenv.2022.156990 PMID:35764147

Fu, B., Lao, Z., Liang, Y., Sun, J., He, X., Deng, T., He, W., Fan, D., Gao, E., & Hou, Q. (2022). Evaluating optically and non-optically active water quality and its response relationship to hydro-meteorology using multi-source data in Poyang Lake, China. *Ecological Indicators*, *145*, 109675. doi:10.1016/j.ecolind.2022.109675

Galkate, R. V., Jain, S., Jaiswal, R. K., Pandey, R. P., Lohani, A. K., Yadav, S., & Yadava, R. N. (2023). Application of GIS and Remote Sensing Tools in Assessment of Drought Using Satellite and Ground-Based Data. In *Application of Remote Sensing and GIS in Natural Resources and Built Infrastructure Management* (pp. 105–121). Springer International Publishing.

Gallant, A. L., Sadinski, W. J., Brown, J. F., Senay, G. B., & Roth, M. F. (2018). Challenges in Complementing Data from Ground-Based Sensors with Satellite-Derived Products to Measure Ecological Changes in Relation to Climate—Lessons from Temperate Wetland-Upland Landscapes. *Sensors (Basel)*, *18*(3), 880. doi:10.339018030880 PMID:29547531

Gao, W., Zheng, C., Liu, X., Lu, Y., Chen, Y., Wei, Y., & Ma, Y. (2022). NDVI-based vegetation dynamics and their responses to climate change and human activities from 1982 to 2020: A case study in the Mu Us Sandy Land, China. *Ecological Indicators*, *137*, 108745. doi:10.1016/j.ecolind.2022.108745

García-Galiano, S. G., Cánovas-García, F., & Giraldo-Osorio, J. D. (2023). Editorial for Special Issue: "Remote Sensing of Hydrological Processes: Modelling and Applications.". *Remote Sensing (Basel)*, *15*(5), 1466. doi:10.3390/rs15051466

García-Palacios, P., Gross, N., Gaitán, J., & Maestre, F. T. (2018). Climate mediates the biodiversity–ecosystem stability relationship globally. *Proceedings of the National Academy of Sciences of the United States of America*, *115*(33), 8400–8405. doi:10.1073/pnas.1800425115 PMID:30061405

Gatti, L. V., Basso, L. S., Miller, J. B., Gloor, M., Gatti Domingues, L., Cassol, H. L. G., Tejada, G., Aragão, L. E. O. C., Nobre, C., Peters, W., Marani, L., Arai, E., Sanches, A. H., Corrêa, S. M., Anderson, L., Von Randow, C., Correia, C. S. C., Crispim, S. P., & Neves, R. A. L. (2021). Amazonia as a carbon source linked to deforestation and climate change. *Nature*, *595*(7867), 388–393. doi:10.103841586-021-03629-6 PMID:34262208

Gaume, E., Bain, V., Bernardara, P., Newinger, O., Barbuc, M., Bateman, A., Blaškovicová, L., Blöschl, G., Borga, M., Dumitrescu, A., Daliakopoulos, I., Garcia, J., Irimescu, A., Kohnova, S., Koutroulis, A., Marchi, L., Matreata, S., Medina, V., Preciso, E., & Viglione, A. (2009). A collation of data on European flash floods. *Journal of Hydrology (Amsterdam)*, *367*(1-2), 70–78. doi:10.1016/j.jhydrol.2008.12.028

Ghebrezgabher, M. G., Yang, T., Yang, X., & Sereke, T. E. (2020). Assessment of NDVI variations in responses to climate change in the Horn of Africa. *The Egyptian Journal of Remote Sensing and Space Sciences*, *23*(3), 249–261. doi:10.1016/j.ejrs.2020.08.003

Giddings, L., & Soto, M. (2005). Standardized Precipitation Index Zones for México. *Atmosfera6, 18*, 33–56.

Giurma, I. (1997). *Reservoirs silting*. Bucharest, Romania: H*G*A* Publishing House.

Glago, F. J. (2021). Flood disaster hazards; causes, impacts and management: a state-of-the-art review. *Natural hazards-impacts, adjustments and resilience*, 29-37.

Goliger, A. M., & Milford, R. V. (1998). A review of worldwide occurrence of tornadoes. *Journal of Wind Engineering and Industrial Aerodynamics*, *74–76*, 111–121. doi:10.1016/S0167-6105(98)00009-9

Gomes, V., Queiroz, G., Ferreira, K. (2020). An Overview of Platforms for Big Earth Observation Data Management and Analysis. *Remote Sensing, 12*(8), 1253. MDPI AG. . doi:10.3390/rs12081253

Goyal, M. K., Gupta, A. K., & Gupta, A. (Eds.). (2022). *Hydro-Meteorological Extremes and Disasters*. Springer Nature. doi:10.1007/978-981-19-0725-8

Grecu, M., & Krajewski, W. (2000). A large-sample investigation of statistical procedures for radar-based short-term quantitative precipitation forecasting. *Journal of Hydrology (Amsterdam)*, *2000*(239), 69–84. doi:10.1016/S0022-1694(00)00360-7

Gregor, M. (2010). *TLM 2.1 User's Manual* [User Manual]. Department of Hydrogeology, Faculty of Natural Science, Comenius University. https://hydrooffice.org/Files/UM%20TLM.pdf

Grieser, J., Gommes, R., & Cofield, S. (2006), On the Estimation of Monthly Precipitation Fields in Afghanistan, A communication uploaded on Researchgate by the authors. https://www.researchgate.net/publication/267723218_On_the_estimation_of_monthly_precipitation_fields_in_Afghanistan#fullTextFileContent

Grotch, S. L., & MacCracken, M. C. (1991). The use of general circulation models to predict regional climatic change. *Journal of Climate*, *4*(3), 286–303. doi:10.1175/1520-0442(1991)004<0286:TUOGCM>2.0.CO;2

Grumezea, N., Kleps, C., & Tusa, C. (1990). *Evolutia Nivelului Si Chimismului Apei Freatice Din Amenajarile De Irigatii In Interrelatie Cu Mediul Inconjurator*. Intreprinderea poligrafica Oltenia.

Guastavino, S., Piana, M., Tizzi, M., Cassola, F., Iengo, A., Sacchetti, D., Solazzo, E., & Benvenuto, F. (2022). Prediction of severe thunderstorm events with ensemble deep learning and radar data. *Scientific Reports*, *12*(1), 20049. doi:10.103841598-022-23306-6 PMID:36414648

Guo, A., Chang, J., Wang, Y., Huang, Q., Guo, Z., & Li, Y. (2019). Uncertainty analysis of water availability assessment through the Budyko framework. *Journal of Hydrology (Amsterdam)*, *576*, 396–407. doi:10.1016/j.jhydrol.2019.06.033

Guo, L., Sun, X., Fu, P., Shi, T., Dang, L., Chen, Y., & Zeng, C. (2021). Mapping soil organic carbon stock by hyperspectral and time-series multispectral remote sensing images in low-relief agricultural areas. *Geoderma*, *398*, 115118. doi:10.1016/j.geoderma.2021.115118

Guri, P. K., Champati Ray, P. K., & Patel, R. C. (2015). Spatial prediction of landslide susceptibility in parts of Garhwal Himalaya, India, using the weight of evidence modeling. *Environmental Monitoring and Assessment*, *187*(6), 1–25. doi:10.100710661-015-4535-1 PMID:25944750

Gustard, A., & Demuth, S. (2008). *Manual on low-flow estimation and prediction. Open Hydrology Report, Vol. No. 50*. World Meteorological Organization.

GusyevM.HasegawaA.MagomeJ.KuribayashiD.SawanoH.LeeS. (2015, December 4). *Drought assessment in the Pampanga River basin, the Philippines – Part 1: Characterizing the role of dams in historical droughts with standardized indices*. doi:10.13140/RG.2.1.4931.4321

Haile, G. G., Tang, Q., Li, W., Liu, X., & Zhang, X. (2020). Drought: Progress in broadening its understanding. *WIREs. Water*, *7*(2), e1407. doi:10.1002/wat2.1407

Halofsky, J. E., Peterson, D. L., & Harvey, B. J. (2020). Changing wildfire, changing forests: The effects of climate change on fire regimes and vegetation in the Pacific Northwest, USA. *Fire Ecology*, *16*(1), 1–26. doi:10.118642408-019-0062-8

Hamdard, M. H., Soliev, I., Rasouli, H., Kløve, B., & Belhassan, K. (2022). Groundwater Quality Assessment in Chak Karstic Sedimentary Basin, Wardak Province, Afghanistan. *Central Asian Journal of Water Research*, *8*(2), 110–127. Advance online publication. doi:10.29258/CAJWR/2022-R1.v8-2/110-127.eng

Hamidi, E., Peter, B. G., Muñoz, D. F., Moftakhari, H., & Moradkhani, H. (2023). Fast Flood Extent Monitoring With SAR Change Detection Using Google Earth Engine. *IEEE Transactions on Geoscience and Remote Sensing*, *61*, 1–19. doi:10.1109/TGRS.2023.3240097

Hamid, J., & Alireza, P. (2022). A comprehensive review of remote sensing platforms, sensors, and applications in nut crops. *Computers and Electronics in Agriculture*, *197*, 106844. doi:10.1016/j.compag.2022.106844

Hapuarachchi, H. P., & Wang, Q. J. (2008). A review of methods and systems available for flash flood forecasting. *Report for the Bureau of Meteorology*. Bureau of Meteorology.

Hasnat, G. T., Kabir, M. A., & Hossain, M. A. (2018). Major environmental issues and problems in South Asia, particularly Bangladesh. Handbook of environmental materials management, 1-40.

Hassan, F. H., & Azelan, N. A. (2019). Comparing the performance of machine learning algorithms in a flood prediction model with real data sets. *International Journal of Advanced Trends in Computer Science and Engineering, 8*(1.4 S1), 152–157. doi:10.30534/ijatcse/2019/2381.42019

Hasson, S., Saeed, F., Böhner, J., & Schleussner, C.-F. (2019). Water availability in Pakistan from Hindukush-Karakoram-Himalayan watersheds at 1.5°C and 2°C Paris Agreement Targets. *Advances in Water Resources, 131*, 103365. doi:10.1016/j.advwatres.2019.06.010

Hayder, I. M., Al-Amiedy, T. A., Ghaban, W., Saeed, F., Nasser, M., Al-Ali, G. A., & Younis, H. A. (2023). An Intelligent Early Flood Forecasting and Prediction Leveraging Machine and Deep Learning Algorithms with Advanced Alert System. *Processes (Basel, Switzerland), 11*(2), 481. doi:10.3390/pr11020481

Hayes, M., Svoboda, M., Wall, N., & Widhalm, M. (2009). *THE LINCOLN DECLARATION ON DROUGHT INDICES: Universal Meteorological Drought Index Recommended.* CORE. https://core.ac.uk/reader/20047170

Hazaymeh, K., & Hassan, Q. K. (2015). Fusion of MODIS and Landsat-8 surface temperature images: A new approach. *PLoS One, 10*(3), e0117755. doi:10.1371/journal.pone.0117755 PMID:25730279

Hazaymeh, K., & Hassan, Q. K. (2016). Remote sensing of agricultural drought monitoring: A state of art review. *AIMS Environmental Science, 3*(4), 604–630. doi:10.3934/environsci.2016.4.604

He, B., Xing, M., & Bai, X. (2014). A Synergistic Methodology for Soil Moisture Estimation in an Alpine Prairie Using Radar and Optical Satellite Data. *Remote Sensing (Basel), 6*(11), 10966–10985. doi:10.3390/rs61110966

He, G., Chen, Y., Fang, G., & Li, Z. (2022). Hydrometeorological Forecast of a Typical Watershed in an Arid Area Using Ensemble Kalman Filter. *Water (Basel), 14*(23), 3970. doi:10.3390/w14233970

Heilmayr, R., Echeverría, C., & Lambin, E. F. (2020). Impacts of Chilean forest subsidies on forest cover, carbon and biodiversity. *Nature Sustainability, 3*(9), 701–709. doi:10.103841893-020-0547-0

Henao Salgado, M. J., & Zambrano Nájera, J. (2022). Assessing flood early warning systems for flash floods. *Frontiers in Climate, 4*, 787042. doi:10.3389/fclim.2022.787042

Hepites, S. (1906). Secetele în România. *Buletinul Societății Geografice Române, XXVII.* dSpace. http://dspace.bcu-iasi.ro/static/web/viewer.html?file=http://dspace.bcu-iasi.ro/bitstream/handle/123456789/54680/BCUIASI_CARTE_IIIB-3.944_1906.pdf?sequence=2&isAllowed=y

Herath, M., Jayathilaka, T., Hoshino, Y., & Rathnayake, U. (2023). Deep Machine Learning-Based Water Level Prediction Model for Colombo Flood Detention Area. *Applied Sciences (Basel, Switzerland), 13*(4), 2194. doi:10.3390/app13042194

Hill, G. W., Hales, T. A., & Aldridge, B. N. (n.d.). *Flood Hydrology Near Flagstaff, Arizona*. Research Gate.

Hisdal, H., Tallaksen, L., Clausen, B., Peters, E., & Gustard, A. (2004). Hydrological drought characteristics, Hydrological drought, Processes and estimation methods for streamflow and groundwater. In *Developments in Water Science, 48* (pp. 139–198). Elsevier Science B. V.

Hoffmann, R., Muttarak, R., Peisker, J., & Stanig, P. (2022). Climate change experiences raise environmental concerns and promote Green voting. *Nature Climate Change, 12*(2), 148–155. doi:10.103841558-021-01263-8

Hofmann, J., & Schüttrumpf, H. (2019). Risk-based early warning system for pluvial flash floods: Approaches and foundations. *Geosciences, 9*(3), 127. doi:10.3390/geosciences9030127

Holden, P. B., Rebelo, A. J., Wolski, P., Odoulami, R. C., Lawal, K. A., Kimutai, J., Nkemelang, T., & New, M. G. (2022). Nature-based solutions in mountain catchments reduce impact of anthropogenic climate change on drought streamflow. *Communications Earth & Environment, 3*(1), 51. doi:10.103843247-022-00379-9

Hosking, J. R. M., & Wallis, J. R. (2005). *Regional Frequency Analysis: An Approach Based on L-Moments*. Cambridge University Press.

Hosseini, F. S., Choubin, B., Mosavi, A., Nabipour, N., Shamshirband, S., Darabi, H., & Haghighi, A. T. (2020). Flash-flood hazard assessment using ensembles and Bayesian-based machine learning models: Application of the simulated annealing feature selection method. *The Science of the Total Environment, 711*, 135161. doi:10.1016/j.scitotenv.2019.135161 PMID:31818576

Huang, F., Shangguan, W., Li, Q., Li, L., & Zhang, Y. (2023). Beyond prediction: An integrated post-hoc approach to interpret complex model in hydrometeorology. *Environmental Modelling & Software, 167*, 105762. doi:10.1016/j.envsoft.2023.105762

Huang, J., Zhang, G., Zhang, Y., Guan, X., Wei, Y., & Guo, R. (2020). Global desertification vulnerability to climate change and human activities. *Land Degradation & Development, 31*(11), 1380–1391. doi:10.1002/ldr.3556

Huang, Z., Liu, X., Jin, M., Ding, C., Jiang, J., & Wu, L. (2016). Deriving the Characteristic Scale for Effectively Monitoring Heavy Metal Stress in Rice by Assimilation of GF-1 Data with the WOFOST Model. *Sensors (Basel), 16*(3), 340. doi:10.339016030340 PMID:26959033

Huang, Z., Yuan, X., & Liu, X. (2021). The key drivers for the changes in global water scarcity: Water withdrawal versus water availability. *Journal of Hydrology (Amsterdam), 601*, 126658. doi:10.1016/j.jhydrol.2021.126658

Hubert, P. (2000). The segmentation procedure as a tool for discrete modeling of hydrometeorological regimes. *Stochastic Environmental Research and Risk Assessment, 14*(4), 297–304. doi:10.1007/PL00013450

Ibarreche, J., Aquino, R., Edwards, R. M., Rangel, V., Pérez, I., Martínez, M., Castellanos, E., Álvarez, E., Jimenez, S., Rentería, R., Edwards, A., & Álvarez, O. (2020). Flash flood early warning system in Colima, Mexico. *Sensors (Basel), 20*(18), 5231. doi:10.339020185231 PMID:32937798

ICHARM. (2023). Integrated Flood Analysis System (IFAS). *International Centre for Water Hazard and Risk Management under the auspices of UNESCO*. PWRI. https://www.pwri.go.jp/icharm/research/ifas/ifas.html

International Institute for Infrastructural, Hydraulic and Environmental Engineering (IHE). (1995). DUFLOW – Manual.

IPCC. (2013). The Physical Science Basis. In T. F. Stocker, D. Qin, G.-K. Plattner, M. Tignor, S. K. Allen, J. Boschung, A. Nauels, Y. Xia, V. Bex, & P. M. Midgley (Eds.), *Climate Change 2013: Contribution of Working Group I to the Fifth Assessment Report of the Intergovernmental Panel on Climate Change*. Cambridge University Press.

IRDR International Centre of Excellence on Risk Interconnectivity and Governance on Weather. (2022). *Peril Classification and Hazard Glossary* (IRDR DATA Publication No. 1; p. 28). International Science Council. doi:10.24948/2022.05

Jajarmizadeh, M., Lafdani, E. K., Harun, S., & Ahmadi, A. (2015). Application of SVM and swat models for monthly streamflow prediction, a case study in South of Iran. *KSCE Journal of Civil Engineering, 2015*(19), 345–357. doi:10.100712205-014-0060-y

Jakovljević, G., & Govedarica, M. (2018). Water body extraction and flood risk assessment using LIDAR and open data. In Climate change management (pp. 93–111). doi:10.1007/978-3-030-03383-5_7

Jentsch, A., Kreyling, J., & Beierkuhnlein, C. (2007). A new generation of climate-change experiments: Events, not trends. *Frontiers in Ecology and the Environment, 5*(7), 365–374. doi:10.1890/1540-9295(2007)5[365:ANGOCE]2.0.CO;2

Jiang, L., Bao, A., Guo, H., & Ndayisaba, F. (2017). Vegetation dynamics and responses to climate change and human activities in Central Asia. *The Science of the Total Environment*, *599*, 967–980. doi:10.1016/j.scitotenv.2017.05.012 PMID:28505889

Jiao, W., Wang, L., & McCabe, M. F. (2021). Multi-sensor remote sensing for drought characterization: Current status, opportunities and a roadmap for the future. *Remote Sensing of Environment*, *256*, 112313. doi:10.1016/j.rse.2021.112313

John, F. England, Cohn, T. A., Faber, B. A., Stedinger, J. R., Jr, W. O. T., Veilleux, A. G., Kiang, J. E., & Robert R. Mason, J. (2019). Guidelines for determining flood flow frequency—Bulletin 17C. In Techniques and Methods (4-B5). U.S. Geological Survey. doi:10.3133/tm4B5

Joseph, S., Sahai, A. K., Sharmila, S., Abhilash, S., Borah, N., Chattopadhyay, R., Pillai, P. A., Rajeevan, M., & Kumar, A. (2015). North Indian heavy rainfall event during June 2013: Diagnostics and extended range prediction. *Climate Dynamics*, *44*(7-8), 2049–2065. doi:10.100700382-014-2291-5

Kakooei, M., & Baleghi, Y. (2020). A two-level fusion for building irregularity detection in post-disaster VHR oblique images. *Earth Science Informatics*, *13*(2), 459–477. doi:10.100712145-020-00449-6

Kakooei, M., Ghorbanian, A., Baleghi, Y., Amani, M., & Nascetti, A. (2022). Remote sensing technology for post-disaster building damage assessment. In *Computers in Earth and Environmental Sciences* (pp. 509–521). Elsevier. doi:10.1016/B978-0-323-89861-4.00047-6

Kar, A. K., Lohani, A. K., Goel, N. K., & Roy, G. P. (2010). Development of flood forecasting system using statistical and ANN techniques in the downstream catchment of Mahanadi basin, India. *Journal of Water Resource and Protection*, *2010*(2), 880.

Kellens, S., Terpstra, T., & De Maeyer, P. (2013). Perception and communication of flood risks: A systematic review of empirical research. *Risk Analysis*, *33*(1), 24–49. doi:10.1111/j.1539-6924.2012.01844.x PMID:22651128

Kelsey, K. C., Pedersen, S. H., Leffler, A. J., Sexton, J. O., Feng, M., & Welker, J. M. (2021). Winter snow and spring temperature have differential effects on vegetation phenology and productivity across Arctic plant communities. *Global Change Biology*, *27*(8), 1572–1586. doi:10.1111/gcb.15505 PMID:33372357

Kennon, F. W. (1955). Magnitude and frequency of summer floods in western New Mexico and eastern Arizona. In *Open-File Report* (pp. 55–82). U.S. Geological Survey. doi:10.3133/ofr5582

Keyantash, J. A., & Dracup, J. A. (2004). An aggregate drought index: Assessing drought severity based on fluctuations in the hydrologic cycle and surface water storage. *Water Resources Research*, *40*(9), 2003WR002610. doi:10.1029/2003WR002610

Khan, I., Lei, H., Shah, A. A., Khan, I., & Muhammad, I. (2021). Climate change impact assessment, flood management, and mitigation strategies in Pakistan for a sustainable future. *Environmental Science and Pollution Research International*, *28*(23), 29720–29731. doi:10.100711356-021-12801-4 PMID:33566292

Khan, N., Sachindra, D. A., Shahid, S., Ahmed, K., Shiru, M. S., & Nawaz, N. (2020). Prediction of droughts over Pakistan using machine learning algorithms. *Advances in Water Resources*, *139*, 103562. Advance online publication. doi:10.1016/j.advwatres.2020.103562

Khan, R., & Gilani, H. (2021a). Global drought monitoring with drought severity index (DSI) using Google Earth Engine. *Theoretical and Applied Climatology*, *146*(1-2), 411–427. doi:10.100700704-021-03715-9

Khan, R., & Gilani, H. (2021b). Global drought monitoring with big geospatial datasets using Google Earth Engine. *Environmental Science and Pollution Research International*, *28*(14), 17244–17264. doi:10.100711356-020-12023-0 PMID:33394397

Khunrattanasiri, W. (2023). Application of Remote Sensing Vegetation Indices for Forest Cover Assessments. In *Concepts and Applications of Remote Sensing in Forestry* (pp. 153–166). Springer.

Kia, M. B., Pirasteh, S., Pradhan, B., Mahmud, A. R., Sulaiman, W. N. A., & Moradi, A. (2012). An artificial neural network model for flood simulation using GIS: Johor River Basin, Malaysia. *Environmental Earth Sciences*, *67*(1), 251–264. doi:10.100712665-011-1504-z

Kim, S., Matsumi, Y., Pan, S., & Mase, H. (2016). A real-time forecast model using an artificial neural network for after-runner storm surges on the Tottori Coast, Japan. *Ocean Engineering*, *2016*(122), 44–53. doi:10.1016/j.oceaneng.2016.06.017

Kim, S., & Singh, V. P. (2013). Flood forecasting using neural computing techniques and conceptual class segregation. *Journal of the American Water Resources Association*, *2013*(49), 1421–1435. doi:10.1111/jawr.12093

Koffler, D., & Laaha, G. (2013). *LFSTAT - Low-Flow Analysis in R, Geophysical Research Abstracts* (Vol. 15). EGU General Assembly.

Kogan, F. N. (1995). Droughts of the Late 1980s in the United States as Derived from NOAA Polar-Orbiting Satellite Data. *Bulletin of the American Meteorological Society*, *76*(5), 655–668. doi:10.1175/1520-0477(1995)076<0655:DOTLIT>2.0.CO;2

Kolarik, N., Roopsind, A., Pickens, A., & Brandt, J. (2023). A satellite-based monitoring system for quantifying surface water and mesic vegetation dynamics in a semi-arid region. *Ecological Indicators*, *147*, 109965. doi:10.1016/j.ecolind.2023.109965

Koons, P. (1989). The topographic evolution of collisional mountain belts: A numerical look at the Southern Alps, New Zealand. *American Journal of Science*, *289*(9), 1041–1069. doi:10.2475/ajs.289.9.1041

Kotal, S. D., Roy, S. S., & Bhowmik, S. R. (2014). Catastrophic heavy rainfall episode over Uttarakhand during 16–18 June 2013–observational aspects. *Current Science*, 234–245.

Kovats, R. S., Valentini, R., Bouwer, L. M., Georgopoulou, E., Jacob, D., Martin, E., Rounsevell, M., & Soussana, J. F. (2014) Europe. In: Barros VR, Field CB, Dokken DJ, Mastrandrea MD, Mach KJ, Bilir TE, Chatterjee M, Ebi KL, Estrada YO, Genova RC, Girma B, Kissel ES, Levy AN, MacCracken, Mastrandrea PR, White LL (eds) Climate change 2014: impacts, adaptation, and vulnerability. Part B: regional aspects. Contribution of Working Group II to the Fifth Assessment Report of the Intergovernmental Panel on Climate Change

Kseňak, Ľ., Pukanská, K., Bartoš, K., & Blišťan, P. (2022). Assessment of the Usability of SAR and Optical Satellite Data for Monitoring Spatio-Temporal Changes in Surface Water: Bodrog River Case Study. *Water (Basel)*, *14*(3), 299. doi:10.3390/w14030299

Kuenzer, C., Klein, I., Ullmann, T., Foufoula-Georgiou, E., Baumhauer, R., & Dech, S. (2015). Remote Sensing of River Delta inundation: Exploiting the potential of coarse Spatial Resolution, Temporally-Dense MODIS Time Series. *Remote Sensing (Basel)*, *7*(7), 8516–8542. doi:10.3390/rs70708516

Kumar, R., Devrani, R., Dangwal, A., Deshmukh, B., & Dutt, S. (2022). Extreme Hydrological Event-Induced Temporal Variation in Soil Erosion of the Assiganga River Basin, NW Himalaya. *Advances in Remote Sensing Technology and the Three Poles*, 230-246.

Kumar, P., Debele, S. E., Sahani, J., Rawat, N., Marti-Cardona, B., Alfieri, S. M., & Zieher, T. (2021). An overview of monitoring methods for assessing the performance of nature-based solutions against natural hazards. *Earth-Science Reviews*, *217*, 103603. doi:10.1016/j.earscirev.2021.103603

Kumar, S., Tiwari, M. K., Chatterjee, C., & Mishra, A. (2015). Reservoir inflow forecasting using ensemble models based on neural networks, wavelet analysis, and bootstrap method. *Water Resources Management*, *2015*(29), 4863–4883. doi:10.100711269-015-1095-7

Kundzewicz, Z. W., & Robson, A. (2000). Detecting Trend and Other Changes in Hydrological Data: Vol. WMO/TD-No. 1013.

Kushiyama, Y., & Matsuoka, M. (2019). Time Series GIS Map Dataset of Demolished Buildings in Mashiki Town after the 2016 Kumamoto, Japan Earthquake. *Remote Sensing (Basel)*, *11*(19), 2190. doi:10.3390/rs11192190

Laaha, G., Gauster, T., Delus, C., Vidal, J.-P., & the UNESCO FRIEND-Water low flow and drought group Team (2016). *The European 2015 drought from a hydrological perspective*, *EGU (European Geosciences Union)*, General Assembly, Vienna, Austria.

Laaha, G. (2023). A mixed distribution approach for low-flow frequency analysis – Part 1: Concept, performance, and effect of seasonality. *Hydrology and Earth System Sciences*, *27*(3), 689–701. doi:10.5194/hess-27-689-2023

Laaha, G., & Blöschl, G. (2006). Seasonality indices for regionalizing low flows. *Hydrological Processes*, *20*(18), 3851–3878. doi:10.1002/hyp.6161

Laaha, G., Demuth, S., Hisdal, H., Kroll, C. N., van Lanen, H. A. J., Nester, T., Rogger, M., Sauquet, E., Tallaksen, L. M., Woods, R., & Young, A. (2013). Prediction of low flows in ungauged basins. In G. Bloeschl, M. Sivapalan, T. Wagener, A. Viglione, & H. Savenije (Eds.), *Runoff Prediction in Ungauged Basins: Synthesis across Processes, Places and Scales* (pp. 163–188). Cambridge University Press. doi:10.1017/CBO9781139235761.011

Laaha, G., Parajka, J., Viglione, A., Koffler, D., Haslinger, K., Schöner, W., Zehetgruber, J., & Blöschl, G. (2016). A three-pillar approach to assessing climate impacts on low flows. *Hydrology and Earth System Sciences*, *20*(9), 3967–3985. doi:10.5194/hess-20-3967-2016

Lacroix, P., Handwerger, A. L., & Bièvre, G. (2020). Life and death of slow-moving landslides. *Nature Reviews. Earth & Environment*, *1*(8), 404–419. doi:10.103843017-020-0072-8

Lafdani, E. K., Nia, A. M., Pahlavanravi, A., Ahmadi, A., & Jajarmizadeh, M. (2013). Research article daily rainfall-runoff prediction and simulation using ANN, ANFIS, and conceptual hydrological MIKE11/NAM models. Int. J.nEng. *Technol.*, *2013*(1), 32–50.

Laiolo, P., Gabellani, S., Campo, L., Silvestro, F., Delogu, F., Rudari, R., & Puca, S. (2016). Impact of different satellite soil moisture products on the predictions of a continuous distributed hydrological model. *International Journal of Applied Earth Observation and Geoinformation*, *48*, 131–145. doi:10.1016/j.jag.2015.06.002

Laonamsai, J., Julphunthong, P., Saprathet, T., Kimmany, B., Ganchanasuragit, T., Chomcheawchan, P., & Tomun, N. (2023). Utilizing NDWI, MNDWI, SAVI, WRI, and AWEI for Estimating Erosion and Deposition in Ping River in Thailand. *Hydrology*, *10*(3), 70. doi:10.3390/hydrology10030070

Lave, J. & Avouac, J.P. (2001). Fluvial incision and tectonic uplift across the Himalayas of central Nepal. *Journal Of Geophysical Research, 106*(26).

Lazar, Gh. I. (2007). Computer-assisted numerical modeling of currents with free level in the regulated regime, Timisoara, Romania: Politechnica Publishing House.

Leal, K. B., Robaina, L. E. S., & De Lima, A. S. (2022). Coastal impacts of storm surges on a changing climate: A global bibliometric analysis. *Natural Hazards*, *114*(2), 1455–1476. doi:10.100711069-022-05432-6

Lee, A. F. S., & Heghinian, S. M. (1977). A Shift Of The Mean Level In A Sequence Of Independent Normal Random Variables—A Bayesian Approach—. *Technometrics*, *19*(4), 503–506. doi:10.1080/00401706.1977.10489592

Lee, S.-J., Kim, N., & Lee, Y. (2021). Development of Integrated Crop Drought Index by Combining Rainfall, Land Surface Temperature, Evapotranspiration, Soil Moisture, and Vegetation Index for Agricultural Drought Monitoring. *Remote Sensing (Basel)*, *13*(9), 1778. doi:10.3390/rs13091778

Lee, T., Singh, V. P., & Cho, K. H. (2021). *Deep Learning for Hydrometeorology and Environmental Science*. Springer. doi:10.1007/978-3-030-64777-3

Lehnebach, R., Campioli, M., Gričar, J., Prislan, P., Mariën, B., Beeckman, H., & Van den Bulcke, J. (2021). High-resolution X-ray computed tomography: A new workflow for the analysis of xylogenesis and intra-seasonal wood biomass production. *Frontiers in Plant Science*, *12*, 698640. doi:10.3389/fpls.2021.698640 PMID:34421949

Lemenkova, P., & Debeir, O. (2023). Multispectral Satellite Image Analysis for Computing Vegetation Indices by R in the Khartoum Region of Sudan, Northeast Africa. *Journal of Imaging*, *9*(5), 98. doi:10.3390/jimaging9050098 PMID:37233317

Lenton, T. M., Held, H., Kriegler, E., Hall, J. W., Lucht, W., Rahmstorf, S., & Schellnhuber, H. J. (2008). Tipping elements in the Earth's climate system. *Proceedings of the National Academy of Sciences of the United States of America*, *105*(6), 1786–1793. doi:10.1073/pnas.0705414105 PMID:18258748

Lhotka, O., Kyselý, J., & Farda, A. (2018). Climate change scenarios of heat waves in Central Europe and their uncertainties. *Theoretical and Applied Climatology*, *131*(3-4), 1043–1054. doi:10.100700704-016-2031-3

Li, P., Li, D., Sun, X., Chu, Z., Xia, T., Zheng, B. (2022). Application of Ecological Restoration Technologies for the Improvement of Biodiversity and Ecosystem in the River. *Water*, *14*(9), 1402. MDPI AG. . doi:10.3390/w14091402

Li, J., & Wang, S. (2018). Using SAR-Derived Vegetation Descriptors in a Water Cloud Model to Improve Soil Moisture Retrieval. *Remote Sensing (Basel)*, *10*(9), 1370. doi:10.3390/rs10091370

Li, L., Xu, H., Chen, X., & Simonovic, S. (2010). Streamflow forecast and reservoir operation performance assessment under climate change. *Water Resources Management*, *2010*(24), 83–104. doi:10.100711269-009-9438-x

Linsley, R. K. Jr, Kohler, M. A., & Paulhus, J. L. H. (1975). *Hydrology for Engineers* (2nd ed.). McGraw-Hill. http://www.esalq.usp.br/lepse/imgs/conteudo_thumb/Understanding-the-Drought-Phenomenon-The-Role-of-Definitions.pdf

Li, P., Stuart, E. A., & Allison, D. B. (2015, November 10). Multiple Imputation: A Flexible Tool for Handling Missing Data. *Journal of the American Medical Association*, *314*(18), 1966–1967. doi:10.1001/jama.2015.15281 PMID:26547468

Li, P., Wang, J., Liu, M., Xue, Z., Bagherzadeh, A., & Liu, M. (2021). Spatio-temporal variation characteristics of NDVI and its response to climate on the Loess Plateau from 1985 to 2015. *Catena*, *203*, 105331. doi:10.1016/j.catena.2021.105331

Liu, J. G. (2013). *REMOTE SENSING | Passive Sensors*. Reference Module in Earth Systems and Environmental Sciences. doi:10.1016/B978-0-12-409548-9.02956-0

Liu, W., Feng, Q., Engel, B. A., Yu, T., Zhang, X., & Qian, Y. (2023). A probabilistic assessment of urban flood risk and impacts of future climate change. *Journal of Hydrology (Amsterdam)*, *618*, 129267. doi:10.1016/j.jhydrol.2023.129267

Llasat-Botija, M., & Llasat, M. C. (2022). 6. FLOODUP, a citizen science project to increase flood risk awareness and collective knowledge. *Information Technologies and Social Media: New Scientific Methods for the Anthropocene*, 106.

Lloyd-Hughes, B. (2014). The impracticality of a universal drought definition. *Theoretical and Applied Climatology*, *117*(3), 607–611. doi:10.100700704-013-1025-7

Loarie, S. R., Duffy, P. B., Hamilton, H., Asner, G. P., Field, C. B., & Ackerly, D. D. (2009). The velocity of climate change. *Nature*, *462*(7276), 1052–1055. doi:10.1038/nature08649 PMID:20033047

Lopac, N., Jurdana, I., Brnelić, A., & Krljan, T. (2022). Application of laser systems for detection and ranging in the modern road transportation and maritime sector. *Sensors (Basel)*, *22*(16), 5946. doi:10.339022165946 PMID:36015703

Lopez, T., Al Bitar, A., Biancamaria, S., Güntner, A., & Jäggi, A. (2020). On the use of satellite remote sensing to detect floods and droughts at large scales. *Surveys in Geophysics*, *41*(6), 1461–1487. doi:10.100710712-020-09618-0

Lorenc, A. C. (1995). *Atmospheric data assimilation*. Meteorological Office.

Maddox, R. A., Zhang, J., Gourley, J. J., & Howard, K. W. (2002). Weather radar coverage over the contiguous United States. *Weather and Forecasting*, *2002*(17), 927–934. doi:10.1175/1520-0434(2002)017<0927:WRCOTC>2.0.CO;2

Maftei, C. (2013, October 17). *Hydrological Drought Assessment in the Dobrogea Region*. Kos 2013- Facets of Uncertainty" Statistical Hydrology (STAHY '13), Kos Island, Greece.

Maftei, C., & Barbulescu, A. (2008). Statistical analysis of climate evolution in Dobrudja region. *Proceedings of the World Congress on Engineering*. Research Gate.

Maftei, C., Barbulescu, A., Buta, C., & Serban, C. (2007). Change points detection and variability analysis of some precipitation series. *Recent Researches in Computational Techniques, Non-Linear Systems and Control*, (pp. 978–1).

Maftei, C., & Barbulescu, A. (2012). Statistical analysis of precipitation time series in the Dobrudja region. *Mausam (New Delhi)*, *63*(4), 553–564. doi:10.54302/mausam.v63i4.416

Maftei, C., Barbulescu, A., & Carsteanu, A. A. (2016). Long-range dependence in the time series of Taiţa River discharges. *Hydrological Sciences Journal*, *61*(9), 1740–1747. doi:10.1080/02626667.2016.1171869

Maftei, C., Buta, C., Draghici, G., & Filip, C. (2019). Analysis of changes in hydrometeorological variables of Techirghiol Lake. *IOP Conference Series. Earth and Environmental Science*, *344*(1), 012025. doi:10.1088/1755-1315/344/1/012025

Maftei, C., Dobrica, G., Cerneaga, C., & Buzgaru, N. (2021). Drought Land Degradation and Desertification—Case Study of Nuntasi-Tuzla Lake in Romania. In A. Vaseashta & C. Maftei (Eds.), *Water Safety, Security and Sustainability: Threat Detection and Mitigation* (pp. 583–597). Springer International Publishing., doi:10.1007/978-3-030-76008-3_25

Maftei, C., & Papatheodorou, K. (2016). Mathematical Models Used for Hydrological Floodplain Modeling. In *Extreme Weather and Impacts of Climate Change on Water Resources in the Dobrogea Region* (pp. 69–100). IGI Global.

Mahdizadeh Gharakhanlou, N., & Perez, L. (2022). Spatial Prediction of Current and Future Flood Susceptibility: Examining the Implications of Changing Climates on Flood Susceptibility Using Machine Learning Models. *Entropy (Basel, Switzerland)*, *24*(11), 1630. doi:10.3390/e24111630 PMID:36359720

Mahto, S. S., & Mishra, V. (2020). Dominance of summer monsoon flash droughts in India. *Environmental Research Letters*, *15*(10), 104061. doi:10.1088/1748-9326/abaf1d

Makanda, K., Nzama, S., Kanyerere, T. (2022). Assessing the Role of Water Resources Protection Practice for Sustainable Water Resources Management: A Review. *Water, 14*(19), 3153. MDPI AG. . doi:10.3390/w14193153

Malgary, F. (1987). Mineralogy book. Geosciences faculty Kabul University, Kabul Afghanistan.

Marchi, L., Borga, M., Preciso, E., & Gaume, E. (2010). Characterization of selected extreme flash floods in Europe and implications for flood risk management. *Journal of Hydrology (Amsterdam)*, *394*(1–2), 118–133. doi:10.1016/j.jhydrol.2010.07.017

Marzban, C., Liu, J., & Tissot, P. (2022). On Variability due to Local Minima and K-Fold Cross Validation. *Artificial Intelligence for the Earth Systems*, *1*(4), e210004. doi:10.1175/AIES-D-21-0004.1

Matese, A., Toscano, P., Di Gennaro, S. F., Genesio, L., Vaccari, F. P., Primicerio, J., Belli, C., Zaldei, A., Bianconi, R., & Gioli, B. (2015). Intercomparison of UAV, aircraft, and satellite remote sensing platforms for precision viticulture. *Remote Sensing (Basel)*, *7*(3), 2971–2990. doi:10.3390/rs70302971

Matthies-Wiesler, F., & Quevauviller, P. (Eds.). (2022). *Hydrometeorological Extreme Events and Public Health*. John Wiley & Sons. doi:10.1002/9781119259350

Mavhura, E. (2019). Systems analysis of vulnerability to hydrometeorological threats: An exploratory study of vulnerability drivers in Northern Zimbabwe. *International Journal of Disaster Risk Science*, *10*(2), 204–219. doi:10.100713753-019-0217-x

McKee, T. B., Doesken, N. J., & Kleist, J. (1993). *The Relationship of Drought Frequency and Duration to Time Scales*. 179–184.

Meis, M., Llano, M. P., & Rodriguez, D. (2022). A statistical tool for a hydrometeorological forecast in the lower La Plata Basin. *International Journal of River Basin Management*, 1–12.

Micu, D. M., Amihaesei, V. A., Milian, N., & Cheval, S. (2021). Recent changes in temperature and precipitation indices in the Southern Carpathians, Romania (1961–2018). *Theoretical and Applied Climatology*, *144*(1-2), 691–710. doi:10.100700704-021-03560-w

Ministry of Environment and Water Management. (2005). HG 1854/2005. Floods risk management national strategy. Bucharest, Romania.

Modarres, R. (2007). Streamflow drought time series forecasting. *Stochastic Environmental Research and Risk Assessment*, *21*(3), 223–233. doi:10.100700477-006-0058-1

Mohamadi, S., Sammen, S. S., Panahi, F., Ehteram, M., Kisi, O., Mosavi, A., Ahmed, A. N., El-Shafie, A., & Al-Ansari, N. (2020). Zoning map for drought prediction using integrated machine learning models with a nomadic people optimization algorithm. *Natural Hazards*, *104*(1), 537–579. doi:10.100711069-020-04180-9

Mohsan, S. A. H., Othman, N. Q. H., Li, Y., Alsharif, M. H., & Khan, M. A. (2023). Unmanned aerial vehicles (UAVs): Practical aspects, applications, open challenges, security issues, and future trends. *Intelligent Service Robotics*, *16*, 109–137. doi:10.100711370-022-00452-4 PMID:36687780

Molnar, P. A. (1990). Review of the seismicity and rates of active under thrusting deformation at the Himalayas. *J. Himalayan Geol.*, *1*, 131–154.

Mondini, A. C., Guzzetti, F., Chang, K.-T., Monserrat, O., Martha, T. R., & Manconi, A. (2021). Landslide failures detection and mapping using Synthetic Aperture Radar: Past, present and future. *Earth-Science Reviews*, *216*, 103574. doi:10.1016/j.earscirev.2021.103574

Montgomery, D. R. (1994). Valley incision and the uplift of mountain peaks. *J. Geophys. Res.*, *99*(13), 913-13, 921.

Moon, W. M., Staples, G., Kim, D., Park, S.-E., & Park, K.-A. (2010). RADARSAT-2 and Coastal Applications: Surface Wind, Waterline, and Intertidal Flat Roughness. *Proceedings of the IEEE*, *98*(5), 800–815. doi:10.1109/JPROC.2010.2043331

Moore, G., McGuire, K., Troch, P., Barron-Gafford, G. (2015). Ecohydrology and the Critical Zone: Processes and Patterns across Scales. *Developments in Earth Surface Processes*, 239–266. . doi:10.1016/B978-0-444-63369-9.00008-2

Moore, D. O. (1976). *Estimating peak discharge from small drainages in Nevada according to basin areas within elevation zones* (Vol. 3). Nevada Highway Department Hydrologic.

Morán-Tejeda, E., Fassnacht, S. R., Lorenzo-Lacruz, J., López-Moreno, J. I., García, C., Alonso-González, E., & Collados-Lara, A. J. (2019). Hydro-meteorological characterization of major floods in Spanish mountain rivers. *Water (Basel)*, *11*(12), 2641. doi:10.3390/w11122641

Morecroft, M. D., & Keith, S. A. (2009). Plant ecology as an indicator of climate and global change. In *Climate Change* (pp. 297–305). Elsevier. doi:10.1016/B978-0-444-53301-2.00016-6

Muhadi, N. A., Abdullah, A. F., Bejo, S. K., Mahadi, M. R., & Mijic, A. (2020). The Use of LiDAR-Derived DEM in Flood Applications: A review. *Remote Sensing (Basel)*, *12*(14), 2308. doi:10.3390/rs12142308

Mukul, S. A., Alamgir, M., Sohel, M. S. I., Pert, P. L., Herbohn, J., Turton, S. M., Khan, M. S. I., Munim, S. A., Reza, A. H. M. A., & Laurance, W. F. (2019). Combined effects of climate change and sea-level rise project dramatic habitat loss of the globally endangered Bengal tiger in the Bangladesh Sundarbans. *The Science of the Total Environment*, *663*, 830–840. doi:10.1016/j.scitotenv.2019.01.383 PMID:30738263

Mulvaney, T. J. (1851). On the use of self-registering rain and flood gauges in making observations of the relations of rainfall and flood discharges in a given catchment. *Proceedings of the Institution of Civil Engineers of Ireland*, *4*, 18–33.

Munsell Color Co. (2023). *Soil color charts*. Munsell Color Company, Inc., USA.

Muzhoffar, D. A. F., Sakuno, Y., Taniguchi, N., Hamada, K., Shimabukuro, H., & Hori, M. (2023). Automatic Detection of Floating Macroalgae via Adaptive Thresholding Using Sentinel-2 Satellite Data with 10 m Spatial Resolution. *Remote Sensing (Basel)*, *15*(8), 2039. doi:10.3390/rs15082039

Nabi, G., Ullah, S., Khan, S., Ahmad, S., & Kumar, S. (2018). China-Pakistan Economic Corridor (CPEC): Melting glaciers—a potential threat to ecosystem and biodiversity. *Environmental Science and Pollution Research International*, *25*(4), 3209–3210. doi:10.100711356-017-1151-3 PMID:29290058

Nakata, T., (1972). Geomorphic history and crustal movements of the foothills of the Himalayas, Sci. Rep. Tohoku Univ. Ser.

Nalbantis, I., & Tsakiris, G. (2009). Assessment of Hydrological Drought Revisited. *Water Resources Management*, *23*(5), 881–897. doi:10.100711269-008-9305-1

Naseem, B., Ajami, H., Liu, Y., Cordery, I., & Sharma, A. (2016). Multi-objective assessment of three remote sensing vegetation products for streamflow prediction in a conceptual ecohydrological model. *Journal of Hydrology (Amsterdam)*, *543*, 686–705. doi:10.1016/j.jhydrol.2016.10.038

Nasiri, V., Sadeghi, S. M. M., Bagherabadi, R., Moradi, F., Deljouei, A., & Borz, S. A. (2022b). Modeling wildfire risk in western Iran based on the integration of AHP and GIS. *Environmental Monitoring and Assessment*, *194*(9), 644. doi:10.100710661-022-10318-y PMID:35930117

Nasiri, V., Sadeghi, S. M. M., Moradi, F., Afshari, S., Deljouei, A., Griess, V. C., Maftei, C., & Borz, S. A. (2022a). The Influence of Data Density and Integration on Forest Canopy Cover Mapping Using Sentinel-1 and Sentinel-2 Time Series in Mediterranean Oak Forests. *ISPRS International Journal of Geo-Information*, *11*(8), 423. doi:10.3390/ijgi11080423

National Oceanic and Atmospheric Administration, National Severe Storms Laboratory. (n.d.). *Flood Forecasting*. NOAA. https://www.nssl.noaa.gov/education/svrwx101/floods/forecasting/

National Oceanic and Atmospheric Administration. (2000). *What are flash floods?* NOAA.

Nayak, M. A., & Ghosh, S. (2013). Prediction of extreme rainfall events using weather pattern recognition and support vector machine classifier. *Theoretical and Applied Climatology*, *2013*(114), 583–603. doi:10.100700704-013-0867-3

Neumann, R. B., Moorberg, C. J., Lundquist, J. D., Turner, J. C., Waldrop, M. P., McFarland, J. W., Euskirchen, E. S., Edgar, C. W., & Turetsky, M. R. (2019). Warming effects of spring rainfall increase methane emissions from thawing permafrost. *Geophysical Research Letters*, *46*(3), 1393–1401. doi:10.1029/2018GL081274

Nguyen, T. N., Pham, C. X., Nguyen, H. Q., & Dang, T. N. B. (2019). Establishing an early warning system for flash floods in Hoang Su Phi District, Ha Giang Province, Vietnam. *Singapore Journal of Tropical Geography*, *40*(2), 312–333. doi:10.1111jtg.12276

Oram, N. J., De Deyn, G. B., Bodelier, P. L., Cornelissen, J. H., van Groenigen, J. W., & Abalos, D. (2020). Plant community flood resilience in intensively managed grasslands and the role of the plant economic spectrum. *Journal of Applied Ecology*, *57*(8), 1524–1534. doi:10.1111/1365-2664.13667

Otkin, J. A., Svoboda, M., Hunt, E. D., Ford, T. W., Anderson, M. C., Hain, C., & Basara, J. B. (2018). Flash Droughts: A Review and Assessment of the Challenges Imposed by Rapid-Onset Droughts in the United States. *Bulletin of the American Meteorological Society*, *99*(5), 911–919. doi:10.1175/BAMS-D-17-0149.1

Oyedotun, T. D. T., Moonsammy, S., Renn-Moonsammy, J., Renn-Moonsammy, D. M., Nedd, G. A., Bishop, J., & Burningham, H. (2022). Data consolidation for coastal management: The development of the Guyana coastal database platform. *Interdisciplinary Environmental Review*, *22*(2), 143–162. doi:10.1504/IER.2022.125019

Paetzold, S. (2005). *Arbeitung zur Durchfuerung von Boden Analyses* (10th ed.). Institut fuer Bodenkunde, University Bonn.

Palmer, W. C. (1965). Meteorological Drought. *Weather Bureau Research Paper, 45*, 1–58.

Pamela, S. F., Hyun-Han, K., Meththika, V., & Joo-Heon, L. (2023). Modeling, challenges, and strategies for understanding impacts of climate extremes (droughts and floods) on water quality in Asia: A review. *Environmental Research*, *225*, 115617. doi:10.1016/j.envres.2023.115617 PMID:36871941

Pandey, A. C., Kaushik, K., & Parida, B. R. (2022). Google Earth Engine for large-scale flood mapping using SAR data and impact assessment on agriculture and population of Ganga-Brahmaputra basin. *Sustainability (Basel)*, *14*(7), 4210. doi:10.3390u14074210

Panwar, V., & Sen, S. (2019). Economic impact of natural disasters: An empirical re-examination. *Margin - the Journal of Applied Economic Research*, *13*(1), 109–139. doi:10.1177/0973801018800087

Papacharalampous, G., & Tyralis, H. (2022). Time series features for supporting hydrometeorological explorations and predictions in ungauged locations using large datasets. *Water (Basel)*, *14*(10), 1657. doi:10.3390/w14101657

Park, S., Berenguer, M., Sempere-Torres, D., Baugh, C., & Smith, P. (2017). Toward seamless high-resolution flash flood forecasting over Europe based on radar nowcasting and NWP: An evaluation with case studies. *Geophysical Research Abstracts*, *19*, EGU2017–EGU12158.

Partnership (GWP). G. W., & Organization (WMO), W. M. (2013). *Flood Mapping*. WMO. https://library.wmo.int/records/item/37083-flood-mapping

Patro, S., Chatterjee, C., Mohanty, S., Singh, R., & Raghuwanshi, N. S. (2009). Flood inundation modeling using MIKE FLOOD and remote sensing data. *Photonirvachak (Dehra Dun)*, *37*(1), 107–118. doi:10.100712524-009-0002-1

Pelletier, J. D., & Turcotte, D. L. (1997). Long-range persistence in climatological and hydrological time series: Analysis, modeling, and application to drought hazard assessment. *Journal of Hydrology (Amsterdam)*, *203*(1), 198–208. doi:10.1016/S0022-1694(97)00102-9

Perera, D., Seidou, O., Agnihotri, J., Mehmood, H., & Rasmy, M. (2020). Challenges and technical advances in flood early warning systems (FEWSs). *Flood impact mitigation and resilience enhancement.*

Perera, D., Agnihotri, J., Seidou, O., & Djalante, R. (2020). Identifying societal challenges in flood early warning systems. *International Journal of Disaster Risk Reduction, 51*, 101794. doi:10.1016/j.ijdrr.2020.101794

Petrescu (Maftei). C. (2002). *Etudes concernant les écoulements superficiels modélisation spatialisée de l'écoulement sur le bassin versant de Voinesti-Roumanie* [PhD Thesis, Université de Montpellier II].

Petrişor, A.-I., Sirodoev, I., & Ianoş, I. (2020). Trends in the national and regional transitional dynamics of land cover and use changes in Romania. *Remote Sensing (Basel), 12*(2), 230. doi:10.3390/rs12020230

Petropoulos, G. P., & Islam, T. (Eds.). (2017). *Remote sensing of hydrometeorological hazards.* CRC Press. doi:10.1201/9781315154947

Pettitt, A. N. (1979). A Non-Parametric Approach to the Change-Point Problem. *Applied Statistics, 28*(2), 126–135. doi:10.2307/2346729

Phillips, B. C., Yu, S., Thompson, G. R., & De Silva, N. (2005). *1D and 2D modeling of urban drainage systems using XP-SWMM and TUFLOW.* In 10th international conference on urban drainage (pp. 21-26). Copenhagen, Denmark.

Piao, S., Ciais, P., Huang, Y., Shen, Z., Peng, S., Li, J., Zhou, L., Liu, H., Ma, Y., Ding, Y., Friedlingstein, P., Liu, C., Tan, K., Yu, Y., Zhang, T., & Fang, J. (2010). The impacts of climate change on water resources and agriculture in China. *Nature, 467*(7311), 43–51. doi:10.1038/nature09364 PMID:20811450

Piao, S., Liu, Q., Chen, A., Janssens, I. A., Fu, Y., Dai, J., Liu, L., Lian, X., Shen, M., & Zhu, X. (2019). Plant phenology and global climate change: Current progresses and challenges. *Global Change Biology, 25*(6), 1922–1940. doi:10.1111/gcb.14619 PMID:30884039

Piri, J., Abdolahipour, M., & Keshtegar, B. (2023). Advanced Machine Learning Model for Prediction of Drought Indices using Hybrid SVR-RSM. *Water Resources Management, 37*(2), 683–712. doi:10.100711269-022-03395-8

Pons, L. J. (1992). Natural resources. In *Conservation Status of the Danube Delta* (Vol. 4, pp. 23–36). IUCN.

Popovici, N. (1991). *Soil Erosion Control Practice.* Politechnical Institute Lithography.

Potter, S., Harrison, S., & Kreft, P. (2021). The benefits and challenges of implementing impact-based severe weather warning systems: Perspectives of weather, flood, and emergency management personnel. *Weather, Climate, and Society, 13*(2), 303–314. doi:10.1175/WCAS-D-20-0110.1

Pourshanbeh, Y., Khoorani, A., & Jafari, L. (2022). The Impact of Climate Change on Jiroft Spring Wheat Yield Using Radiative Forcing Scenarios. *Land Ecology, 1*(1), 1-13. https://dorl.net/dor/20.1001.1.28211782.1401.1.1.1.3

Pouyan, S., Rahmanian, S., Amindin, A., & Pourghasemi, H. R. (2022). Spatial and seasonal modeling of the land surface temperature using random forest. In *Computers in Earth and Environmental Sciences* (pp. 221–234). Elsevier. doi:10.1016/B978-0-323-89861-4.00035-X

Pradhan, B., Moneir, A. A. A., & Jena, R. (2018). Sand dune risk assessment in Sabha region, Libya using Landsat 8, MODIS, and Google Earth Engine images. *Geomatics, Natural Hazards & Risk, 9*(1), 1280–1305. doi:10.1080/19475705.2018.1518880

Prăvălie, R., Patriche, C. V., Sirodoev, I., Bandoc, G., Dumitraşcu, M., & Peptenatu, D. (2016). Water deficit and corn productivity during the post-socialist period. Case study: Southern Oltenia drylands, Romania. *Arid Land Research and Management, 30*(3), 239–257. doi:10.1080/15324982.2015.1091399

Prăvălie, R., Sîrodoev, I., Nita, I.-A., Patriche, C., Dumitraşcu, M., Roşca, B., Tişcovschi, A., Bandoc, G., Săvulescu, I., Mănoiu, V., & Birsan, M.-V. (2022). NDVI-based ecological dynamics of forest vegetation and its relationship to climate change in Romania during 1987–2018. *Ecological Indicators*, *136*, 108629. doi:10.1016/j.ecolind.2022.108629

Praveen, B., & Sharma, P. (2019). A review of literature on climate change and its impacts on agriculture productivity. *Journal of Public Affairs*, *19*(4), e1960. doi:10.1002/pa.1960

Qian, C., Qiang, H., Wang, F., & Li, M. (2021). Optimization of Rocky Desertification Classification Model Based on Vegetation Type and Seasonal Characteristic. *Remote Sensing (Basel)*, *13*(15), 2935. doi:10.3390/rs13152935

Quesada-Román, A., & Villalobos-Chacón, A. (2020). Flash flood impacts of Hurricane Otto and hydrometeorological risk mapping in Costa Rica. *Geografisk Tidskrift*, *120*(2), 142–155. doi:10.1080/00167223.2020.1822195

Quiroga, G., Erice, G., Aroca, R., Zamarreño, Á. M., García-Mina, J. M., & Ruiz-Lozano, J. M. (2020). Radial water transport in arbuscular mycorrhizal maize plants under drought stress conditions is affected by indole-acetic acid (IAA) application—. *Journal of Plant Physiology*, *246-247*, 153115. doi:10.1016/j.jplph.2020.153115 PMID:31958683

Rafaqat, W., Iqbal, M., Kanwal, R., & Weiguo, S. (2022). Evaluation of Wildfire Occurrences in Pakistan with Global Gridded Soil Properties Derived from Remotely Sensed Data. *Remote Sensing (Basel)*, *14*(21), 5503. doi:10.3390/rs14215503

Rao, P., Gupta, K., Roy, A., & Balan, R. (2021). Spatio-temporal analysis of land surface temperature for identification of heat wave risk and vulnerability hotspots in Indo-Gangetic Plains of India. *Theoretical and Applied Climatology*, *146*(1-2), 567–582. doi:10.100700704-021-03756-0

Rasheed, M. W., Tang, J., Sarwar, A., Shah, S., Saddique, N., Khan, M. U., Imran Khan, M., Nawaz, S., Shamshiri, R. R., Aziz, M., & Sultan, M. (2022). Soil Moisture Measuring Techniques and Factors Affecting the Moisture Dynamics: A Comprehensive Review. *Sustainability (Basel)*, *14*(18), 11538. doi:10.3390u141811538

Rasouli H., (2021b). Analysis of Groundwater Quality in Jabal Sarage and Charikar Districts, Parwan, Afghanistan. *Journal of Geological Research, 3*. doi:10.30564/jgr.v3i4.3717

Rasouli H., Qureshi R., Belhassan K., (2021c). Investigations on River Sediments in Chak Sedimentary Basin, Wardak Province, Afghanistan. *Journal of Geological Research, 3*.. doi:10.30564/jgr.v3i4.3574

Rasouli, H. (2019). A Study on Some River Sediments, Hydrology and Geological Characteristics in Chak Sedimentary Basin, Wardak, Afghanistan. *International Journal of Geology, Earth & Environmental Sciences*, *9*(2).

Rasouli, H. (2019). A Study on Some River Sediments, Hydrology and Geological Characteristics In Chak Sedimentary Basin, Wardak, Afghanistan. *International Journal of Geology, Earth & Environmental Sciences.*

Rasouli, H. (2020c). Well Design and Stratigraphy of Sheerkhana Deep Well in Chak District, Wardak, Afghanistan. *International Journal of Geology, Earth & Environmental Sciences.* https://www.cibtech.org/jgee.htm2020.

Rasouli, H. (2020c). Well, Design and Stratigraphy of Sheer Khana Deep Well in Chak District, Wardak, Afghanistan. *International Journal of Geology, Earth & Environmental Sciences, 10*(2) May-August, pp.54-68/Rasouli

Rasouli, H. (2022a). Climate Change Impacts on Water Resource and Air Pollution in Kabul Sub-basins, Afghanistan. *Advances in Geological and Geotechnical Engineering Research, 04*(1). https://ojs.bilpublishing.com/index.php/agger

Rasouli, H. (2022a). Climate Change Impacts on Water Resource and Air Pollution in Kabul Sub-basins, Afghanistan. *Advances in Geological and Geotechnical Engineering Research, 04*. https://ojs.bilpublishing.com/index.php/agger

Rasouli, H., Kayastha, R. B., Bikas C. B., Ahuti S., Arian, H., Armstrong, R., (2015). Estimation of Discharge from Upper Kabul River Basin, Afghanistan Using the Snowmelt Runoff Model. *Journal of Hydrology and Meteorology, 9.* . doi:10.3126/jhm.v9i1.15584

Rasouli, H. (2020a). *Application of soil physical and chemical parameters and it's Comparing in Kabul Sedimentary basins.*, doi:10.24327/ijrsr.2020.1102.5095

Rasouli, H. (2022b). Methods of Well Construction Complication, Design, and Developing for Sixteen Observation and Test Wells at the Eight Locations of Zarang district, Nimroz, Afghanistan. *International Journal of Earth Science Knowledge and Applications*, *4*(3), 426–448.

Rasouli, H. (2022b). Methods of Well Construction Complication, Design, and Development for Sixteen Observation and Test Wells at the Eight Locations of Zarang district, Nimroz, Afghanistan. *International Journal of Earth Science Knowledge and Applications*, *4*(3), 426–448.

Rasouli, H. Sarwari, M. H., Khairuddin R.· Said A. H., (2020a). Geological Study of Tangi Mahipar Mountain Range along Kabul Jalalabad Road. *Afghanistan*. doi:10.4236/ojg.2020.1010044

Rasouli, H., & Safi, A. G. (2021a). Geological, Soil and Sediment Studies in Chelsaton Sedimentary Basin, Kabul, Afghanistan. *International Journal of Geosciences*, *12*(2), 170–193. doi:10.4236/ijg.2021.122011

Rasouli, H., & Vaseashta, A. (2023). Investigation of Physicochemical Properties of Qalay Abdul Ali Soil, Kabul, Afghanistan. *Advances in Geological and Geotechnical Engineering Research*, *5*(3), 55–68. doi:10.30564/agger.v5i3.5773

Rasouli, H., Vaseashta, A., & Belhassan, K. (2023). Mechanical Analysis of Khair Abad Village, Surskhrud District, Nangarhar Province, Afghanistan. *International Journal of Earth Sciences Knowledge and Applications.*, *5*(1), 103–118. http://www.ijeska.com/index.php/ijeska

Rasouli, H., Vaseashta, A., & Hamdard, M. H. (2023). Sedimentological Study of Chack Hydropower Reservoir, Wardak, Afghanistan. *International Journal of Earth Sciences Knowledge and Applications*, *5*(1), 21–32. http://www.ijeska.com/index.php/ijeska

Razali, N., Ismail, S., & Mustapha, A. (2020). Machine learning approach for flood risk prediction. *IAES International Journal of Artificial Intelligence*, *9*(1), 73–80. doi:10.11591/ijai.v9.i1.pp73-80

Reliefweb. (2022). *Practical Action's approach in disseminating localized flood early warning in Bangladesh.* ReliefWeb. https://reliefweb.int/report/bangladesh/practical-actions-approach-disseminating-localized-flood-early-warning-bangladesh

Richardson, A. D., Keenan, T. F., Migliavacca, M., Ryu, Y., Sonnentag, O., & Toomey, M. (2013). Climate change, phenology, and phenological control of vegetation feedbacks to the climate system. *Agricultural and Forest Meteorology*, *169*, 156–173. doi:10.1016/j.agrformet.2012.09.012

Rivero, R. M., Mittler, R., Blumwald, E., & Zandalinas, S. I. (2022). Developing climate-resilient crops: Improving plant tolerance to stress combination. *The Plant Journal*, *109*(2), 373–389. doi:10.1111/tpj.15483 PMID:34482588

Roshan, G., Sarli, R., & Grab, S. W. (2021). The case of Tehran's urban heat island, Iran: Impacts of urban "lockdown" associated with the COVID-19 pandemic. *Sustainable Cities and Society*, *75*, 103263. doi:10.1016/j.scs.2021.103263 PMID:36568531

Roy, D. P., Wulder, M. A., Loveland, T. R., Woodcock, C. E., Allen, R. G., Anderson, M. C., & Zhu, Z. (2014). Landsat-8: Science and product vision for terrestrial global change research. *Remote Sensing of Environment*, *145*, 154–172. doi:10.1016/j.rse.2014.02.001

Ruleman, C. A., Crone, A. J., Machette, M. N., Haller, K. M., & Rukstales, K. S. (2007). *Geological Map and database of probable and possible Quaternary faults in Afghanistan.* U.S Survey Open-File Report 2007–1103. US Survey.

Sadek, M., & Li, X. (2019). Low-Cost solution for assessment of urban flash flood impacts using Sentinel-2 satellite images and fuzzy analytic hierarchy process: A case study of Ras Ghareb City, Egypt. *Advances in Civil Engineering, 2019*, 1–15. doi:10.1155/2019/2561215

Saghafian, B., Meghdadi, A. R., & Sima, S. (2015). Application of the WEPP model to determine sources of run-off and sediment in a forested watershed. *Hydrological Processes, 29*(4), 481–497. doi:10.1002/hyp.10168

Saleh, A., Salama, M. S., Elhag, M., & Elagib, N. (2020). Assessing the accuracy of machine learning algorithms in mapping flood inundation using Sentinel-1 SAR imagery and Google Earth Engine platform. *Water (Basel), 12*(5), 1405. doi:10.3390/w12051405

Salinas, J. L., Laaha, G., Rogger, M., Parajka, J., Viglione, A., Sivapalan, M., & Blöschl, G. (2013). Comparative assessment of predictions in ungauged basins – Part 2: Flood and low flow studies. *Hydrology and Earth System Sciences, 17*(7), 2637–2652. doi:10.5194/hess-17-2637-2013

Salmi, T., Määttä, A., Pia, A., Ruoho-Airola, T., & Amnell, T. (2002). Detecting Trends of Annual Values of Atmospheric Pollutants by the Mann-Kendall Test and Sen's Solpe Estimates the Excel Template Application MAKESENS. *Publications on Air Quality, 31.*

Samansiri, S., Fernando, T., & Ingirige, B. (2022). Advanced Technologies for Offering Situational Intelligence in Flood Warning and Response Systems: A Literature Review. *Water (Basel), 14*(13), 2091. doi:10.3390/w14132091

Sample, M., Thode, A. E., Peterson, C., Gallagher, M. R., Flatley, W., Friggens, M., Evans, A., Loehman, R., Hedwall, S., Brandt, L., Janowiak, M., & Swanston, C. (2022). Adaptation Strategies and Approaches for Managing Fire in a Changing Climate. *Climate (Basel), 10*(4), 58. doi:10.3390/cli10040058

Sarker, M. N. I., Peng, Y., Yiran, C., & Shouse, R. C. (2020). Disaster resilience through big data: Way to environmental sustainability. *International Journal of Disaster Risk Reduction, 51*, 101769. doi:10.1016/j.ijdrr.2020.101769

Sati, S. P., & Gahalaut, V. K. (2013). The fury of the floods in the north-west Himalayan region: The Kedarnath tragedy. *Geomatics, Natural Hazards & Risk, 4*(3), 193–201. doi:10.1080/19475705.2013.827135

Sati, V. P. (2013). Extreme weather-related disasters: A case study of two flash floods hit areas of Badrinath and Kedarnath Valleys, Uttarakhand Himalaya, India. *Journal of Earth Science and Engineering, 3*(8), 562.

Scheip, C. M., & Wegmann, K. W. (2021). HazMapper: A global open-source natural hazard mapping application in Google Earth Engine. *Natural Hazards and Earth System Sciences, 21*(5), 1495–1511. doi:10.5194/nhess-21-1495-2021

Schumann, G., Giustarini, L., Tarpanelli, A., Jarihani, B., & Martinis, S. (2022). Flood Modeling and Prediction Using Earth Observation Data. *Surveys in Geophysics*, 1–26.

Schwanghart, W., Worni, R., Huggel, C., Stoffel, M., & Korup, O. (2016). Uncertainty in the Himalayan energy–water nexus: Estimating regional exposure to glacial lake outburst floods. *Environmental Research Letters, 11*(7), 074005. doi:10.1088/1748-9326/11/7/074005

Schwank, J., Escobar, R., Girón, G. H., & Morán-Tejeda, E. (2014). Modeling of the Mendoza River watershed as a tool to study climate change impacts on water availability. *Environmental Science & Policy, 43*, 91–97. doi:10.1016/j.envsci.2014.01.002

Seleiman, M. F., Al-Suhaibani, N., Ali, N., Akmal, M., Alotaibi, M., Refay, Y., Dindaroglu, T., Abdul-Wajid, H. H., & Battaglia, M. L. (2021). Drought Stress Impacts on Plants and Different Approaches to Alleviate Its Adverse Effects. *Plants*, *10*(2), 259. doi:10.3390/plants10020259 PMID:33525688

Selvarajah, H., Koike, T., Rasmy, M., Tamakawa, K., Yamamoto, A., Kitsuregawa, M., & Zhou, L. (2021). Development of an integrated approach for the assessment of climate change impacts on the hydro-meteorological characteristics of the Mahaweli river basin, Sri Lanka. *Water (Basel)*, *13*(9), 1218. doi:10.3390/w13091218

Sene, K., & Sene, K. (2016). Flash floods. *Hydrometeorology: forecasting and applications*, 273-312.

Sene, K. (2012). *Flash floods: forecasting and warning*. Springer Science & Business Media.

Seneviratne, S. I., Nicholls, N., Easterling, D., Goodess, C. M., Kanae, S., Kossin, J., Luo, Y., Marengo, J., McInnes, K., Rahimi, M., Reichstein, M., Sorteberg, A., Vera, C., & Zhang, X. 2012: Changes in climate extremes and their impacts on the natural physical environment. In: Managing the Risks of Extreme Events and Disasters to Advance Climate Change Adaptation. Cambridge University Press. doi:10.1017/CBO9781139177245.006

Senthil Kumar, A., Sudheer, K., Jain, S., & Agarwal, P. (2005). Rainfall-runoff modeling using artificial neural networks: Comparison of network types. *Hydrological Processes*, *2005*(19), 1277–1291. doi:10.1002/hyp.5581

Seo, D. J., & Breidenbach, J. (2002). Real-time correction of spatially nonuniform bias in radar rainfall data using rain gauge measurements. *Journal of Hydrometeorology*, *2002*(3), 93–111. doi:10.1175/1525-7541(2002)003<0093:RTCOSN>2.0.CO;2

Seo, Y., Kim, S., Kisi, O., & Singh, V. P. (2015). Daily water level forecasting using wavelet decomposition and artificial intelligence techniques. *Journal of Hydrology (Amsterdam)*, *2015*(520), 224–243. doi:10.1016/j.jhydrol.2014.11.050

Serban, C., & Maftei, C. (2011). Thermal analysis of climate regions using remote sensing and grid computing. *ArXiv Preprint ArXiv:1101.4474.*

Serban, C., Maftei, C., & Barbulescu, A. (2010). Assessment of evapotranspiration using remote sensing data and grid computing and application. *WSEAS Transactions on Computers*, *9*(11), 1245–1254.

Shafaei, M., & Kisi, O. (2017). Predicting river daily flow using wavelet-artificial neural networks based on regression analyses in comparison with artificial neural networks and support vector machine models. *Neural Computing & Applications*, *2017*(28), 15–28. doi:10.100700521-016-2293-9

Shamal, S., & Rasouli, H. (2018). Comparison between pH, EC, $CaCO_3$, and mechanical analysis of Qala Wahid and Company Areas soil, Kabul, Afghanistan. *International Journal of Scientific Research*, *8*(5). Advance online publication. doi:10.21275/ART20197318

Sharma, K., Anand, D., Sabharwal, M., Tiwari, P. K., Cheikhrouhou, O., & Frikha, T. (2021). A disaster management framework using internet of things-based interconnected devices. *Mathematical Problems in Engineering*, *2021*, 1–21. doi:10.1155/2021/9916440

Sheffield, J., Andreadis, K., Wood, E., & Lettenmaier, D. (2009). Global and continental drought in the second half of the twentieth century: Severity-area-duration analysis and temporal variability of large scale events. *Journal of Climate*, *22*(8), 1962–1981. doi:10.1175/2008JCLI2722.1

Shen, F., Yang, L., Zhang, L., Guo, M., Huang, H., & Zhou, C. (2023). Quantifying the direct effects of long-term dynamic land use intensity on vegetation change and its interacted effects with economic development and climate change in jiangsu, China. *Journal of Environmental Management*, *325*, 116562. doi:10.1016/j.jenvman.2022.116562 PMID:36308967

Sherman, l., (1932b). The relation of hydrographs of runoff to size and character of drainage basins. *Eos, Transactions American Geophysical Union, 13*(1), 332–339. doi:10.1029/TR013i001p00332

Sherman, I. (1932a). Streamflow from Rainfall by Unit-Graph Method. *Engineering News Record, 108*, 501–505.

Sheskin, D. J. (2000). *Parametric and Nonparametric Statistical Procedures* (2nd ed.). Chapman & Hall.

Shew, A. M., & Ghosh, A. (2017). USING MULTI-TEMPORAL REMOTE SENSING DATA TO ANALYZE THE SPATIO-TEMPORAL PATTERNS OF DRY SEASON RICE PRODUCTION IN BANGLADESH. *ISPRS Annals of the Photogrammetry, Remote Sensing and Spatial Information Sciences, IV-4*(W2), 4. doi:10.5194/isprs-annals-IV-4-W2-61-2017

Shoaib, M., Shamseldin, A. Y., & Melville, B. W. (2014). Comparative study of different wavelet-based neural network models for rainfall–runoff modeling. *Journal of Hydrology (Amsterdam), 2014*(515), 47–58. doi:10.1016/j.jhydrol.2014.04.055

Shu, C., & Ouarda, T. (2008). Regional flood frequency analysis at ungauged sites using the adaptive neuro-fuzzy inference system. *Journal of Hydrology (Amsterdam), 2008*(349), 31–43. doi:10.1016/j.jhydrol.2007.10.050

Singh, A., Gaurav, K., Meena, G. K., & Kumar, S. (2020). Estimation of Soil Moisture Applying Modified Dubois Model to Sentinel-1; A Regional Study from Central India. *Remote Sensing (Basel), 12*(14), 2266. doi:10.3390/rs12142266

Singh, D., Horton, D. E., Tsiang, M., Haugen, M., Ashfaq, M., Mei, R., & Diffenbaugh, N. S. (2014). Severe precipitation in Northern India in June 2013: Causes, historical context, and changes in probability. *Bulletin of the American Meteorological Society, 95*(9), S58–S61.

Smakhtin, V. U., & Hughes, D. A. (2004). *Review, Automated Estimation and Analyses of Drought Indices in South Asia* (No. 83; Working Paper 83). International Water Management Institute.

Smakhtin, V. U. (2001). Low flow hydrology: A review. [Elsevier.]. *Journal of Hydrology (Amsterdam), 240*(3-4), 147–186. doi:10.1016/S0022-1694(00)00340-1

Solomon, S., Qin, D., Manning, M., Chen, Z., Marquis, M., & Averyt, K. B. (Eds.). (2007). *Contribution of Working Group I to the Fourth Assessment Report of the Intergovernmental Panel on Climate Change*. Cambridge University Press.

Son, N. T., Chen, C. F., Chen, C. R., Masferrer, M. G. M., & Recinos, L. E. M. (2018). Multitemporal Landsat-MODIS fusion for cropland drought monitoring in El Salvador. *Geocarto International, 34*(12), 1363–1383. doi:10.1080/10106049.2018.1489421

Souriau, M. E., Ruddiman, W. F., & Froelich, P. N. (1988). Influence of late Cenozoic Mountain building on ocean geochemical cycles. *Geology, I6*, 649–653.

Sousa, D., & Small, C. (2023). Which Vegetation Index? Benchmarking Multispectral Metrics to Hyperspectral Mixture Models in Diverse Cropland. *Remote Sensing (Basel), 15*(4), 971. doi:10.3390/rs15040971

Spinoni, J., Naumann, G., Vogt, J. V., & Barbosa, P. (2015). (2015). The biggest drought events in Europe from 1950 to 2012. *Journal of Hydrology. Regional Studies, 3*, 509–524. doi:10.1016/j.ejrh.2015.01.001

Stagge, J. H., Tallaksen, L. M., Kohn, I., Stahl, K., & van Loon, A. (2013). *A European Drought Reference Database: Design and Online Implementation*. (DROUGHT-R&SPI, technical Report; No. 12). EU. https://edepot.wur.nl/291635

Stahl, K., Vidal, J.-P., Hannaford, J., Tijdeman, E., Laaha, G., Gauster, T., & Tallaksen, L. M. (2020). The challenges of hydrological drought definition, quantification, and communication: an interdisciplinary perspective. *Proc. IAHS*. IEEE. 10.5194/piahs-383-291-2020

Stančikaitė, M., Zernitskaya, V., Kluczynska, G., Valūnas, D., Gedminienė, L., Uogintas, D., & Šeirienė, V. (2021). The Lateglacial and Early Holocene vegetation dynamics: New multi-proxy data from the central Belarus (Vol. 630). Quaternary International. , PP 121-136. doi:10.1016/j.quaint.2021.05.004

Stanescu, V.A., & Drobot, R. (2002). *Non-structural measures for flood management.* H*G*A* Publishing House.

Staudniger, M., Stahl, K., & Seibert, J. (2014). A drought index accounting for snow. *Water Resources Research, 50*(10), 7861–7872. doi:10.1002/2013WR015143

Stringer, L. C., Mirzabaev, A., Benjaminsen, T. A., Harris, R. M. B., Jafari, M., Lissner, T. K., Stevens, N., & Tirado-von der Pahlen, C. (2021). Climate change impacts on water security in global drylands. *One Earth, 4*(6), 851–864. doi:10.1016/j.oneear.2021.05.010

Subiyantoro, A., Van Westen, C. J., Den Bout, B. V., Yuniawan, R. A., & Mulyana, A. R. (2022, November). Semi-automatic Landslide Detection Using Google Earth Engine, a Case Study in Poi Village, Central Sulawesi. In *2022 IEEE International Conference on Aerospace Electronics and Remote Sensing Technology (ICARES)* (pp. 1-4). IEEE. 10.1109/ICARES56907.2022.9993507

Sudhishri, S., Kumar, A., & Singh, J. K. (2016). Comparative evaluation of neural network and regression-based models to simulate runoff and sediment yield in an outer Himalayan watershed. *J. Agric. Sci. Technol., 2016*(18), 681–694.

Sulaiman, J., & Wahab, S. H. (2017). Heavy rainfall forecasting model using artificial neural network for flood-prone areas. In It Convergence and Security 2017. Springer: Singapore.

Summerfield, M. A., & Hulton, N. J. (1994). Hulton, Natural controls of fluvial denudation rates in Major world drainage basins. *Journal of Geophysical Research, 99*(B7), 13,871–13,883. doi:10.1029/94JB00715

Sundararajan, K., Garg, L., Srinivasan, K., Bashir, A., Kaliappan, J., Ganapathy, G., Selvaraj, S., & Meena, T. (2021). A Contemporary Review on Drought Modeling Using Machine Learning Approaches. *Computer Modeling in Engineering & Sciences, 128*(2), 447–487. doi:10.32604/cmes.2021.015528

Sutanto, S. J., & Van Lanen, H. A. J. (2020). Hydrological Drought Characteristics Based on Groundwater and Runoff Across Europe, *Proc. IAHS*. IEEE. 10.5194/piahs-383-281-2020

Tallaksen, L., & van Lanen, H. A. J. (2004). *Hydrological drought. Processes and estimation methods for streamflow and groundwater.* Elsevier. https://research.wur.nl/en/publications/hydrological-drought-processes-and-estimation-methods-for-streamf

Tallaksen, L. M., & Van Lanen, H. A. J. (2004). *Hydrological drought: processes and estimation methods for streamflow and groundwater. Developments in water science*. Elsevier.

Tamiminia, H., Salehi, B., Mahdianpari, M., Quackenbush, L., Adeli, S., & Brisco, B. (2020). Google Earth Engine for geo-big data applications: A meta-analysis and systematic review. *ISPRS Journal of Photogrammetry and Remote Sensing, 164*, 152–170. doi:10.1016/j.isprsjprs.2020.04.001

Tansar, H., Babur, M., & Karnchanapaiboon, S. L. (2020). Flood inundation modeling and hazard assessment in Lower Ping River Basin using MIKE FLOOD. *Arabian Journal of Geosciences, 13*(18), 1–16. doi:10.100712517-020-05891-w

Tayebi, A., Kasmaeeyazdi, S., Tinti, F., & Bruno, R. (2023). Contributions from experimental geostatistical analyses for solving the cloud-cover problem in remote sensing data. *International Journal of Applied Earth Observation and Geoinformation, 118*, 103236. doi:10.1016/j.jag.2023.103236

Tayfur, G., Singh, V. P., Moramarco, T., & Barbetta, S. (2018). Flood hydrograph prediction using machine learning methods. *Water (Basel), 10*(8), 968. doi:10.3390/w10080968

Taylor, W. (2000). *Change-Point Analysis: A Powerful New Tool For Detecting Changes.* Variation. https://www.variation.com/cpa

Tehrany, M. S., Pradhan, B., & Jebur, M. N. (2013). Spatial prediction of flood susceptible areas using rule-based decision tree (DT) and a novel ensemble bivariate and multivariate statistical models in GIS. *Journal of Hydrology (Amsterdam), 2013*, 504. doi:10.1016/j.jhydrol.2013.09.034

Telesca, L., Aromando, A., Faridani, F., Lovallo, M., Cardettini, G., Abate, N., Papitto, G., & Lasaponara, R. (2022). Exploring Long-Term Anomalies in the Vegetation Cover of Peri-Urban Parks Using the Fisher-Shannon Method. *Entropy (Basel, Switzerland), 24*(12), 1784. doi:10.3390/e24121784 PMID:36554193

Telesca, L., Lovallo, M., Lopez-Moreno, I., & Vicente-Serrano, S. (2012). Investigation of scaling properties in monthly streamflow and Standardized Streamflow Index (SSI) time series in the Ebro basin (Spain). *Physica A, 391*(4), 1662–1678. doi:10.1016/j.physa.2011.10.023

Tellman, B., Sullivan, J., Kettner, A., Brakenridge, G. R., Slayback, D. A., Kuhn, C., & Doyle, C. (2016, December). Developing a global database of historic flood events to support machine learning flood prediction in Google Earth Engine. In AGU Fall Meeting Abstracts (Vol. 2016, pp. H13Q-06). Research Gate.

Temimi, M., Leconte, R., Chaouch, N., Sukumal, P., Khanbilvardi, R., & Brissette, F. (2010). A combination of remote sensing data and topographic attributes for the spatial and temporal monitoring of soil wetness. *Journal of Hydrology (Amsterdam), 388*(1-2), 28–40. doi:10.1016/j.jhydrol.2010.04.021

Teodoro, A. C., & Duarte, L. (2022). The synergy of remote sensing and geographical information systems in the management of natural disasters. In *Nanotechnology-Based Smart Remote Sensing Networks for Disaster Prevention* (pp. 217–230). Elsevier. doi:10.1016/B978-0-323-91166-5.00023-9

Thomas, D. M., & Benson, M. A. (1970). Generalization of streamflow characteristics from drainage-basin characteristics. In *Water Supply Paper*. U.S. Govt. Print. Off. (Original work published 1975), doi:10.3133/wsp1975

Thuiller, W. (2007). Climate change and the ecologist. *Nature, 448*(7153), 550–552. doi:10.1038/448550a PMID:17671497

Tigkas, D., Vangelis, H., & Tsakiris, G. (2013). *The Drought Indices Calculator.* DrinC.

Torabi, O., & Karimi, N. shesh angosht, S., Rashtbari, M., & Sarbazvatan, A. H. (2023). Investigating the effect of radar images in classifying land use classes in machine learning-based algorithms. *Land Ecology, 2*(2), 137-150. https://dorl.net/dor/20.1001.1.28211782.1402.1.1.4.0

Torge, T., Georg, H., & Thomas, H., (2003). *Hydrogeology of the Kabul Basin Part I: Geology, aquifer characteristics, climate, and hydrography.* Foreign office of the Federal Republic of Germany.

Traganos, D., Aggarwal, B., Poursanidis, D., Topouzelis, K., Chrysoulakis, N., & Reinartz, P. (2018). Towards Global-Scale Seagrass Mapping and Monitoring using Sentinel-2 on Google Earth Engine: The case study of the Aegean and Ionian Seas. *Remote Sensing (Basel), 10*(8), 1227. doi:10.3390/rs10081227

Trambadia, N. K., Patel, D. P., Patel, V. M., & Gundalia, M. J. (2023). Flood inundation assessment in the data-scarce region using hydrodynamic models and google earth engine-A case of Ghed region, Ozat River basin, Gujarat, India. Research Gate.

Tripathy, P., & Malladi, T. (2022). Global Flood Mapper: A novel Google Earth Engine application for rapid flood mapping using Sentinel-1 SAR. *Natural Hazards, 114*(2), 1341–1363. doi:10.100711069-022-05428-2

Tucker, C. J., Pinzon, J. E., Brown, M. E., Slayback, D. A., Pak, E. W., Mahoney, R., Vermote, E. F., & El Saleous, N. (2005). An extended AVHRR 8-km NDVI dataset compatible with MODIS and SPOT vegetation NDVI data. *International Journal of Remote Sensing*, *26*(20), 4485–4498. doi:10.1080/01431160500168686

Tu, H., Wang, X., Zhang, W., Peng, H., Ke, Q., & Chen, X. (2020). Flash flood early warning coupled with hydrological simulation and the rising rate of the flood stage in a mountainous small watershed in Sichuan Province, China. *Water (Basel)*, *12*(1), 255. doi:10.3390/w12010255

Tünnemeier. T. (2005). Hydrogeology of Kabul Basin Part 1. Geology, Aquifer characteristics, climate, and hydrography (BGR). Afghanistan.

Tylkowski, J., & Hojan, M. (2019). Time decomposition and short-term forecasting of hydrometeorological conditions in the south Baltic Coastal zone of Poland. *Geosciences*, *9*(2), 68. doi:10.3390/geosciences9020068

U.S. Army Corps of Engineers. (1993). *Hydrologic Frequency Analysis*. US Army.

US Army Corps of Engineers, Hydrologic Engineering Centers. (2000). *Hydrologic Modeling System HEC-HMS Technical Reference Manual*. US Army.

US Army Corps of Engineers, Hydrologic Engineering Centers. (2010). HEC-HMS Quick Start Guide version 3.5. US Army.

US Department of Agriculture, Agricultural Research Service (USDA-ARS). (1995). *WEPP User's Manual*. Purdue University.

USACE. (1992). HEC-FFA: Flood Frequency Analysis. User's Manual [HEC-FFA Frequency]. US Army Corps of Engineers, Hydrologic Engineering Center.

Valipour, M., Bateni, S. M., Dalezios, N. R., Almazroui, M., Heggy, E., Şen, Z., & Angelakis, A. N. (2020). Hydrometeorology: Review of Past, Present and Future Observation Methods. *Hydrology*.

Van Beek, C., & Tóth, G. (2012). *Risk assessment methodologies of soil threats in Europe : status and options for harmonization for risks by erosion, compaction, salinization, organic matter decline, and landslides*. Publications Office of the European Union. https://data.europa.eu/doi/10.2788/47096

Van Der Knijff, J. M., Younis, J., & De Roo, A. P. J. (2010). LISFLOOD: A GIS-based distributed model for river basin scale water balance and flood simulation. *International Journal of Geographical Information Science*, *24*(2), 189–212. doi:10.1080/13658810802549154

Van der Plas, G. W., De Cort, G., Petek-Sargeant, N., Wuytack, T., Colombaroli, D., Lane, P. J., & Verschuren, D. (2019). Distinct phases of natural landscape dynamics and intensifying human activity in the central Kenya Rift Valley during the past 1300 years. *Quaternary Science Reviews*, *218*, 91–106. doi:10.1016/j.quascirev.2019.06.009

Van Dijk, A. I. J. M., Brakenridge, G. R., Kettner, A. J., Beck, H. E., De Groeve, T., & Schellekens, J. (2016). River gauging at a global scale using optical and passive microwave remote sensing. *Water Resources Research*, *52*(8), 6404–6418. doi:10.1002/2015WR018545

Van Hoek, M., Jia, L., Zhou, J., Zheng, C., & Menenti, M. (2016). Early Drought Detection by Spectral Analysis of Satellite Time Series of Precipitation and Normalized Difference Vegetation Index (NDVI). *Remote Sensing (Basel)*, *8*(5), 422. doi:10.3390/rs8050422

Van Huijgevoort, M. H. J., Van Lanen, H. A. J., Teuling, A. J., & Uijlenhoet, R. (2014). Identification of changes in hydrological drought characteristics from a multi-GCM driven ensemble constrained by observed discharge. *Journal of Hydrology (Amsterdam)*, *512*, 421–434. doi:10.1016/j.jhydrol.2014.02.060

Van Lanen, H. A. J., Laaha, G., Kingston, D. G., Gauster, T., Ionita, M., Vidal, J.-P., Vlnas, R., Tallaksen, L. M., Stahl, K., Hannaford, J., Delus, C., Fendekova, M., Mediero, L., Prudhomme, C., Rets, E., Romanowicz, R. J., Gailliez, S., Wong, W. K., Adler, M.-J., & Van Loon, A. F. (2016). Hydrology needed to manage droughts: The 2015 European case. *Hydrological Processes*, *30*(17), 3097–3104. doi:10.1002/hyp.10838

Van Loon, A. F., & Van Lanen, H. A. J. (2012). A process-based typology of hydrological drought. *Hydrology and Earth System Sciences*, *16*(7), 1915–1946. doi:10.5194/hess-16-1915-2012

Vanama, V. S. K., Mandal, D., & Rao, Y. S. (2020). GEE4FLOOD: Rapid mapping of flood areas using temporal Sentinel-1 SAR images with Google Earth Engine cloud platform. *Journal of Applied Remote Sensing*, *14*(3), 034505–034505. doi:10.1117/1.JRS.14.034505

Vaseashta, A. (2015). Life Cycle Analysis of Nanoparticles Risk, Assessment, and Sustainability. DEStech Publications, Inc.

Vaseashta, A. (2021a). Introduction to Water Safety, Security, and Sustainability. In A. Vaseashta & C. Maftei (Eds.), *Water Safety, Security and Sustainability. Advanced Sciences and Technologies for Security Applications*. Springer. doi:10.1007/978-3-030-76008-3_1

Vaseashta, A. (2022). Applying Resilience to Hybrid Threats in Infrastructure, Digital, and Social Domains Using Multisectoral, Multidisciplinary, and Whole-of-Government Approach. NATO Science for Peace and Security Series - D. *Information and Communication Security.*, *61*, 42–59. doi:10.3233/NICSP220017

Vaseashta, A. (2022). Nexus of Advanced Technology Platforms for Strengthening Cyber-Defense Capabilities. In I. Djugumanov & M. Hadji-Janev (Eds.), *NATO Science for Peace and Security Series – E: Human and Societal Dynamics*. IOS Press. doi:10.3233/NHSDP220003

Vaseashta, A., Duca, G., & Travin, S. (Eds.). (2022). *Handbook of Research on Water Sciences and Society* (Vols. 1–2). IGI Global., doi:10.4018/978-1-7998-7356-3

Vaseashta, A., Gevorgyan, G., Kavaz, D., Ivanov, O., Jawaid, M., & Vasović, D. (2021b). Exposome, Biomonitoring, Assessment, and Data Analytics to Quantify Universal Water Quality. In A. Vaseashta & C. Maftei (Eds.), *Water Safety, Security and Sustainability. Advanced Sciences and Technologies for Security Applications*. Springer. doi:10.1007/978-3-030-76008-3_4

Veneros, J., & García, L. (2021). Application of the Standardized Vegetation Index (SVI) and Google Earth Engine (GEE) For Drought Management In Peru. *Tropical and Subtropical Agroecosystems*, *25*(1). Advance online publication. doi:10.56369/tsaes.3784

Vespremeanu-Stroe, A., Preoteasa, L., Hanganu, D., Brown, A., Bîrzescu, I., Toms, P., & Timar-Gabor, A. (2013). The impact of the Late Holocene coastal changes on the rise and decay of the ancient city of Histria (southern Danube delta). *Quaternary International*, *293*, 245–256. doi:10.1016/j.quaint.2012.11.039

Vicca, S., Gilgen, A. K., Serrano, M. C., Dreesen, F., & Dukes, J. (2012). Urgent need for a common metric to make precipitation manipulation experiments comparable. *The New Phytologist*, *195*(3), 518–522. doi:10.1111/j.1469-8137.2012.04224.x PMID:22734795

Vicente-Serrano, S., Cabello, D., Tomás-Burguera, M., Martín-Hernández, N., Beguería, S., Azorin-Molina, C., & Kenawy, A. (2015). Drought Variability and Land Degradation in Semiarid Regions: Assessment Using Remote Sensing Data and Drought Indices (1982–2011). *Remote Sensing (Basel)*, *7*(4), 4391–4423. doi:10.3390/rs70404391

Walker, D., Forsythe, N., Parkin, G., & Gowing, J. (2016). Filling the observational void: Scientific value and quantitative validation of hydrometeorological data from a community-based monitoring programme. *Journal of Hydrology (Amsterdam)*, *538*, 713–725. doi:10.1016/j.jhydrol.2016.04.062

Wang, J., Ling, Z., Wang, Y., & Zeng, H. (2016). Improving the spatial representation of soil moisture by integration of microwave observations and the temperature–vegetation–drought index derived from MODIS products. *Isprs Journal of Photogrammetry and Remote Sensing, 113*, 144. doi:10.1016/j.isprsjprs.2016.01.009

Wang, B., Liu, J., Kim, H. J., Webster, P. J., & Yim, S. Y. (2012). Recent change of the global monsoon precipitation (1979–2008). *Climate Dynamics, 39*(5), 1123–1135. doi:10.100700382-011-1266-z

Wang, M., He, G., Ishwaran, N., Hong, T., Bell, A., Zhang, Z., Wang, G., & Wang, M. (2020). Monitoring vegetation dynamics in East Rennell Island World Heritage Site using multi-sensor and multi-temporal remote sensing data. *International Journal of Digital Earth, 13*(3), 393–409. doi:10.1080/17538947.2018.1523955

Wang, S., Li, R., Wu, T., Zhao, L., Wu, X., Hu, G., Yao, J., Ma, J., Liu, W., Jiao, Y., Xiao, Y., Yang, S., Shi, J., & Qiao, Y. (2022). Evaluating the Impact of Soil Enthalpy upon the Thawing Process of the Active Layer in Permafrost Regions of the Qinghai–Tibet Plateau Using CLM5.0. *Remote Sensing (Basel), 15*(1), 249. doi:10.3390/rs15010249

Wang, W., Samat, A., Abuduwaili, J., De Maeyer, P., & Van de Voorde, T. (2023). Machine learning-based prediction of sand and dust storm sources in arid Central Asia. *International Journal of Digital Earth, 16*(1), 1530–1550. doi:10.1080/17538947.2023.2202421

Wang, W., Samat, A., Abuduwaili, J., Ge, Y., De Maeyer, P., & Van de Voorde, T. (2022). A novel hybrid sand and dust storm detection method using MODIS data on GEE platform. *European Journal of Remote Sensing, 55*(1), 420–428. doi:10.1080/22797254.2022.2093278

Wang, X., Rodríguez, D. A., & Mahendra, A. (2021). Support for market-based and command-and-control congestion relief policies in Latin American cities: Effects of mobility, environmental health, and city-level factors. *Transportation Research Part A, Policy and Practice, 146*, 91–108. doi:10.1016/j.tra.2020.12.004 PMID:34295022

Wang, Y. (2015). Advances in remote sensing of flooding. *Water (Basel), 7*(11), 6404–6410. doi:10.3390/w7116404

Wang, Y., Pineda-Munoz, S., & McGuire, J. L. (2023). Plants maintain climate fidelity in the face of dynamic climate change. *Proceedings of the National Academy of Sciences of the United States of America, 120*(7), e2201946119. doi:10.1073/pnas.2201946119 PMID:36745797

Wang, Y., Zhang, F., Li, X., Johnson, V. C., Tan, M. L., Kung, H.-T., Shi, J., Bahtebay, J., & He, X. (2023). Methodology for Mapping the Ecological Security Pattern and Ecological Network in the Arid Region of Xinjiang, China. *Remote Sensing (Basel), 15*(11), 2836. doi:10.3390/rs15112836

Wang, Y., Zhou, X., Ao, Z., Xiao, K., Yan, C., & Xin, Q. (2022). Gap-Filling and Missing Information Recovery for Time Series of MODIS Data Using Deep Learning-Based Methods. *Remote Sensing (Basel), 14*(19), 4692. doi:10.3390/rs14194692

Wang, Z., Lai, C., Chen, X., Yang, B., Zhao, S., & Bai, X. (2015). Flood hazard risk assessment model based on random forest. *Journal of Hydrology (Amsterdam), 2015*(527), 1130–1141. doi:10.1016/j.jhydrol.2015.06.008

Wanniarachchi, S., & Sarukkalige, R. (2022). A Review on Evapotranspiration Estimation in Agricultural Water Management: Past, Present, and Future. *Hydrology, 9*(7), 123. doi:10.3390/hydrology9070123

Water Cadastre Atlas of Romania. (1992).

Water Resources Council (U.S.) (Ed.). (1977). Guidelines for determining flood flow frequency (Rev. June 1977). Water Resources Council, Hydrology Committee.

Water Resources Council (U.S.). (1981). *Guidelines for determining flood flow frequency* (Rev September 1981). Water Resources Council, Hydrology Committee. https://library.wmo.int/records/item/35804-guide-to-hydrological-practices-volume-i

Weilnhammer, V., Schmid, J., Mittermeier, I., Schreiber, F., Jiang, L., Pastuhovic, V., Herr, C., & Heinze, S. (2021). Extreme weather events in europe and their health consequences–A systematic review. *International Journal of Hygiene and Environmental Health, 233*, 113688. doi:10.1016/j.ijheh.2021.113688 PMID:33530011

Wei, X., Shao, M., Fu, X., Ågren, G. I., & Yin, X. (2011). The effects of land use on soil N mineralization during the growing season on the northern Loess Plateau of China. *Geoderma, 160*(3-4), 590–598. doi:10.1016/j.geoderma.2010.11.007

Wei, Y., Sun, S., Liang, D., & Jia, Z. (2022). Spatial–temporal variations of NDVI and its response to climate in China from 2001 to 2020. *International Journal of Digital Earth, 15*(1), 1463–1484. doi:10.1080/17538947.2022.2116118

WHO. (2009). *WHO Handbook on Indoor Radon: A Public. Health Perspective.* World Health Organization.

WHO. (2011). *Uranium in drinking water. Background document, for development of WHO Guidelines for drinking-water quality.* WHO.

Wilhite, D. A., & Glantz, M. H. (1985). Understanding the Drought Phenomenon: The Role of Definitions. *Water International, 10*(3), 111–120. doi:10.1080/02508068508686328

Williams, A. P., Abatzoglou, J. T., Gershunov, A., Guzman-Morales, J., Bishop, D. A., Balch, J. K., & Lettenmaier, D. P. (2019). Observed impacts of anthropogenic climate change on wildfire in California. *Earth's Future, 7*(8), 892–910. doi:10.1029/2019EF001210

WMO, & GWP. (2014). *National Drought Management Policy Guidelines: A template for action.* MWO. https://www.droughtmanagement.info/literature/IDMP_NDMPG_en.pdf

WMO. (2017). *Guidelines on Best Practices for Achieving User Readiness for New Meteorological Satellites.* WMO. https://library.wmo.int/doc_num.php?explnum_id=3553

WMO. (2018). *Guidelines on Satellite Skills and Knowledge for Operational Meteorologists.* WMO. https://library.wmo.int/doc_num.php?explnum_id=5799

WMO. (2023a). *Atlas of Mortality and Economic Losses from Weather, Climate, and Water Extremes (1970–2019).* WMO. https://storymaps.arcgis.com/stories/8df884dbd4e849c89d4b1128fa5dc1d6

WMO. (2023b). *Guidelines on the Definition and Characterization of Extreme Weather and Climate Events.* World Meteorological Organization. https://library.wmo.int/records/item/58396-guidelines-on-the-definition-and-characterization-of-extreme-weather-and-climate-events

WMO. (n.d.-a). *Climate Data Management System Specifications.* WMO. https://library.wmo.int/records/item/51447-climate-data-management-system-specifications

WMO. (n.d.-b). *Guide to Hydrological Practices, Volume I.* WMO. https://library.wmo.int/records/item/35804-guide-to-hydrological-practices-volume-i?offset=2

World Bank (Ed.). (2018). *From uneven growth to inclusive development: Romania's path to shared prosperity.* World Bank Group. 10.1596/9781464813177

World Meteorological Organization (WMO). (2008). Guide to Hydrological Practices (WMO-No. 168), Volume I: Hydrology – From Measurement to Hydrological Information (Edition 2008, updated in 2020). WMO.

World Meteorological Organization (WMO). (2020). *Vision for the WMO Integrated Global Observing System in 2040* (2019 edition). WMO.

World Meteorological Organization (WMO). (2021). *Manual on the WMO Integrated Global Observing System (WMO-No. 1160): Annex VIII to the WMO Technical Regulations* (2021 edition). WMO.

World Meteorological Organization -WMO, & Global Water Partnership - GWP. (2016). *Handbook of Drought Indicators and Indices* (M. Svoboda & B. Fuchs, Eds.). WMO. https://public.wmo.int/en/resources/library/handbook-of-drought-indicators-and-indices

World Meteorological Organization -WMO, & Global Water Partnership - GWP. (2016). *Handbook of Drought Indicators and Indices* WMO. https://public.wmo.int/en/resources/library/handbook-of-drought-indicators-and-indices

World Meteorological Organization -WMO. (2021). *WMO Atlas of Mortality and Economic Losses from Weather, Climate and Water Extremes (1970–2019) (WMO-No. 1267)*. WMO.

Wu, C., & Chau, K. W. (2010). Data-driven models for monthly streamflow time series prediction. *Engineering Applications of Artificial Intelligence*, *2010*(23), 1350–1367. doi:10.1016/j.engappai.2010.04.003

Wu, G., Fang, Y., Jiang, Q., Cui, M., Li, N., Ou, Y., Diao, Z., & Zhang, B. (2023). Early identification of strawberry leaves disease utilizing hyperspectral imaging combing with spectral features, multiple vegetation indices and textural features. *Computers and Electronics in Agriculture*, *204*, 107553. doi:10.1016/j.compag.2022.107553

Wu, X., Lu, G., & Wu, Z. (2023). Remote Sensing Technology in the Construction of Digital Twin Basins: Applications and Prospects. *Water (Basel)*, *15*(11), 2040. doi:10.3390/w15112040

Xie, Z., Huete, A., Cleverly, J., Phinn, S., McDonald-Madden, E., Cao, Y., & Qin, F. (2019). Multi-climate mode interactions drive hydrological and vegetation responses to hydroclimatic extremes in Australia. *Remote Sensing of Environment*, *231*, 111270. doi:10.1016/j.rse.2019.111270

Xu, X., Li, J., & Tolson, B. A. (2014). Progress in integrating remote sensing data and hydrologic modeling. *Progress in Physical Geography*, *38*(4), 464–498. doi:10.1177/0309133314536583

Yacouby, R., & Axman, D. (2020, November). Probabilistic extension of precision, recall, and f1 score for more thorough evaluation of classification models. In *Proceedings of the first workshop on evaluation and comparison of NLP systems* (pp. 79-91). IEEE. 10.18653/v1/2020.eval4nlp-1.9

Yamazaki, F., & Matsuoka, M. (2007). Remote sensing technologies in post-disaster damage assessment. *Journal of Earthquake and Tsunami*, *1*(3), 193–210. doi:10.1142/S1793431107000122

Yang, F., Koukoula, M., Emmanouil, S., Cerrai, D., & Anagnostou, E. N. (2023). Assessing the power grid vulnerability to extreme weather events based on long-term atmospheric reanalysis. *Stochastic Environmental Research and Risk Assessment*, *37*(11), 1–16. doi:10.100700477-023-02508-y

Yang, Y., Anderson, M. C., Gao, F., Wood, J. D., Gu, L., & Hain, C. (2021). Studying drought-induced forest mortality using high spatiotemporal resolution evapotranspiration data from thermal satellite imaging. *Remote Sensing of Environment*, *265*, 112640. doi:10.1016/j.rse.2021.112640

Yasmin, T., Khamis, K., Ross, A., Sen, S., Sharma, A., Sen, D., Sen, S., Buytaert, W., & Hannah, D. M. (2023). Brief communication: Inclusiveness in designing an early warning system for flood resilience. *Natural Hazards and Earth System Sciences*, *23*(2), 667–674. doi:10.5194/nhess-23-667-2023

Yevjevich, V. (1967). *An Objective Approach to Definitions and Investigations to Continental Hydrologic Droughts*. Colorado State University.

Yevjevich, V. M. (1969). An objective approach to definitions and investigations of continental hydrologic droughts. *Journal of Hydrology (Amsterdam)*, *7*(3), 353. doi:10.1016/0022-1694(69)90110-3

Yin, H., & Li, C. (2001). Human impact on floods and flood disasters on the Yangtze River. *Geomorphology*, *41*(2-3), 105–109. doi:10.1016/S0169-555X(01)00108-8

You, J.-H., Oh, S., Park, J.-E., Song, H., & Kim, Y.-K. (2021). A novel LiDAR sensor alignment inspection system for automobile productions using 1-D photodetector arrays. *Measurement*, *183*, 109817. doi:10.1016/j.measurement.2021.109817

Young, A. R., Round, C. E., & Gustard, A. (2000). Spatial and temporal variations in the occurrence of low flow events in the UK. *Hydrology and Earth System Sciences*, *2000*(4), 35–45. doi:10.5194/hess-4-35-2000

Yousefi, E., Sayadi, M. H., & Chamenhpour, E. (2022). Google Earth Engine platform to calculate the hydrometeorology and hydrological water balance of wetlands in arid areas and predict future changes. *Journal of Applied Research in Water and Wastewater*, *9*(1), 52–68.

Zanchetta, A. D., & Coulibaly, P. (2020). Recent advances in real-time pluvial flash flood forecasting. *Water (Basel)*, *12*(2), 570. doi:10.3390/w12020570

Zeppetello, L. R. V., Battisti, D. S., & Baker, M. B. (2022). The physics of heat waves: What causes extremely high summertime temperatures? *Journal of Climate*, *35*(7), 2231–2251. doi:10.1175/JCLI-D-21-0236.1

Zhang, B., Salem, F. K. A., Hayes, M. J., & Tadesse, T. (2022). *Quantitative Assessment of Drought Impacts Using XGBoost based on the Drought Impact Reporter.* arXiv preprint arXiv:2211.02768.

Zhang, C., Wang, K., Yue, Y., Qi, X., & Zhang, M. (2023). Assessing Regional Ecosystem Conditions Using Geospatial Techniques—A Review. *Sensors (Basel)*, *23*(8), 4101. doi:10.339023084101 PMID:37112442

Zhang, H., Lan, Y., Suh, C. P.-C., Westbrook, J., Clint Hoffmann, W., Yang, C., & Huang, Y. (2013). Fusion of remotely sensed data from airborne and ground-based sensors to enhance detection of cotton plants. *Computers and Electronics in Agriculture*, *93*, 55–59. doi:10.1016/j.compag.2013.02.001

Zhang, K., Wang, Q., Chao, L., Ye, J., Li, Z., Yu, Z., Yang, T., & Ju, Q. (2019). Ground observation-based analysis of soil moisture spatiotemporal variability across a humid to semi-humid transitional zone in China. *Journal of Hydrology (Amsterdam)*, *574*, 903–914. doi:10.1016/j.jhydrol.2019.04.087

Zhang, M., Yang, X., Cleverly, J., Huete, A., Zhang, H., & Yu, Q. (2022). Heat wave tracker: A multi-method, multi-source heat wave measurement toolkit based on Google Earth Engine. *Environmental Modelling & Software*, *147*, 105255. doi:10.1016/j.envsoft.2021.105255

Zhang, X., Shi, P., Liu, X., Liu, J., & Lu, Y. (2020). Analyzing the spatiotemporal pattern of extreme precipitation using Google Earth Engine: A case study in China. *Journal of Hydrology (Amsterdam)*, *590*, 125384. doi:10.1016/j.jhydrol.2020.125384

Zhang, Y., Wang, Y., Chen, Y., Xu, Y., Zhang, G., Lin, Q., & Luo, R. (2021). Projection of changes in flash flood occurrence under climate change at tourist attractions. *Journal of Hydrology (Amsterdam)*, *595*, 126039. doi:10.1016/j.jhydrol.2021.126039

Zhang, Y., & Ye, A. (2021). Machine learning for precipitation forecasts postprocessing: Multimodel comparison and experimental investigation. *Journal of Hydrometeorology*, *22*(11), 3065–3085.

Zheng, X., Streimikiene, D., Balezentis, T., Mardani, A., Cavallaro, F., & Liao, H. (2019). A review of greenhouse gas emission profiles, dynamics, and climate change mitigation efforts across the key climate change players. *Journal of Cleaner Production*, *234*, 1113–1133. doi:10.1016/j.jclepro.2019.06.140

Zhou, J., Chen, J., Chen, X., Zhu, X., Qiu, Y., Song, H., Rao, Y., Zhang, C., Cao, X., & Cui, X. (2021). Sensitivity of six typical spatiotemporal fusion methods to different influential factors: A comparative study for a normalized difference vegetation index time series reconstruction. *Remote Sensing of Environment*, *252*, 112130. doi:10.1016/j.rse.2020.112130

Zhou, X., Geng, X., Yin, G., Hänninen, H., & Hao, F. (2020). Legacy effect of spring phenology on vegetation growth in temperate China. *Agricultural and Forest Meteorology*, *281*, 107845. doi:10.1016/j.agrformet.2019.107845

Zribi, M., Albergel, C., & Baghdadi, N. (2020). Editorial for the Special Issue "Soil Moisture Retrieval using Radar Remote Sensing Sensors.". *Remote Sensing (Basel)*, *12*(7), 1100. doi:10.3390/rs12071100

About the Contributors

Radu Muntean is an Associate Professor at the Faculty of Civil Engineering, Transilvania University of Brasov, Romania. He has a master's degree in Rehabilitation and consolidation of buildings and a Ph.D. in Civil Engineering. The main fields of interest and research expertise areas are in sustainability of constructions, waste and recovered materials, ecological building materials, timber and concrete structures, management and economics in the constructions sector, circular economy issues.

Ashok Vaseashta currently serves as the CRO, Executive Director of Research and Chair of IRB and ORI. Concurrently, he serves as Chaired Professor of Nanotechnology at the Academy of Sciences of Moldova, Academician at EMAAS, Senior Strategic Research Advisor for several organizations, and Professor at Riga Technical University and Transylvania University of Brasov. Inspired by nature and guided by societal necessities, he strives for innovations to address global challenges such as environment, human safety and security, and sustainability using advanced technological solution platforms. He is a scholar, dedicated futurist and visionary leader. He received PhD from the Virginia Tech, VA in 1990 followed by Kobe post-doctoral fellowship. He served the U.S. Department of State in two rotations, as strategic S&T advisor and U.S. diplomat. His research interests span nanotechnology, environmental/ecological science and safety and security. He has authored over 250 research publications, edited/authored eight books on nanotechnology, and presented many keynote and invited lectures worldwide. He serves on the editorial board of several highly reputed international journals.

Adrian Alexandru Aldea is Researcher at the National Institute of Hydrology and Water Management. He attended Technical University of Construction Bucharest, Faculty of Hydrotechnics from 2007 – 2011. In 2013 he received the Master of Science degree, specialization Hydraulic Engineering. His research activity is focused on: hydrology and hydrological-hydraulic modelling, photogrammetry, hydrometry, water management.

Alina Bărbulescu graduated from the University of Craiova (Romania), Faculty of Mathematics, and from the Petre Andrei University of Iaşi (Romania), Faculty of Law. After an MSc in Mathematics at Bucharest University (Romania), she got a Ph.D. in Mathematics from Al. I. Cuza University of Iasi (Romania), a Ph.D. in Cybernetics and Economic Statistics from the Academy of Economic Studies Bucharest (Romania), and a Ph.D. in Civil Engineering, with Magna cum Laude, from the Technical University of Civil Engineering, Bucharest (Romania). She got the habilitation in Civil Engineering in 2014, with the thesis Modeling the spatially distributed precipitations, and in Cybernetics and Economics Statistics in 2019, with the thesis Contribution to statistical analysis and modeling in economics and finance. She is a Professor at Transilvania University of Braşov, Department of Civil Engineering. Before her current position, she taught at the Ovidius University of Constanţa, Romania, Higher Colleges of Technology, U.A.E, and was a visiting professor at Abu Dhabi University. Her research interests span applied statistics, econometrics, hydrological modeling, and water and air pollution. She has authored over 200 articles, 30 books, and book chapters, was a coeditor of 20 special issues of scientific journals and conference proceedings, and presented invited lectures at international conferences. She is a member of IAHS, IAENG, SSMR, and a topic editor at different journals, including Water, Atmosphere (Clarivate Analytics).

Silvia Chelcea is a Senior Researcher at the National Institute of Hydrology and Water Management, in Bucharest, Romania. She attended the Faculty of Mathematics at the University of Bucharest in 1995. Silvia has a PhD in Geography since 2012. She has been working at the National Institute of Hydrology and Water Management for more than 25 years, currently being part of the Danube River and the Black Sea Hydrology, Hydrometry Department. Her research activity is focused on hydrological extremes, hydrological characteristics of drought events and the impact of climate change on the hydrological regime. This involves the employment of approaches such as: statistical analysis of water and sediments variability, computation and analysis of hydrological indices and indicators, trends. She is also the author/co-author of scientific articles published in national and international ISI-listed journals.

Gabriel Dobrica was born in Medgidia City, Constanta, Dobrogea, Romania in 1988. He received the B.S. degree in civil engineering from "Ovidius" University of Constanta, Romania, in 2012, also degree in geodetic engineering from University of Agronomic Sciences and Veterinary Medicine of Bucharest, Romania, in 2014 and the M.S. degree in civil engineering from "Ovidius" University of Constanta, Romania, in 2014. He is currently pursuing the Ph.D. degree in Civil engineering and installations at "Ovidius" University of Constanta, Doctoral School of Applied Sciences, Romania. From 2016 up to present, he is working at the National Administration of "Romanian Waters". His Assessment of Hydrological Drought in the Watershed of Lake Nuntasi-Tuzla, Constanta County, this is also the title of the research topic of the doctoral thesis. Mr. Gabriel Dobric˘a was member of the target group of the project "Academic excellence and entrepreneurial values scholarship system to ensure opportunities for training and development of entrepreneurial skills of doctoral and postdoctoral students—ANTREPRENORDOC", in 2019.

Khaled Hazaymeh received his PhD in earth observation (Remote Sensing) from the University of Calgary in Canada in 2016. In 2004 he was graduated from the Department of geography in Yarmouk University in Jordan and was granted a scholarship to pursue his graduate studies. He is currently an associate professor in the department of geography at Yarmouk University. He writes and presents widely on issues of environmental modeling and remote sensing development techniques and is the author of more 20 research papers and book chapters since 2016.

Dinesh Kumar is working as an Assistant Professor in the Department of Environmental Science, Central University of Jammu

Shivali Kundan is presently enrolled as a PhD researcher in the Department of Environmental Science, Central University of Jammu.

Carmen Maftei is Professor at the Civil Building Services Department and Vice dean for Academic Affairs at the Transilvania University of Brasov, Romania. She attended Polytechnic Institute of Iasi where she majored in Land Reclamation in 1988 and from 1996-1998 she attends a specialization in Water resources and protection of water resources, equivalent of Master of Science. In 2002 she finished the PhD studies in Civil Engineering. Her research activity is focused on: (i) hydrology and hydrological modelling.

Ayodeji Samuel Makinde is a lecturer and researcher in the Department of Computer Science, Faculty of Science, Edo State University Uzairue, Edo State, Nigeria. He is a multidisciplinary professional with over 7 years of professional experience cutting across Data Science, Machine Learning, Artificial Intelligence, Business Intelligence, Data Visualization, and Web development using Django. He currently works as the Ag. Director of Information and Communication Technology at Edo State University Uzairue. Academically, he holds a PhD in Computer Science from the Federal University of Agriculture Abeokuta, MSc in Information and Communication Science and Engineering from the Nelson Mandela African Institution of Science and Technology, Arusha, Tanzania, where he was awarded the best Master's Student in Research and Innovation of his faculty. He also holds a BSc. in Computer Science from the Federal University of Agriculture Abeokuta. His research interests include Data Science, Artificial Intelligence, Blockchain, IoT, E-learning and Business Intelligence. He has several publications in reputable journals.

Ayodeji S. Makinde holds a PhD in Computer Science with bias in Artificial Intelligence and Software systems. He is currently a lecturer in the Department of Computer Science, Edo University Iyamho Nigeria. He is a member of several learned societies.

Anna Msigwa is a lecturer in hydrology and water resources engineering at the department of Water Environmental Sciences and Engineering at the Nelson Mandela African Institution of Science and Technology (NM-AIST), Tanzania. Msigwa is a hydrologist and water management expert, researching water resource management and climate-smart agricultural water management strategies using remote sensing technologies.

Radu Muntean is an Associate Professor at the Faculty of Civil Engineering, Transilvania University of Brasov, Romania. He has a master's degree in Rehabilitation and consolidation of buildings and a Ph.D. in Civil Engineering. The main fields of interest and research expertise areas are in sustainability of constructions, waste and recovered materials, ecological building materials, timber and concrete structures, management and economics in the constructions sector, circular economy issues.

Zahid Nabi is a PhD researcher in the Department of Environmental Science, Central University of Jammu

Sorour Rahmanian is a plant ecologist.

Cristina Serban is an Associate Professor at Faculty of Mathematics and Informatics, specialized în evolutionary computation, databases and remote sensing.

Maria Cristina Trifu is Senior Researcher at the National Institute of Hydrology and Water Management, head of the Danube River and Black Sea Hydrology, Hydrometry Department, Romania. She attended Polytechnic Institute of Bucharest and from 1994 - 1996 she attends a specialization in water resources and protection of water resources, equivalent of Master of Science. In 2001 she received the Ph. D. from Pierre et Marie Curie University, Paris VI, in the specialization Geoscience and Natural Resources. Her research activity is focused on: hydrology and hydrological modelling, hydrometry, nutrient water pollution, water management.

Neha Verma is enrolled as a PhD researcher in the Department of Environmental Science, Central University of Jammu.

Mohammad Zeitoun received his PhD in geography (climatology) from the University of Jordan in Jordan in 2013. He was graduated from the department of geography in University of Jordan. He is currently an associate professor and the head of the department of geography at Yarmouk University. He was the chairman of the accreditation department at Yarmouk University between 2019-2022. He writes and presents widely on issues of environmental modeling and climatology related studies including climate change and environmental hazards and is the author of many research papers in these fields.

Index

www.ingramcontent.com/pod-product-compliance
Lightning Source LLC
LaVergne TN
LVHW082052011026
849162LV00019B/1435

* 9 7 8 1 6 6 8 4 8 7 7 1 6 *